AF494764

Polymer Matrix Composites

Soviet Advanced Composites Technology Series

Series editors: J. N. Fridlyander, Russian Academy of
Sciences, Moscow, Russia
I. H. Marshall, University of Paisley,
Paisley, UK

This series forms a unique record of research, development and application of composite materials and components in the former Soviet Union. The material presented in each volume, much of it previously unpublished and classified until recently, gives the reader a detailed insight into the theory and methodology employed and the results achieved, by the Soviet Union's top scientists and engineers in relation to this versatile class of materials.

Titles in the series

1. **Composite Manufacturing Technology**
   Editors: A. G. Bratukhin and V. S. Bogolyubov

2. **Ceramic- and Carbon-matrix Composites**
   Editor: V. I. Trefilov

3. **Metal Matrix Composites**
   Editor: J. N. Fridlyander

4. **Polymer Matrix Composites**
   Editor: R. E. Shalin

5. **Fibre Science and Technology**
   Editor: V. I. Kostikov

6. **Composite Materials in Aerospace Design**
   Editors: G. I. Zagainov and G. E. Lozino-Lozinski

Polymer Matrix Composites

Edited by

R. E. Shalin

All-Russian Scientific Institute for Aviation Materials, Moscow

SPRINGER-SCIENCE+BUSINESS MEDIA, B.V.

Contributors

G.M. Gunyaev
All-Russia Scientific Institute of Aviation Materials Scientific and Industrial Association,
  Moscow

I.P. Khoroshilova
All-Russia Scientific Institute of Aviation Materials Scientific and Industrial Association,
  Moscow

B.A. Kiselev
All-Russia Scientific Institute of Aviation Materials Scientific and Industrial Association,
  Moscow

G.P. Mashinskaya
All-Russia Scientific Institute of Aviation Materials Scientific and Industrial Association,
  Moscow

B.V. Perov
All-Russia Scientific Institute of Aviation Materials Scientific and Industrial Association,
  Moscow

V.D. Protassov
Scientific and Technological Institute of Special Machine-Building, Khot'kovo

R.E. Shalin
All-Russia Scientific Institute of Aviation Materials Scientific and Industrial Association,
  Moscow

T.G. Sorina
All-Russia Scientific Institute of Aviation Materials Scientific and Industrial Association,
  Moscow

E.B. Trostyanskaya
The Tsiolkovsky Institute of Aircraft Technology, Moscow

Preface

Polymeric composites based on continuous glass, carbon and organic fibres have found wide application in many branches of modern engineering. High-strength and high-modulus reinforced plastics possess a unique combination of mechanical, technological and service properties. They differ from other traditional structural materials in the fact that the design of composite components and articles cannot be carried out separately from the design and development of the initial materials, beginning with the selection and preparation of raw materials. The above-mentioned features and also the factor of the free variation of composite properties over the whole complex of technical and theoretical problems associated with composite development, design and production, i.e. the study of their properties, reveal their potential.

In the modern scientific and technical literature, editions of complex character that provide the specialist with the necessary information are rather rare.

This book (Volume 4 of the series) gives the basic principles of the development of high-strength high-modulus composites in Russia. A wide spectrum of problems has been considered, associated with the synthesis of the initial components (binders and reinforcing fillers), the development of formulations, the study of the properties of glass-, carbon- and organic-fibre-reinforced plastics, and the peculiarities of hybrid composites design. Definite types of reinforced materials widely used in the Russian aerospace industry and other branches of engineering and national economy are considered.

The data presented can serve as a useful information source for scientists, industrial engineers and designers working in this advanced materials science field, and also for representatives of industry and business dealing with the production and application of micro-reinforced and hybrid composite articles.

R.E. Shalin
Moscow, Russia

1

Polymeric matrices in fibre-reinforced composite materials

E.B. Trostyanskaya

1.1 INTRODUCTION

In spite of the fact that polymeric composite materials have developed simultaneously with plastics production, the most accurate definition of the term 'composite material' and the scientific problems associated with this new trend in materials science were formulated in 1967 by L.Y. Broutman and R. Krock.

Among the various multiphase materials that could be referred to as composites, the most important are materials intended for the production of articles that are subjected to mechanical loads. Here fibres manufactured of different materials are the phase responsible for the strength and rigidity of the material as a whole. Fibres of different nature are used in a matrix material, or spherical or scaly particles are added at the same time in order to increase composite functionality. Any combination of performance properties can be achieved in a material by suitable selection of the components.

Component unity is provided by a continuous phase (matrix), which fulfils the following functions in a composite material:

1. Provides shape and size stability of the article.
2. Fixes a given distribution of the fibres relative to each other.
3. Gives the material deformation stability under thermal and mechanical loading up to the level given by the performance specifications.
4. Redistributes the external action on all the elementary fibres in a composite material, including broken and distorted ones.
5. Gives the material outdoor resistance and stability in the medium for which it is intended.

The selection of the material is made by taking into account the need to keep the following inherent properties at the given performance specifications:

1. Hardness and monolithic character.

2. Strength in respect of mechanical and thermal loading.
3. Elongation at a particular strain.
4. Outdoor stability, waterproofing and stability in media as shown in the performance specifications.

A material can be the matrix of a composite material if it provides:

1. Limited complete wettability of elementary fibres in a fibrous filler (plait, thread, tape, textile) at a temperature that is lower than the melt, thermal degradation or structural disorientation temperature of the fibres.
2. Lack of chemical reaction between matrix and fibres, which could modify the shape and fibre bulk properties.
3. Strong coupling over the whole surface of the elementary fibres as a result of chemical and polar interactions preserving a clear phase boundary.
4. Material monolithization and moulding of articles in regimes preventing thermo-degradation and mechanical fracture or destruction of elementary fibres and disturbing their mutual arrangement (distribution).
5. Continuity of an uniform matrix distribution over the whole interstitial space at a filling degree up to 60–65 vol % and diameter of elementary fibres of 7–10 μm, i.e. under conditions when the contact zone reaches 400–500 mm^2 in 1 mm^3 of material.

Polymers for technical use are artificially created materials. If polymer synthesis is carried out in consequent stages of composite material production and fabrication of the corresponding articles, then the contradictory requirements for the technological and performance properties of the matrix can be eliminated.

Composite material production involves the following stages:

1. Prepreg manufacture, i.e. impregnation of fibrous filler with the melt or matrix polymer solution. A uniform film of polymer of 6–8 μm thickness remains after solvent removal at the end of the process of prepreg preparation at the elementary fibre surface.
2. Lay-out of the packet and blank winding from prepreg with a given fibre distribution (with a given scheme of strengthening).
3. Packet monolithization into the material with simultaneous moulding of the article as a result of polymeric film softening at the surface of the fibres and densification of the packet.

The solution or the melt of the polymer must have a low viscosity at this stage to wet the whole surface of elementary fibres in a fibrous filler, to distribute polymeric film uniformly and to achieve rapidly an equilibrium structure of adsorption layers at the fibre surface.

After the end of the impregnation, the viscosity of the polymer in the prepreg must be raised so that the film does not peel off from the fibre surface but attained compliance is necessary for lay-out of the prepreg on the mould.

In monolithization and moulding of the blank at the selected pressure and elevated temperature, the melt viscosity of the polymeric film at the surface of the fibres must provide · their coalescence into a monolithic matrix. Without pressing out the filler, the matrix vitrifies, and fixes the article's shape and the mutual distribution of the fibres.

Only step-by-step synthesis can provide a repeated change of matrix viscosity at the prepreg and article manufacture stage. This allows one also to give the matrix the properties required to satisfy performance requirements. Any synthesis stage must be coupled with the material and article manufacture stage, possibly lengthening its holding time at each stage. At the end of the chemical conversions, the polymer must become stable to deformation, tough and thermostable. These properties correspond to a greater degree at the end of the process to high-crosslink-density polymers, semicrystalline linear polymers with elevated stiffness of the macromolecules and low-crosslink-density polymers with elevated stiffness of interjunction fragments.

1.2 HIGH-CROSSLINK-DENSITY POLYMERIC MATRICES

According to an idealized treatment, a high-crosslink-density polymer is considered to consist of multiblock chains with strong chemical crosslinks (chemical junctions) and polar groups in the units introducing an additional polar interaction ('physical junctions') in the polymeric network. The idealized polymer network density is characterized by the molecular weight of the fragment between adjacent chemical junctions (M_c) and the number of chemical junctions (N_c) in unit volume [1–3]. It is supposed that a high-crosslink-density polymer remains solid until the beginning of thermodegradation, has a minimal (for polymers) thermal expansion coefficient, preserves its mechanical characteristics, does not exhibit cold flowability, is impermeable to low-molecular-weight compounds, is insoluble and does not swell. Only in the region where the physical junctions disappear can some change in high crosslink density take place. The intensity of such a change is defined by the M_c value and the rigidity of the interjunction fragment.

Synthesis of high-crosslink-density polymers includes several stages:

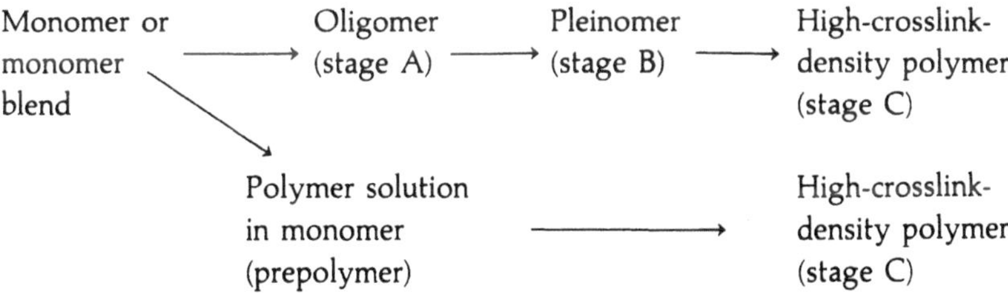

Transition from stage A to stage C, or from the prepolymer stage, is usually called 'curing'. The initial product of the synthesis (stage A) or prepolymer consists of fractions with different numbers of units in the macromolecules, different branching, different numbers of functional groups and their different mutual distributions. Some quantities of the initial substances not entering the reaction and synthesis by-products remain in the initial composition. Such a mixture of complex composition is usually called 'the resin'. The resin at stage A is characterized by conventional indices of molecular weight and percentage content of functional groups that can take part in the cure reaction, and information about melt or solution viscosity and time of attaining these indices on storage (shelf-life) is given.

In the cure process, addition of multifunctional oligomer molecules is accompanied by an increase in the number of branches of different length. The branches are as a rule in interaction with each other, and lead to ring formation. The ring density increases continuously. The probability of intramolecular reactions increases with the medium viscosity, decreasing number of units in the oligomeric medium and increasing flexibility [4].

Networks arise when branched molecules are combined with each other even by a single branch ('gel point'). But long before this stage a new phase arises in the resin as the finest gels ('microgels') form, with a diameter of 20–30 μm and with no clear separation from the liquid phase [5]. As the cure reaction advances, liquid phase viscosity increases and the size of microgels increases too. The reaction proceeds immediately at the microgel surface and inside. As a result, both microgel size and density increase. The microscopic flow increases and increase of size of disperse phase particles is interrupted. The prepolymer turns into an elastic (low-crosslink-density) polymer and then into a glass-like (high-crosslink-density) polymer from the viscous flow state [6–8].

The system gets to the gel point from solubility to the limited swelling. The functional groups decrease and then their mobility is interrupted. The polymer becomes vitreous at the selected synthesis temperature. The cure process ceases, though there are reactive functional groups in the polymer. Increasing the temperature above the glass transition point can increase the degree of conversion of the functional groups. The reaction proceeds until the glass transition appears again, now at a higher temperature. The polymer network density ceases to increase as the reaction temperature attains the temperature of rapid degradation processes.

The curing process can be followed by transmittance (optical density), viscosity or elasticity modulus increase [9–11], by the increase in the number of functional groups in reaction products and by swelling ratio.

The kinetics of the cure reaction is described by an S-like curve with acceleration at the stage of microgel formation as the functional-group concentration increases in microvolumes and the isothermal character of the reaction is broken.

Physical junctions arise as a result of polar interactions between units of adjacent chains, and influence cure kinetics in the curing process. Medium viscosity increases and the reaction reaches the diffusional regime already in its initial stages. To increase the reaction rate, the resin is heated to a temperature at which the influence of polar interaction decreases.

Below the gel point, reaction rate decreases sharply, probably due to topological restrictions [12,13]. The amorphous structure is fixed at the final stage by vitrification and the sharply expressed uniformity of chemical junction density in the polymer network. Maximum density of chemical junctions is located in microvolumes – microgels. The latter are surrounded by and combined with an intergel region of weaker chemical junctions between chain fragments, strong ruptures therein and different lengths of fragments. The structure of a high-crosslink-density polymer resembles that of a polymeric material with a discrete distribution of polymer filler particles in a polymeric matrix and strong chemical interaction between phases [14].

The sharp distinction of real polymeric networks from idealized ones leads to a description of their structure by M_c and N_c values that has a conventional character. The properties of real high-crosslink-density polymers have the same distinction. The heterogeneity becomes apparent in low strength and high brittleness. The major portion of the defects in the interlayer region become apparent in an increase of thermal expansion coefficient, in forced elasticity in long-term loading and even some plasticity (cold flowability), in decreasing modulus of elasticity and heat stability, and in marked changing of temperature, gas permeability and swelling, even if these are limited. The structure of a high-crosslink-density polymer is estimated by the degree of heterogeneity and sizes of microgel particles. The temperature of disappearance of polar and hydrogen-bond interactions is shown for each polymer.

If the average length of an interjunction fragment in the microgel region exceeds the mechanical segment length, then the polymer becomes elastic after the physical junctions disappear. The degree of elasticity is determined by the flexibility of interjunction fragments. A polymer with smaller length of interjunction fragments than the length of mechanical segments remains vitreous until thermodegradation of the chemical junction proceeds.

The reversible decrease of the indices of the properties up to a level inadmissible by the performance specifications is implied by the term 'thermal stability'. The thermal stability of a high-crosslink-density polymer increases with chemical junction density and decreasing number of structural defects in the intergel region. It is supposed that the density of chemical junctions in a high-crosslink-density polymer that exhibits thermostability attains 10^{21} junctions/cm^3. It is difficult to determine this value experimentally owing to the heterogeneity of polymer network structure and the large amount of defects in the intergel region. The density of intermolecular packing of atoms decreases simultaneously and the conformation set of adjacent units shortens. Therefore, local rigidity increases and steric restrictions are exhibited to a greater degree.

The heterogeneity of a polymeric network causes a local overtension. Stress concentration arises near any defect, and this brings about local plastic deformation and breaking of the most stressed fragments. The rupture of several chemical bonds causes subcrack formation. Where these coalesce, cracks arise, which propagate along the boundaries of microheterogeneities.

Prolonged treatment can increase the uniformity of the polymeric film structure and decrease the free volume in the polymeric network and polar interaction strength.

Forced elasticity becomes apparent in high-crosslink-density polymers on slow or prolonged loading. This process is accompanied by the rupture of defect regions of a network polymer and overstressed chemical bonds. Therefore it is irreversible. 'Neck' formation precedes sample destruction.

On fast loading, forced elasticity is observed as the background of chemical junction fragmentation and fast crack propagation [15, 16].

The greater the polymer network density in relation to chemical and physical junctions, the greater is the polymer matrix brittleness.

The conversion of functional groups is often incomplete owing to steric and diffusional restrictions and to frequent terminations of the chains between junctions and irregular arrangement of the chemical junctions, and to the length of interjunction fragments exceeding the mechanical segment length. The glass transition temperature in these systems is below the thermodegradation temperature.

Above the glass transition temperature a polymer is in a limited elastic state. Though the area of chain packing and the strength of polar interaction increase with more irregular distribution of chemical junctions, the influence of polar interaction and segment mobility of interjunction fragments decrease near the glass transition temperature.

Below the glass transition temperature, relaxation transitions are caused because the mobility of short chain segments or lateral substituents becomes apparent. Above the secondary relaxation transition temperature (T_β) the fracture viscosity of high-crosslink-density polymer (brittleness temperature) increases to some degree. To decrease the secondary transition temperature, it is reasonable to introduce groups or atoms into interjunction fragments of crosslinked polymer to facilitate the mobility of adjacent units or to lower the concentration of polar groups.

When limited glass transition temperature is above the thermodegradation temperature, the degree of conversion can be followed by T_β increase [17−19].

Parameters characterizing α and β relaxation transition are functions of cohesive energy density. Therefore, these depend to some degree on thermal expansion coefficients, density, modulus of elasticity and strength. However, analytical calculations of the location of α and β transitions on the temperature scale using the contributions of oligomer unit structure and chemical junction structure are limited. As the cure process proceeds, the heterogeneity of the structure increases, units of new composition and new polar groups arise and the interjunction fragment packing density changes. If the cure conditions for the same composition are the same, relaxation transition temperatures increase with chemical junction concentration. The peak of the curve of the dependence of mechanical and dielectric losses on temperature is lowered simultaneously and the plateau of the maximum is expanded.

The presence of physical junctions in a polymeric network is revealed visually in monotonically changing property indices with temperature. As the temperature increases, the junction strength, brittleness and modulus of elasticity decrease but elongation at break increases. As polar interaction disappears, the fracture energy increases sharply, but deformation stability decreases in an analogous manner. Further behaviour of high-crosslink-density polymer is determined by the chemical junction density and the rigidity of interjunction fragments.

If functional groups arise in an oligomer at stage A that do not react with each other, the presence of another component − a curing agent − is required for the cure process. The functional groups of the curing agent react with oligomer functional groups, and this leads to the formation of branched macromolecules. Functional groups take part in intra- and intermolecular addition reactions. Low-molecular-weight multifunctional compounds usually serve as curing agents. Information about lifetime, melt or solution viscosity and cure reaction regimes relates to a resin with a selected curing agent.

One can stop the synthesis of a high-crosslink-density polymer at stage A or prepolymer stage. Then, using the low viscosity of the resin or prepolymer melt, one can impregnate the fibrous filler. A low-viscosity formulation penetrates into the interfibre space, wets the surface of elementary fibres (monofilaments) and spreads into a uniform thin film. In the adsorption zone the structure of adsorbed composition is created.

A continuous plait of fibres, tape or textile (fabric) is impregnated by resin in impregnating baths. The process is carried out by immersion of filler into melt or solution, moving the former through guide rollers. Excess resin or prepolymer is pressed out in the gap between the pressing rollers, which are fixed above the impregnating bath. Better impregnation of the filler and more uniform resin distribution are achieved when the tape or textile passes between two rollers rotating towards each other, and resin is distributed on their surface.

Impregnation is followed by a preliminary cure stage, which is the transition to stage B. The reaction of chemical conversion of the resin is realized in horizontal driers.

This stage is intended to remove the solvent and accompanying low-molecular-weight components and to increase polymer molecular weight. As molecular weight increases, separation of the resin from the fibres on storage of the prepreg is prevented, stickiness is lowered, shrinkage decreases (when the polymeric film is cured in a mould) and the duration of moulding of an article decreases.

If low-molecular-weight compounds are evolved in the cure process (polycondensation), then these substances are removed. The degree of conversion of chemical processes is determined by the polymer melt viscosity that is necessary for the selected conditions (temperature and pressure) of monolithization and moulding of the material. As the mechanism of cure is polycondensation, the prepreg is precured to the pregel stage to provide minimum formation of low-molecular-weight compounds in a closed mould. A deep precure increases polymer melt viscosity. Mould filling and monolithization of the material are conducted in this case under a pressure above 20–25 MPa.

It is desirable to conduct precure in wells equipped with lamp heating. Ultraviolet radiation provides heating for the deep layers of prepreg and unhindered evaporation of low-molecular-weight substances from the film surface.

If the impregnation composition does not contain solvent and cure proceeds without evolution of low-molecular-weight substances (polymerization, polyaddition), then the precure stage is unnecessary and the prepreg goes to lay-out of the packet or winding the half-finished product (liquid-phase winding), and then on to material monolithization, carried out at low pressure (up to 1–5 MPa).

Precure duration and temperature are determined in each particular case by resin composition, rate of cure reaction and holding time at intermediate stages.

The material is kept in the moulding equipment (instrument) until the article gains sufficient shape stability to remove it from the heated mould without fear of warping. The final cure of deep layers and lowering of the stresses caused by non-uniform (due to article thickness) cure are realized outside the forming instrument (thermal treatment or additional cure).

Cure in a closed mould proceeds under conditions that are not favourable for chemical reactions. The composition has low viscosity and low heat conductivity. Heat transfer is realized from the mould walls only. The exothermicity of the reaction is determined by resin composition and process mechanism. The temperature gradient in the bulk of the material depends on article thickness, heat conductivity and filler heat capacity and on the amount of low-molecular-weight substances formed in the cure process.

The fibre surface influences substantially the cure process of thin resin films [20]. The effect of selective sorption of components becomes apparent in an adsorption zone. Sorption changes the composition of reacting components, and the effect of accelerating or retarding the cure reaction is observed [21–23]. It was stated that the film surface can also take part in cure due to its reactive functional groups [24–26].

As a result of incomplete wettability of the fibre surface, components of air can remain thereon. These components can influence the cure reaction. The latter promotes the heterogeneity of the polymeric phase in a composite material and increases tension in the layers that adjoin directly to the films or in the transient region, where the influence of adsorption forces is less perceptible.

The resin cure process in contact with the fibres can be followed, which will be used in the composition of composite material. It is convenient to use 'the torsion method'. Here the sample is a thread (plait) manufactured from supposed filler. The sample is wetted with resin composition, which transforms to polymeric matrix after cure. The period of the damped torsion oscillations can be used to follow the rigidity increase as liquid composition cure proceeds, in a microsample, imitating the cure process at selected temperatures [9, 27].

The degree of conversion at the final stage of cure can be estimated by the number of functional groups that remain in the polymer.

If the reaction proceeds without evolution of low-molecular-weight substances and has sharply expressed exothermic effect, then the degree of conversion in the cure process can be estimated by the differential thermal analysis (DTA) method or by differential scanning calorimetry (DSC) [28].

The limited degree of conversion of functional groups and therefore limited glass transition temperature are for many high-crosslink-density polymers outside the limits of polymer thermodegradation. In this case the reaction is terminated at an intermediate stage of cure characterized by a glass transition temperature that corresponds to a given stage of conversion (T_g). The conventional glass transition temperature corresponds to the cure temperature or exceeds it by 20–30°C [29].

The resin or prepolymer cure reaction can be retarded by lowering the temperature and introducing reaction inhibitors. The holding time of resin or prepolymer at an intermediate stage is characterized by the lifetime index of the composition as a whole.

Composite material is moulded long before the transition of the matrix to the gel state. The transition of the polymer from viscous flow state to plastic one is completed in a forming instrument and the slow cure stage commences. The residence time of the material in the mould is determined by its transition to a size-stable state at the temperature of curing. Maximum possible degree of reaction completeness is obtained at the final stage of the technological process – thermotreatment stage (postcure). The stability of property parameters is increased and tension in the material is reduced.

The brittleness of high-crosslink-density polymers is the main shortcoming of these materials, as cracks in a thin polymeric matrix upset hermeticity and uniformity of redistribution of stresses on the filler. Therefore, modification of high-crosslink-density polymers is directed preferably to increasing the fracture energy and deformability index under loading.

It is proposed to decrease the physical junction concentration in a network polymer to increase the fracture energy. The copolymer of polybutadiene and styrene (which fulfils the function of a curing agent) is an example of these polymeric matrices. However, elevated impact strength is combined with low thermostability and low strength of adsorption interaction with fibres [30–32].

If traditional plasticizers are introduced into the initial composition and these screen some of the oligomer polar groups, then, as cure proceeds, the defectiveness of the network polymer increases, especially in the intermicrogel region, the density of chemical junctions decreases and the free volume increases [33]. This becomes apparent in lowering the glass transition temperature, elasticity modulus and hardness, and in increasing the thermal expansion coefficient [34]. The main part of the plasticizer is displaced into the zone of contact with the fibres' surface, lowering the cohesive strength, and into the boundary with the face of the mould.

To prevent seizure of plasticizer, the method of 'structural plasticization' is used. A flexible-chain oligomer compatible with the main prepolymer or resin and containing reactive functional groups in the end units is selected. This oligomer is necessary for taking part in cure. Non-polar flexible-chain oligomers are inserted into interjunction fragments. Structural modification does not reduce the degree of conversion of the functional groups, but makes interjunction fragments longer and increases their flexibility, and reduces the number of physical junctions in the network polymer. The fracture energy increases, but heat resistance and modulus of elasticity of the network polymer are reduced.

Some other mechanism of structural plasticization operates in the case when non-polar oligomers have a functional group in one end of the monomeric unit. A monofunctional oligomer taking part in a cure reaction interrupts the branching of macromolecules, and thereby reduces the number of intramolecular cycles (i.e. microgel density) but simultaneously increases the number of breaks in the microgel region. Such a method of plasticization lowers the structural heterogeneity of the network polymer, and a larger volume of the material is drawn into resisting crack propagation. The heat resistance and modulus of elasticity decrease to a lesser degree. Structural plasticizers that are involved in resin composition in the initial stage fulfil the function of lowering the viscosity by dilution and facilitating wetting of fibrous filler ('active diluents'). After curing the plasticized polymeric matrix, the tension reduces in the zone adjacent to the fibre surface.

To increase the resistance of high-crosslink-density polymers to impact loading while preserving inherent deformation heat resistance and elasticity modulus, the elasticization process is used. An elasticizer is introduced into the initial composition. It is a flexible-chain pleinomer or polymer compatible with the initial composition. The end units of the elasticizer molecule contain functional groups that can take part in the curing reaction of general oligomer. Differing from structural plasticizers, an elasticizer evolves in a cured polymer as a fine dispersed elastic or plastic phase that is chemically bonded with the high-crosslink-density polymer [35–37]. Elasticizer particles of 1–2 μm are distributed uniformly in a glass-like matrix with clear phase boundary [38]. The glass-like matrix provides rigidity and heat resistance, and the elastic one promotes dissipation of impact energy and changes the character of crack propagation. The elasticization effect depends on the phase structure of cured composition, the shape and size of elastic phase particles, its volume fraction, its deformation strength properties and the adhesive strength with the matrix.

Elasticizer is introduced into a resin as powder, viscous liquid or monomer mixture, forming a disperse phase of elastic polymer during resin cure. It is necessary that the elasticizer should be as stable against thermodegradation as the basic polymer.

Heterogeneous structures of cured polymer with elastic inclusions and therefore with optimum elasticizer effect are attained only for definite differences between the solubility parameters of elasticizer in the polymer to be cured. This is achieved by selection of elasticizer macromolecule units and molecular weight. This allows one to increase 10-fold the fracture energy of a high-crosslink-density polymer and the stress intensity coefficient by 1.5–3 times, while preserving the heat resistance and modulus of elasticity inherent to the basic polymer. If phase separation is not very good or there is a too fine dispersion of elasticizer in the polymeric matrix, the presence of flexible-chain polymer becomes apparent to a greater degree as a plasticization effect. As a consequence, the fracture energy increases less sharply and is followed by heat resistance and elasticity modulus decrease.

The elasticization effect for high-crosslink-density polymer is reduced in a thin matrix film distributed between elementary fibres in the composite material.

Apparently, deformation is hindered in a thin film. Besides that, the most overstressed zone in a fibrous composite material is the matrix layer absorbed by the fibre surface. The presence of elasticizer in this layer is unlikely [38].

It was shown experimentally that the fibre surface sorbs preferentially the more polar components of the polymer to be cured and the less polar plasticizer or elasticiser are not so readily sorbed. As a result, the danger zone remains brittle as before and the volume and character of the elasticizer distribution change in the polymeric matrix.

To prevent crack propagation in the fibre contact zone with a polymeric matrix, it is necessary to introduce a thin (20–30 nm) layer of elastic polymer on the fibre surface. The elastic boundary layer must preserve strong cohesion with fibres by lowering tension in the polymeric matrix [39–42].

Resin cure processes are accompanied by sharp (by 15–20%) volume change. To lower shrinkage of a polymeric film, various methods are proposed. Preference is given to cure processes that are not accompanied by formation of low-molecular-weight substances; the stage of preliminary cure is made longer outside the forming instrument; or a thin powder of mineral filler is dispersed in a resin. A linear polymer incompatible with the network polymer can also be added.

All the above methods have some limitations. Removal of low-molecular-weight components from the initial composition as well as use of deep preliminary cure increases resin viscosity at the stage of article moulding. This requires higher pressures for monolithization. The resin viscosity increases sharply in the presence of a filler or linear polymer powder in the resin at the stage of fibre wetting. The powder does not penetrate into the interfibre space, aggregates and slightly upsets matrix monolithization. To lower the sticking of fine powder particles and composition viscosity, filler particles are manufactured in spherical shape. To decrease material mass, the particles are made hollow. Such a filler locates between plaits or in the interlayer region, lowers polymeric film shrinkage and increases the compression strength of the film. Single-crystal fibres promote considerable increase of shear strength and elasticity modulus, though they increase the resin composition viscosity [43].

To lower shrinkage in the cure process, linear polymer is introduced into resins that is incompatible with the network polymer. During cure it separates into the intergel region, increasing density and decreasing shrinkage. High-molecular-weight linear polymer addition increases resin viscosity sharply. Therefore, it is introduced only in small amounts (3–4%) into low-viscosity compositions of prepolymers, combining these with mineral filler powders.

The substances are introduced into curing resin composition in small amounts to increase polymeric matrix stability to external influences ('antipyrenes', 'antirads', photostabilizers) or to obtain a desirable decorative effect.

All these components taken together, i.e. resin, curing agent, accelerator or retarder of cure process, elasticizer or plasticizer and modifiers for changing a particular polymer property, are called a 'binder' of the composite material.

Among the great variety of curing binders, epoxy, phenol–formaldehyde, poly-imide and oligoester binders are used widely as matrices for the construction of composite materials [44, 45].

1.2.1 Epoxy matrices

Among the various curing compositions, epoxy binders satisfy to the greatest degree the requirements (especially technological ones) for fibrous composite material matrices. Epoxy resins are the main component. Resins also contain monomers and oligomers with different numbers of units and degree of branching. Epoxy groups are located in all units or in part of the end units, and hydroxyl groups in intermediate units and part of the end units. To make designations and analysis simpler, epoxy resins are called by their monomeric composition and are depicted as oligomers of linear structure with epoxy groups in the end units.

Each type of resin is characterized by average molecular weight, total number of epoxy and hydroxyl groups and viscosity at 20 or 50°C.

Below, the most widespread types of epoxy resins are given, which are identified by monomer structure.

Diglycidyl ether or bisphenol A (DPhP) ('dian resins')

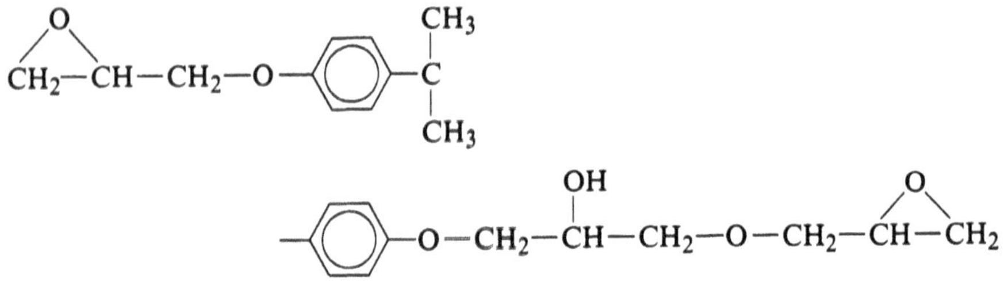

	Epoxy group content (%)	Viscosity (Pa s)
ED-24	23.5	6–10 at 20°C
ED-24N	23.5	4–6 at 20°C
ED-22	21.1–22.3	7–12 at 20°C
ED-20	19.9–22.0	12–25 at 20°C
ED-16	16.0–18.0	3–20 at 50°C
ED-14	13.9–15.9	20–40 at 50°C
UP-64	6.0–9.0	9–14 at 50°C

Epon-1001, DER-661: softening temperature, 65–75°C.
DER-332, DER-300, Epon-825, DER-331, Epon-826: epoxy group content, 17.1–20%.

Diglycidyl ether of bisphenol F (DPhM)

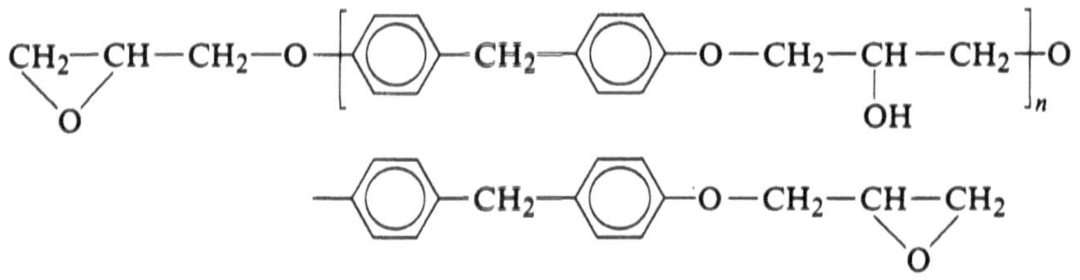

DGEBPh, XD-7818: epoxy group content, 15.8–16.5%; viscosity, 5–8 Pa s.

Diglycidyl ether of aniline

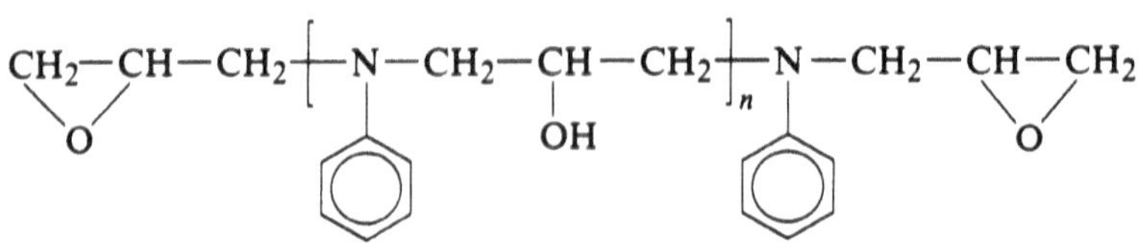

BA: epoxy group content, 31%; viscosity, 0.3–0.5 Pa s.

Diglycidyl ether or resorcinol

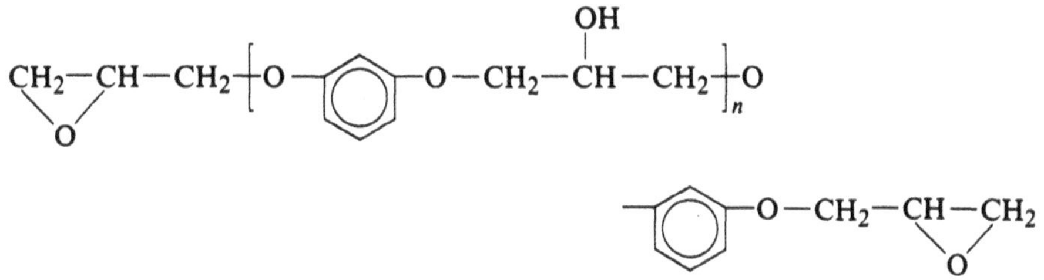

RES, UP-637 EPE-1359: epoxy group content, 30%; viscosity, 1–0.5 Pa s.

Diglycidyl ether or methyltetrahydrophthalic acid

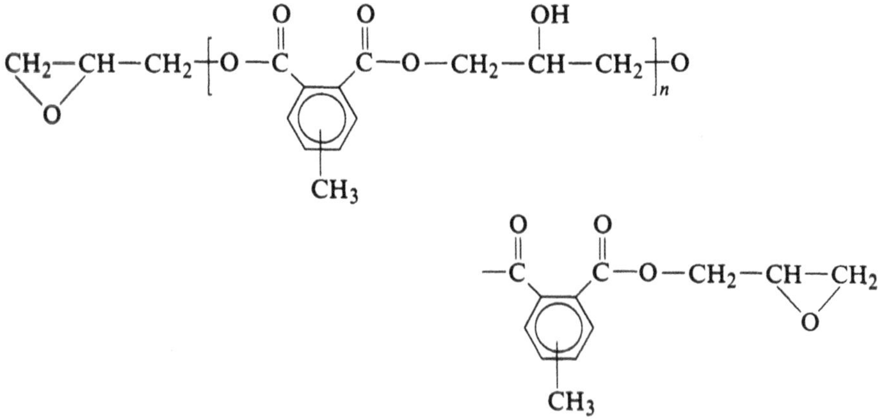

UP-640: viscosity, 10–1.5 Pa s.

Triglycidyl ether of p-aminophenol ('triepoxide') and its oligomers

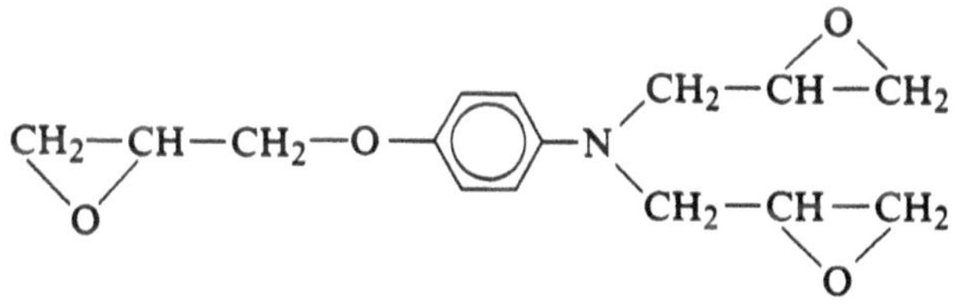

UP-610: epoxy group content, 33–40%; viscosity at 40°C, 1.0–2.0 Pa s.
ERL-0510: viscosity, 0.55–0.85 Pa s.

Triglycidyl isocyanurate and its oligomers
EC, EC-N, EC-K: epoxy group content, 30–38%; softening temperature, 95–115°C.

Tetraglycidyldianilinemethane and its oligomers ('tetraepoxide')

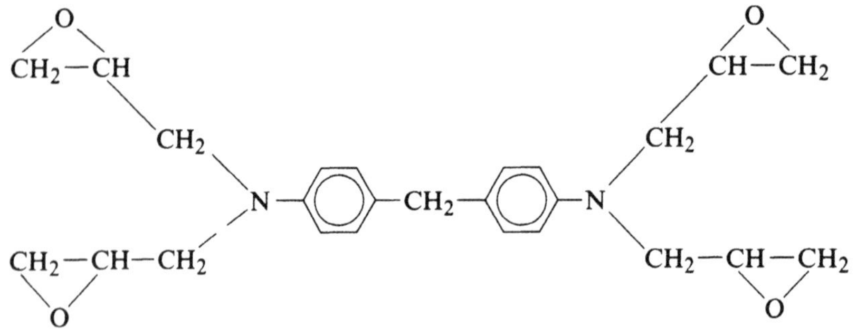

EHD: epoxy group content, 26–30%; viscosity at 50°C, 12–13 Pa s.
Araldite MY-720: viscosity at 50°C, 10.5–15 Pa s; and low-viscosity trademark, 3.5–6.0 Pa s at 50°C (with elevated monomer content).

Polyglycidyl ethers of novolac resin

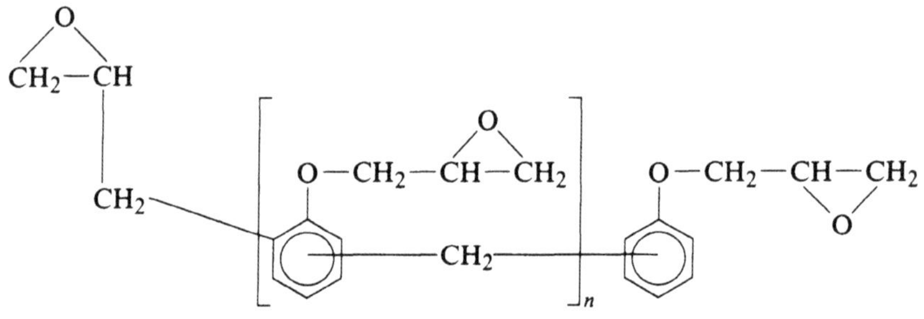

EN-6, 5-N, ETPh, UP-643; epoxy group content 17–23%; softening temperature, 40–50°C.
DEN-431: viscosity, 1.1–1.7 Pa s at 52°C.
DEN-438: viscosity, 20–50 Pa s at 52°C.
XD-785500: viscosity, 69–77 Pa s.
DEN-439: viscosity, 4–10 Pa s.
ERR-0100: softening temperature, 85–100°C.

Dioxide of the cyclic acetal

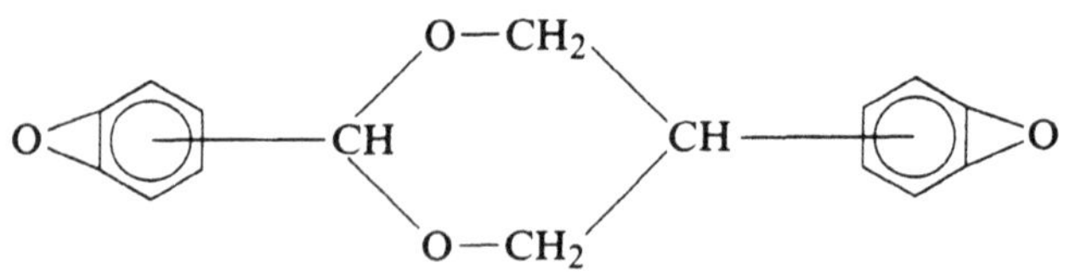

UP-612: epoxy group content, 27–29%; viscosity, 6–10 Pa s at 40°C.

Dioxide of the ester of tetrahydrobenzyl alcohol and sebacic acid

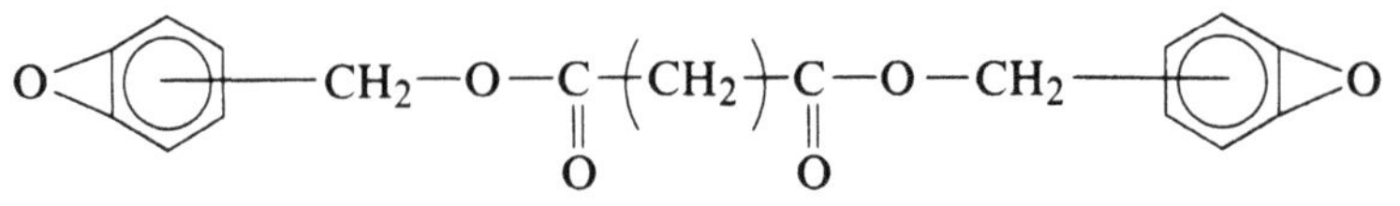

$$O\!\!<\!\!\bigcirc\!\!>\!\!-CH_2-O-\underset{\underset{O}{\|}}{C}\!-\!(CH_2)\!-\!\underset{\underset{O}{\|}}{C}-O-CH_2-\!\!<\!\!\bigcirc\!\!>\!\!O$$

UP-648: epoxy group content, 18–20%; viscosity, 0.4–0.5 Pa s.

Tetraoxide of tetrahydrobenzoic acid and pentaerythritol

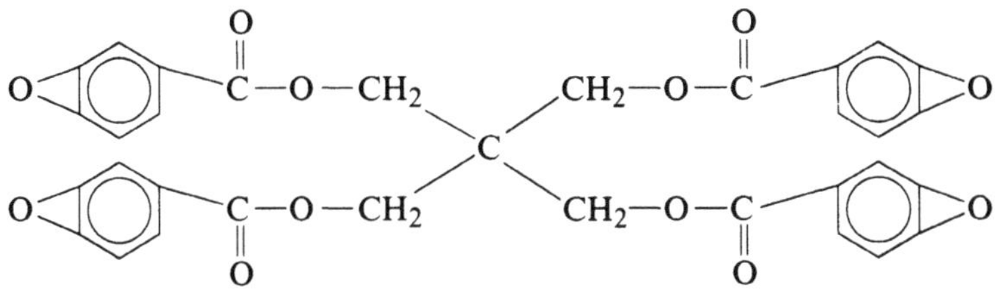

$$O\!\!<\!\!\bigcirc\!\!>\!\!-\underset{\underset{O}{\|}}{C}-O-CH_2 \qquad CH_2-O-\underset{\underset{O}{\|}}{C}\!\!<\!\!\bigcirc\!\!>\!\!O$$
$$\overset{}{C}$$
$$O\!\!<\!\!\bigcirc\!\!>\!\!-\underset{\underset{O}{\|}}{C}-O-CH_2 \qquad CH_2-O-\underset{\underset{O}{\|}}{C}\!\!<\!\!\bigcirc\!\!>\!\!O$$

To illustrate the multicomponent nature of epoxy resins, the content of the most widespread resin is given in Table 1.1 as an example of a material prepared by reaction of epichlorohydrin with p, p'-dioxydiphenylpropane (bisphenol A), which is called 'dian resin' (ED resin).

Table 1.1 Oligomeric diepoxide fraction content in 'dian' resins with various molecular weights

Average epoxy resin molecular weight	Functional-group content (%)		Oligomer fraction content (%)				Aggregate state
	Epoxy groups	OH groups	Monomer	Dimer	Trimer	Rest[a]	
350–400	23–21.5	0.1–0.8	92–85	8–15	2–30	0	Liquid at 40°C
400–600	21.5–14.5	0.8–2.5	85–50	15–20	8–10	5–10	Liquid at 40°C
600–800	14.5–10.0	2.5–4.6	50–20	12–16	8–11	45–50	Viscous liquid at 100°C
1000–1400	8.0–6.0	5.1–6.0	13–8	7–9	8–10	70–75	Solid, $T_{soft}=$ 50–55°C
1400–1800	6.0–4.0	6.0–6.5	6–4	6–8	8–10	80–85	Solid, $T_{soft}=$ 70–85°C
1800–3500	4.0–2.0	6.5–6.8	4–2	3–5	5–8	83–90	Solid, $T_{soft}=$ 85–100°C

[a] The remaining amount: mono- and diepoxides with different number of units and the products of hydroxyl group interaction in the central (middle) units of oligomers with epichlorohydrin or epoxyoligomers.

Depending on the relationship of monomer and low-unit oligomers to high-unit oligomers, the viscosity and composition molecular weight change. The lower the resin molecular weight, the greater is the number of epoxy groups and the lower is the number of hydroxyl groups.

Epoxy resins are stable on storage, and they are soluble in acetone, ethanol, methyl ethyl ketone, toluene and xylol. Low-viscosity epoxy resins can be prepared by displacing diphenylolisopropane by diphenylolmethane. The epoxy-group content increases in the resin simultaneously (until 24%). The viscosity of the mixture of resin with triethylenetetramine equals 4.8 Pa s at 20°C [46]. To lower the initial composition viscosity, different resins are mixed. For example, low-viscosity 'tri-epoxide' resin or diepoxyresorcinol resin are added to higher-viscosity resin ('tetra-epoxide').

The initial epoxy resin viscosity decreases to a greater degree if a small amount (10–15 mass %) of liquid or monoepoxide are added. Liquid diepoxides taking part in a cure reaction lengthen interjunction fragments and bring about greater flexibility ('structural modification').

The brittleness of the cured polymer decreases as well as the modulus of elasticity and heat resistance. Liquid monoepoxides are 'active diluents' in the cure process and interrupt branching, introducing structural defects into microgels and terminating interjunction fragments. The tension decreases as well as polymer brittleness. The heat resistance and modulus of elasticity decrease less sharply than in 'structural modification'. Some types of epoxy resin diluents are given in Table 1.2. Examples of compositions with low viscosity are given in Table 1.3.

In contact with substances containing proton donor functional groups, epoxy resins take part in polyaddition reaction:

$$H-R'-H + CH_2-CH-R''-CH-CH_2 \longrightarrow$$

$$-R'CH-CH-R''-CH-CH_2-$$
$$OH \qquad OH$$

Amines, phenols, alcohols and acids belong to the substances having proton donor groups. All these can be curing agents for epoxy resins if the number of mobile hydrogen atoms provides 'chemical junction' formation [47].

Reaction of viscid resins with low-molecular-weight curing agents even in the final stages is accompanied by less steric and diffusional restrictions as compared with oligomeric curing agents. Their reactivity is used therefore to a greater degree.

Cure of epoxy resins is accelerated by catalysts weakening the epoxy ring stability. Lewis acids can serve as catalysts, and tertiary amines and phosphorus too [48–50]:

$$R_3N^{\delta+}\cdots\cdots{}^{\delta-}O \begin{array}{c} CH \\ | \\ CH_2 \end{array} \longrightarrow R_3N^{\delta+}\cdots\cdots{}^{\delta-}O-CH-\!\!\sim\!\!\sim$$
$$CH_2-\!\!\sim\!\!\sim$$

Table 1.2 Active diluents and 'structural plasticizers' of epoxy resins

Monomer structure and name	Epoxy-group content (%)	Viscosity (Pa s)
CH_2—CH—CH_2—O—(phenyl) (epoxide ring on CH_2—CH) Glycidyl ether of phenol (PEG)	22	0.003–0.006
CH_2—CH—CH_2—O—(phenyl)—CH_3 (epoxide ring on CH_2—CH) Glycidyl ether of cresol	20.2	0.005–0.05
CH_2—CH—CH_2—O—$(CH_2)_7$—CH_3 (epoxide ring) Glycidyl ether of octanol	11.0	0.010
CH_2—CH—CH_2—O—$(CH_2)_3$—CH_3 (epoxide ring) Glycidyl ether of butanol	21.5	0.002–0.003
CH_2—CH—CH_2—O—(phenyl)—C$(CH_3)_3$ (epoxide ring) Glycidyl ether of t-butylphenol	16.0	0.015–0.030
CH_2—CH—CH_2—O—$(CH_2)_4$—O—CH_2—CH—CH_2 (epoxide rings on both ends) Diglycidyl ether of 1,4-butane diol (RD-2, DEB)	20.7	0.01–0.25
CH_2—CH—$CH_2$$\left[\text{O}-CH_2-\text{CH}(CH_3)\right]_n$O—$CH_2$—CH—$CH_2$ (epoxide rings on both ends) Diglycidyl ether of poly(propylene glycol)		
$\quad$ $n \simeq 4$, DER-736	27.0	0.03–0.06
$\quad$ $n \simeq 9$, DER-732	15.3	0.055–0.1
CH_2—CH—$CH_2$$\left[\text{O}-CH_2-CH_2\right]_n$O—$CH_2$—CH—$CH_2$ (epoxide rings on both ends) with CH_3 branch 		
$\quad$ $n = 1$	28.5	0.06
$\quad$ $n = 2$ Diglycidyl ether of ethyleneglycol	24.0	0.07
$\quad$ $n = 3$	19.0	0.09

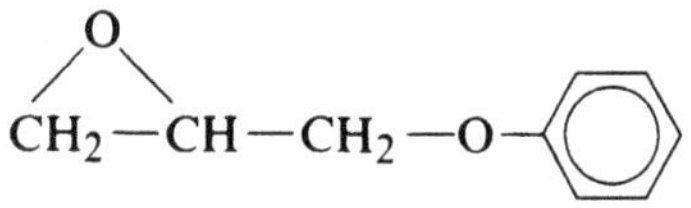

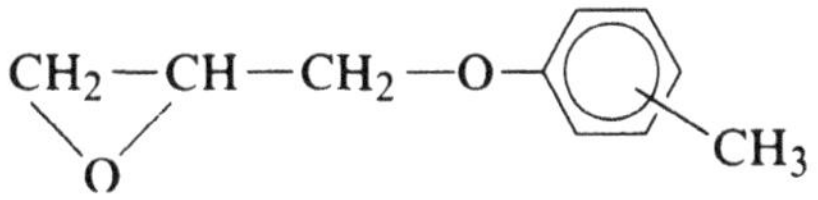

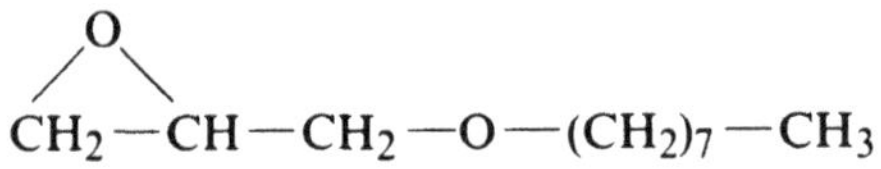

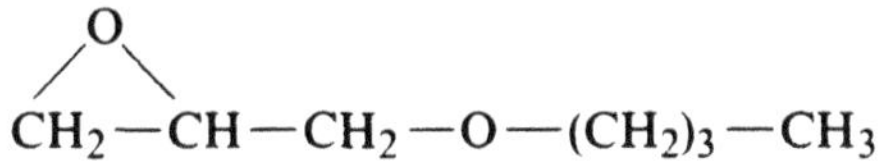

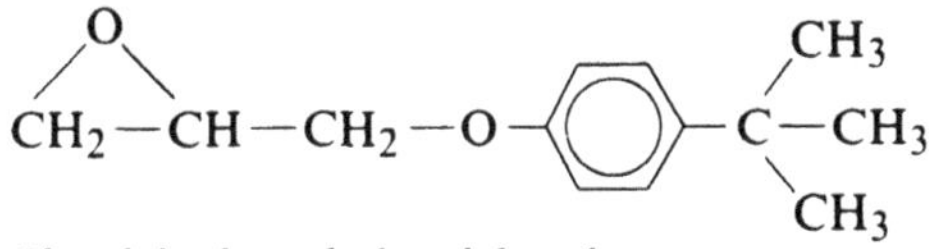

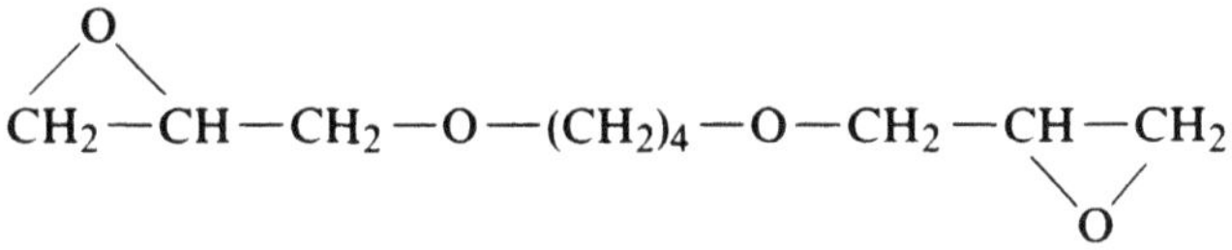

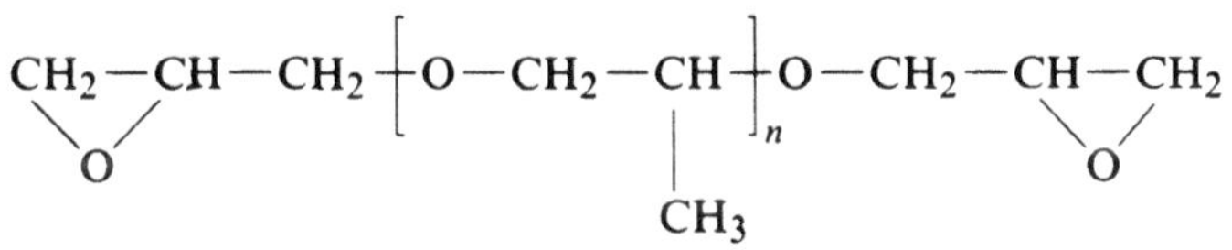

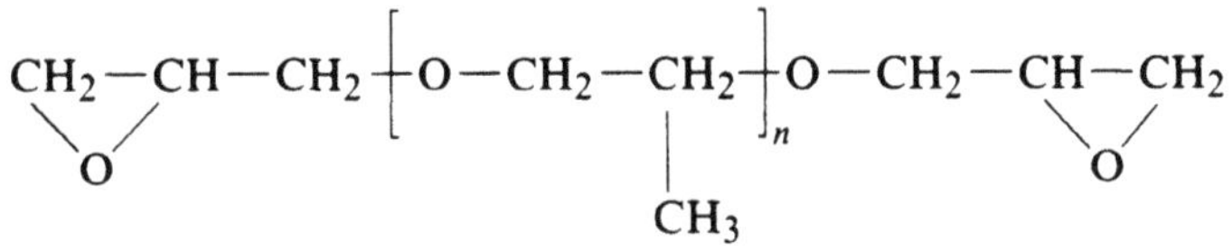

Table 1.3 Epoxy resin compositions with low viscosity

Resin	Ratio (%)	Viscosity (Pa s)
Diglycidyl ether of bisphenol A	80	1.20 at 20°C
Diglycidyl ether of 1,4-butanediol	20	–
Diglycidyl ether of bisphenol A	30	–
Diglycidyl ether of linoleic acid	50	0.54 at 20°C
Diglycidyl ether of n-amylglycol	20	–
Diglycidyl ether of bisphenol A	85	–
Monoglycidyl ether of phenol	15	0.50 at 60°C
Epoxynovolac resin	70	30 at 60°C
Dian resin ED-22	30	–
Tetraepoxide	50	–
Epoxynovolac resin	30	36.0 at 50°C
Diglycidyl ether of 1,4-butanediol	20	–

As polarity increases, the electron density shift strengthens and rupture becomes easier.

To increase the lifetime of a prepreg with a binder that contains a curing catalyst, the latter is introduced in salt form ('latent' catalyst). As the temperature increases, the salt complexes are destroyed, and the catalytic activity is regenerated. Lewis salts (e.g. boron trifluoride) form complexes with monoamine. After the complex is destroyed, the monoamine takes part in the cure reaction as a lengthener of interjunction fragments. The catalyst is introduced as a solution, for example, a solution in diethylene glycol.

Table 1.4 Accelerators of cure of epoxy resins

Accelerator	Structure	Temperature
Amine complexes with BF_3		
Ethylamine	$C_2H_5H_2N{:}F_3B$	$T_m = 85-93°C$
Aniline	$C_6H_5H_2N{:}BF_3$	$T_{degrad} = 120°C$
Benzylamine	$C_6H_5{-}CH_2{-}H_2N{:}BF_3$	$T_m = 125-127°C$
p-Toluidine	$CH_3C_6H_4H_2N{:}BF_3$	$T_m = 45°C$
Quaternary bases		
Triethylbenzylammonium chloride	$[(C_2H_5)_3(C_6H_5CH_2)N]^+ Cl^-$	
Trimethylbenzylammonium chloride	$[(CH_3)_3(C_6H_5CH_2)N]^+ Cl^-$	
Benzyltriphenylphosphonium chloride	$[(C_6H_5)_3(C_6H_5CH_2)P]^+ Cl^-$	
Methyltributylphosphonium chloride	$[CH_3(C_4H_9)_3P]^+ Cl^-$	
Salts		
Manganese triacetylacetonate		
Chromium acetylcetonate		
Zinc dioleate or distearate		

Examples of catalysts used for accelerating epoxy resin cure processes are given in Table 1.4.

(a) Curing epoxy resins

Cure with amines

To cure epoxy resins, aliphatic and aromatic amines with two or more amino groups can be used (see Table 1.5). To distribute curing agent uniformly in resin, a common solvent is used, which is removed after wetting fibrous filler, or the resin and curing

Table 1.5 Amine curing agents for epoxy resins

Amine name and structure	T_m (°C)	Viscosity (Pa s)
Diethylenetriamine (DETA) $H_2NCH_2-CH_2-\underset{\underset{H}{\mid}}{N}-CH_2-CH_2NH_2$	–	0.0055–0.0085
Triethylenetetramine (TETA) $H_2NCH_2-CH_2-\underset{\underset{H}{\mid}}{N}-CH_2-CH_2-\underset{\underset{H}{\mid}}{N}-CH_2-CH_2NH_2$	–	0.020–0.023
m-Phenylenediamine (MPDA)	60	–
4,4′-Methylenedianiline (4,4′-diaminophenylmethane)(MDA) $H_2N-C_6H_4-CH_2-C_6H_4-NH_2$	170–180	–
4,4′-Diaminodiphenyl ether (DDPE) $H_2N-C_6H_4-O-C_6H_4-NH_2$	–	–
2,6-Diaminopyridine (DAP)	121	–
Dicyanodiamide $H_2N-\underset{\underset{HN}{\parallel}}{C}-\underset{\underset{H}{\mid}}{N}-C\equiv N$	207–209	–

agent are heated to curing agent melting temperature, mixed and melt is distributed on the fibre. In the latter case composition viscosity begins to increase at prepreg preparation stage.

Therefore liquid amines or eutectic mixtures of amines (Table 1.6) are used preferably as curing agents. To facilitate compatibility of high-viscosity resins with curing agents, low-melting-point or liquid amine adducts with glycidyl ethers were proposed: oxyethylated diethylenetetramine (UP-0619), oxyethylated polyethylene-polyamine (UP-0622), adduct of ED-20 with excess of diethylenetriamine (UP-616), adduct of ED-22 with excess of diethylenetriamine (UP-620), and monocyanoethylated diethylenetriamine (UP-0633M) (viscosity 0.115–1.5 Pa s).

Curing rate increases with increase of amine basicity. The reaction of epoxy resins with aliphatic amines proceeds at a great rate at 20°C. Until the gel point the rate increases continuously as a result of the exothermal nature of the process ('cold cure'). Outside the limits of gel formation, viscosity increases sharply and increase of temperature is required up to 100–120°C.

The lifetime of epoxy resin mixture with aliphatic amine curing agent could be increased if the amine is first encapsulated. At elevated pressure and under heating, the thin film of the capsule is destroyed and the cure process begins. Despite the substantial increase of composition lifetime, heating agent encapsulation does not have widespread application. That is associated with the necessity to form prepregs at pressures sufficient for capsule destruction but with uniform distribution of curing agent in the resin, which causes increased uniformity of the polymer cured [51].

The rate constants of epoxy group reaction with the first hydrogen atom of amino groups is higher than with the second one. In curing by aromatic amines, the difference in rate constants attains 10-fold order.

Table 1.6 Eutectic mixtures of aromatic amines

Amine	Content (%)	Viscosity (Pa s)
1. MPDA	33.3	
2. MDA	33.3	5
3. Isopropylphenylamine	33.4	
1. MDA	40	
2. Diaminodiethyldiphenylmethane	60	2–5
1. MPDA	40	1.5
2. MDA	60	

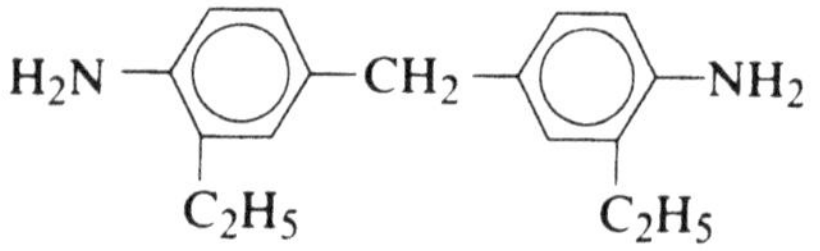

$$H_2N-\!\!\bigcirc\!\!-CH_2-\!\!\bigcirc\!\!-NH_2$$

with C_2H_5 substituents on each ring

Therefore, the length of interjunction fragment in excess of amine curing agent increases in a network polymer, the glass transition temperature decreases and so do heat resistance and elasticity modulus. Thus, in cured 'dian' resin with stoichiometric amount of m-phenylenediamine, molecular weight of interjunction fragment $M_c = 370$ and $T_g = 170°C$. If the curing agent is used in two-fold quantity, then M_c increases up to 3000 and T_g decreases to 100°C despite complete epoxy-group conversion [52].

For excess of epoxy groups in relation to amino groups, the number of defects in a network polymer increases, which causes glass transition temperature lowering [53]. It was shown on the example of epoxy 'dian' resin curing by m-phenylenediamine.

The glass transition temperature of a polymer increases from 71 to 170°C with increase of number of amine groups from 50 to 100%. After that, as it increases to 170%, glass transition decreases to 102°C.

The amount of amine groups in the curing agent influences analogously cured tetraepoxide structure. Thus the viscosity increases up to the transition to the solid state to which T_g corresponds. If complete conversion of functional groups is impossible due to intensive degradation, then T_g attained as a result of cure will be below the temperature limit possible for the given system. In epoxy polymers the limited T_g is a success as a rule to be attained before thermodegradation begins.

At 20°C the lifetime of a mixture of epoxy resins with aromatic amines is limited to 10–12 days and becomes shorter if cure catalyst, even in complex form, is contained in the composition. At 180–200°C maximum cure rate can be obtained. As the reaction mass viscosity increases sharply, the rate of intramolecular cyclization and therefore defectiveness of polymer structure increases sharply. This forces one to carry out the reaction step by step, increasing the temperature and staying some time at each stage. Some information about the cure process of 'dian' resin by aromatic amines is given in Table 1.7.

Table 1.7 Curing 'dian' resins by aromatic amines

Amine	Temperature (°C)	Rate of gelformation (h)		Temperature of maximum cure rate (°C)	T_g (°C)
		at 130°C	at 175°C		
m-Phenyl- enediamine	70	2.75	1.25	180	155
4,4'-Diamino- diphenyl- methane	110	9.35	1.90	200	185
4,4'-Diamino- diphenyl ether	130	9.43	1.95	200	190
4,4'-diamino- diphenyl- sulphone	120	8.70	5.80	220	208

Displacement of the aromatic amine by an aliphatic one leads to lowering of the glass transition temperature of the cured polymer and increase of elasticity. The greater the number of methylene groups in the aliphatic amine, the sharper becomes the difference in properties of network polymer cured by aromatic amine from that cured by aliphatic amine.

Increase of number of methylene groups in linear aliphatic diamine from 2 to 12 ensures the decrease of glass transition temperature of epoxy polymer from 150 to 55°C, bending strength from 98 to 67 MPa and tensile strength from 56 to 40 MPa; impact strength increases from 8 to $30\,kJ\,m^{-2}$ and elongation at break from 3.5 to 7.0%.

The tendency to intramolecular cyclization of tetraepoxide resins in reactions with aromatic amines is especially high. So, in reaction with diaminodiphenylsulphone, the number of epoxy groups taking part in the cyclization reaction reaches 75%. The tendency to cyclization increases as viscosity increases. If cure begins at 130°C and a slow, step-by-step, temperature increase is used, then it could support the minimum (for a given composition) viscosity and reproducibility of the structure of cured polymer is possible. At elevated temperature of cure, both secondary hydrogen atoms of amino groups and epoxy groups react with hydroxyl groups and hydroxyl groups react with each together, and the polymer density decreases and melt flow index increases [54].

A composition intended for liquid-phase winding or impregnation of filler must have viscosity not above 6 Pa s and it must increase more than two-fold for 6–8 h. In that case resins with elevated monomer content are used. This causes polymeric network density and brittleness increase (elongation decreases up to 1.2–1.8).

A particular place among amine curing agents is held by dicyanodiamide as the lifetime of its mixture with epoxy resins is several weeks. The melting temperature of dicyanodiamide is above 200°C and it is insoluble in the usual solvents for epoxy resins. To achieve more uniform distribution of curing agent in resin, dicyanodiamide is dissolved in water and introduced into resin emulsion. The emulsion is distributed on fibrous filler and resin film is dried to remove water. The cure is carried out at 150–175°C up to $T_g = 120$°C (VES-21 composition). The maximum rate of curing 'dian' resin by dicyanodiamide is observed at 205–210°C (activation energy is $137\,kJ\,mol^{-1}$) [55].

Cure with anhydrides

To increase the polymer network density of chemical junctions, one must draw into the cure reaction both epoxy and hydroxyl groups including those formed after destroying epoxy rings. For this purpose, acid anhydrides are used as curing agents, and therefore, the reaction of hydroxyl groups with anhydrides is not accompanied with water evolution [56]:

The carboxylic groups formed react with the epoxide rings and a chemical junction arises from the ester group:

$$\text{structure} \quad + \quad \text{structure} \longrightarrow \text{structure}$$

The most frequently used anhydride curing agents are given in Table 1.8. The chemical junctions that are formed are less flexible than the amine ones, are more stable to thermo-oxidative degradation and are destroyed as a result of hydrolysis, especially in alkaline media. The cure by anhydrides proceeds at a temperature above 160°C and requires a prolonged stay at elevated temperatures. Therefore, the reaction is usually carried out in the presence of accelerators [57].

'Dian' resin ED-20 is cured with isomethyltetrahydrophthalic anhydride (iso-MTGPhA) for 18–20 h at 200°C. The molecular weight of interjunction fragment reaches $M_c = 830$ (determined by swelling). In the presence of manganese triacetyl-acetonate (MTA) curing is complete after 5 h at 160°C and $M_c = 536$ [58]. In the presence of BF_3 complex with benzylamine, the same composition has a lifetime at 20°C of 30 days. At 150°C a gel forms after several minutes. At 150–170°C the process can be completed after 3 h but the polymer has clearly defined heterogeneity and as a result high brittleness. Therefore the reaction was conducted at 80, 130 and 150°C for 4 h each. The mechanical properties indices increase in this case and T_g attains the value 170°C [59].

Phosphonium bases are even more active in epoxy-group reaction with acid anhydrides [60].

High temperature of curing causes activation of epoxy-group reaction with hydroxyl groups:

$$\text{structure} \quad + \quad \text{structure} \longrightarrow \text{structure}$$

and of hydroxyl groups with each together. Therefore chemical junction ether links arise in polymer network structure. As medium acidity increases, its concentration increases too.

Among the anhydride curing agents, low-melting-point and liquid curing agents are preferable and liquid eutectic mixture too. They are distributed in viscous resin, lowering the mixture viscosity. Moisture in the resin bulk or at the filler surface changes sharply with the progress of the reaction. In this case a portion of the anhydride groups are converted into acid ones, which react with epoxy groups, as the activation energy for this process is much lower than that of the etherification

Table 1.8 Anhydride curing agents for epoxy resins

Anhydride name	Structure	T_m (°C)
Phthalic anhydride	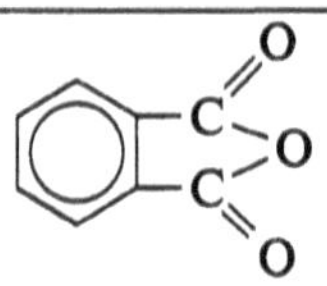	130
Hexahydrophthalic anhydride	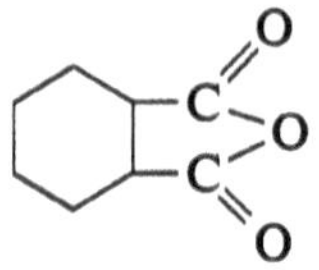	40
Methylhexahydrophthalic anhydrid	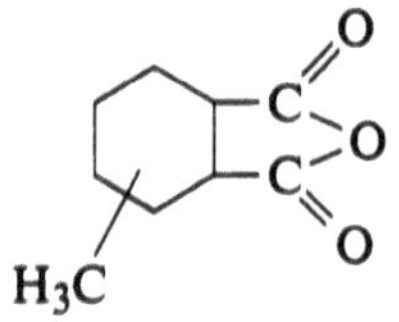	63–64
Endic anhydride	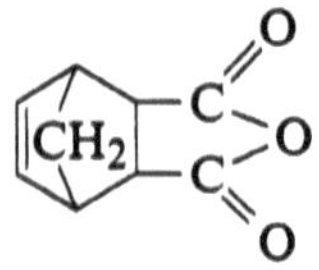	163
Methylendic anhydride	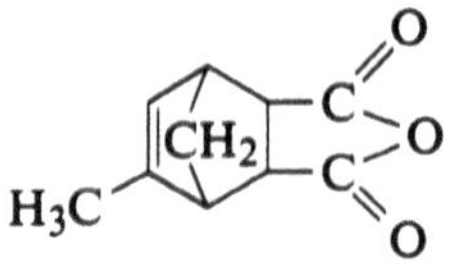	Liquid at 20°C, viscosity 0.2 Pa s
Tetrahydrophthalic anhydride	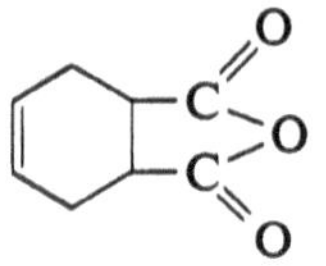	97–101
Methyltetrahydrophthalic anhydride	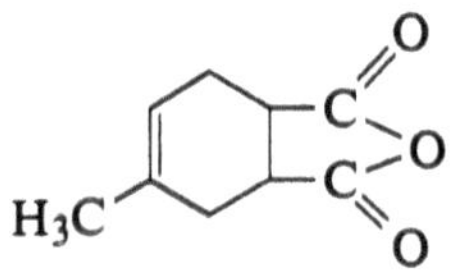	4; at 25°C, viscosity 0.06 Pa s
Trimellitic anhydride	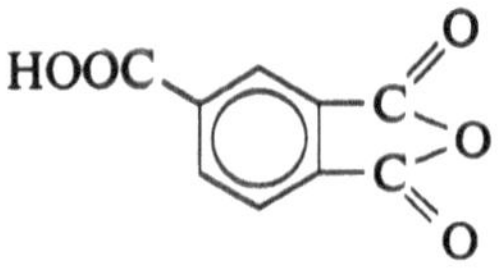	161–164

(Continued)

Table 1.8 (*Contd.*)

Anhydride name	Structure	T_m (°C)
Dodecylsuccinic anhydride	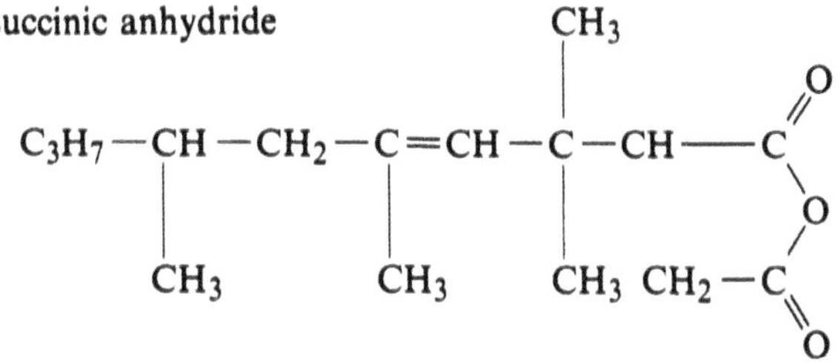 (see structure below)	At 25°C, viscosity 0.2 Pa s
Eutectic mixture: phthalic anhydride 15%, hexahydrophthalic anhydride 85%		Liquid
Eutectic mixture: maleic anhydride 40%, hexachlorendic anhydride 60%		Liquid
Mixture of methylendic anhydride isomers		Viscosity 0.17–0.27 Pa s
Mixture of methyltetra-hydrophthalic anhydride isomers		Viscosity 0.06 Pa s

Structure for dodecylsuccinic anhydride:

$$C_3H_7-\underset{\underset{CH_3}{|}}{CH}-CH_2-\underset{\underset{CH_3}{|}}{C}=CH-\underset{\underset{CH_3}{|}}{C}-CH-C\overset{O}{\underset{O}{\diagdown}}$$

process. If primary amino groups are in the binder or at the fibre surface, then acid anhydrides are converted into amides.

At the metal or mineral glass surface, curing epoxy resins in the adsorption layer facilitates the structural heterogeneity of network polymer. The greater the free-energy of the solid surface, the sharper are the heterogeneities, and microgel size attains 100–500 μm [61].

The rigidity of chemical junctions created by the reaction of hydroxyl and epoxy groups with anhydrides becomes apparent in the elevated brittleness of the cured polymer. The brittleness temperature of the polymer based on ED-20 resin cured by isomethyltetrahydrophthalic anhydride is + 96.5°C, the glass transition temperature is + 131°C and the residual stress in the polymers is between 2.2 and 4.1 MPa depending on temperature and reaction time [62].

Cycloepoxides are cured by acid anhydrides with minimum shrinkage stresses and their glass transition temperature is higher than with other curing agents. Thus the glass transition temperature in the polymer prepared by cure with tetracyclooxide of tetrahydrophthalic acid attains 224°C, and elongation at break is 3%.

Cure by isocyanates
Like the acid anhydrides, isocyanates enter into reaction with hydroxyl groups of epoxy resins at the first stage. Therefore the number of hydroxyl groups in the resin determines the initial stage of the reaction [63]:

first stage

$$\sim\!\!\!-OH + O\!=\!C\!=\!N\!-R \longrightarrow \sim\!\!\!-O-\underset{\underset{H}{|}}{\overset{\overset{O}{\|}}{C}}-N-R\sim$$

second stage

$$\sim\!\!\!-O-\underset{\underset{H}{|}}{\overset{\overset{O}{\|}}{C}}-N-R + CH_2-CH\underset{\diagdown O \diagup}{}\sim \longrightarrow \sim\!\!\!-O-\overset{\overset{O}{\|}}{\underset{\underset{CH_2-\underset{\underset{OH}{|}}{CH}\sim}{\underset{|}{N-R}}}{C}}$$

Despite the hydrogen atom of the urethane group being less reactive than the amino one, in reaction with the epoxy group the lifetime of resin mixture with atomatic diisocyanate at room temperature equals 2–3 days only. At 100–120°C gel point commences in 40–60 min.

To increase composition lifetime and polymer network density, isocyanate groups in the curing agent are blocked by amines, phenols, alcohols and lactams [64]. For instance, the complex toluylene diisocyanate and diaminodiphenylsulphone is used as a curing agent. The lifetime of epoxy resin with this complex mixture is 10–12 days. At 120–160°C the complex decomposition rate increases and both components – isocyanate and amine – are involved in cure reactions. Compared with the epoxy resin cured only by diaminodiphenylsulphone, the polymer glass transition temperature increases by 20–40°C, elasticity modulus increases and thermal expansion coefficient decreases. The presence of urethane groups in the structure of cured polymer becomes apparent in increasing the strength of adhesive interaction with filler fibres.

The stability of isocyanate complex increases if isocyanate groups in the complex are blocked by aliphatic amines or phenols. Toluylene diisocyanate complex with diethylamine is stable up to 147°C. If the complex is dissolved in piperidine, whose activity in reaction with epoxy groups is low, then the mixture of the curing agent with the resin can be kept for 14–18 days. The composition cures completely after 3 h at 90°C and 5 h at 150°C, the glass transition temperature being at the level characteristic for cure by aliphatic amines and does not exceed 120°C.

Isocyanate adducts with phenols are more stable and less toxic. The adducts are completely destroyed above 150°C, but the cure process requires higher temperature and longer duration.

Cure by compounds containing hydroxyl or phenol groups
The rate constant of the reaction of epoxy groups with the hydroxyl groups of alcohols and phenols is considerably lower than the rate constant of the reaction with amines. Therefore resin mixture with a curing agent containing alcoholic or phenolic groups can be kept for a longer time at room temperature. The reaction proceeds at sufficiently high rate at 180–200°C. Amines and quaternary bases can serve as reaction catalysts.

The reaction proceeds without evolution of low-molecular-weight substances. Hydrogen atom migration from hydroxyl groups of curing agent to epoxy groups of resin proceeds in the addition process:

$$\sim\!\!\!-OH + CH_2-CH-\!\!\!\sim \quad\longrightarrow\quad -O-CH_2-\underset{\underset{\textstyle OH}{|}}{CH}-\!\!\!\sim$$

The reaction of epoxy groups with the hydroxyls that arise in the preceding stage proceeds simultaneously.

The chemical junction of polymeric network differs from all those described above by greater flexibility and oxidative and hydrolytic stability [65].

Three atomic phenols, novolac and resol resins are used as curing agents. The viscosity of the mixture of 'dian' resins with the novolac one reaches 450 Pa s.

A liquid hardener is often used to cure epoxy resins. This agent is prepared by re-esterification of tetrabutyltitanate with triethylamine. The curing agent is called 'tetraminotitanate'. Hydroxyl groups take part in the reaction with epoxy groups, amine groups accelerating the reaction. EDT-10 composition containing 10 wt% of tetraminotitanate and 10 wt% of diglycidyl ether of diethylene glycol in 100 wt% of ED-20 resin has a liquid consistency and is recommended for the impregnation processes of collected filler or for liquid-phase winding. The mixture remains liquid for 5 days. The epoxy number is decreased from 22.4% to 15.5% after 10 days of storage. At 75°C the epoxy number is decreased to 7.5% after 11 h. The composition is cured for 1 h at 100°C, 3 h at 120°C and 2 h at 160°C [66].

There are chemical junctions in the network polymer structure as the ester of titanic acid. These junctions are destroyed gradually as a result of hydrolysis.

(b) Structure and properties of cured epoxy resins

The properties of epoxy resins depend on the degree of structural heterogeneity, the polymer network density near chemical junctions, the rigidity of the units (originating from the chemical junctions) and their chemical resistance, and the polymer network density (originating from hydrogen bonds). The resin containing the monomer and oligomers with small number of units is easy to handle in the initial stage in terms of its viscosity; however, after curing, the polymer becomes more brittle compared with the polymer prepared by curing the resin with the average number of 4–8 units in the oligomer. Among the 'dian' resins, the most preferable are those with minimum molecular-weight scatter (narrow molar-mass distribution). If the molecular-weight distribution of the oligomers in the resin is narrow, then even at a small number of units the regularity of the distribution of chemical junctions in the network polymer increases, the tension of interjunction fragments decreases and polymer fracture energy increases.

Hydrogen bonds in the polymer between adjacent hydroxyl groups or between hydroxyl groups and amine considerably influence the properties of cured polymer. As the temperature and humidity increase, the strength of hydrogen bonds falls and polymer properties change. In low-molecular-weight models, epoxy–amine polymer hydrogen bond strength is about $6-7$ kcal mol^{-1}. In epoxy polymers cured by amines, hydrogen bond strength goes down to 4 kcal mol^{-1}, which is evidence of less regulation in interjunction fragment packing [67]. In 'dian' resins cured by aliphatic amines down to $H_2N(CH_2)NH_2$, the temperature of disappearing hydrogen bonds coincides with the glass transition temperature. As the number of $-CH_2-$ groups in diamine increases, hydrogen bonds are preserved down to $132°C$ in spite of the fact that the polymer is still outside the limits of the glass state. In polymers cured by aromatic amines and especially by acid anhydrides, the hydrogen bonds are weakened by rigid chemical junctions and disappear long before the glass transition temperature [68].

The distribution of functional groups in a molecule of curing agent influences the relationship between intra- and intermolecular reactions and therefore influences polymer structural heterogeneity. It becomes apparent in polymer resistance to mechanical deformation. For 'dian' resin cured by alkyldiamine $H_2N(CH_2)_nNH_2$ and polyalkylpolyamine

$$H_2N-(CH_2)_n-N-(CH_2)_m-NH_2$$
$$|$$
$$H$$

for example, a monotonic decrease of the absorbed energy and an increase of initiation energy on mechanical fracture of the polymer as the numbers n and m increase in diamine curing agent was revealed [69]. Apparently this is caused by the formation of more flexible chemical bonds between polymer network fragments and increasing hydrogen bond strength. The presence of additional functional groups in the curing agent increases the network polymer defectiveness and this leads to the sharp decrease of initiation energy of fracture.

In epoxy resins cured by ethylene-, tetramethylene- or hexamethylenediamines, the fracture initiation energy is 329 ± 38, 489 ± 83 or 575 ± 50 J m^{-2}, respectively; use of diethylenetriamine, triethylenetetramine or tetraethylenepentamine instead leads to the decrease of the initiation energy of fracture to 130 ± 2, 141 ± 29 or 136 ± 20 J m^{-2}, respectively.

Tri- and tetraepoxides react intermoleculary mainly with the curing agent. The structural microheterogeneity and overtension of interjunction fragments can be seen. This becomes apparent in increased brittleness, lower strength of hydrogen bonds, water absorption increase and absence of complete reversibility of the properties on drying a polymer swollen in water.

The polymer network density is the greatest near the chemical junctions of a novolac cured by amines. The glass transition temperature is above the degradation temperature, and it remains to discuss the completeness of the reaction by second relaxation transition T_β. A sharp increase of the viscosity and steric restrictions impede the participation of all epoxy groups in the cure reaction. The process proceeds preferably by intermolecular interaction, leading to increased microheterogeneity [70].

Polymer fracture proceeds in the vicinity of intermicrogel regions. The stress intensity factor on extension K_{Ic} for 'dian' resins cured by aromatic amines is usually inside the limits $0.55-0.62\,\mathrm{MPa\,m}^{1/2}$ (according to other data it is inside the limits $0.82-1.1\,\mathrm{MPa\,m}^{1/2}$). In cured tetraepoxides K_{Ic} decreases up to $0.45-0.50\,\mathrm{MPa\,m}^{1/2}$ at 20°C. The cured cycloaliphatic epoxy resins, for which K_{Ic} is inside the limits $0.74-0.78\,\mathrm{MPa\,m}^{1/2}$, are destroyed in more viscous ways [71].

The contributions of hydrogen bonds and structural heterogeneity to indices of properties of cured epoxy polymers become apparent especially visually as water diffuses. Water is absorbed by polar groups of the polymer, destroying the physical junctions, i.e. it displays itself as a plasticizer. As the polymer network tension and the number of crazes increase, by means of swelling pressure water can cause the destruction of chemical bonds in overstressed chemical junctions or interjunction fragments. Finally, the hydrolysis of the chemical junction can proceed if the temperature increases [72]. Plasticizing influence of water is eliminated on drying. The initial indices of the properties are recovered in the dry polymer. Mechanical or hydrolytic destruction of chemical bonds leads to irreversible changes in polymer due to onset of microcracking [73]. The higher the polymer structural heterogeneity, the higher are the water diffusion constant (diffusivity) and equilibrium water uptake.

In 'dian' resin cured by 'tetraminotitanate' (EDT-10 polymer) water uptake can attain 8%. All the water is absorbed by the polar groups. As a result, the intensity of relaxation processes increases sharply. For dry sample T_g equals 110°C, for the sample with 1.5% water uptake T_g decreases to 70°C and with 5% water uptake to 58°C. The initial T_g value is restored after drying [74].

In 'dian' resin cured by triethyleneamine (10%) equilibrium water absorption is 3.3% and glass transition temperature decreases reversibly by 33°C. In the polymer from tetraepoxide cured by diaminediphenylsulphone (50%) equilibrium water uptake attains 6.5% and glass transition temperature decreases by 102°C [75]. Thermal expansion coefficient increases from $99 \times 10^{-6}\,°\mathrm{C}^{-1}$ to $170 \times 10^{-6}\,°\mathrm{C}^{-1}$, elasticity modulus decreases by 30% and tensile strength by 15–17% [76]. As observed, the polymer heterogeneity and water uptake increase if the cure temperature of tetraepoxide by aromatic amine and duration at that temperature increase [77]. In the polymer cured at 130°C the equilibrium water uptake does not exceed 3.5%, and in that cured at 230°C it increases to 6.0%.

As water content increases, the polymer volume increases too. In 'dian' resin cured by *m*-phenylenediamine the increase of water content from 2 to 6% causes the glass transition temperature depression to change from 40 to 125°C, and the polymer volume increases up to 4.5%.

The structure of the thin polymeric film that connects elementary fibres (monofilaments) strengthening the binder can be changed considerably, being influenced by adsorption forces of solid surface, substances present at the surface and the inherent functional groups arising from the apprete (sizing, coupling or keying agent) layer.

The influence of the fibre surface extends as far as 25–70 mm [78] into the boundary polymeric binder, and this surface considerably influences the shear strength of the composite material, its deformation stability on heating and crack resistance on loading and moistening.

The surface of the glass fibres promotes increased size of cured resin microgel in the adsorption zone. The concentration of epoxy groups is greater than that of amine ones in the adsorption zone, and ionic polymerization with ring closure is initiated. Silanol appretes with amine, acidic and phenolic groups take part in epoxy resin cure, and as a result the apprete layer density at the fibre surface increases, attaining a thickness up to 20 nm [20, 79]. To decrease residual stresses in epoxy polymer at the boundary with the glass fibre, it was proposed to use the apprete representing comb-like block copolymer with molecular weight of 3000–7000 prepared from polydimethylsiloxane and poly(ethylene oxide) (lateral branchings) (PEO). At layer thickness of 20–25 nm, residual stresses in epoxy–amine film decrease by 48% (determined using the model of a rod polymer) and in cured epoxynovolac film by 52.5% [79].

Overtension of polymeric film at the carbon fibre surface becomes especially apparent with moistening. The initial properties are not restored completely after drying. Apparently a polymeric film is retained at the carbon fibre surface preferably by hydrogen bonds [80]. On moistening, the strength of the interaction declines, shear strength decreases by 30–35% and tensile strength by 55–60%.

In the moistened state, the deformation stability of carboplastic based on 'dian' resin cured by aromatic amine decreases up to 120°C.

To decrease the tension of the boundary layer, it was proposed to distribute a thin layer of epoxy resin on carbon fibres (up to 0.8–1.0%). This layer functions as a grease and apprete. The curing agent migrates partially into this layer from the composition found at the filler in the prepreg preparation process. The shear strength of carboplastic reaches 75.6 MPa. If film surface oxidation is carried out first, the shear strength of epoxy carboplastic increases up to 82.6 MPa. However information is available that the boundary layer from partially cured epoxy resin has increased brittleness. On loading, cracks arise preferentially in this layer and propagate, thus reducing the viscosity of plastic fracture [81].

In wetting polyaramide fibres by epoxy binder, diffusion of most low-molecular-weight components of the liquid phase into the fibre skin is observed. On plastic loading, especially perpendicular to the fibres, the skin is separated from the core. The shear strength in the boundary layer does not exceed 40 MPa and the tension of the resin film in it is 20 MPa. Moistening of organoplastic changes the strength and elasticity modulus insignificantly down to 70°C but creep increases remarkably [82].

Chemical resistance of epoxy polymers is determined by the presence of methylene groups —CH_2—, hydroxyls, imine (in the case of cure by amines) or ester groups (in the case of cure by anhydrides) [83]. As the reagents penetrate slowly into the bulk of a vitreous polymer, chemical reactions arise preferably in surface layers, and the rate of destruction of the article is determined by its shape, pore and crack number and duration of contact with the reagent.

The oxidation reaction is characteristic for epoxy polymers. Epoxy resins cured by amines are the most unstable in oxidative media. As oxidation proceeds, the brittleness of polymer increases and tensile strength decreases. The oxygen index of epoxy polymers is within the limits 20–26%. Combustion (burning) is accompanied by intensive smoke formation. To decrease flammability, chlorine or bromine atoms are introduced into the structure of the epoxy resin or antipyrene (fireproofing or fire-retardant) is added to the binder composition.

Acid resistance of epoxy polymer articles increases sharply with a protective film of poly(vinylidene fluoride), which is pressed strongly onto the epoxy polymer surface.

Epoxy polymers are rapidly destroyed in nitric acid. Several seconds after immersion the sample surface becomes green and then as a result of degradation the polymer goes into solution.

Epoxy polymers swell in both water and polar solvents. Swelling ratio at 20°C is inside limits 0.75–3.5% and increases sharply with temperature. The majority of the solvent is dissolved in the polymer, acting as a plasticizer.

Thermodegradation of epoxy polymers proceeds at a remarkable rate above 220°C. In an inert medium above 250°C, degradation into fragments begins; cake residue does not usually exceed 23% and grows to 35% only in separate cases.

The thermal linear expansion coefficient of epoxy polymers of various composition is inside the limits $(50–90) \times 10^{-6}\,°C^{-1}$, specific heat 0.3–0.5 kcal $kg^{-1}\,°C^{-1}$, heat conductivity 0.2–0.225 kcal $m^{-1}\,h^{-1}\,°C^{-1}$, volume electrical resistance 1×10^{15}–$8 \times 10^{16}\,\Omega\,cm$, dielectric loss tangent 0.010–0.30 and dielectric constant 3.2–4.5.

The behaviour of epoxy resins at low temperatures was investigated with EDT-10 resin as an example [84]. As temperatue decreases from 293 to 4.2 K, elasticity modulus and yield stress at break increase from 3000 to 9170 and from 75 to 137 MPa, respectively, elongation at break decreases from 4.5 to 1.5% and heat conductivity changes from 0.2236 to 0.125 W $m^{-1}\,°C^{-1}$.

(c) Elasticization of epoxy binders

To increase the stability of cured epoxy polymer to crack formation, elasticizer is introduced into the epoxy binder composition. Pleinomers with flexible chains or polymers that attach to the network polymer with block-copolymer formation are used as elasticizers. Elastic blocks are aggregated in cured polymer into a separate phase consisting of particles of colloidal size distributed uniformly in a high-crosslink-density matrix and attached to it strongly. The glass-like matrix of epoxy polymer provides rigidity and heat resistance, and the elastic phase provides elevated crack resistance and viscosity of fracture. Three methods are proposed for elasticization of epoxy binders.

According to the first method an elastomeric latex is dispersed in the epoxy resin. Latex particles as a result of oxidation contain epoxy or carboxyl groups, which fix their distribution in the coprecipitation process [85–87].

The second method is of more widespread use. It consists of the dissolution of flexible-chain pleinomers in an epoxy binder. Pleinomer macromolecules contain functional groups in the end units, which participate in the cure reaction. The phase structure of cured polymer depends on the difference of solubility parameters of epoxy resin and pleinomer and pleinomer molecular weight [88, 89].

According to the third method, monomers or oligomers forming elasticizer solution in it are introduced into the epoxy resin composition. The polymer separates in the process of epoxy resin cure into an independent phase as fine particles chemically attached to the matrix [90].

A latex of butadiene–acrylonitrile copolymer is used usually in the first method. The latex is vulcanized and oxidized down to formation of epoxy group at the particle surface. Then latex is mechanically distribued in epoxy binder and water is removed. To achieve optimum effect in increasing crack resistance, it is sufficient to introduce 2.5–4% of rubber in relation to epoxy resin. The crack resistance index G_{Ic} increases 10-fold, stress intensity factor K_{Ic} by 2.5–3 times and impact strength by 30–40%. Elasticity modulus and heat resistance remain at the same level.

Other methods of elasticization are desired because of non-uniform distribution of rubber particles in resin and their insufficiently strong cohesion with the matrix.

On dispersion of linear pleinomer or polymers with functional groups in the end units, it is necessary to achieve clear phase separation at the stage of resin cure and elasticizer particle size not below 1.5 μm.

If microheterogeneous structure is formed in the process of cure of epoxy–elastomeric composition, then the flexible-chain modifier functions as a plasticizer, promoting a sharp decrease of heat resistance and rigidity and increasing the fracture viscosity insignificantly. If an elastomer in a cure process separates into an independent phase, forming disperse inclusions (1–2 μm), then an elastization effect – sharp increase of crack resistance and fracture viscosity with small change of heat resistance and rigidity – becomes apparent. As the elastomer contains carboxyl or amine groups in the end units of macromolecules, it takes part in the cure reaction and its particles are attached strongly to the polymeric matrix. The regime of curing considerably influences the elasticization process. Careful maintenance of cure regimes is required in each particular case to introduce the elasticizer and to provide attachment of elasticizer to the resin and phase separation with a necessary dispersion of elasticizer. Filler presence restricts maintenance of these conditions [87, 91].

Such a method of elasticization has widespread application. This is related to the use of tetraepoxy and triepoxy resins. The low viscosity of these resins facilitates elasticizer dispersion in them, which is added to decrease the brittleness of the polymers. Polyepoxides based on tetraepoxides are exploited up to 200°C. Therefore their elasticizers require sufficient heat resistance. Poly(ether sulphones) with molecular weight 7000–20000 are usually used and they have carboxyl or amine end-groups. Elasticizer is introduced in amounts 10–25% in relation to epoxy resin [92, 93].

Tetraepoxide crack resistance increases by 14–18 times as a result of elasticization by polysulphone and cure by diaminodiphenylsulphone. In glass plastics this increase is equal to 4.4, and in carboplastics to 7 times.

The synthesis of polyurethane plasticizer directily in an epoxy resin is an example of the third method. Oligoether (for instance, poly(tetramethylene glycol) with molecular weight 1000–2000), triol and diisocyanate are introduced into the resin. Epoxypolyurethane dispersion is mixed with the curing agent and curing catalyst. The heterogeneous structure achieved earlier is fixed in the cure process; the polymer gains a clearly expressed macroheterogeneous structure with a bimodal distribution of inclusions by size (0.5–1.0 μm).

Changing component ratio can regulate the phase structure of elasticized polymer and its properties in a given direction and within wide limits. In epoxypolyurethane systems containing 20% polyurethane, impact strength increases from 7.1 to 16.5 kJ m^{-2}, K_{Ic} from 0.89 to 2.29 MPa m$^{1/2}$ and G_{Ic} from 240 to 1790 J m^{-2}, rigidity and heat resistance being unchanged [94].

In fibrous composite materials the thickness of polymeric film in the interfibre space is small and the elasticization effect becomes apparent to a considerably lower degree.

Data on the properties of 'dian' resins cured by m-phenylenediamine are given in Table 1.9 depending on elasticization method.

The fracture energy of interlayer shear increases considerably in glass plastics and carboplastics without a sharp change in elasticity modulus. Fracture proceeds step by step, a notched split remaining on which fractured fibres are seen. These observations were made on epoxynovolac resin elasticized by butadiene–acrylonitrile copolymer (NBR) and cured by dicyanodiamide [95].

For wet winding the compositions UP-25AC, UP-2202 and UP-2215 with 'active diluent' are recommended.

For dry winding and pressing, the binders UP-2216, UP-2220 and UP-2130 are recommended; for the filler packet vacuum impregnation, compositions EDT-10 and EDT-10P are suitable.

The binders of type ETPh, ENPhB, EPh, 5-211-B and UP-2130 are used in combination with carbon and organic fibres. Packets are monolithized by pressing or autoclave method.

Binders suitable for exploitation of articles above 170°C contain epoxynovolac, triglycidyl-N-aminophenol or tetraglycidyldiaminodiphenylmethane resin, and diaminodiphenylsulphone, a mixture of m-phenylenediamine and aminodiphenylmethane, or dicyanodiamide as curing agent.

Examples of epoxy binder compositions for composite materials moulding by

Table 1.9 Influence of elasticization method and type of elasticizer on the mechanical properties of 'dian' resin cured by m-phenylenediamine

		Properties of cured polymer			
Elasticizer	*Elasticizer content (%)*	*Fracture energy (kJ m^{-2})*	*Impact strength (kJ m^{-2})*	*Elasticity modulus (MPa)*	*Tensile strength (MPa)*
Without elasticizer	–	0.12	7.19	3300	87
Polybutadiene latex	10	0.79	5.22	3200	33
Polybutadiene latex	2.5	0.41	10.28	3600	35
Polyurethane elasticizer	15	2.6	21.27	2590	69.4
'Liquid' rubber: copolymer of butadiene and acrylonitrite (18%)	15	3.43	10.3	2300	56

autoclave method are as follows:

1. Tetraepoxide + epoxynovolac + diluent 1,4-diglycidyl ether of butanediol + diaminodiphenylmethane + accelerator.
2. Epoxynovolac resin + diaminodiphenylmethane + 1,4-diglycidyl ether of butanediol.
3. Tetraepoxide + 1,4-diglycidyl ether of butanediol + adduct of 'dian' resin with diethyltoluylenediamine.

The properties of several compositions are given in Table 1.10.

1.2.2 Polyimide matrices

(a) Some principles of creating polyimide matrices

These matrices are necessary for composite materials that function for long periods at temperatures of 300–360°C. They must maintain shape stability and tensile strength under such severe conditions. Heterocyclic polymers [96] and aromatic heterocyclic polyimides of linear [97, 99] or network [98] structure are investigated in more detail. They belong to the class of thermostable polymers. Degradation in an inert medium begins above 800°C accompanied by decrease in mass and properties indices.

Table 1.10 Properties of epoxy binders of various compositions

Properties	ED-10	ETPh +MPDA	EA +TGPhA	UP-610	EN-6 +DADPhM +PEG	EXOD +DDPhS	EPON 826 +DEBD +MPhDA +MDA	DGEBPh +MNA	ED-23 +maleic anhydride	UP-652 +MPhDA	MY-720 +DADPhS	ED-20 +MPhDA	UP-632	UP-652	ED-20 +DADPhS	Tetracyclo-aliphatic +TGPhA
Density (g cm^{-3})	1.23	1.23	1.22	1.23	1.22	1.24	1.21	1.22	1.23	1.23	1.283	1.23	–	1.24	–	1.24
Stress at break (MPa)																
extension	75	62	40	80	61	85	85	–	80	97	85	92	40	97	143	47
compression	160	136	126	130	165	200	111	72.4	130	163	–	142	136	163		169
bending	116	67	95	120	78	120	140	–	120	158	–	125	60	158	150	77
Shear strength (MPa)	51	34	26	120	70	–	–	–	–	–	–	77	–	–	–	–
Elasticity modulus on extension (MPa)	2900	3150	3900	4000	3600	5500	3500	3450	–	4100	5400	3300	–	–	4300	–
Elongation at break (%)	4.5	2.0	1.7	3.0	2.0	2.5	8.0[a]	2.7[b]	3.8	2.0	4.8	3.0–5.3	0.9	2.0	7.0	3.0
Impact strength (kJ m^{-2})	16	7	5	12	13	12	–	–	20	4.8	–	7.2	7.3	4.8	–	–
Linear thermal expansion coefficient (10^6 $^\circ$C^{-1})	90–105	–	–	–	–	117	68	–	99	–	–	70	–	–	–	–
Glass transition or melting temperature, T_g or T_m ($^\circ$C)	T_g 130	T_m 180–200	T_m 180	T_m 160	T_m 150	T_m 180–200	T_g 126–133	–	T_m 120	T_m 193	T_m 240	T_m 110	T_m 176	T_m 193	T_m 150	T_m 225

[a] Compression.
[b] Extension.

The polyimide prepared from pyromellitic acid anhydride and diaminodiphenoxide:

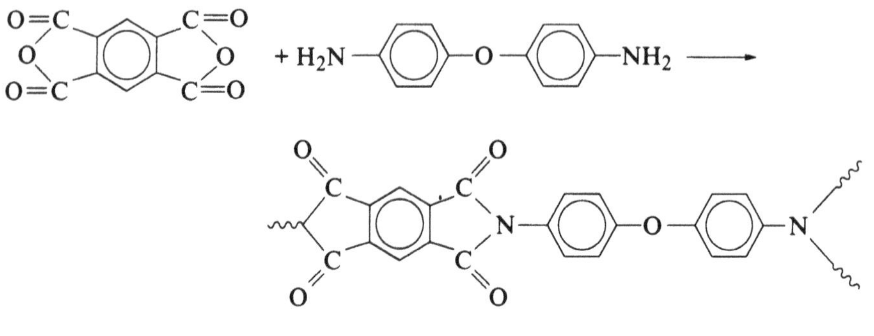

lost up to 52% mass at 870°C [100]. An intense exothermic peak was fixed caused by thermo-oxidative degradation inside the limits 350–380°C, depending on poly(aryl amide) composition.

Mass losses of the same polyimide in dry air at 400°C reach 1.5% and 5.7% at 425°C. The main component among the gaseous products is carbon monoxide, its content being 78%.

In prolonged use of poly(aryl amide) at 250°C and above, degradation processes arise that slowly bring about continuous decrease of property indices. The activation energy calculated by strength indices change is of $160 \pm 12\,\text{kJ}\,\text{mol}^{-1}$ for polyimides of different composition.

Branching and network formation increase at 275–375°C with simultaneous formation of chemical bonds between the units of adjacent macromolecules:

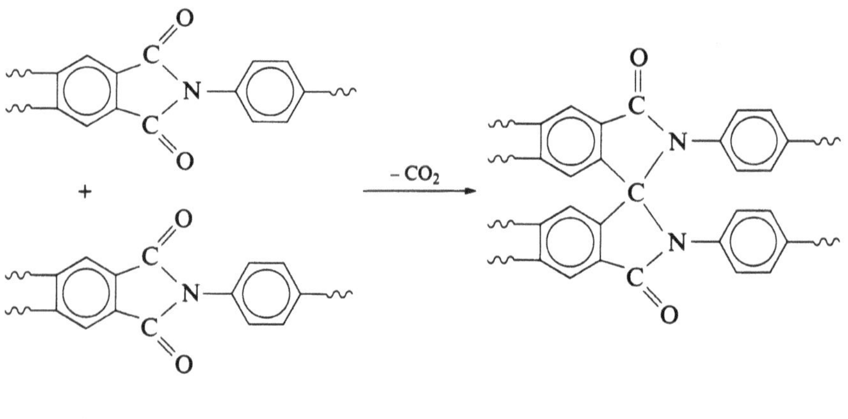

Degradation processes in treated moistened polymer (relative humidity of the medium 60–70%) are accompanied by more intensive gaseous evolution. As a temperature of 400°C is attained, mass losses reach 2.6%. This argues that thermodegradation of the moistened polyimide is accompanied by hydrolysis in the units with incomplete cyclization [101]:

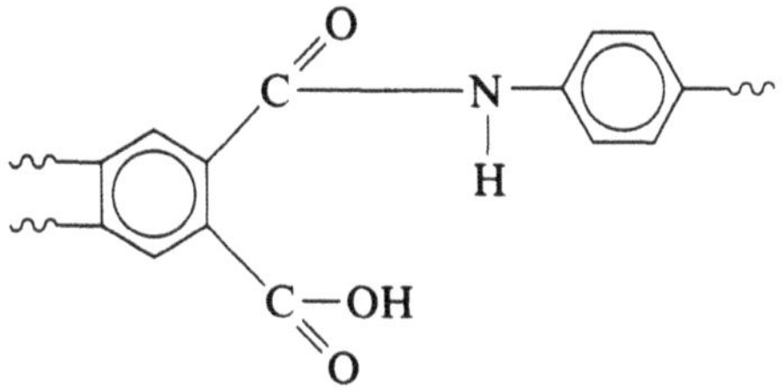

To decrease the tension of the network polyimide the units

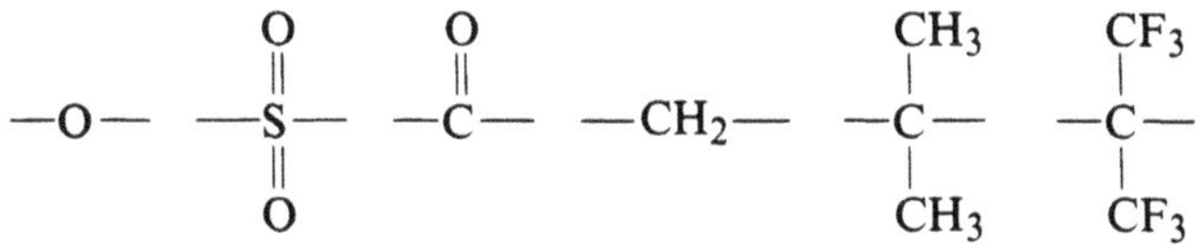

are introduced into the composition of interjunction fragments.

To complete the network polyimide synthesis with prepreg and article manufacture stages, the process is divided into several stages. Such a division can be different [102, 98].

First version of dismemberment process

Two-stage process

1. Poly(amino acid) synthesis:

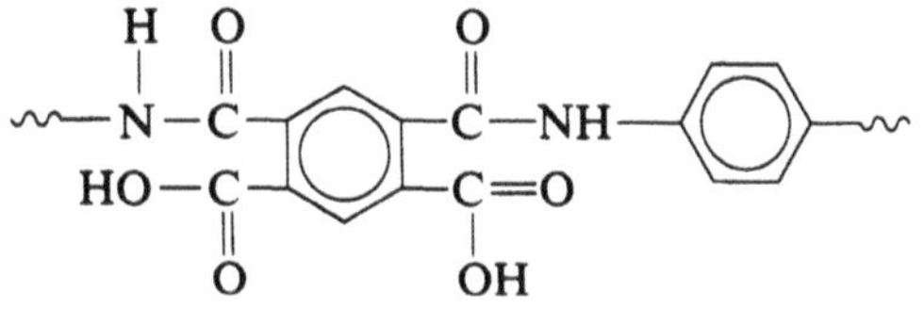

2. Poly(amino acid) imidization and network polymer formation with the units:

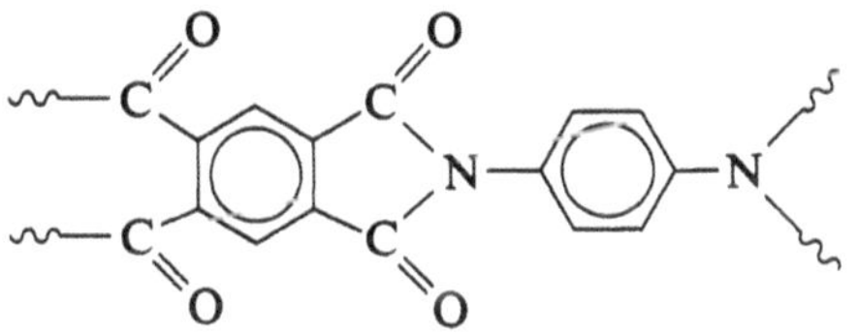

or

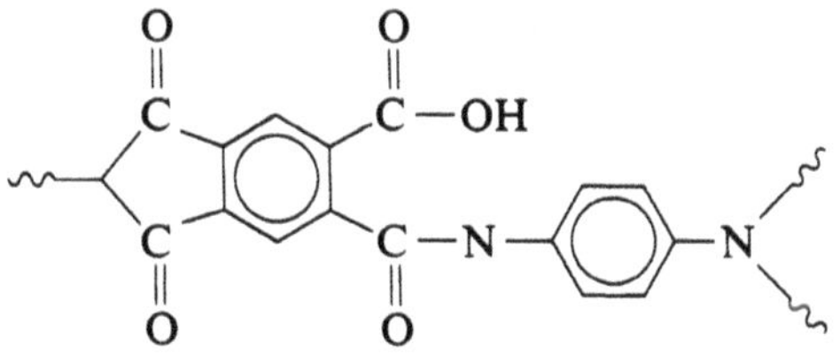

Three-stage process

1. Synthesis of telofunctional oligo(amino acids), for instance:

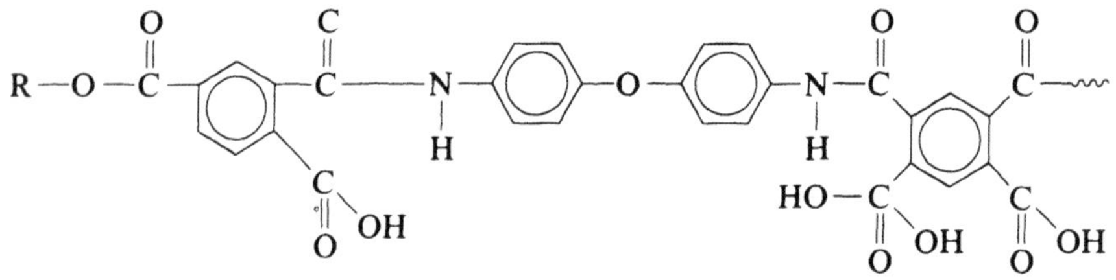

2. Imidization of telofunctional oligo(amino acids):

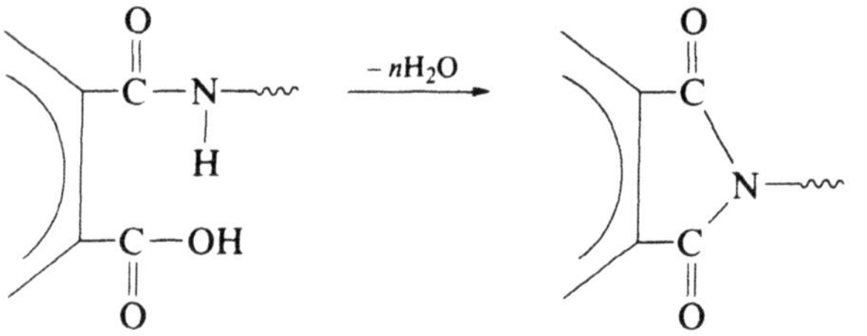

3. Cure of telofunctional oligoimides:

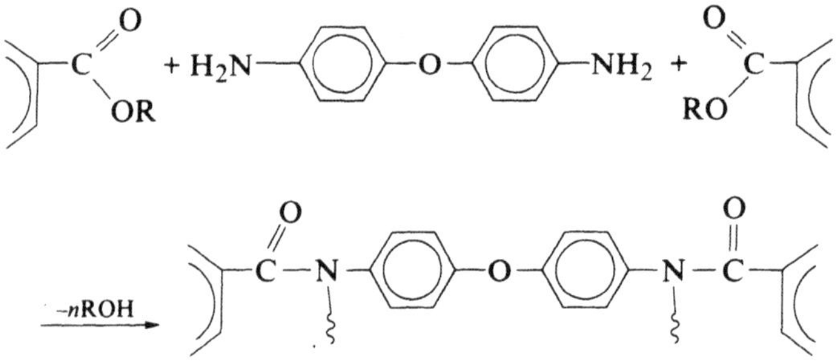

Second version of dismemberment process

Two-stage synthesis

1. Synthesis of oligoimide with end maleimide units (bismaleimides, BMI):

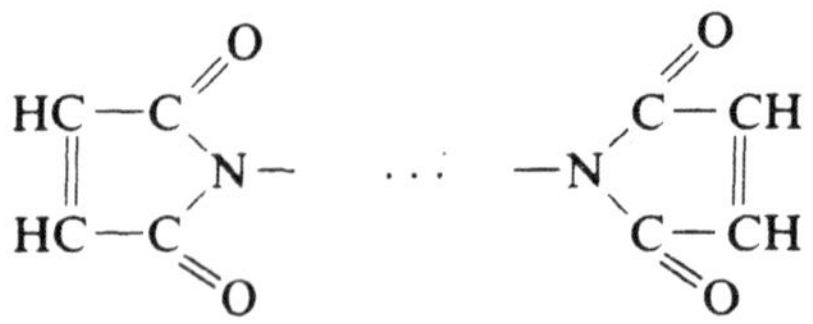

2. Cure of bismaleimide into network polyimide:

$$\ce{^^^^^CH=CH^^^^^} + \ce{H2N-R'-NH2} + \ce{^^^^^CH=CH^^^^^}$$

$$\longrightarrow$$

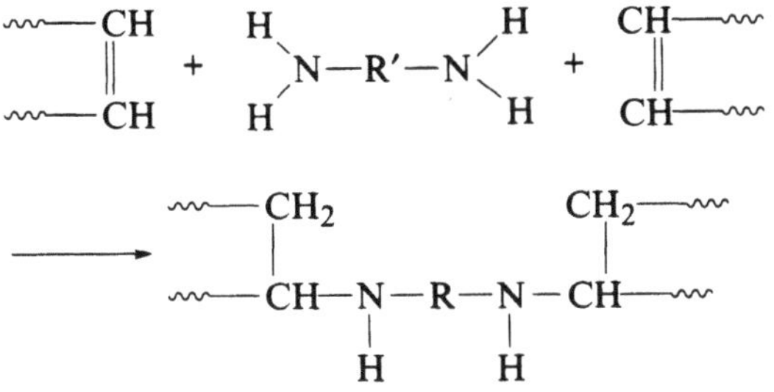

Three-stage synthesis

1. Codissolution of dianhydride, diamine and monoanhydride of unsaturated acid and oligomide formation, for instance:

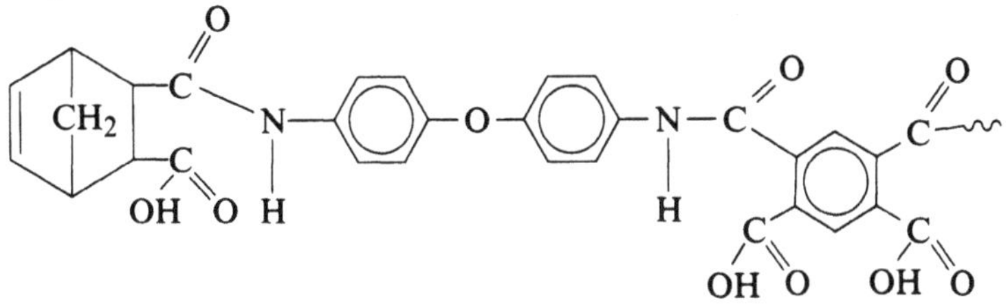

2. Cyclization of oligoamide into oligoimide.
3. Cure of oligoimide into network polyimide by pyrolytic polymerization:

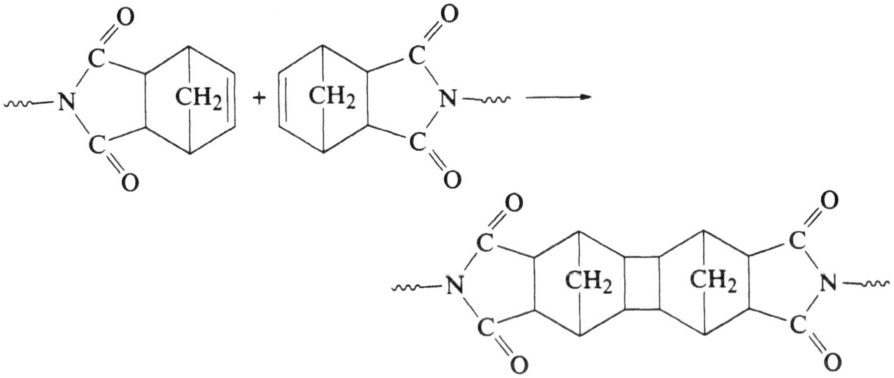

Among the various compounds that can be used in polyimide synthesis, the following have commercial significance:

Tetracarbonic acids

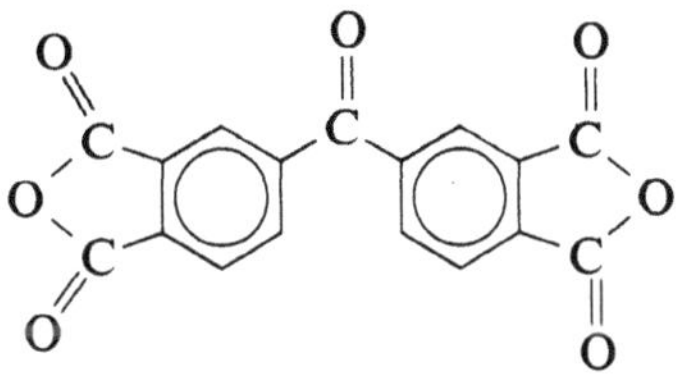

Benzophenone acid
dianhydride (DABPA)

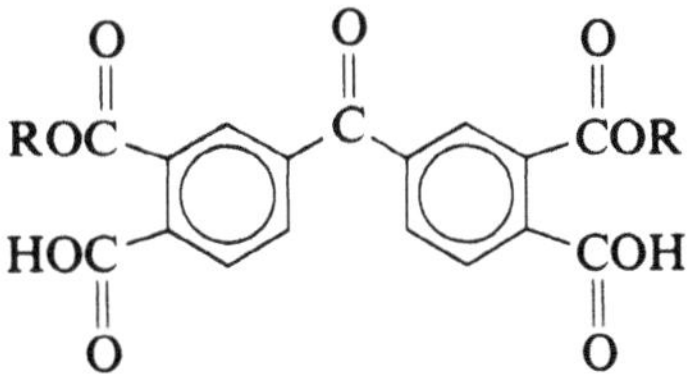

Benzophenone acid
diester (DEBPK)

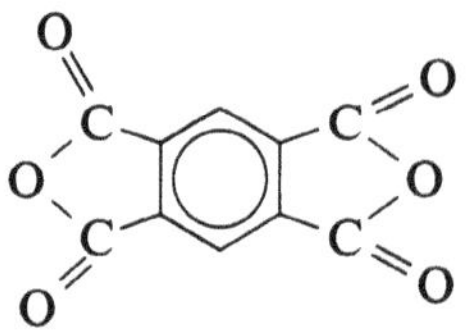

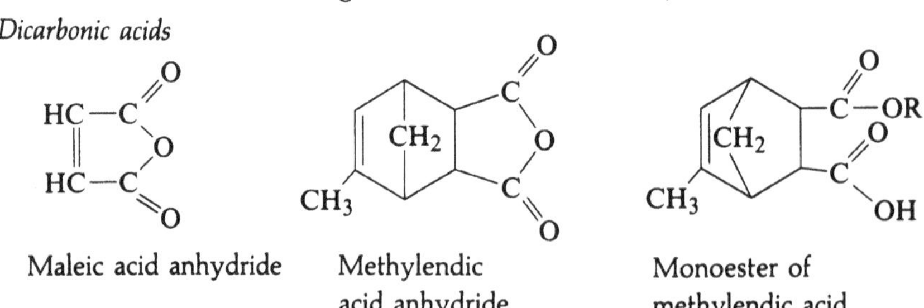

Pyromellitic acid
dianhydride

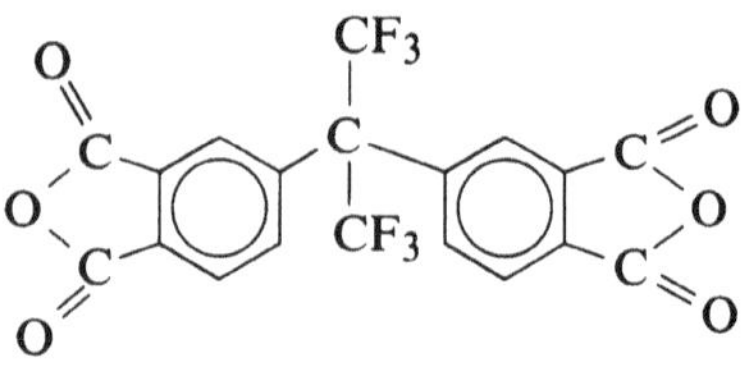

2,2-Bis(3,4-dicarbodiphenyl)hexafluoro-
propane dianhydride

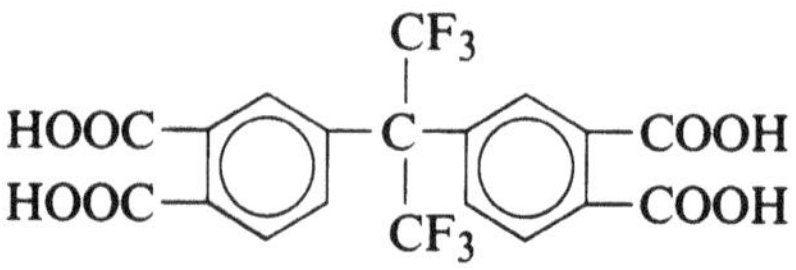

2,2-Bis(3,4-carboxyphenyl)hexafluoro-
propane tetraacid (6FTR)

Diamines

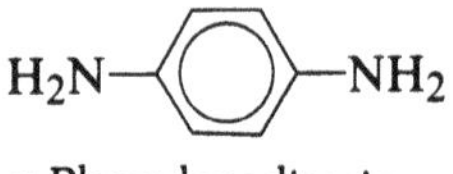

p-Phenylenediamine

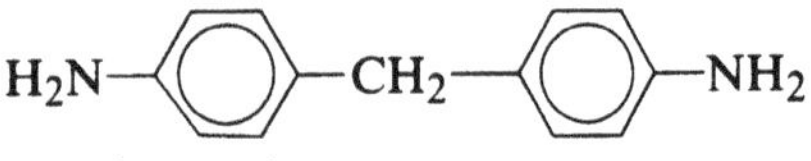

Dianilinemethane

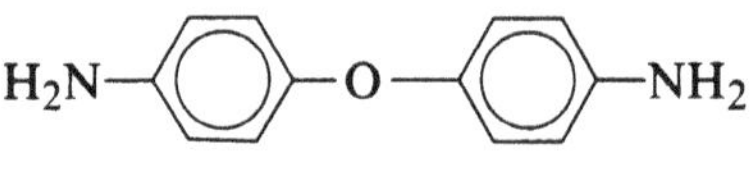

Dianiline oxide

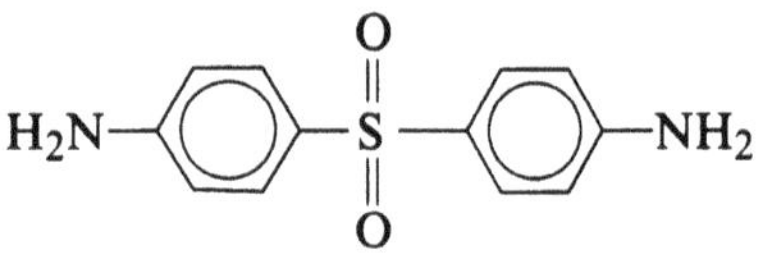

Dianilinesulphone

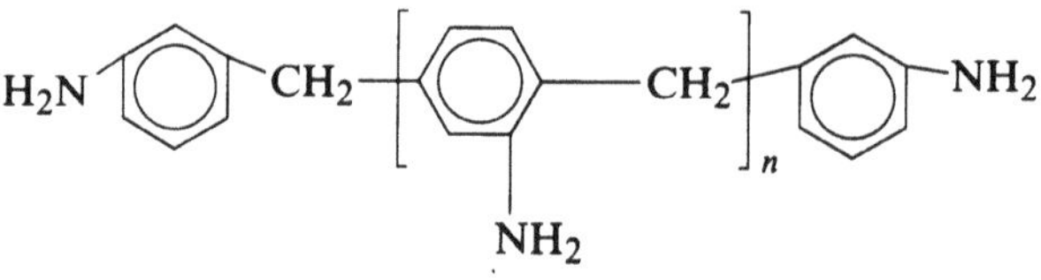

Oligoaramine (Jeffamine AP-22)

Dicarbonic acids

Maleic acid anhydride

Methylendic
acid anhydride

Monoester of
methylendic acid

(b) Condensation polyimides

First version of dismemberment of network polymer
According to the two-stage synthesis, stoichiometric quantities of dianhydride or
diester of tetracarbonic acid and diamine are dissolved in polar solvent (*N*-methyl-

pyrrolidone, dimethylformamide or dimethylacetamide) and heated to 50–60°C to form poly(amino acid).

To use the solution further with more ease polymer concentration in solution should not exceed 20%. The viscosity of such a solution in the methylpyrrolidone–xylol mixture is inside the limits 0.3–2.0 Pa s. The solution viscosity increases on storage and after 5–7 days it is difficult to distribute it in the interfibrous space of the filler as a thin film (SP-6 (Russia), Skybond 705 (USA)).

Thermo-oxidative degradation stabilizers are sometimes added to the solution. The fibrous filler surface is wetted by poly(amino acid) solution and the solvent is removed from the prepreg. Partical imidization of macromolecule units in poly(amino acid) proceeds simultaneously. The process is accompanied by water and alcohol release, decreasing polymer film solubility, and increasing softening temperature and melt viscosity.

If the prepreg is intended to mould articles by the autoclave method, then drying is interrupted when 12–16% of volatile substances remain in the prepreg (based on polymeric matrix). If the article is moulded by pressing, then drying continues until the quantity of volatiles removed on complete imidization reaches 8–9%.

Moulding is carried out at 300–320°C at increasing pressure (from 69 to 170 MPa). The material is maintained under pressure for 30 min and cooled to 60°C. The articles undergo sizing outside the mould, increasing the temperature step by step up to 320–350°C. The thermal treatment lasts for 12–14 h.

The packet from the prepreg is monolithized in vacuum at slow temperature increase up to 180°C, then the pressure is increased up to 0.7 MPa, the article is maintained at this pressure for 60 min and cooled to 60°C. Thermal treatment of the articles is conducted in the same manner as for articles prepared at high pressure [103].

In the initial stages, especially in the presence of solvent, imidization is accompanied preferentially with intramolecular ring formation:

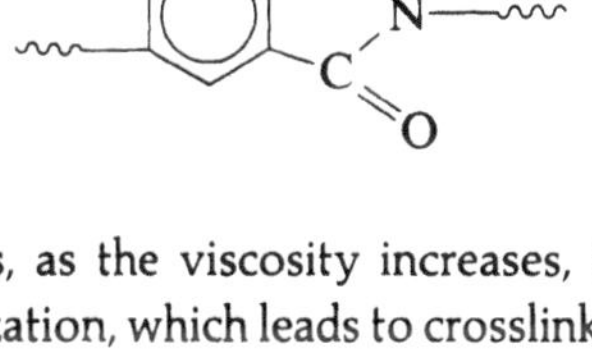

In the subsequent stages, as the viscosity increases, the units of adjacent macromolecules take part in imidization, which leads to crosslinked polymer formation [104]:

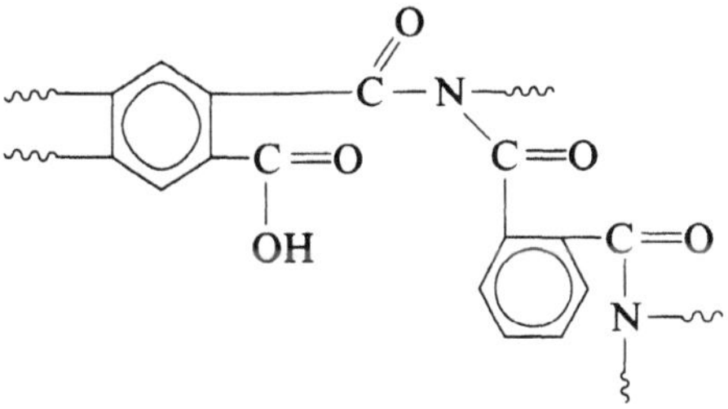

Random distribution of chemical junctions stops the overtension of chemical bonds, and 12–15% amide units remain in the polymer. These groups are less stable to thermodegradation and hydrolysis compared with imide groups.

Second version of network polymer synthesis

This is divided into two stages: poly(amino acid) and network polyimide. For instance, conjugation of SP-97 or Skybond 705 binder is associated with a number of technological difficulties. Poly(amino acids) are soluble only in polar solvents that are difficult to remove and are toxic. Yet at concentration 16–18%, solution viscosity attains 2.0 Pa s. On storage, viscosity increases quickly and after 6–7 days gelation of the composition begins. Removal of toxic solvent proceeds not only at the drying stage but during article formation too. Finally, the great volume of evolving volatiles in the formation process impedes the selection of methods and regimes providing material monolithization.

To eliminate partially the above-mentioned difficulties, a three-stage synthesis of polyimides was proposed: synthesis of oligo(amino acid) with end functionalities; imidization of oligo(amino acid) into oligoimide while preserving the end functionalities; and polycondensation of oligoimide into network polyimide by means of curing agent.

The binder that is synthesized by the three-stage process must retain the solubility at oligo(amino acid) stage after imidization into oligoimide. It is desirable that solvent mixtures contain the minimum quantity of toxic solvents and solutions of concentration 40–60% have low (up to 7.0 Pa s) viscosity. This is achieved by oligo(amino acid) molecular-weight decrease and selecting the functional groups in the end units, for example:

$$C_4H_9-O-C(=O)-C_6H_3(-C(=O)OH)-C(=O)-C_6H_3(-C(=O)OH)-C(=O)-NH-C_6H_4-NH-C(=O)-C_6H_3(-C(=O)OH)-C(=O)-C_6H_3(-C(=O)OH)-C(=O)-O-C_4H_9$$

The oligomers are soluble in methylpyrrolidone–methanol mixture. At oligomer concentration in solution of 40–60%, solution viscosity does not exceed 7.0 Pa s and remains invariable for 3–6 months (SP-95, API, BRI (Russia); Skybond 700, 703, 705, 709, 710 binders). Oligo(amino acids) that contain 6FTA units are known under the trademark NR-150.

Aromatic amines are used as curing agents. The rate of the oligoimide–curing agent reaction up to 120°C is low, but it proceeds with increasing rate at a temperature of 200°C and above.

After wetting the fibrous filler with the solution of oligo(amino acid) and curing agent, the prepreg is dried by stepped increase of the temperature from 120°C up to 177°C to remove the solvent and to imidize the oligomer. Some 8.0–8.5% of volatile substances remain in the prepreg. Oligoimide loses its solubility at this stage, but it can be converted into the plastic state and pressed at 208–270°C and a pressure of 69–170 MPa, temperature being increased up to 316°C and then forming comes to an end. Polyimide can be dissolved in N-methylpyrrolidone, and converts into the plastic state at 175–177°C.

If oligo(amino acid) contains benzophenone acid units, then the carbonyl group is also included in the reaction with amines. This creates additional chemical junctions of the network polyimide and a large increase of melt viscosity long before the end of the cure process:

The possibility of this reaction proceeding in poly(amino acids) containing 6FTA units (NR-150 B) is excluded. The polymer reserves plasticity to the end stage of imidization too. Prepregs can be dried to a deeper stage of imidization and therefore articles can be moulded with lower porosity (1–2% instead of 3–5% in articles from benzophenone acid).

The quantity of low-molecular-weight substances evolving over the moulding period equals 3–4%. To remove low-molecular-weight compounds completely and thermostabilize the material, the article is maintained at 300°C for 60 min. The following moulding regime is recommended: prepreg heating at 12–15 MPa at a rate of $60°C\,h^{-1}$ up to 140°C; holding under pressure at 140°C for 75 min; temperature increase up to 315°C and exposure for 60 min. The bending strength of glass cloth laminate at 20°C is 500 MPa, whereas at 315°C it is 200 MPa.

At 2% porosity, carboplastic based on NR-150 B binder (pressed at 17.2 MPa) at 343°C has a bending strength of 500 MPa (56% of initial value), shear strength of 32 MPa (64% of initial value) and elasticity modulus under bending of 11 700 MPa (83% of initial value) [105].

The porosity of polyimide articles cured by polycondensation reaction leads to moisture and air (oxygen) penetration to the fibre surface. The consequence of this is breaking of the binder film from the fibres or fibre fracture. The tear-off of the film becomes especially apparent in plastics strengthened by carbon fibres. Therefore polyimide binders cured by polycondensation process are used preferably in combination with mineral fibres.

In one-direction glass fibres, losses at 370°C for 100 h attain 5%, and bending stress at break decreases from 590 to 240 MPa (in pressed articels) and up to 170 MPa (in articles monolithized in an autoclave).

The matrix film of NR-150 B is much more monolithic compared with the film of Skybond or SP or API. It protects carbon fibres reliably from oxidation, and NR-150 B binder can be used in combination with carbon fibres at elevated temperature for a long time.

(c) Addition polyimides

If one decreases oligoimide molecular weight and replaces polycondensation by stepped polyaddition or by pyrolytic polymerization, one can exclude completely the presence of polar solvents at the stage of filler impregnation and evolution of low-molecular-weight substances at the article moulding stage.

For curing by polyaddition mechanism, low-melt-temperature imides (called bismaleinimides (BMI)) are synthesized ($T_{soft} = 80-105$°C) from aromatic diamines and maleic acid anhydride:

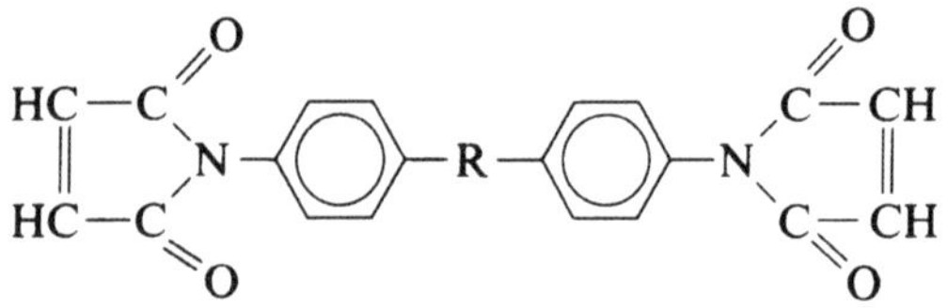

where R is **X**

or

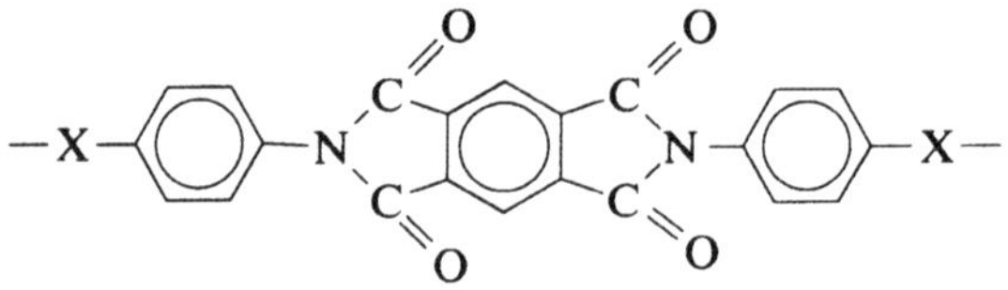

or

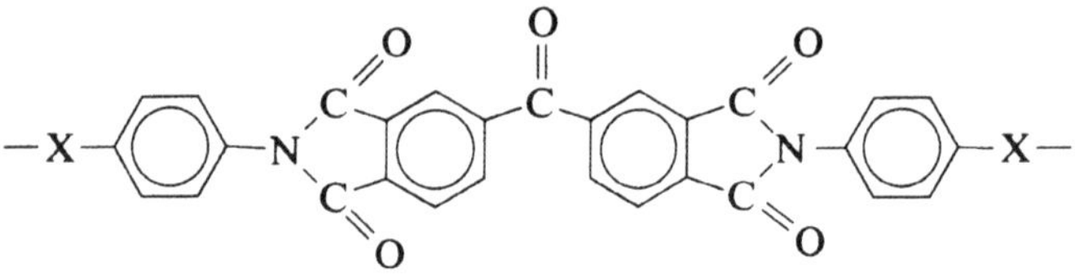

X in diamine can be

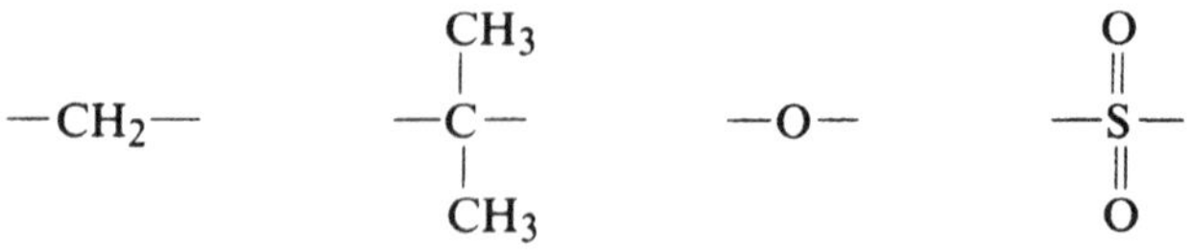

Cure in such a case is the process of joining oligoimides with each other by aromatic diamine. The reaction proceeds at 130–1240°C and is accompanied by pyrolytic polymerization:

$$
\sim\!\!\text{CH}\!\!=\!\!\text{CH}\!\!\sim + \text{H}_2\text{N}-\!\!\!\bigcirc\!\!\!-\text{X}-\!\!\!\bigcirc\!\!\!-\text{NH}_2 + \sim\!\!\text{CH}\!\!=\!\!\text{CH}\!\!\sim
$$

$$
\longrightarrow \sim\!\!\text{CH}_2-\text{CH}-\text{N}-\!\!\!\bigcirc\!\!\!-\text{X}-\!\!\!\bigcirc\!\!\!-\text{N}-\text{CH}-\text{CH}_2\!\!\sim
$$

and simultaneously:

$$
\sim\!\!\text{CH}\!\!=\!\!\text{CH}\!\!\sim + \sim\!\!\text{CH}\!\!=\!\!\text{CH}\!\!\sim \longrightarrow \sim\!\!\text{CH}-\text{CH}\!\!\sim \quad\text{or}\quad -\text{CH}-\text{CH}-\text{CH}-\text{CH}-
$$

To use the curing agent completely, the reaction is carried out stepwise. At the first stage the material is maintained at 135–145°C. At the second stage the temperature is increased up to 230–250°C. It was proposed to maintain the material in autoclave moulding at 80°C for 1 h, at 177°C for 4 h and at 204°C for 4 h. Overcure outside the autoclave is carried out at 220–260°C. Polyimide glass transition temperature is within the limits of 158–300°C depending on type of curing agent, BMI and curing conditions. Polymer mass loss at 250°C for 200 h is up to 0.8% and for 1000 h up to 3.0% [106] (Imilone, Kerimide-60, PAIS, Mialow). Thermodegradation activation energy is 134–156 kJ mol^{-1}. Fracture stress on extension at 250°C for 400 h falls by 13% and for 1000 h by 32%. Elasticity modulus decreases for 1000 h by 8%.

Cured bismaleinimides differ advantageously from epoxide binders by higher water resistance, fibre resistance (oxygen index 60–80) and long-term wear resistance.

Carbon plastics retain 60–90% of their initial mechanical properties indices, but the brittleness of these materials is very high. Elongation at break is 0.45–1.0%, fracture toughness $G_{\text{Ic}} = 25\text{–}30\,\text{J m}^{-2}$, stress intensity factor $K_{\text{Ic}} = 0.4\,\text{N m}^{-3/2}$ and shear strength is 26–29 MPa [107].

If low-molecular-weight aromatic diamines are replaced by polyamine of type Jeffamine AP-22, then stress intensity factor of BMI increases up to $0.6\,\text{N m}^{-3/2}$.

Different methods were proposed to elasticize BMI by thermostable linear polymers. Linear polyimides with end amine groups were proposed as elasticizers. BMI and linear polyimide are dissolved in dimethylformamide. The fibrous filler is impregnated by the solution and dried at 150°C. Melt viscosity at this stage is 105 Pa s at 200°C. The packet is monolithized under a pressure of 4 MPa and at 200°C for 2 h, then overcured at 245°C. Polyimide glass transition temperature is within the limits 309–333°C depending on the ratio BMI/linear polyimide. Shear strength increases up to 55–70 MPa [108].

BMI is elasticized both by polysulphone with end amino groups (molecular weight 7000–10 000), which fulfils the function of curing agent simultaneously [109], and by oligodimethylsiloxane with end amino groups [110].

The impact strength of BMI increases considerably on decreasing the number of chemical junctions. To preserve thermostability at the 200–250°C level, oligomeric fragments of poly(ether sulphone) or poly(ether ketone) are created between maleinimide end units [111], for example:

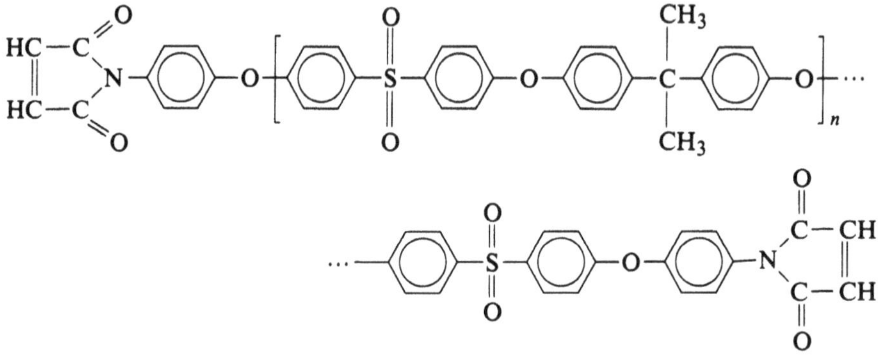

where $n = 1, 2$

The cure by diamine is completed at 270°C after 30 min. At $n = 1$, polyimide has $T_g = 239$°C. Mass losses at the heating rate 10°C min^{-1} up to 400°C are equal to 2%. Introducing oligo(ether ketone) with molecular weight 7500 can increase the K_{Ic} value up to $3.8\,\mathrm{N\,m}^{-3/2}$.

To eliminate polar solvents at the stage of impregnation of fibrous filler and to increase the heat resistance of the cured polyimide, all synthetic stages (mixing the monomers, oligo(amino acid) synthesis, cyclization into oligoimide, cure with poly-imide, formation) are conducted at the filler surface by pyrolytic polymerization (PMR-15, PMR-10, Lack-100, APM-2, API-3).

The initial components of the monomer mixture are acid dianhydrides or tetracarbonic acid diesters, for example:

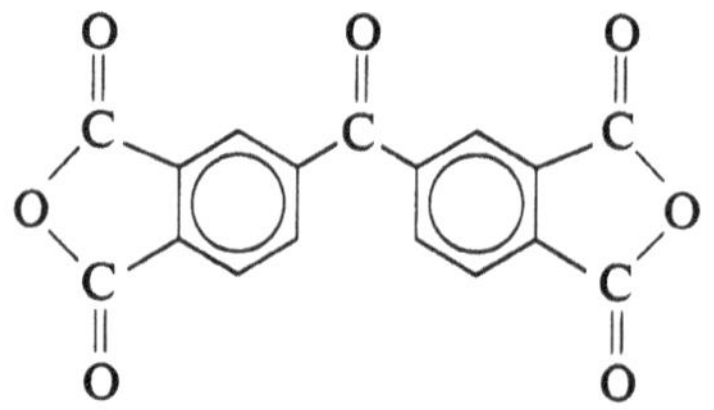

DBTK 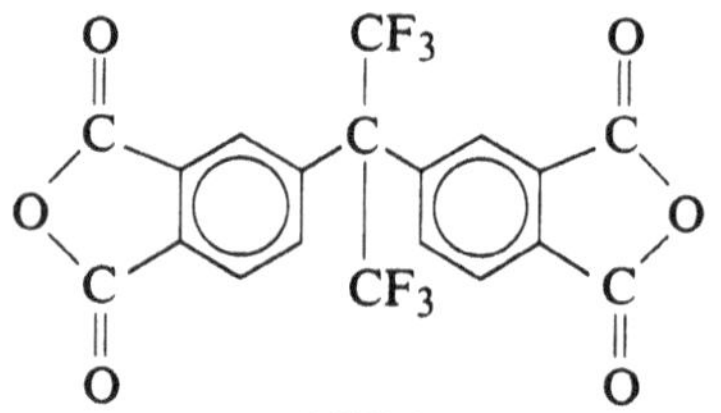   6 FTA

monoesters of endic and methylendic acid, for example:

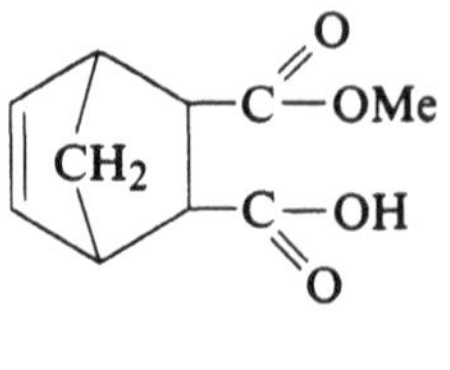

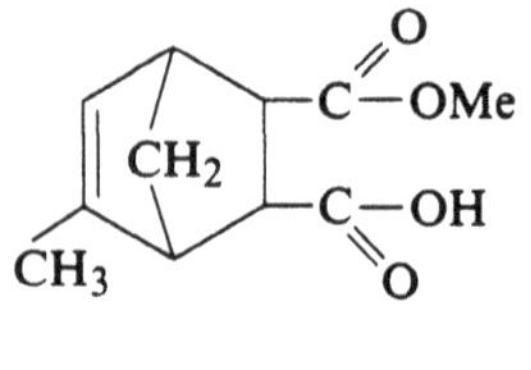

and aromatic diamine, for example:

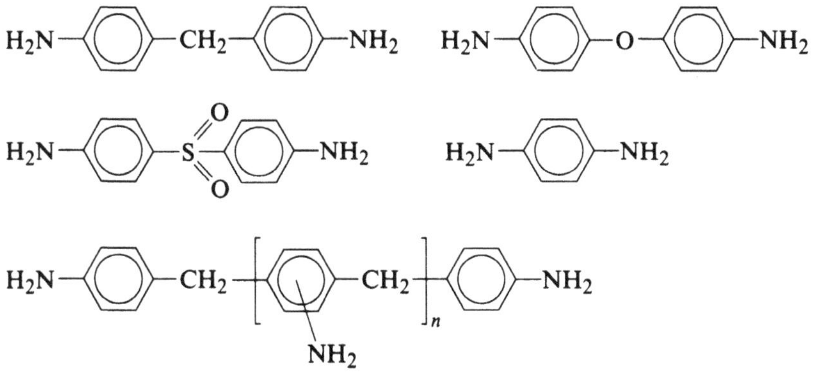

All three components are dissolved in methanol or ethanol at the ratio of n moles of dianhydride to $(n + 1)$ moles of diamine to 2 moles of methylendic ester [112, 113]. The concentration of monomer mixture in solution can be 75–80%. The solution is distributed in the fibrous filler surface and the amino acid is prepared simultaneously with solvent removal. The further product is oligoimide prepared at 100–120°C and having the following structure:

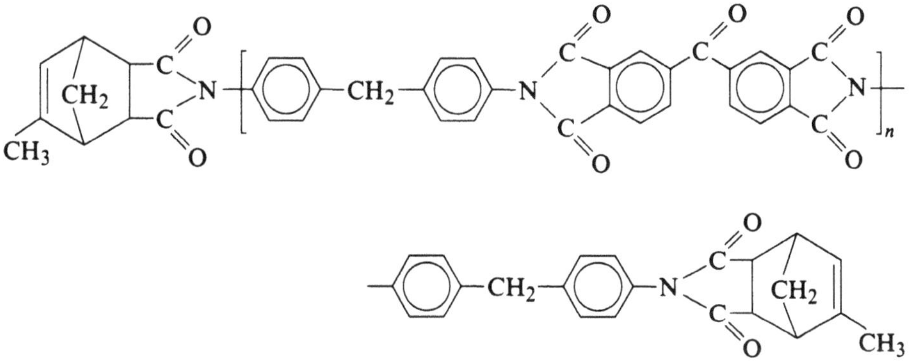

It is desirable to adjust the component ratio to $N/I = 2.08$. The average molecular weight of oligoimide in such a case is 1500 (PMR-15, APJ). If one maintains each of the stoichiometric ratios of the components, then the molecular weight of the oligoimide equals 1000 (PMR-10). Flowability of its melt is above the PMR-15 melt flowability, but is lower than thermo-oxidative degradation resistance of the network polyimide prepared.

Pyrolytic polymerization proceeds at 250–330°C. The packet is pressed at the initial stage (at 230–240°C), the temperature is increased at the second stage up to 316°C and the pressure reaches 69 MPa. The article is maintained outside the mould at 316°C for 16 h.

Selecting a different tetracarbonic acid dianhydride, replacing it by 6FTA anhydride or 6FTA diester, changing the diamine and using a diamine mixture or aromatic oligoimide of type Jeffamine AP-22 can all influence the crosslinked polyimide properties.

The activation energy of polyimide thermodegradation equals $170-180\,kJ\,mol^{-1}$ (in epoxy polymers it is $120\,kJ\,mol^{-1}$). Especially high thermoresistance indices were attained by combining 6FTA acid with *p*-phenylenediamine and methylendic anhydride (PMR-11 polyimide). The plastic can be maintained at 312°C for 1000 h without remarkable change of the shear strength. Its glass transition temperature is 390°C and mass losses are 10% at 330°C for 100 h.

On complete cure of crosslinked polyimides of type:

the following relaxation transitions were found [114]:

X	Y	$T_\alpha(°C)$	$T_\beta(°C)$	$T_\gamma(°C)$
–	–	None	275	-20
—O—	–	None	220	None
–	—O—	320	200	None
—O—	—O—	385	175	-70
—CO—	—O—	308	180	None

To compare the behaviour of polymeric matrices based on epoxide, maleinimide and imide binders, information is given about their heat resistance. The temperature at which the mass losses are 5% after 2000 h is 230°C for polyimides, 160°C for BMI and 90°C for epoxypolymers; for 5000 h, 276, 180 and 104°C, respectively; and for 500 h, 318, 199 and 128°C, respectively.

The polyimide that contains dianilinemethane units:

was tested at 1000 cycles of cooling to 0°C and heating to 232°C. It was stated that the onset of microcracking in the surface layer of samples is caused by —CH$_2$— group oxidation to carboxyl ones [115].

It is difficult to prepare monolithic samples from unfilled polyimides. Information about their mechanical properties is contradictory as these samples have lot of pores and latent defects.

In carboplastics based on polyimide binders, the carbon fibre is subject to oxidation in air at 335°C on the butt end of the panel. Then it is destroyed at the zone at the boundary with the fibre. The higher the sodium or potassium content in the carbon fibre, the more remarkable is the oxidation at sites that are not protected by polyimide. After 316°C for 600 h in an air flow at a flow rate of $900\,cm^3\,min^{-1}$, the shear strength in a carboplastic containing 55 vol.% of T-300 fibres decreases from 96.5 to 24 MPa.

The results of an investigation of polyimide behaviour at temperatures down to $-269°C$ are worthy of mention. The investigation of degree of properties change was carried out by the acoustic method. At -93 to $-113°C$ the γ relaxation transition was revealed in polyimides, which is apparently related to onset of oscillations of phenylene groups in diamidodiphenyl ether units. The fracture energy of polyimide changes very little down to $-269°C$ (it is much less than in polyepoxides). The changes of Poisson's ratio are unremarkable (from 0.36 to 0.41). This shows the small free-volume change in polyimide on cooling to $-269°C$ [116]. Thermal expansion coefficient changes from $61 \times 10^{-6}\,°C^{-1}$ (at 20°C) to $61 \times 10^{-6}\,°C^{-1}$ (at $-253°C$).

To decrease the oligoimide cure temperature and to eliminate the step-by-step character of the temperature increase, which is especially difficult for article pressing, different methods have been proposed. Among them the use of active diluents and replacement of pyrolytic polymerization process by a step-by-step addition reaction using the curing agent are known.

Furfural or furfurayl alcohol are proposed as active diluents. The linear furan polymer is formed at the amino acid imidization stage and this facilitates form filling with imidoplastic [117]. The cleavage of double bonds in furan units proceeds at the pyrolytic polymerization stage at 250–270°C. These units are apparently included into oligoimide cure reaction. The cure of oligoimide begins in the presence of polyfuran at 180–200°C and is completed at 300°C without substantial change of polyimide thermostability. The binder (API-3) is distributed on the filler surface, heated at $5°C\,min^{-1}$ up to 185°C and maintained for 1 h at this temperature. The packet from the prepreg is pressed (moulded) at 10–15 MPa and at 300°C with maintenance under pressure for 50–60 min [118].

Divinylbenzene is also proposed as a diluent. The temperature is increased in such a case step-by-step up to 177°C and the sample is then overcured at 250°C [119].

The temperature of curing decreases also in such a case when pyrolytic polymerization is replaced by stepped polyaddition with participation of a curing agent. But it is accompanied by decrease of polymer heat stability and thermodegradation stability.

BMI as binder component is used in combination with glass and carbon fibres. Oligoimide binders are used in combination with carbon, basalt and thermostable silica fibres. Polyaramide fibres are destroyed in the presence of the monomers used for oligoimide synthesis and in the process of their cure.

The grease (lubricant) is removed in thermostable imidocarboplastic manufacture from the carbon fibre surface by burning off at 300°C for 3 h in a nitrogen atmosphere.

A thermostable grease (lubricant) was proposed recently that can serve as a sizing layer too at the prepreg preparation stage without lessening imidocarboplastic thermostability. Such a grease can be prepared by polymerization from water solutions of monomer mixture where monomers penetrate into fibre surface microdefects and form the polymer film on its surface (fibre mass increase by 1.0–1.5%). The copolymer is stable up to 320°C, and contains functional groups participating in polyimide formation reaction [120]. Polymerization is initiated by the electrochemical method, continuously movable carbon tape being used as the anode [121], or radical polymerization initiators. The presence of such a film eliminates the necessity for preliminary electrochemical oxidation of the carbon fibre, which is very important in carboplastics intended for exploitation at high temperatures.

It was proposed to eliminate the high brittleness characteristic of polyimides by introduction of elasticizer [122]: the block copolymer of polyimide (from benzophenone acid anhydride and diaminodiphenylsulphone) with polydimethylsiloxane. The binder fracture energy increases two-fold in the presence of block copolymer without decreasing the elasticity modulus and heat resistance.

1.2.3 Phenol–formaldehyde matrices

(a) Curing phenolic resins

The chemistry of various phenol–formaldehyde resins has been described in detail [123–127]. The initial substances usually used in phenolic resin synthesis are given in Table 1.11. Resin synthesis involves two stages:

1. Aldehyde addition to phenol, forming a mixture of isomeric mono-, di- and trisubstituted methylphenols:

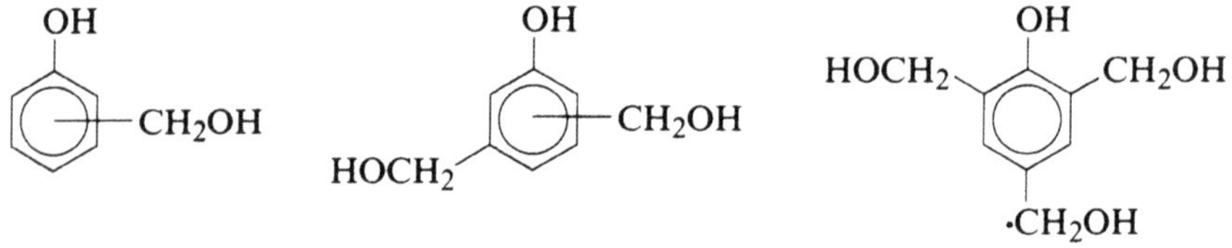

2. Polycondensation of the isomeric mixture to highly crosslinked polymer:

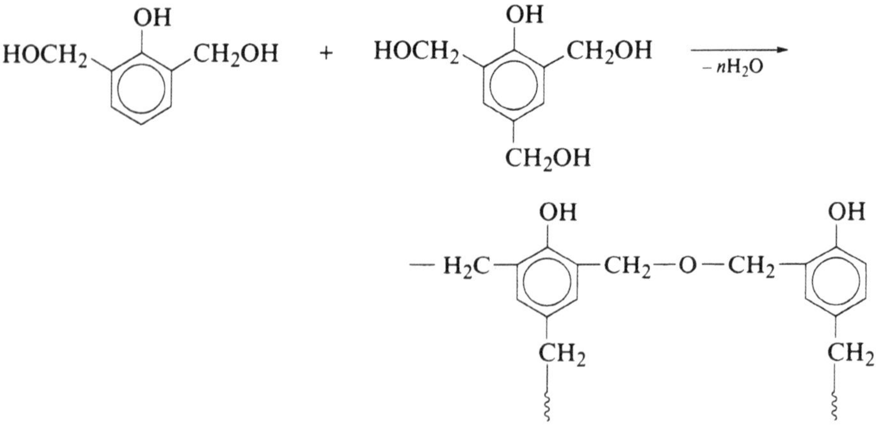

Table 1.11 Initial components of phenol–formaldehyde resin synthesis

Substance	Chemical formula	Physical constants	
		$T_m(^\circ C)$	$T_b(^\circ C)$
Phenol		40	181.8
o-Cresol		30.9	191.0
p-Cresol		34.7	201.9
Mixture of cresol isomers (o, m, p)		Liquid	–
Mixture of xylenol isomers		Liquid	–
Resorcinol		110	281.0
Bisphenol A		157.3	
Formaldehyde (formalin – as a water solution)	CH_2O	-92	-21
Trioxan		62	115
Acetaldehyde	CH_3CHO	-123	48.8

The addition reaction (aldehyde to phenol) is intensified in alkaline medium. Thus according to [128] in the presence of NaOH and at molar ratio phenol/formaldehyde = 1/ 1.4 the isomer mixture contains:

$$
\begin{array}{ll}
10\text{--}15\% & \text{of } o\text{-methylolphenol} \\
35\text{--}40\% & \text{of } p\text{-methylolphenol} \\
30\text{--}35\% & \text{of 2,4-dimethylolphenol} \\
4\text{--}8\% & \text{of 2,4,6-trimethylolphenol} \\
5\text{--}10\% & \text{of phenol}
\end{array}
$$

Polycondensation of such a complex mixture leads to oligomer formation, differing in degree of branching and unit number even at the initial stage. Several representatives of this complex mixture were identified in [129–131]. In acid media the polycondensation rate increases and a deeper conversion of functional groups is obtained. Each stage of methylol derivative addition to each other is accompanied by water evolution:

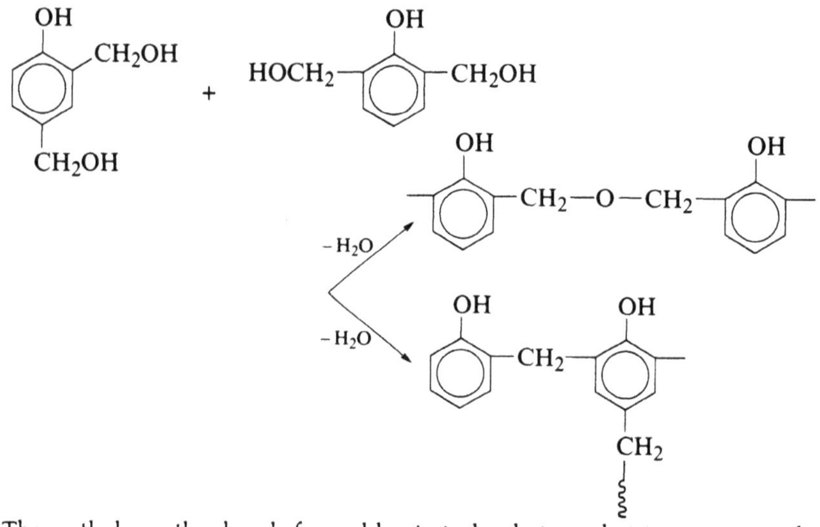

The methylene ether bonds formed begin to be destroyed at temperatures above 140°C and the formaldehyde evolved can add again to some of the phenolic units:

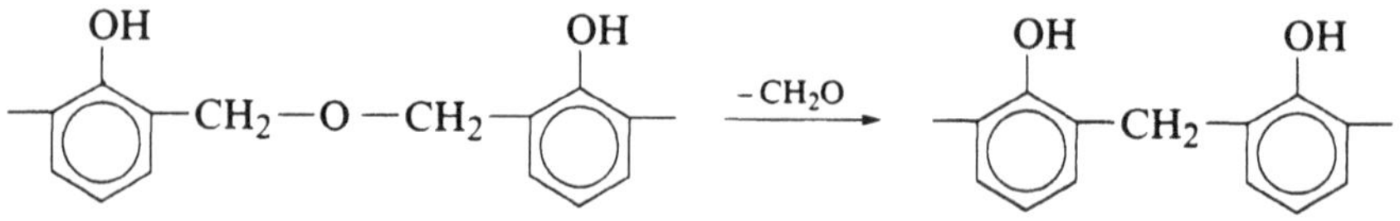

At 180–200°C, methylene ether bonds between phenolic units are transformed completely into methylene ones.

The rate of the polycondensation reaction, especially in neutral and acidic media, changes sharply with temperature. This allows the process to be stopped at any interrmediate stage by decreasing the temperature and to be continued again at the same rate by increasing the temperature up to the initial level. Polycondensation of methylolphenols is stopped at the stage that is called 'resol resin' or stage A. At stage A the oligomer mixture attains average molecular weight 500–700. The resin is solid, brittle, amber to brown in colour, softens at 60–75°C, and is soluble in alcohol, acetone and ethanol–benzene mixture. Resol resin viscosity and solubility are not changed sharply after 14 days storage at 20°C, nor after 35 days at 10°C.

The resin transforms at 130°C from B stage to G stage after 8–10 min. Using different phenols or their mixtures, changing phenol/formaldehyde ratio, and varying the catalyst and its amount can alter the resol resin composition, its lifetime and polymer properties at final cure stage – resite or polymer of stage C. As oligomer addition proceeds, macromolecule branching and intramolecular cyclization probabilities increase. At stage C the polymer represents a typical heterogeneous high-crosslink-density polymer with microgel size of 80–90 nm [132], with irregularly distributed interjunction fragments differing in length. Some of the methylol groups that did not enter the reaction due to sharply increasing viscosity at the selected polycondensation temperature remain in the polymer. Increasing the temperature of the reaction, an additional amount of functional groups can be drawn into polycondensation. As polymer network density increases, the glass transition temperature increases. Its limited value is outside the limits of thermo-oxidative degradation. It is easier to follow the reaction by secondary relaxation transition temperature change (T_β), which is at 45–47°C depending on degree of cure.

In the case of excess of phenol to formaldehyde (1/0.75–0.85) the methylolphenol formation reaction comes to an end as soon as all the methylol groups formed enter the reaction. Average molecular weight of the novolac resin does not exceed 1000 and it contains preferably oligomers with 5–6 units. It is supposed that oligomers with such a number of units create rings strengthened by inter-unit hydrogen bonds between adjacent hydroxyl groups [131]. The softening temperature of the resin is 82–95°C. Free phenol content equals 2–5%.

Table 1.12 Curing agents for novolac resins

Curing agent	Structure	Physical state
Hexamethylenetetramine ('hexa')	$(CH_2)_6N_3$	Solid, decomposition temperature above 115°C
Furfural	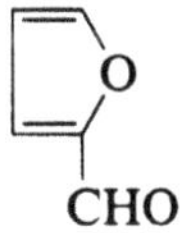 (furan ring with CHO group)	Liquid, boiling temperature 162°C
Difurfurylideneacetone (adduct of furfural and acetone) (FAM)	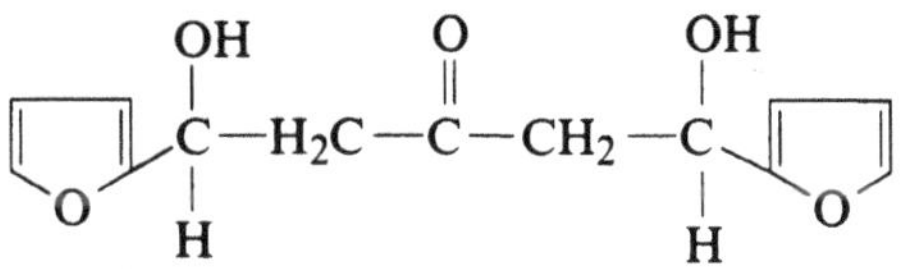 (difurfurylideneacetone structure)	Viscous liquid

To convert novolac resin into resite, it is necessary to introduce a curing agent (Table 1.12). Hexamethylenetetramine or furfural are usually used as curing agents in combination with a small amount of FAM.

Hexamethylenetetramine ('hexa') undergoes decomposition above 115°C and reacts with oligomers of novolac resin, creating connecting units of different composition between them:

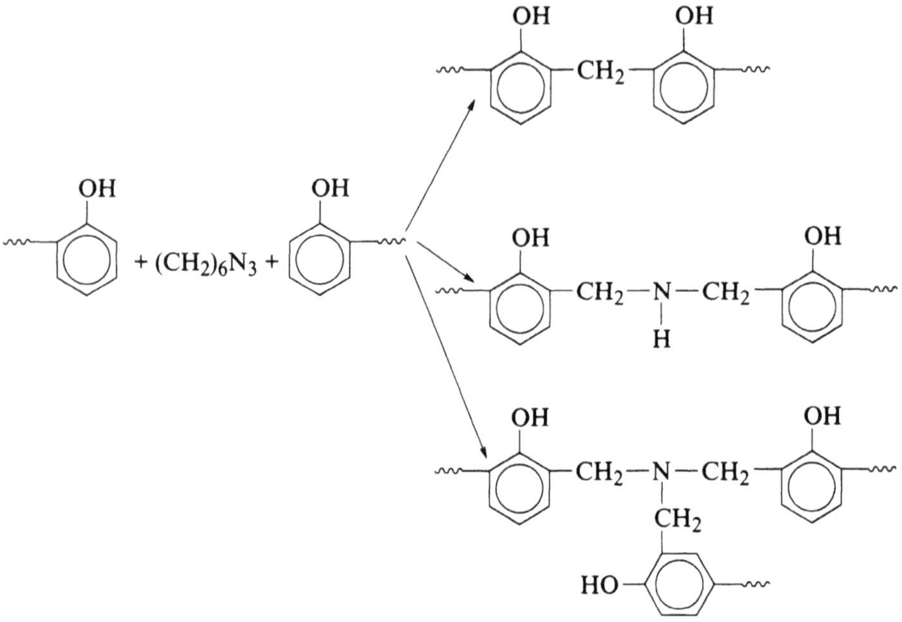

At 130°C transition of the resin from stage B to stage C proceeds after 16–20 min. The process is accompanied by ammonia evolution.

Furfural dissolves novolac resin at the stage of prepreg manufacture and can be used as a binder diluent. In the cure process 'hexa' (by means of furfural) reacts partially with novolac resin oligomers:

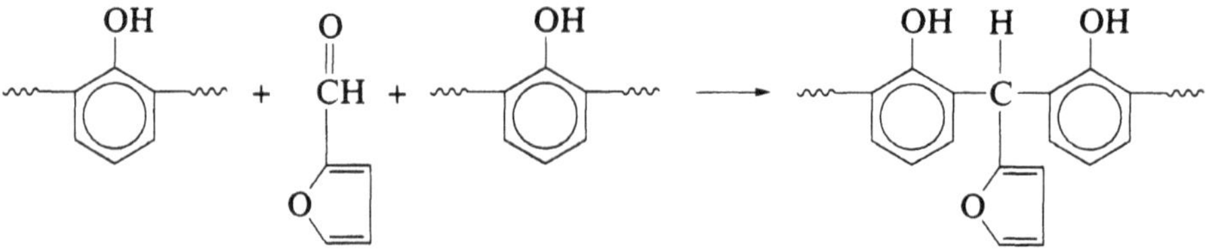

and simultaneously forms furan resin, which builds into a polymeric network by chemical means. The ammonia evolved accelerates furfurol polymerization.

Information is available that one can prepare novolac resin by polycondensation of phenol with acetaldehyde. These resins, cured by hexamethylenetetramine, have superior dielectric properties to and higher strength than phenol–formaldehyde novolacs [133].

Resol resins are synthesized as powder (SPhP), ethanolic solution (LBS), water emulsion (SPhG) and microspheres of 40–80 µm diameter (SPh-460, SPh-480).

Resol resin is mixed with modifiers and a cure catalyst is introduced. The melt, solution or water emulsion of resol resin, novolac resin and 'hexa' in furfural solution or solution of novolac resin in FAM are used for fibrous filler impregnation.

Resol resin curing into resite proceeds at the surface of the fibres. The process is accompanied by evolution of water, ammonia, formaldehyde and phenol incompletely reacted. To minimize of evolution of volatiles at the stage of moulding articles, the resin at the surface of prepregs is preliminarily cured. Preliminary cure was carried out up to resin transition to stage B (resitol stage). At this stage polymer loses solubility and its viscosity increases sharply. At 140–160 °C resitol viscosity equals 3×10^5 to 3×10^6 Pa s. To monolithize the packet from the prepreg requires a pressure of 10–15 MPa at 140–160°C. Transition of resitol into resite at 160°C is accompanied by the formation of 2.0–2.5% of volatile substances only 98% of them being water [134].

Furfural solution of novolac resin (PhN composition) is used for the liquid-phase process of impregnation of fibrous filler, packet assembly and monolithization at pressures up to 1.0 MPa.

To accelerate curing, acidic catalysts are introduced into resolic resins, preferably weak organic acids, avoiding a sharp decrease of lifetime of the semifinished product.

At pH 1–3 the cure reaction can be carried out at 75–110°C. Activation energy equals 100 kJ mol^{-1}. In alkaline medium (pH 11.3) the same degree of curing can be attained at 150–170°C. At 20°C the process proceeds very slowly. The heat of reaction exceeds by 3–4 times the heat of reaction in acidic medium. In acidic medium polycondensation proceeds preferably between methylene ether bonds. In alkaline medium methylol groups react preferably with the hydrogen atoms of the benzene ring, forming methylene links between oligomers. The reaction proceeds more slowly than in acidic media, but the degree of formaldehyde utilization is higher in polycondensation reaction and a higher polymer network density near chemical junctions is achieved [135].

Phenol–formaldehyde resins are cured in closed moulds at 150–160°C at a rate of 0.8–1.0 min per millimetre of article thickness. There is no necessity for stepped temperature increase.

(b) Properties of cured phenolic resins

Resites are amorphous, high-crosslink-density polymers of heterogeneous structure. In an inert medium they preserve hardness up to 400°C.

Their elasticity modulus changes little up to 150–170°C, then decreases sharply and becomes nearly constant again. It is far from α transition, as the activation energy of this relaxation transition is 20–30 kJ mol^{-1} only. It is assumed that resite glass transition temperature is above 350°C. Below 400°C in an inert medium only 3% gaseous products are evolved. This is water evolved as the polycondensation process continues. Thermal gradation rate attains maximum value at 430–475°C, but the polymer remains glass-like. Water, carbon monoxide, carbon dioxide, methane and phenol are evolved at this stage. Above 600°C remarkable shrinkage of the material commences. Density and heat conductivity increase. At the final stage of carbonization, resite transforms into glass carbon. Yield of glass carbon ('cake') attains maximum value 71.4% (instead of 54–56%) in the case when resolic resin is synthesized at phenol/formaldehyde ratio of 1/3 in the presence of ammonia (2% based on phenol). Before curing, the introduction of toluenesulphonic acid into resolic resin and cure at 200°C are recommended [136, 137].

The thinner the resol film, the more it retains its strength after carbonization. At a thickness of 90 μm, carbonized resite film preserves its tensile strength of 32 MPa. If the film thickness is increased to 650 μm, then the strength after carbonization decreases to 19 MPa [138].

According to thermogravimetric analysis data, resite is stable in nitrogen atmosphere up to 400°C. The dependences of compression strength of pressed samples of phenol–formaldehyde resins of different composition on mass loss as a result of thermo-oxidative degradation are given in [139].

Resite heterogeneity and lowered flexibility of interjunction fragments leads to increased brittleness of polymer. It was shown [140] that resite impact strength increases from $2.8–3.5\,kJ\,m^{-2}$ up to $9.5\,kJ\,m^{-2}$ and bending strength from 70 MPa up to 165 MPa, if polymer is synthesized not from oligomer mixture of random composition but from the trimer having the following structure:

$$HO-H_2C-\underset{CH_2OH}{\overset{OH}{\bigcirc}}-CH_2-\underset{CH_3}{\overset{OH}{\bigcirc}}-CH_2-\underset{CH_2OH}{\overset{OH}{\bigcirc}}-CH_2OH$$

It was stated in these studies that hydrogen bonds between adjacent polymer network fragments are destroyed completely at 140°C. At lower temperatures the contribution of hydrogen bonds is so great that elasticity modulus, hardness and thermomechanical properties change little with resin composition change and cure conditions. Above 140°C, property indices of polymer become functions of chemical junction frequency only, and defects caused by resin manufacture and cure become especially remarkable. Cure of resol resins in the presence of ionogenic oxides leads to some phenolic units transforming into phenolate ones and hydrogen bonds being substituted by coordinate bonds. Replacement of hydrogen atoms, for example, in hydroxyl groups of phenol by magnesium ions increases the elasticity modulus two-fold, the hardness 30%, compression strength 8% and heat resistance 35°C. However, the replacement of hydrogen atoms in resin by metal ions decreases the reactivity of methylol groups and increases sharply the melt viscosity. It can cause a considerable decrease of polymer network density near chemical junctions [141].

Resin properties can be changed by replacing phenol with cresol isomer mixture or with resorcinol or by copolycondensation of methylolphenols with aniline, melamine, *m*-aminophenol or cresol. The presence of cresol units decreases brittleness of resite, aniline or melamine, and increases heat stability, water resistance and dielectric properties [143].

The rate of novolac resin cure increases sharply if oligomer end units are *m*-aminophenols [142]. In the presence of polybasic acids (phosphoric, boric, silicic), orthonovolacs are esterified intramolecularly, for example:

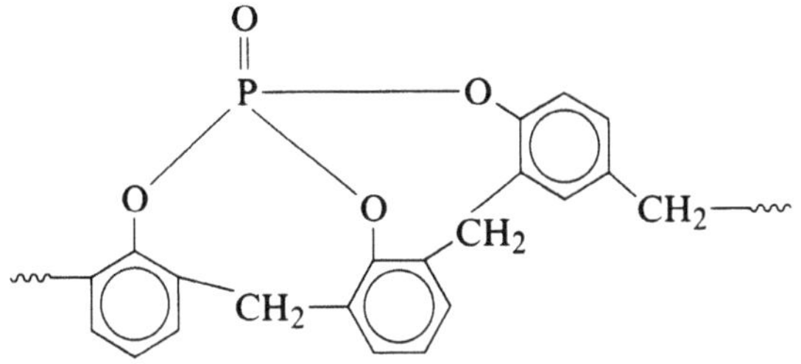

Displacing up to 50% of hydroxyl groups in novolac resin by ester bonds can preserve its curability by means of 'hexa', but increase heat stability and fire retardancy. Cured novolac losses for 4 h at 400°C in air are 60–90%. If it contains 6% phosphorus as phenylphosphoric acid ester, this mass loss does not exceed 28% [144].

The thick polymeric network of resite prevents water diffusion to polar groups. But by penetrating into the polymer, bulk water breaks the hydrogen bonds in it and is absorbed by hydroxyl groups. After completion of hydrogen-bond rupture, water diffusion in polymer obeys the laws characteristic for the processes of inert-gas and vapour sorption. The lower the polymeric network density, the greater is the degree of aggregation of water and the more intensive is the increase in sorption isotherm at the final stage. The diffusion coefficient of water into resite is $0.8 \times 10^{-6} \, \text{cm}^2 \text{h}^{-1}$, and relative elongation at uniform swelling is 0.68%. At a relative humidity of 70% the effect caused by polymer plasticization is overlapped by swelling pressure, which can bring about the mechanical destruction of a thin film. The greater the polymer network density, the lower is its resistance to swelling pressure. Resite structure heterogeneity promotes increase of polymer network tension, which leads to increased water sorption. Low-molecular-weight fractions present in resite are hydrated in the swelling process and sharply increase the swelling pressure [145].

Equilibrium water uptake attains 3.8–4.48% depending on polymer network density [146].

In alkaline medium resite undergoes chemical conversion as a result of hydrogen displacement of OH groups in phenolic units by metal ions from solution – phenolic unit conversion into phenolate is not accompanied by the direct fracture of the polymeric network but can lead to mechanical destruction of the film as a result of the increased swelling pressure. On increasing the alkali concentration, the degree of conversion of phenolic units into phenolate and electrolyte diffusion rate pass through maxima at NaOH concentration of 4 N. Maximum conversion degree equals $6 \times 10^{-3} \, \text{mol g}^{-1}$, and the diffusion coefficient is $3.4 \times 10^{-6} \, \text{cm}^2 \text{h}^{-1}$. Degree of hydration of phenolate units decreases continuously as alkali concentration increases. Mechanical fracture of the resite is most probable in 2 N alkali solution.

In acidic and oxidative media, resites are quite stable. The oxygen index is within the limits of 45–80% O_2. Burning is accompanied by a little fume formation, the fumes being non-toxic. On being removed from the fire, these articles no longer burn.

Resite is stable to strong irradiation. At a dose of 108 rad, strength indices decrease by 20–25%. In the presence of glass fibres, the radiation stability increases.

Limited information on the values of the properties indices for resites of different composition and cure conditions are given in Table 1.13 and data about resites frequently used as binders for polymeric materials are given in Table 1.14.

The behaviour of cured samples was investigated at low temperatures, phenol–aniline–formaldehyde resin elasticized by polyvinylbutyral (P-2M) being an example.

As temperature decreases from 293 to 4.2 K, tensile modulus of elasticity increases from 2800 to 5000 MPa, breaking stress at extension from 55 to 82 MPa, and breaking elongation decreases from 2.0 to 0.9%.

The polycondensation reaction continues up to 350°C and is accompanied by evolution of low-molecular-weight substances (3–7 mass%). Thermo-oxidative degradation of resites begins from 375°C, and up to 50% of initial mass is removed at 600°C as degradation products.

Phenol–formaldehyde binders of different composition are used in combination with asbestos, basaltic, glass and carbon fibres. Asbestos fibre increases article strength in bending for exploitation in acidic media and at temperatures up to 250°C. Basaltic fibres are used to replace asbestos fibres in articles that depend for their performance on friction junctions at elevated temperatures [147]. Glass fibres are used for strengthening components of general use. The advantages of phenolic resin are

Table 1.13 Properties of resites of different composition

Properties	Indices
Density (g cm^{-3})	1.19–1.28
Breaking strength (MPa) at	
extension	40–75
compression	100–125
bending	70–110
elongation (%)	0.4–1.0
Bending elasticity modulus (MPa)	3000–5000
Heat resistance (Martens) (°C)	140–180
Impact strength (kJ m^{-2})	2.8–9.5
Linear thermal expansion	
coefficient $(°C^{-1})$	$(40\text{–}80) \times 10^{-6}$
Equilibrium water uptake (%)	3.8–4.48
Dielectric permittivity (10^6 Hz)	3.9–4.1
Dielectric loss tangent	
at 10^6 Hz	0.03–0.043
Oxygen index (% O_2)	45–80

Table 1.14 Composition and properties indices of resites used as binders

| | Composition | | | | |
Properties	Phenol–aniline–formaldehyde (P-300)	Novolac–furfural (FN)	Phenol–aniline–formaldehyde + poly(vinylbutyral) (P-2M)	Phenol–formaldehyde (LBS-4)	Phenol–epoxy alloy
Density (g cm^{-3})	1.19	1.25	1.2	1.23	1.29
Linear thermal expansion coefficient at 20°C (°C^{-1})	75×10^{-6}	41.8×10^{-6}	51×10^{-6}	56×10^{-6}	72×10^{-6}
Elasticity modulus at extension (MPa)	4100	3180	2800	3600	3150
Breaking stress (MPa) at					
extension	42	50	55	40	62
bending	45	78	84	80	74
compression	85	95	–	–	136
Relative elongation at break (%)	0.45	1.6	2.0	0.5	1.23
Impact strength (kJ m^{-2})	3.8	8.2	6.4	3.4	7.5

especially large in combination with carbon fibres. High heat resistance, low flammability and stability in oxidative media and in elevated humidity conditions favour use of carbophenoplastics as compared with carboepoxyplastics. Carbophenoplastic carbonization is the basis of carbon–carbon composite manufacture, as matrix pyrolysis allows one to convert it into glass carbon with volume up to 71.4% without the article changing shape.

Compositions reinforced by chopped fibres are moulded into articles by pressing. Resol resins sometimes modified by linear polymers in order to increase the brittleness of the matrix serve as binder in most cases. In the case of long fibres a solution of novolac resin in furfural (FN) or in FAM (polyfuran) in combination with the curing agent are used. In this case the prepreg is manufactured by the 'liquid-phase' method and monolithization and cure of semifinished product can be realized under low pressures.

Even greater attention has been paid to phenolic binders when pultrusion moulding of profiles strengthened by long fibres. These are suitable for this moulding method because of their low viscosity and prolonged lifetime of impregnation compositions, rapidly increasing viscosity in the moulding process and the need for a short overcure at the exit from the forming instrument.

In the presence of mineral fillers the conditions of phenol–formaldehyde resin cure are changed. It was stated that, on increasing surface alkalinity of mineral fibres, the rate and degree of curing the phenol–formaldehyde resins decreases remarkably [148]. If the surface is covered by silane appreture [finishing coating], this behaviour disappears and the indices of plastic mechanical properties increase. However, a too thin film of low-molecular-weight appreture cannot eliminate the stresses arising in resite film in contact with mineral fibres [149].

It was shown on the models of glass rod–cured resin film that the stresses attain up to 27.6 MPa. In glass plastics and basaltoplastics the resite layer directly contacting the fibre and penetrating into the thin layer of low-molecular-weight appreture is the reason for crack formation. To decrease matrix tension at the boundary layer to a level preventing microcrack formation during cooling of the moulded article, it was proposed to use a comb-like copolymer of polydimethylsiloxane and poly(ethylene oxide) (KEP) with molecular weight of 7000 as appreture. This polymer is a detergent. On being introduced into phenol–formaldehyde resin, it creates a transition layer of thickness 150–200 Å from the fibre to the polymeric film. The stresses in the boundary layer decrease to 19.3 MPa and phenol glass or phenol basaltoplastic preserves matrix integrity even after multiple changes of temperature from $-70°C$ to $+200°C$ [150].

To prevent crack formation in carbon fibres and resite matrix, it was proposed to create a boundary layer from novolac resin by distributing it in a thin layer from a water emulsion. It is supposed that novolac resin takes part only partially in cure during combination with resol matrix. A phenol carboplastic with uniaxial fibre distribution appreted by novolac, with filling degree of 50%, has a shear strength 37.5 MPa [151].

To lower crack number in the boundary layer that arises during phenol carboplastic pyrolysis, it was proposed to spread a resin made of furfuryl alcohol ('Hitofusan') at the surface of the fibres. The presence of such a layer decreases both the matrix tension and shrinkage that arise during pyrolysis [152]. Taking into account the difficulties related with lubricant removal (the lubricant is spread on the carbon fibre), the complex topography of the surface and the negligible size of pores and cracks, it was proposed to polymerize monomer or monomer mixtures, which satisfactorily wet carbon fibres, at the fibre surface. The polymeric film fills the topographical defects, strengthens the fibre, eliminates the necessity for oxidative activation of the surface and provides the necessary amount of functional groups in order to interact with the binder. The film fulfils the function of a lubricant and of a boundary layer that is water-resistant and heat-stable up to 300°C [153].

(c) Methods of phenol resin modification

Methods of structural modification of phenolic resins and binder composition are directed at replacement of the polycondensation cure process by processes that proceed without evolution of low-molecular-weight compounds, with decrease of brittleness of polymeric matrix, increase of heat stability, imparting to phenolic resin the required combination of physical properties and increase of yield of glass carbon during carbonization.

Under atmospheric conditions at 20°C phenoplastics can be exploited for years. With further temperature increase exploitation time diminishes, and above 250°C phenoplastics retain their properties only for several hours [154].

To prolong phenoplastic performance and to make it more stable above 200°C, it is necessary to decrease the concentration of groups that are oxidized more easily: hydroxyl and methylene groups. For such purposes phenolic units are replaced partially by alkylbenzenic or phenolate ones, which are complexed with residual phenolic units, and methylene links between phenols are replaced by oxygen or nitrogen atoms.

Polymers in which a portion of the phenolic units are replaced by xylenolic ones are more stable to thermo-oxidative degradation, have lower brittleness and lower water uptake, and are therefore more stable indices of dielectric properties [155]:

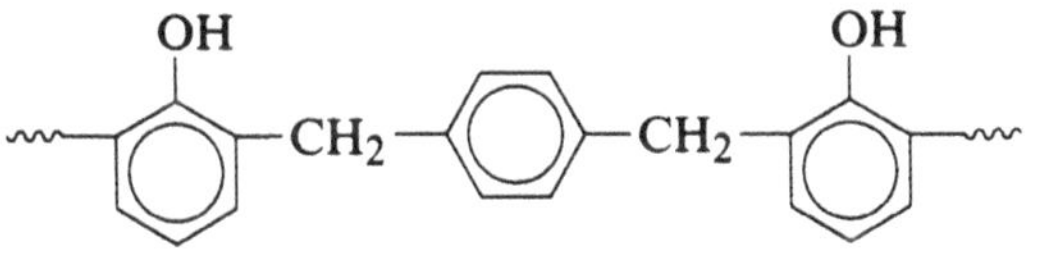

$$\sim\!\!\!\!\sim\!\underset{\displaystyle OH}{\bigcirc}\!-CH_2-\bigcirc-CH_2-\underset{\displaystyle OH}{\bigcirc}\!\sim\!\!\!\!\sim$$

Analogous results are obtained when diphenyl ether units are introduced into network polymer:

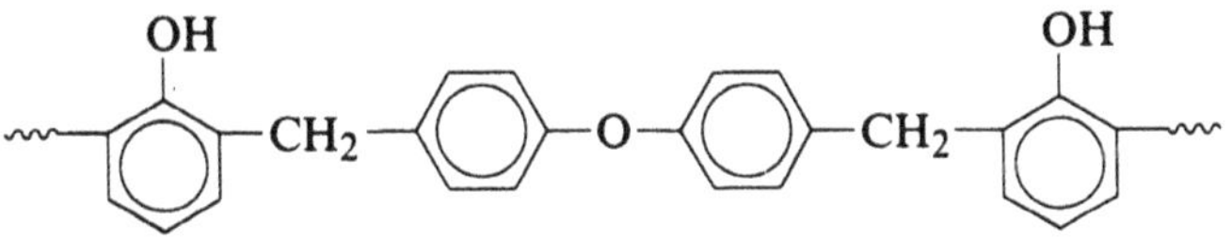

$$\sim\!\!\!\!\sim\!\underset{\displaystyle OH}{\bigcirc}\!-CH_2-\bigcirc-O-\bigcirc-CH_2-\underset{\displaystyle OH}{\bigcirc}\!\sim\!\!\!\!\sim$$

At 300°C mass losses for cured polymer equal 1.4% only; at 500°C, 9.2% of low-molecular-weight substances evolve over 7 min. In an inert medium at 800°C, carbonization of polymer is accompanied by 34% mass loss.

Resol resins containing methylenephenolic and methyleneaniline units (for example, R-300 resin) are obtained by joint polycondensation of phenol and aniline with formaldehyde:

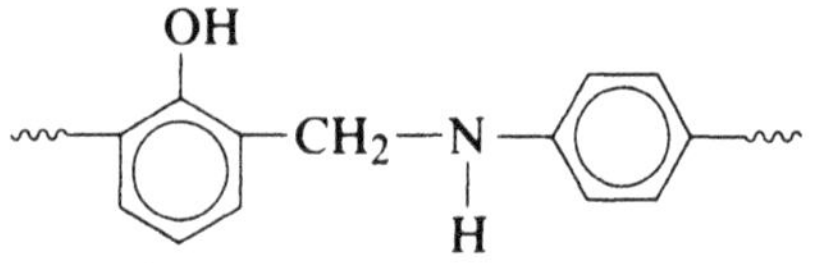

The resin easily wets mineral fibres. In cured polymer, dielectric property and heat stability indices are higher than in a common resite.

Network polymers can be synthesized by resol resin cure from the phenol–trimethylolphenolate–formaldehyde mixture. The presence of metal ions in the polymer structure and the appearance of coordinate bonds between phenolate and phenolic units leads to a polymer with new properties. At low phenolate concentrations in the polymer composition, metal ions are aggregated into clusters of size up to 10 nm; at high concentration they can form domains of size 3–5 μm. The viscosity of metal resolic resins can be decreased by furfural or furfuryl alcohol. The replacement of hydrogen bonds by coordinate bonds is exhibited in polymer heat stability increase by 40–60°C as compared with the common resites, decrease of thermal expansion coefficient, increase of elasticity modulus and appearance of properties that are characteristic for the metal in the polymer composition [156, 157].

To decrease the brittleness of resites, in resol or novolac resin composition polymers of linear composition are introduced: poly(vinyl chloride), polyamide, polyvinylbutyral or butadiene–acrylonitrile copolymer. The presence of linear polymer increases the viscosity of the solution, impedes cure and prevents formation of hydrogen bonds between network polymer units. Therefore fracture energy increase is accompanied by decreasing elasticity modulus and heat stability.

To achieve an elasticizing effect it was proposed to introduce into alcoholic solution acrylonitrile–butadiene copolymer of molecular weight 2600–3000 ('liquid' rubber SKN-26) with carboxylic groups in the end units. To disperse the rubber particles uniformly in the resolic resin, a block copolymer of polybutadiene with polyacrylonitrile was proposed as an emulsifier. In an elasticized resite containing 6% 'liquid' rubber, crack resistance index G_{Ic} increases 5.4-fold and stress intensity factor K_{Ic} increases 3.1 times, whereas heat stability and elasticity modulus do not change to any remarkable extent. The viscosity of the binder at prepreg manufacture stage is lower than in initial resol resin solution [158].

To replace the polycondensation cure mechanism by polyaddition reactions or radical-chain polymerization, new functional groups are introduced into novolac resin structure or other curing agents are selected. Epoxidized novolac resins are used as a binder in moulding materials. On replacing hydroxyl groups in phenolic units by glycidyl ether ones:

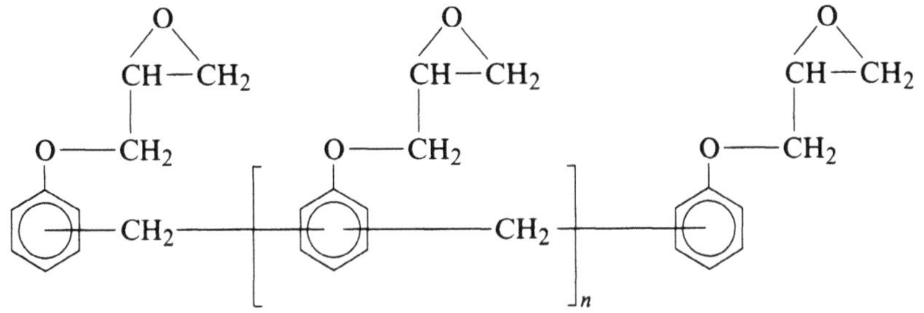

one can replace the polycondensation cure mechanism by polyaddition, if diamine is used as curing agent. Cure proceeds at lower rate and requires a temperature increase up to 180–200°C, but shrinkage decreases on cure and low-molecular-weight products are absent. In cured polymer, the plasticing effect of absorbed moisture is exhibited more sharply and at lower temperature (below 220°C) as compared with common resites. Oxidation processes are developed intensively simultaneously.

The inherent heat stability and stability to thermo-oxidation processes are retained completely in resites if novolac resins are cured by bismaleinimides [159].

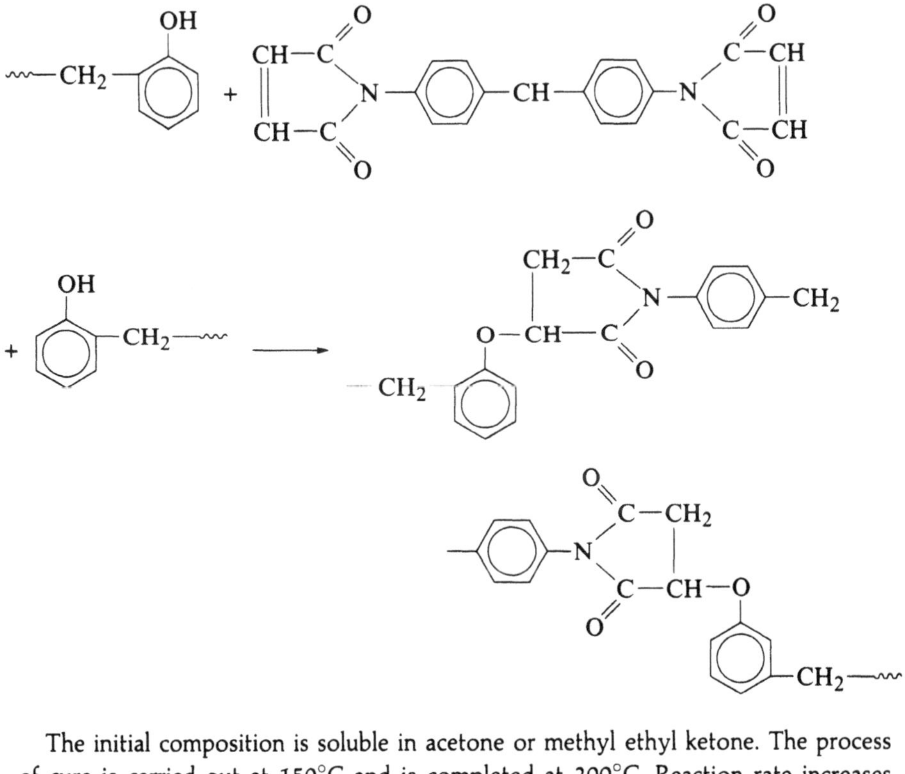

The initial composition is soluble in acetone or methyl ethyl ketone. The process of cure is carried out at 150°C and is completed at 200°C. Reaction rate increases in the presence of bases. Polymer differs from the common resites by higher heat stability (up to 230–250°C). Oxidative degradation begins at 405°C.

1.2.4 Matrices based on unsaturated oligo- and polyesters

Various compositions, the main components of which are unsaturated oligo- and polyesters, that are transformed into network polymer are generalized under the name oligoester binders [160, 161]. These binders are prepared by radical chain polymerization or copolymerization.

Double bonds introduced by unsaturated acids or alcohols participate in polymerization or copolymerization reaction. In the first case maleinic acid is used for prepolymer synthesis (polymaleinates) or methacrylic acid (oligo(ester acrylates)) [162]. Prepolymers containing unsaturated bonds in alcoholic units are prepared by polymerization of allyl alcohol ethers with dibasic unsaturated acids, as a rule with *o*-phthalic acid (diallyl phthalate prepolymers).

Cure process kinetics is described by an S-like curve with clearly expressed activation process stages up to gel formation and slow propagation of the reaction in the gel phase. The reaction develops independently on the various double bonds and macromolecule growth is accompanied with formation of various branchings and often termination as a result of intramolecular cyclization. Even in the early stages, a dispersion of microgel of high-crosslink-density polymer in oligomer or prepolymer medium is formed. Further polymerization develops in microgels, at the microgel surface and in the liquid phase of prepolymer. At a definite stage of conversion, which is characteristic for each composition, clearly expressed separation of microgels from dispersion phase begins and the reaction is localized at high-crosslink-density polymer. Phase inversion proceeds in the final phase. Combined microgels become a continuous phase in which oligomer propagating polymerization is dispersed [163, 164].

After the end of cure, polymer retains a clearly expressed structural heterogeneity with low polymer network density in intergel regions and high density in microgels.

The cure reaction is inhibited by atmospheric oxygen. Being localized in the intergel region, it increases its defectiveness, participating in termination reactions of kinetic chain. Their inhibitory effect is exhibited especially sharply in the surface layers of articles [165].

(a) Polymaleinates

Polymaleinates are esters that are synthesized by esterification of maleic acid anhydrides and dibasic unsaturated acid (or anhydride) with glycol. Molecular weight is controlled by glycol/acid ratio. End units contain hydroxyl groups, average molecular weight being within the limits of 1000–3000. Maleic acid is the *cis* form of the first of two unsaturated dibasic acids with melting point of 130°C. On heating it is isomerized into the thermally stable *trans* form (fumaric acid) with melting point 237°C:

$$
\begin{array}{ll}
\mathrm{HC-COOH} \qquad\qquad & \mathrm{HOOC-CH} \\
\;\;\|\qquad\qquad\qquad & \qquad\quad\| \\
\mathrm{HC-COOH} \qquad\qquad & \mathrm{HC-COOH} \\[4pt]
\text{Maleic acid} \qquad\qquad & \text{Fumaric acid}
\end{array}
$$

Fumaric acid enters less readily into polymerization and copolymerization reactions compared with maleic acid, but as a consequence of unfavourable mutual location of carboxyl groups it does not undergo intermolecular cyclization. The high melt temperature of fumaric acid means that the esterification reaction must be carried out at higher temperature compared with maleic anhydride esterification. The failure of anhydride form increases water yield, promoting process reversibility. Therefore it is preferable to carry out polyester synthesis from maleic acid anhydride and glycol. Isomerization proceeds during synthesis, including the majority of maleinate units in fumarate units. Under standard conditions of synthesis only 10–20% maleinate units remain in the polyester. As fumarate number increases, the hardness of polyester, its softening temperature and melt viscosity increase, intermolecular addition is facilitated and double bonds are involved in this reaction as a result of copolymerization with vinyl derivatives.

The properties of cured polyester prepared from prepolymer can be varied by selection of the glycol and unsaturated dibasic acid.

Usually maleic anhydride esterification is carried out by means of poly(propylene glycol). More elastic and water-resistant polymaleinates are prepared by butylene glycol esterification. More stability to impact loadings can be achieved by replacing the propylene glycol by triethylene glycol. In order to increase stability to hydrolysis while preserving the hardness and rigidity of network polyester, glycols with longer chains are used instead of poly(propylene glycol). These glycols can contain aromatic units, for example, dioxypropylenic ester of diphenylolpropane:

$$HO-\underset{\underset{CH_3}{|}}{CH}-CH_2-CH_2-O-\!\!\bigcirc\!\!-\underset{\underset{CH_3}{|}}{\overset{\overset{CH_3}{|}}{C}}-\!\!\bigcirc\!\!-OCH_2-\underset{\underset{CH_3}{|}}{CH}-OH$$

Units of dibasic aliphatic acids (adipic, sebacic) increase the impact strength of cured maleinate but decrease its heat stability. Units of dibasic aromatic acids (for example, phthalic) increase the rigidity and heat stability of network polymer. If the aromatic acid is halogenated (for example, tetrafluoro- or tetrabromophthalic acid), then the heat stability of the polymer increases.

Polymaleinates that are solid at room temperature soften at 60–80°C. In the melt, oxygen adds to the double bonds of the fumarate units, creating crosslinks between macromolecules and transforming it step by step into a network polymer. The reaction proceeds slowly and is accompanied by various processes. Therefore, polymer network density is low and uncontrolled.

Directed synthesis of network polymer from polymaleinate prepolymer is carried out by copolymerization with monomers. Polymaleinate is soluble in styrene, α-methylstyrene, acrylic and methacrylic acid esters, acrylonitrile, vinyl acetate and diallyl phthalate. At the stage of prepreg preparation, the monomer is used as a diluent of polymaleinate and copolymerizes with it on cure. In network copolymer, chains of several monomer units create crosslinks between prepolymer molecules, drawing in to the reaction the unsaturated bonds of the units of fumarate and partially of maleinate. Of the above monomers, styrene and triethylene glycol dimethacrylate (TGM-3) [166] and more seldom diallyl phthalate [167] are used.

At any ratio of polymaleinate/styrene, homopolymer is absent in the composition. All the styrene takes part in copolymer formation with a number of styrene units in the chains connecting prepolymer molecules [168].

If there is one fumarate unit per mole of styrene, then the solution viscosity is too high, and only 75% of fumarate units are involved in the polymerization process. At a ratio exceeding 1/1.5–2.0 the shrinkage at cure and the indices of mechanical properties of network polymer decrease [169]. In such a case the solution is suitable for fibrous prepolymer impregnation and 95% of unsaturated units of prepolymer are involved in reaction. Volume shrinkage at cure is 7–9%, and the mechanical properties of network polymer achieve the highest indices.

Depending on maleinate/fumarate ratio in prepolymer prepared by maleic anhydride esterification with poly(propylene glycol) and further crosslinked by copolymerization with styrene, the properties of the product are within the following limits: density $1.1–1.46\,\mathrm{g\,cm^{-3}}$; breaking stress at extension 40–60 MPa, compression 80–125 MPa, bending 80–110 MPa; breaking elongation 2.2–6.4%; elasticity modulus at break 2100–3800 MPa; heat stability (at loading $18.5\,\mathrm{kg\,cm^{-2}}$) 40–70°C; oxygen index 24%; dielectric permeability at 60 Hz 3.0–3.4; water uptake (after 24 h) 0.15–0.60%; and shrinkage on curing 7.7–9.6%.

Polymerization is accompanied by intensive heat evolution. At optimum ratio of prepolymer to styrene and at an initial temperature of 100°C, the exothermic peak reaches 200–250°C. The more maleinate units in the prepolymer that are isomerized into fumarate, the higher is the exothermic peak and the shorter is the induction period of copolymerization. Under conditions of intensive heat evolution, monomer evaporability can be minimized. This can be achieved by heat intensity decrease and selection of a monomer with higher vapour pressure. The exothermic peak decreases by 35–40°C if 4–5% of styrene is replaced by α-methylstyrene. The induction period is lengthened simultaneously and does not decrease the double-bond conversion in the prepolymer. Replacing styrene by TGM-3 or diallyl phthalate can eliminate volatility in the cure process and lower polymer shrinkage, but the induction period increases sharply. The conversion of fumarate units is lower. There is much homopolymer TGM-3 or diallyl phthalate in the cure composition. At optimum ratio of prepolymer to styrene, solution viscosity at 20°C is 8000–9000 Pa s, in TGM-3 7000–8000 Pa s, and in diallyl phthalate 5800 Pa s.

The viscosity can be decreased by synthesizing lower-molecular-weight prepolymer. However, the number of structural defects increases in network polymer, which leads to heat stability decrease and lower mechanical properties indices [170]. In order to use the advantages of low-viscosity binder at the stage of prepreg preparation and to preserve the quality of network polymer, chain extenders are introduced into low-molecular-weight polyesters (molecular weight 1000–2500) besides comonomer. Diepoxides or diisocyanates usually serve as chain extenders. The lengthening of macromolecules of polyester proceeds simultaneously in this case.

To prolong shelf-life, an inhibitor is introduced into the binder. The inhibitors used are usually hydroquinone or n-t-butylcatecholamine. Before prepreg manufacture the binder is mixed with copolymerization initiator. Peroxides or peroxide mixtures are used as the initiator. The peroxide is consumed partially in the reaction with the inhibitor but mostly in copolymerization initiation. The more inhibitor that remains in the binder, the lower is the copolymerization exothermic effect, but the longer is the induction period and the lower are copolymer property indices.

The compositions of the peroxides are selected with the aim of realizing a synergistic effect of their cleavage for formation of copolymerization initial radicals or to carry out the stepped process of initiation. At the first stage the molecular weight of polymer film at the prepreg surface increases and copolymerization is completed on moulding the article at the second stage.

The peroxides are ordered into the following series by activity in polymerization initiation of vinyl derivatives: diacetyl peroxides > ketone peroxide derivatives > hydroperoxides > peresters > dialkyl and diaryl peroxides.

The duration of induction period and heat evolution intensity are measures of peroxide activity. The selection of initiator determines the composition conversion into network polymer and its properties (Table 1.15) [171].

Peroxide decomposition temperature and therefore beginning of cure reaction can be decreased by accelerators. Organic acid and transition-metal salts accelerate peroxide cleavage. Cobalt naphthenate is usually used. Such accelerators as vanadium oxide or ferrocenes are even more active. Ferrocene is first dissolved in dibutyl phthalate. The cleavage of peresters and diacetyl peroxides is accelerated by tertiary amines soluble in the copolymer (dimethylaniline, dimethyl-p-toluidine).

In accelerator selection, article moulding conditions are taken into account. If it is proposed to mould the article at a temperature below 30°C, then it is recommended

Table 1.15 Influence of initiator on degree of conversion of poly(maleinate–styrene) and network polymer properties

Initiator	Extractibles (%)	Breaking strength of copolymer at bending (MPa)	Bending flexure (mm)
Benzoyl peroxide	10.22	92.5	9
Dicumyl peroxide	10.64	94.5	9
t-Butyl peroxide	11.35	63.2	11
Benzoyl peroxide + dicumyl peroxide (1/1)	9.47	104.5	8
Benzoyl peroxide + t-butyl peroxide (1/1)	8.42	113.0	7
Benzoyl peroxide + isopropylbenzoyl peroxide (1/1)	9.58	103.5	9

to use methyl ethyl ketone peroxide in combination with cobalt naphthenate. Cure at 30–100°C is initiated by diacetyl peroxide in combination with tertiary amine. To carry out cure above 100°C, t-butylperoxide or hydroperoxide is used and the process is accelerated by dimethylaniline or cobalt naphthenate, respectively.

To achieve deeper conversion in the cure process it was proposed to irradiate the article after moulding has ended. Optimum dose for irradiation of network polymaleinate at overcure stage is 17.5 Mrad. After irradiation, extractibles decrease by 6%, bending strength increases by 56%, elasticity modulus increases by 44% and impact strength increases by 13%. The indices of dielectric properties remain at the earlier level [172].

The properties of polymaleinates cured by different monomers are given in Tables 1.16 and 1.17.

Polymaleinates having Martens heat stability of 130°C (PN-15) are prepared by esterification of maleic anhydride by dipropyl ether of diphenylolisopropane (molecular weight 2000) and curing by diallyl phthalate.

Modifiers in polymaleinate binder composition
Modifiers are intended for viscosity control, increase of shrinkage in copolymerization and temperature of the composition at this stage and for preventing inhibition of the cure process by atmospheric oxygen.

Low viscosity at the initial stage and large shrinkage at the cure stage are characteristic for polyester compositions. The cure process is accompanied by intensive heat evolution. To prevent pressing out of binder from fibre in the moulding process,

Table 1.16 Properties of polymaleinates cured by styrene

	Cured maleinate tradenames			
Properties	*PN-1*	*PN-3S*	*PN-6*	*PM of general use [173]*
Density (g cm^{-3})	1.208	1.205	1.48	1.10–1.496
Breaking strength (MPa) at				
extension	45	50	45	40–90
compression	110	125	110	90–200
bending	105	85	80	–
Brinell hardness (MPa)	180	150	210	
Elasticity modulus at				
bending (MPa)	2800	3500	3900	2000–4400
Relative breaking				
elongation (%)	3.7	2.5	1.8	5.0
Impact strength (kJ m^{-2})	10	7	6	–
Martens temperature (°C)	65	78	75	121
Water uptake after 24 h (%)	0.07–0.14	0.10–0.22	0.94–1.0	0.15–0.60
Dielectric permittivity at 60 Hz	0.36–0.42	–	–	3.0–4.36

Table 1.17 Properties of polymaleinates cured by triethylene dimethacrylate

Property indices	Polymaleinate tradename			
	PN-62	*3CP-3*	*NPS-609*	*PN-40*
Before curing				
Viscosity (Pa s)	6.1–7.0	50–70	3.6–7.1	5.2–7.7
Comonomer content (%)	60	45	55	60
Shelf-life (months)	4	4	4	6
Volume shrinkage at cure	9.2	10	9.6	9.1
After curing				
Breaking stress (MPa) at				
extension	38	50	50	42
bending	56	65	70	85
compression	120	135	140	150
Breaking elongation (%)	1.8	2.6	3.3	3.0
Impact strength (kJ m^{-2})	5.4	2.5	6.0	4.0
Brinell hardness (MPa)	18	22	17	19
Vicat softening temperature (°C)	180	200	165	210

decreased shrinkage and intensive increase of the temperature of the moulded composition mineral powders (chalk, dolomite, kaolin, talc) are introduced. The presence of the powders in the cured binder increases hardness and compression stability, and improves the appearance of the article.

To preserve the low viscosity of the paste (polyester solution in the monomer plus powder filler) at the stage of wetting the fibres and to increase the viscosity in the storage period and prepreg moulding, at the final stage of prepreg manufacture thickeners or thixotropic additives are introduced (Aerosil, bentonite). Magnesia or magnesium hydroxides and calcium hydroxides are used as thickeners. Initial viscosity of the binder increases from 10–40 Pa s by 20–40 times and no pressing out of the binder in the moulding period is observed. The action of oxides and hydroxides is explained by the formation of ionic and coordinate bonds with polyester molecules [174].

Evidence in favour of the above is the sharp decrease of the thickener efficiency in the presence of even negligibly small amounts of water. Thickener efficiency increases in the presence of calcium oxide, phthalic anhydride and lithium chloride [175]. On heating, coordinate bonds are destroyed and paste viscosity decreases.

Powder fillers in combination with thickeners improve article quality considerably, but such imperfections as rough surfaces, inner cracks and voids, and warping of thin-walled articles are not eliminated. These defects are caused additionally by considerable shrinkage of binder in the cure process. The presence of thermoplastics in moulding materials decreases or completely eliminates their residual shrinkage.

The effect is achieved in that case if thermoplastic is incompatible with the copolymer and does not disperse in intergel regions, adding to chain fragments of network polymer by chemical or physical bonds. Cellulose butyrate, poly(vinyl acetate), poly(methyl methacrylate) and poly(ester urethane) are introduced into binder [176, 177]. The presence of linear polymer dispersion in intergel region increases crack stability of cured polymaleinate [178].

The following components are contained in typical binder compositions (in parts by mass):

solution of unsaturated polyester in monomer	20–45
copolymerization initiator	0.2–0.5
mineral powder	20–45
thickener	up to 1
thermoplastic	5–10

Polymaleinates are used usually as binder in composite materials reinforced by glass-fibre filler. Sometimes to increase impact strength some organic fibres are introduced into composite material. Glass fibres are appreted by vinylsilanes [179].

Polymaleinates cured by styrene becomes yellow with time under the action of ultraviolet (UV) radiation. To prevent this effect, styrene as monomer is replaced by methyl methacrylate, or UV absorbers (2-hydroxybenzophenone or hydroxybenzotriazole derivatives) are introduced into the composition of polymaleinate binder. Hermeticity of thin walls of polyester glass plastics is increased by glass scales.

The glass plastics based on polymaleinates are colourless. They can be coloured by adding a dye.

Prepregs with glass-fibre filler are pressed at 125–140°C while maintaining a pressure of 1.0–7.0 MPa for 1–1.5 min as referred to 1 mm of thickness. Manufactured articles containing 40% of glass fibres on the basis of the binder (in which TGM-3 serves as monomer) preserve their indices after 360 h at 130°C at the following levels:

breaking strength (MPa) at extension	79
bending	131
compression	210
impact strength ($kJ\,m^{-2}$)	114

(b) Oligo(ester acrylates)

Oligo(ester acrylates) are prepared by esterification of dibasic saturated acids and acrylic or methacrylic acid with polyatomic alcohols. The oligomer consists of one or several units of unsaturated acid ester and end units of acrylic or methacrylic acid ester. The number of units in the oligomer and molecular degree of branching are determined by component ratio and number of hydroxyl groups in the alcohol. Double bonds are in end units and can polymerize between each other without comonomer participation.

Below, some oligoacrylates in which the dibasic acid is exemplified by phthalic acid, the unsaturated acid by methacrylic acid and with variable polyol are given:

MGF-9   alcohol component = triethylene glycol, ester molecular weight 566, viscosity 1250 Pa s

TMGF-11   alcohol component = glycerin, ester molecular weight 586, viscosity 8000 Pa s

MDF-1   alcohol component = diethylene glycol, ester molecular weight 492, viscosity 6000 Pa s

MGF-1   alcohol component = ethylene glycol, ester molecular weight 390, viscosity 400 Pa s

The higher the oligoester molecular weight, the lower is the shrinkage at cure, but the higher is the viscosity and the lower is the heat stability of network polymer.

MGF-9 polymerizes with a volume shrinkage of 9.6% and has an exothermic peak below 175°C (initial temperature 100°C) and Martens heat stability of network polymer of 45°C.

MGF-11 polymerizes with a volume shrinkage of 8.5% with an exothermic peak below 170°C and network polymer heat stability of 96°C.

Oligo(ester acrylates) were proposed also as curing agents for polymaleinates instead of low-volatility monomers. The compositions are intended for moulding pressing [180] or injection moulding [181].

Of especial importance in oligo(ester acrylates) are the interaction products of methacrylic acid with dihydroxypropylene ester of oligo(dimethyl n-xylyl ether) [182]:

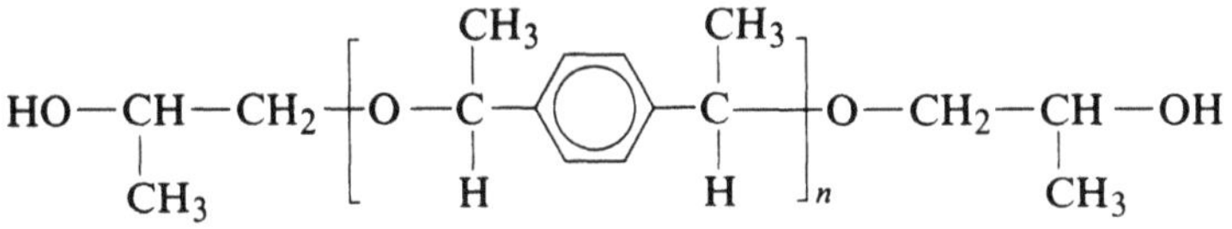

$$\text{HO}-\text{CH}-\text{CH}_2\left[\text{O}-\underset{\underset{\text{H}}{|}}{\overset{\overset{\text{CH}_3}{|}}{\text{C}}}-\langle\text{C}_6\text{H}_4\rangle-\underset{\underset{\text{H}}{|}}{\overset{\overset{\text{CH}_3}{|}}{\text{C}}}\right]_n\text{O}-\text{CH}_2-\text{CH}-\text{OH}$$

(with CH_3 groups on the terminal CH carbons)

$$(n = 1 \text{ to } 5)$$

in a solution of bis(4-vinylphenyl) ether [183].

Depending on unit number, methacrylic acid diester can be liquid or solid, but has low melting temperature. Being an ester of secondary alcohols, the oligomer when polymerized at elevated temperature is destroyed partially and carboxyl and hydroxyl groups arise in the polymer.

Their recombination creates new more thermostable cycloanhydride and cycloether units in the network polymer. The initial viscosity of the oligoester MV-1 at 20°C equals 20 Pa s and decreases to 1–2 Pa s at 150°C. For curing, stepped increase of temperature from 140°C up to 230°C over 8.5 h is required [184].

Depending on cure conditions, polymer property indices are changed within the following limits: breaking strength at extension 73–94 MPa, at compression 126–138 MPa; elongation 9.7–22%; and thermal expansion coefficient $(50–70) \times 10^{-6}\,°\text{C}^{-1}$.

Polymer retains deformation stability up to 250°C. Oligoester MV-1 is used in the manufacture of glass and carbonplastics moulded by liquid-phase winding and monolithization in an autoclave.

(c) Prepolymers of diallyl esters

Double bonds of allyl derivatives are less reactive in polymerization compared with vinyl ones [185].

The process of diallyl ester polymerization is investigated in detail using the example of diallyl sebacate and diallyl phthalate [186, 187]. The reaction, even at initial stages, leads to the formation of polymer of branched structure, which retains solubility in its monomer. The solubility is maintained up to a conversion of 18.6–24%. Molecular weight remains constant at this stage, polymer volume in monomer changing only. Taking into account the ratio of number-average and weight-average molecular weights, which is 5.89, one branch arises on each 12–15 units. So 40–43% of double bonds are retained in the macromolecules. The degree of intramolecular cyclization is high. There is one ring for each 8–10 units. The second stage of polymerization involves interaction of macromolecules with each other and increasing volume of the gel. The number of intramolecular rings increases simultaneously. On increasing the volume of insoluble network polymer, phase inversion occurs. Polymerization of residual monomer is completed in network polymer matrix and density of microgel structures attains the limiting value.

Conversion of double bonds and rate of attaining gel-formation point depend on choice of initiator [188]. If 1.5% of benzoyl peroxide is introduced into the polymer, then a gel fraction is observed after 15 days. Storage time increases in the presence of such inhibitors as phenol, dimethyldi(N-phenylaminophenyl) silane or dimethyldi(N-β-naphthylaminophenoxy)silane. The shelf-life increases up to 12 months in proportion to the amount of inhibitor introduced. Polymerization at 80°C ends after 18 h, at 100°C after 20 min. The presence of inhibitor decreases the polymerization reaction exothermicity. Increasing the shelf-life of the mixture of prepolymer, initiator and inhibitor causes the influence of the latter on polymerization reaction process to decrease.

The conversion degree for double bonds in curing at 140–180°C reaches 95%. Curing at 177°C proceeds for 4 min. Volume shrinkage is near 3% [189].

Thermodegradation of poly(diallyl phthalate) begins from 260°C. Deformation stability at 175°C is retained for a long time (1500 h). The polymer has high indices of dielectric properties; it is colourless and optically transparent.

Breaking strength of the polymer at extension is 21–28 MPa, at compression 150–170 MPa and at bending 50–63 MPa. The temperature of onset of thermodegradation for polyallyl esters depends on acid selection and polymerization conditions. In the presence of benzoyl peroxide (0.5%), t-butyl perbenzoate (1%) and cumyl peroxide (1.0%) polymer was prepared by step-by-step temperature increase from 70°C up to 150°C.

Thermodegradation of poly(diallyl phthalate) begins at 260°C and of poly(triallyl cyanurate) at 300°C. Property indices remain unchanged if poly(diallyl phthalate) is heated for 1500 h at 185°C, poly(diallyl isophthalate) at 190°C and poly(triallyl cyanurate) at 225°C. Thermodegradation rate attains a maximum in polyesters based on phthalic, adipic or sebacic acids at 400°C, and isophthalic acid at 430°C. The source of thermodegradation processes is unconsumed double bonds [190]. Prepolymer is usually combined with glass-fibre, seldom with organic fibres or asbestos.

Diallyl phthalate prepolymer in a mixture with poly(vinyl acetate) (5–10 mass %), chalk (5–10 mass %), benzoyl peroxide, dicumyl peroxide and t-butyl perbenzoate is used as a binder in glass-fibre materials (74%) for manufacturing profiled articles by pultrusion. The prepreg at 25°C retains its lifetime for 180 days, at 50°C for 16 h and at 150°C for 80 s [191].

To provide wetting of fibrous filler it is desired to use prepolymer containing not more than 15% of polymer. To decrease shrinkage in the cure process, mineral powders or a powder of precured polymer are introduced [192]. Using poly(diallyl phthalate) powder containing 90% gelfraction can both decrease shrinkage at moulding and create a monolithic matrix with stable property indices.

To increase fracture viscosity a ternary block copolymer made from one polydivinyl block and two blocks of poly(α-methylstyrene) is introduced [193].

1.3 MATRICES MANUFACTURED OF THERMOSTABLE LINEAR AND LOW-CROSSLINK-DENSITY POLYMERS

1.3.1 Matrices of linear polymers

Structural microheterogeneity of high-crosslink-density polymers and the tension in interjunction chain fragments decrease tensile strength (to 60–110 MPa) and, especially sharply, impact strength. Long before material fracture the matrix loses monolithic character and cleaves partially from the fibres.

Flexible-chain linear polymers (Kuhn segment length of 20–30 Å) have higher resistance to impact and sign-variable loadings but heat stability is low and cold flowability is high. On increasing the structural rigidity of macromolecules, polymer heat stability increases and cold flowability decreases in the glassy state. Macromolecule rigidity (Kuhn segment length above 100 Å) increases by 10-fold if polymer chain units contain phenylene groups, especially in the case of 1,4-addition. For example, in concentrated solution of sulphuric acid, polyarylamide with symmetric phenylene unit addition [194, 195]

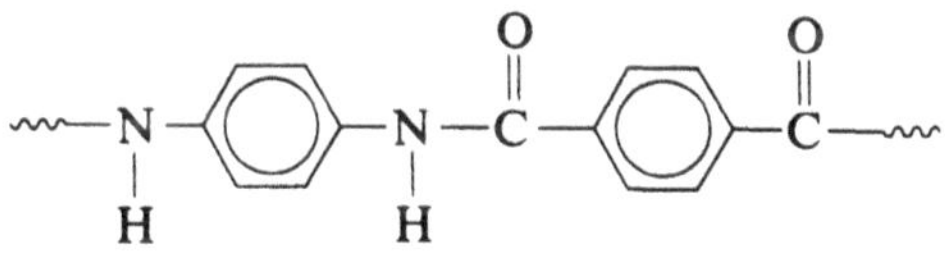

is characterized by Kuhn segment length of 1300 Å, the number of monomeric units in the segment being 200. In the polymer with asymmetric phenylene group units:

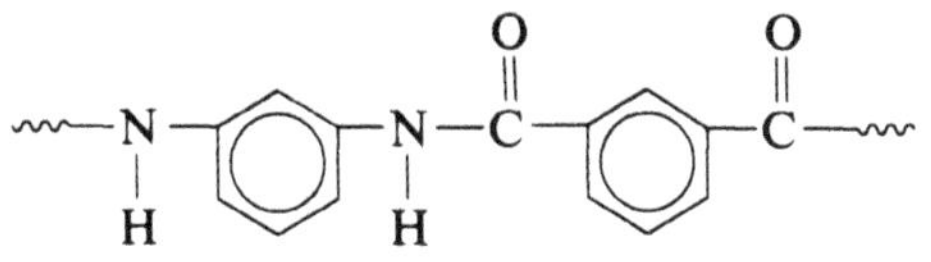

the Kuhn segment length decreases to 50 Å and the number of units in the segment to 8.4.

Glass transition temperature changes from $T_g = 433°C$ to $275°C$, respectively. One can vary widely the structural rigidity of linear polymer macromolecules and therefore the glass transition temperature, melt viscosity and solubility by alternation of phenylene units with polar and flexible 'hinge-like' groups.

Among the various linear polymers with elevated heat stability, structures of the type

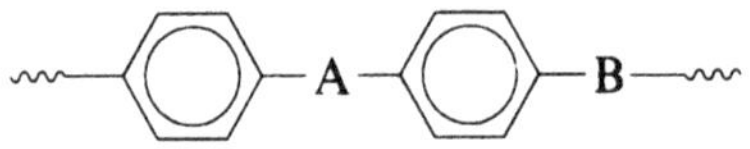

where A are polar groups and B are 'hinge-like' units, have the most widespread application. Examples are as follows:

Polar groups A

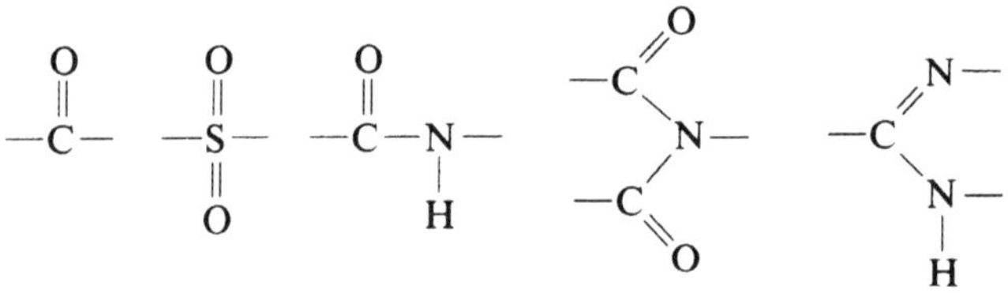

Hinge-like units B

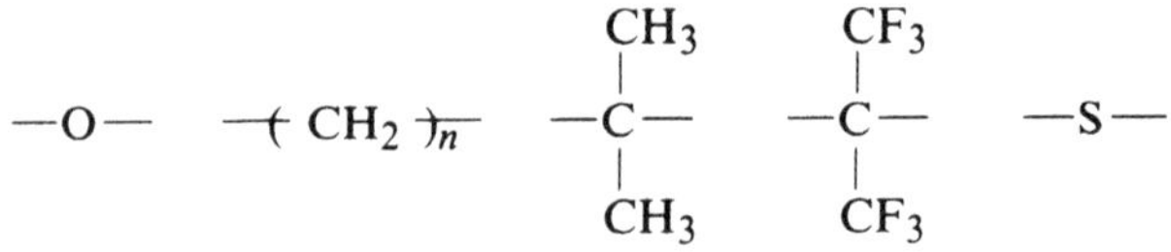

The relationship between polar and 'hinge-like' units can be different. Such linear polymers are usually called 'semirigid' [196]. Glass transition temperature for 'semirigid' linear polymers depends on type and ratio of A and B units, but usually exceeds 180°C. Deformation stability (heat stability) of such polymers is 20–30°C below the glass transition temperature.

To increase heat stability of the polymer, regular structure polymers are synthesized that are capable of crystallizing with crystallinity of 40–60%. Heat stability in such a polymer is retained at the level of 30–40°C above the glass transition temperature [197, 198].

The presence of crystalline phase prevents the cold flow inherent to linear polymers of amorphous structure. When the crystalline phase arises, polymer softening temperature and solubility are decreased even in polar organic solvents. Increased structural polymer rigidity diminishes the rate of achieving equilibrium crystalline structure. Crystallization at the most favourable temperature lasts for 1.5–2.5 h.

The absence of solubility of semicrystalline heat-stable thermoplastics, high softening temperature (above 370°C) and high melt viscosity impede prepreg manufacture.

Polymer does not penetrate into interfiller space of plaits and does not spread over all the relief of fibre surfaces. Polymer melt is usually distributed on fibre plaits by a coextrusion method. On exit from the extruder, the rod made of the plait and polymer distributed on it are crushed into granules. The granules are pressed into the articles by heating the polymer again up to the softening temperature. To decrease material porosity it is proposed to remove grease from strengthening fibres and to impregnate the plaits with dilute polymer solution of polymer having the same unit composition as matrix polymer but lower molecular weight and amorphous structure. After solvent removal, the plait is directed on extrusion in order to be compatible with the matrix polymer.

In the manufacture of large articles, sheet filler (tapes, textile, fabric, thick felt) are alternated with polymer powder layers or previously prepared film of polymer. The packet is pressed at the polymer softening temperature. In this case material porosity attains 6–8 vol % and strength is low. One can decrease porosity caused by weak wettability of elementary fibres by first removing the grease and impregnating the filler with amorphous polymer solution.

The methods of 'fibrous' technology are used to an increasing extent [199, 200]. Matrix polymer is mixed with strengthening fibres into the prepreg prepared in the form of mats or fabrics of different braiding. The packet is pressed at the softening temperature of matrix fibres. In this case one can achieve deeper penetration of melt into the interfibre space of plaits and threads, even decreasing the pressure of moulding. However 'matrix' fibres of heat-stable semicrystalline polymers are moulded from dilute solutions of sulphuric acid or high-boiling-point polar solvents and that feature increases prepreg value sharply.

Table 1.18 gives a list of heat-stable polymers used and recommended for use as matrices for composite fibrous materials. Brief characteristics of each polymer are also given below. Information about property indices and heat stability of linear polymers is given in Table 1.19.

(a) Poly(phenylene oxide)

Poly(phenylene oxide) is an amorphous polymer of molecular weight 30000–40000 that is soluble in benzene, toluene, xylene, tetrahydrofuran and methylene chloride. It is stable to acidic and alkaline media but less stable to oxidizing agents. Mass losses at 250°C for 1.5 h attain 46%. Oxidation proceeds in methylene groups and leads to macromolecules combining with each other. The polymer loses solubility and the ability to change into the viscous flow state, the brittleness increasing. At 100°C for 430 h the gel volume fraction reaches 9%, and at 150°C for 112 h it is 58% [201].

Table 1.18 Heat-stable linear polymers used as matrices in composite materials

Polymer	Abbreviation or trademark	Unit structure	Phase state	Density $(g\,cm^{-3})$	Temperature (°C)			
					T_g	T_m	T^a_{max}	T^b_{onset}
Poly(dimethyl-phenylene oxide)	PPhO	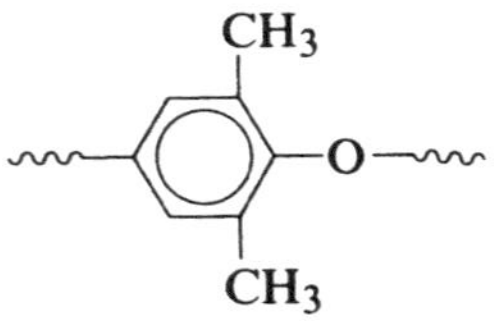 (aromatic ring with two CH_3 groups and $-O-$)	Amorphous	1.06	205–210	–	–	320–380
Poly(ether sulphone)	PES	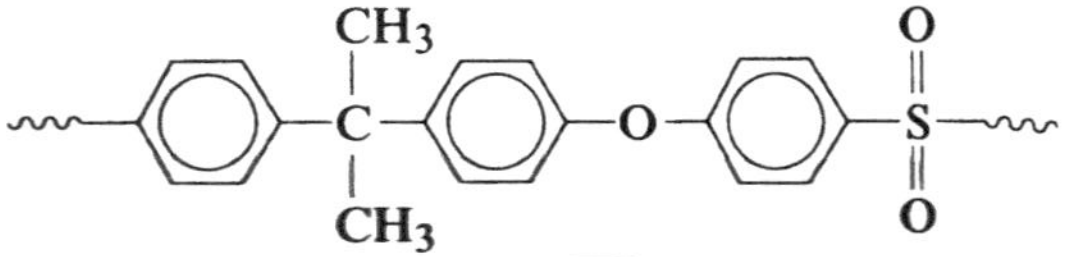 (bisphenol-A / $-O-$ / sulphone $-SO_2-$ unit)	Amorphous	1.35–1.37	200–230	–	–	315
Poly(phenylene-sulphide)	PPhS	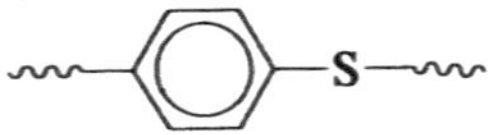 (phenylene $-S-$ unit)	Semicrystalline	1.36	88	277–281	204	315
Poly(ether ether ketone)	PEEK	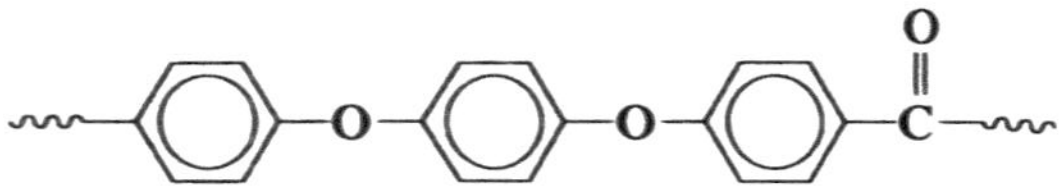 (diphenyl ether / ketone $-CO-$ unit)	Semicrystalline	1.3075	143–150	340–347	302–316	319–330
Poly(ether imide)	Ultem	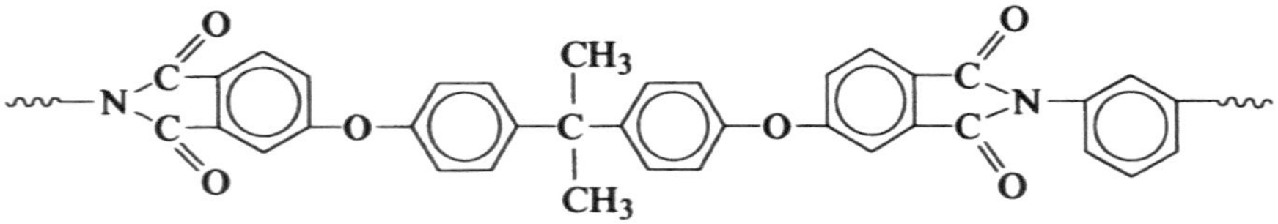 (imide / bisphenol-A ether unit)	Amorphous	1.15–1.27	215	285	–	–

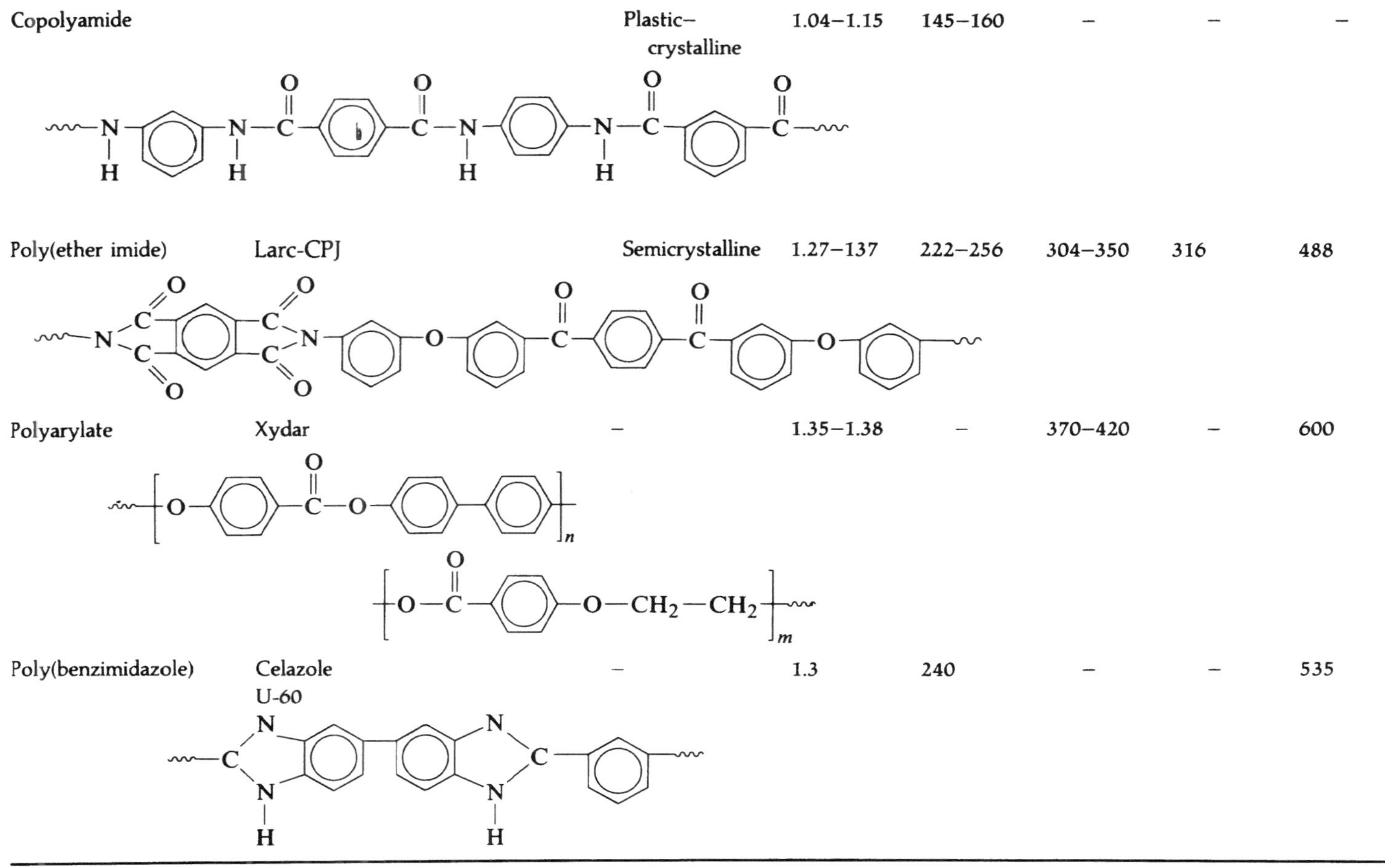

Polymer	Trade name	Physical state	Density	T_a[a]		T_b[b]	
Copolyamide		Plastic–crystalline	1.04–1.15	145–160	–	–	–	–
Poly(ether imide)	Larc-CPJ	Semicrystalline	1.27–137	222–256	304–350	316	488
Polyarylate	Xydar	–	1.35–1.38	–	370–420	–	600
Poly(benzimidazole)	Celazole U-60	–	1.3	240	–	–	535

[a] Temperature of attainment of maximum crystallinity.
[b] Temperature of onset of intensive degradation.

Table 1.19 Properties of heat-stable linear polymers used as matrix in composite material

Polymer name	Breaking stress (MPa)			Elasticity modulus at extension (MPa)	Elonga-tion (%)	Impact strength (kJ m^{-2})	Heat stability, $T_{18.4}$ (°C)	Linear thermal expansion coefficient (10^{-6} °C^{-1})	Water uptake (%)	Dielectric permitti-vity at 60 Hz	Dielectric loss tangent at 60 Hz	Oxygen index (% O_2)
	Extension	Compres-sion	Bending									
Poly(dimethylphenylene oxide)	63–70	–	98–105	2520	50–80	8–10	190	16	0.06	2.58	0.0035	40
Poly(phenylene ether sulphone)	85–90	125	120–129	2800	30–80	90	150–174	47–56	0.43–1.4	3.55	0.0035	41
Poly(phenylene sulphide)	75	–	96	2000	1.4	1.6	135	42	0.1	–	–	47.8
Poly(ether ether ketone)	92–103	–	120–169	3800	40–80	9.6	180	43	0.22	–	–	35–40
Poly(ether imide)	105	140	145	3000–3500	60–80	5.3	200–220	62	1.25	3.15	0.0013	47

Polymer is processed in the presence of antioxidants by injection moulding at 295–340°C and 60–120 MPa into a mould heated to 130°C. The holding time of the material in the cylinder of the moulding machine cannot exceed 10 min. Pressing is carried out at 250–270°C and a pressure of 16–20 MPa.

Poly(phenylene oxide) differs from the usual polymers in its high dielectric properties, which change only slightly up to 200°C and with changing current frequency or medium humidity.

The thermal expansion coefficient has an unusually low value for amorphous polymers (16×10^{-6} °C^{-1}). This leads to the stability of mechanical property indices over a wide range of temperatures. This feature of the polymer was used in the manufacture of articles with dielectric properties for use at cryogenic temperatures or under conditions of sharp temperature falls. The polymer is compatible with polystyrene which fulfils the function of plasticizer. As polystyrene content increases, the softening temperature, melt viscosity and temperature of deformation stability decrease.

In the majority of cases poly(phenylene oxide) is used as a matrix combined with glass fibres. However, there is a proposal to use poly(phenylene oxide) in the manufacture of limitedly filled organoplastics. Plaits or tapes of poly(aryl amide) fibres are impregnated with poly(phenylene oxide) from dilute solution in methylene chloride. Mass gain equals 12–18%. The prepreg is subjected to transverse compression at 250°C and 10 MPa. The fibres are processed into narrow films without loss of oriented structure of the molecules. After assembling the packet and pressing the article at 275°C at a pressure of 7.5 MPa, degree of filling reaches 80–85% [202].

(b) Poly(ether sulphones)

Among the various polymers relating to the poly(ether sulphone) (PES) group, the most widespread application has polymer containing sulphone and isopropyl groups between phenylene units [203–205]:

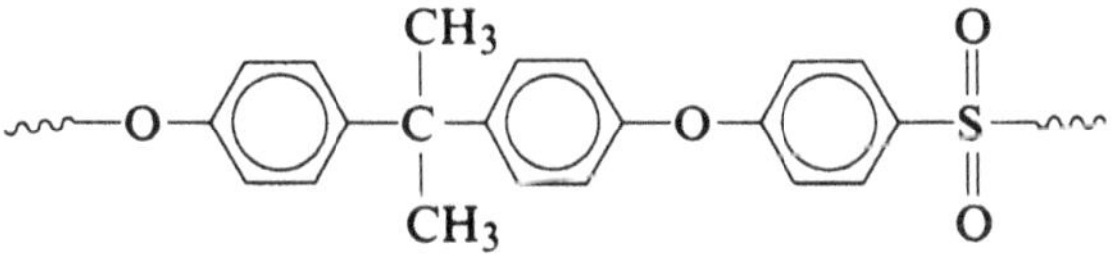

$$\text{---}O\text{---}\langle\bigcirc\rangle\text{---}\underset{\underset{CH_3}{|}}{\overset{\overset{CH_3}{|}}{C}}\text{---}\langle\bigcirc\rangle\text{---}O\text{---}\langle\bigcirc\rangle\text{---}\underset{\underset{O}{\|}}{\overset{\overset{O}{\|}}{S}}\text{---}$$

Molecular weight of polymer is 20 000–55 000. The polymer has an amorphous structure, and is soluble in tetrahydrofuran, methylene chloride, dioxane, dimethylformamide, chlorobenzene and cyclohexane. The high melt viscosity impedes moulding. Injection moulding is carried out at 360–412°C (viscosity 2000 Pa s) into moulds heated to 110–160°C. Shrinkage on cooling is 0.5–0.7%. Sulphur dioxide separation is observed in the moulding process. Intensive degradation begins at 380°C and proceeds with autoacceleration. In the presence of traces of iron, degradation intensity increases sharply, and 0.05% of iron is enough for gel volume fraction at 320°C after 10 min to reach 38%. To retard thermodegradation, thermostabilizers are introduced into the polymer composition. At 140°C for 900 h thermostabilized poly(ether sulphone) loses 3% of mass only [206].

High sensitivity to electron irradiation is a distinguishing property of poly(ether sulphones). On being irradiated by an electron beam it becomes a network polymer as a result of recombination of macroradicals, and its heat stability increases.

Moisture uptake by poly(ether sulphone) after 7 days in humid conditions equals 0.58%.

(c) Poly(phenylene sulphide)

This polymer has semicrystalline structure. Maximum crystallinity (54%) is attained after 2 h heating at 204°C and slow cooling [207]. The polymer is moulded at 320–350°C. Melt viscosity at this temperature equals 2000 Pa s.

Moulding is accompanied by intensive oxidation and gel fraction formation. Only the proper selection of thermostabilizer can retard this process. Lithium carbonate is recommended as stabilizer. It inhibits degradation of poly(phenylene sulphide) if it is heated above 370°C. To thermostabilize in the range 300–370°C poly(ether ketone) or polyamidoimide were proposed. These are introduced in amounts up to 2% [208]. Thermostabilized polymer can stay in the cylinder of the moulding machine for not more than 10 min. As oxidation proceeds, macromolecule branching increases and gel volume fraction increases. This leads to fracture energy decrease. In an inert medium at 350°C only an increase of polymer molecular weight is observed. Above 500°C the polymer carbonizes, mass loss being 50%. It is a characteristic for poly(phenylene sulphide) that it has elevated fire resistance.

On being removed from the fire, the polymer quickly extinguishes. The temperature of inflammability in air is 493°C at minimum fuming [209]. The polymer retains its deformation stability for a long time at 170°C and for a short time at 230°C. The polymer is characterized by its electrical conductivity (electrical resistance $10^{10}\,\Omega$ cm). It can be increased by ion adsorption on polymer macromolecules (electrical resistance decreases to $10^{2}\,\Omega$ cm) [210].

The polymer is used as a matrix in fibrous composite materials, but the melt viscosity is too large and the wettability of fibres too low. Therefore porosity in carboplastics attains 5.9% and shear strength does not exceed 32 MPa. In combination with poly(aryl amide) fibres, the porosity reaches 4.8% and with glass fibre it is 3.8%.

Observations are described that carbon fibres with grease removed intensify polysulphide crystallization in boundary layers, thus increasing plastic shear strength [211].

(d) Poly(ether ether ketone)

This is a semicrystalline polymer with molecular weight near 22 000 [203, 213]. On quick cooling of the melt, completely amorphous polymer with $T_g = 155°C$ and density of $1.264\,\mathrm{g\,cm^{-3}}$ is formed. If polymer was maintained for 24 h at 176°C ('cold crystallization') crystallinity reaches 41%. The crystals are thin with a great number of defects and $T_m = 201$°C; large crystals with $T_m = 334$°C are inserted. Consequent heating at 302°C increases crystallinity up to 50% and irregular crystals disappear. In the amorphous phase the glass transition temperature increases up to 160°C; the crystalline phase melts at 321°C. If the melt experiencing 'cold crystallization' is maintained at 302°C for 29 h, the crystallinity attains 50% again, but among the crystals melting at 321°C crystals arise with a melting point of 331°C. The highest crystallinity (53%) is attained at 316°C. The crystals melt at 338°C, polymer density becoming $1.3075\ \mathrm{g\,cm^{-3}}$ [212]. However, long-term holding of the polymer at 316°C is accompanied by degradation.

In argon at 238°C the polymer can be dissolved in α-chloronaphthalene. At room temperature it is soluble in sulphuric acid only. The polymer does not swell in oil fractions, and is stable to electrolytic action.

Poly(ether ether ketone) (PEEK) can be moulded at 380–400°C. At this temperature, melt viscosity varies from 2000 to 4000 Pa s depending on polymer molecular weight. The polymer preserves its deformation stability for a long time under loading at 145–160°C but only for a short time at 180°C.

Poly(ether ether ketone) can serve as a matrix in carboplastics. If grease is removed from carbon fibre, it accelerates polymer crystallization in boundary layers. In this case plastic porosity does not exceed 2% [214].

The influence of the surface of carbon fibres free of grease on polymer crystallization in boundary layers is supported by the unusually high compression resistance of plastics especially perpendicular to the fibres (250 MPa) and interlaminar shear index of polymer matrix strength (110 MPa).

Elevated carboplastic resistance to crack formation and the capacity to damp impact and cyclic loading is predetermined by the low thermal expansion coefficient of poly(ether ether ketone), its viscous fracture and strong cohesion with carbon fibres. Destruction of the plastic begins with a crack propagating along the fibres even in such a case when a notch in the sample is made on a polymer layer [215, 216].

Polymers in which two ketone groups follow one ether unit

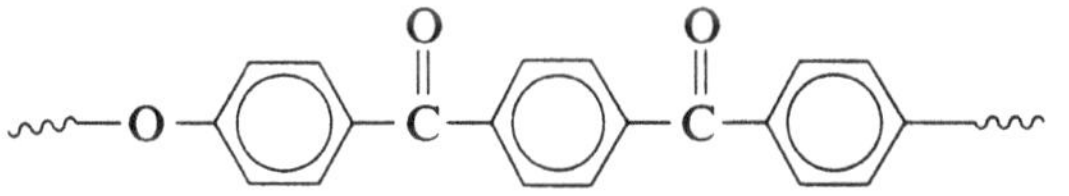

are related to poly(ether ketone ketones) (PEKK). PEKK is more difficult to synthesize than poly(ether ether ketone) but its glass transition temperature is 15°C above and the melting temperature is 11°C below those in poly(ether ether ketone). Melt viscosity is much lower (600–800 Pa s). All this facilitates prepreg manufacture and article moulding.

(e) Poly(aryl amides)

Amorphous and semicrystalline polymers in which the units of diaminophenylene isomers are combined with the units of phthalic, iso- and terephthalic acids belong to aromatic polymeric amides:

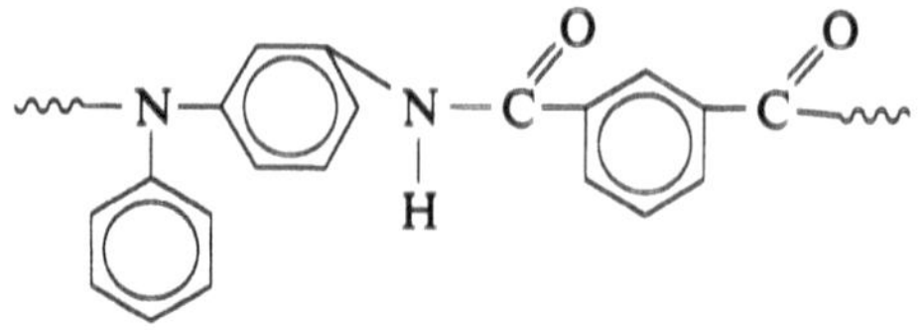

$$T_g = 275\ °C,\ T_{soft} = 380\text{–}390\ °C$$

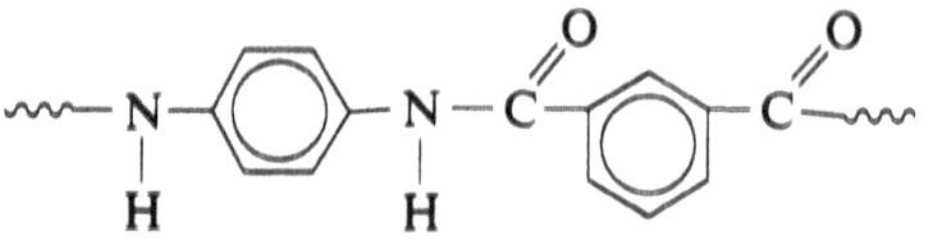

$$T_g = 306\ °C,\ T_{soft} = 490\text{–}500\ °C$$

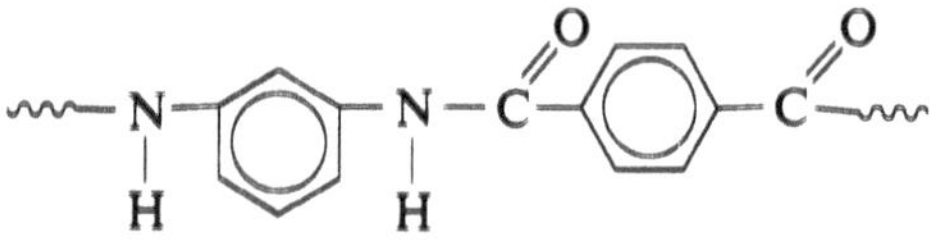

$$T_g = 334\ °C,\ T_{soft} = 490\text{–}500\ °C$$

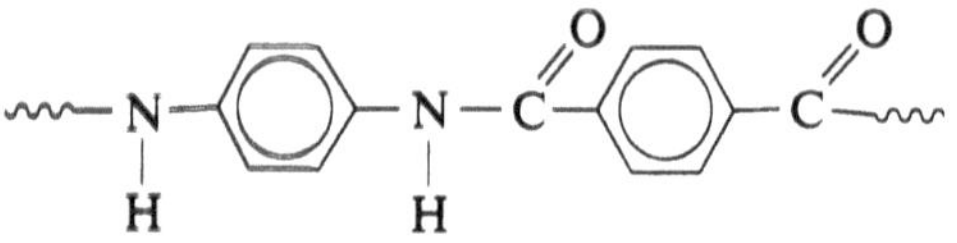

$$T_g = 483\ °C,\ T_{soft} = 530\ °C$$

One can decrease the softening temperature by 80–140°C and facilitate copolymer solubility without decreasing its heat stability. This is achieved by copolycondensation of diamine with a mixture of dicarbonic acid isomers. Resistance to thermo-oxidative degradation is determined by the degree of order of the macromolecular structure. In vacuum, intensive thermodegradation of poly(p-phenylene isophthalamide) begins at 400°C and is completed at 500°C by carbonization with 60% mass loss [217]. Poly(p-phenylene terephthalamide) loses 25% on heating in vacuum up to 540°C. In air at 300°C for 70 h mass losses for poly(m-phenylene terephthalamide) reach 6.8%, for poly(p-phenylene terephthalamide) 10.6%. Oxygen index is within the limits 37–44% depending on unit composition. Water uptake is high and in equilibrium state equals 3.5–7.8%. In acid or alkali solutions, gradual hydrolysis of the polymer is observed.

Copolyamides containing both amide and ester links are more suitable as matrices for fibrous composite materials. Poly(ester amides) of type KS, J-copolyamide, etc., are related to this group. Fibrous filler is impregnated with the polymer from the melt. At 285°C the viscosity of J-copolyamide is near 900 Pa s, which is two-fold lower than the melt viscosity of poly(ether sulphone) and 3.5 times lower than that of poly(ether ether ketone). Shear strength of carboplastic with poly(phenylene sulphide) matrix does not exceed 32 MPa (due to low wettability) and that with poly(ether ether ketone) 54 MPa (due to high melt viscosity). Shear strength of carboplastic (in all cases grease was removed from carbon fibres) based on J-copolyamide varies in the limits 76–86 MPa. Fracture viscosity is 10-fold higher than for epoxycarboplastic and 30–35% higher than for carboplastics based on PEEK or PES.

In combination with aromatic fibres J-copolyamide matrix provides the highest plastic fracture viscosity. Shear strength attains 55 MPa. Prepreg is manufactured by the coextrusion method.

(f) Poly(ether imides)

Linear poly(ether imides) [218] are synthesized in two stages. In the first one poly(amino acid) is synthesized and polymer is distributed from 15% solution onto the surface of fibrous filler. In order for polymer to penetrate into the interfibre space of plaits the filler is first wetted with 2–3% solution of poly(amido acid). The thin underlayer is dried and imidized by intramolecular cyclization of poly(amido acid). The impregnation of prepared fibre filler with 15% solution is repeated three times. Each time the solvent is removed and the distributed layer is imidized. Prepreg is pressed at 350°C and cooled in the mould up to 200°C at a rate of $5°C\,min^{-1}$ [219]. Amorphous polymer (for example, Ultem) can be maintained for a long time under loading at 170°C and briefly at 220°C. Semicrystalline poly(ether imide) (for example, LARC-TPJ) with molecular weight 30 000–35 000, glass transition temperature 222°C and melting temperature 350°C, is pressed at 400°C. At 488°C in air the polymer loses 5% of its initial mass.

The prepolymer – poly(amido acid) – is soluble in dimethylacetamide. To achieve maximum crystallinity the polymer is held in the mould at 316°C. Shear strength of the carboplastic in boiling water decreases by 15% only and recovers completely after drying. The mechanical strength, especially on compression, is characteristic for this polymer (Table 1.19).

Polybenzimidazoles (PBI), for example Celazole U-60:

were proposed as a matrix for carbon and carbon silica fibrous composite materials [220–223].

Like polyimides, polybenzimidazole is synthesized in two stages. In the initial stage, soluble polymer is prepared:

Polyamidoamine is distributed on the surface of fibrous filler and partial intramolecular cyclization is carried out to remove water vapour. The material is moulded at 400°C into thin-walled articles. After the end of cyclization, the polymer loses solubility on heating. The plastic is distinguished by very high surface hardness.

1.3.2 Matrices of low-crosslink-density polymers and block copolymers

The undoubted advantages of heat-stable linear polymers as matrices of fibrous composite materials are clear compared with high-crosslink-density polymers. Storage time of prepregs with matrices based on linear polymers is not limited. These composite materials have higher indices of mechanical properties, especially on impact and cyclic loading: the material preserves monolithic character up to fracture, and it is much easier to vary polymer matrix physical properties. Semicrystalline linear polymers have better heat-stability properties than high-crosslink-density polymers, and are stable to cold flow, solvents and gas permeability.

However, heat-stable linear polymers, especially semicrystalline ones, do not meet the technological requirements of a material for use as a matrix of fibrous composites. The temperature of the transition to the viscous flow state is too high. In the majority of cases it lies outside the limits of the onset of polymer or fibre degradation. Melt viscosity is high, and polymer does not fill completely the interfibre space nor spread over the whole surface of the fibres. In the contact zone near the fibre surface, an adsorption layer of equalibrium structure cannot form. As a result, the most sensitive zone of composite material – the boundary of filler contact with matrix – becomes most variable, which leads to decreasing shear strength indices.

Structural rigidity of macromolecules of heat-stable linear polymers is exhibited in inhibited process of achieving the equilibrium degree of crystallization. This forces one to maintain the material for 2–3 h in the forming instrument at a temperature exceeding the glass transition temperature of the matrix polymer. Finally, high temperatures of matrix polymer softening at the stage of prepreg manufacture and consequent moulding of articles (300–450°C) require new thermostabilizer to inhibit polymer degradation and new equipment.

All these difficulties can be overcome, as in the case of high-crosslink-density matrices, using stepped synthesis of polymers having increased macromolecular rigidity.

In the initial stage, binder can be in the form of oligomer having low melt and solution viscosity in technically acceptable solvents. The functional groups must be in the end units of the oligomer, which react with each other at elevated temperature or with low-molecular-weight bifunctional component – the chain extender [224]. Structural rigidity of oligomeric chain minimizes the probability of intramolecular ring formation in the reaction with the chain extender.

To prevent polymer cold flowing under performance conditions, along with the chain extender (or instead of it) a curing agent (i.e. low-molecular-weight compound with three or more functional groups of the same type) is introduced into the binder composition. On changing the ratio of chain extender to curing agent, the length of the fragments between chemical junctions can be controlled. The following recommendations are an example of stepped synthesis of a heat-stable polymeric matrix.

Amorphous poly(ether ketone ketone) with end maleinimide groups and average molecular weight of 6890 is synthesized [225]:

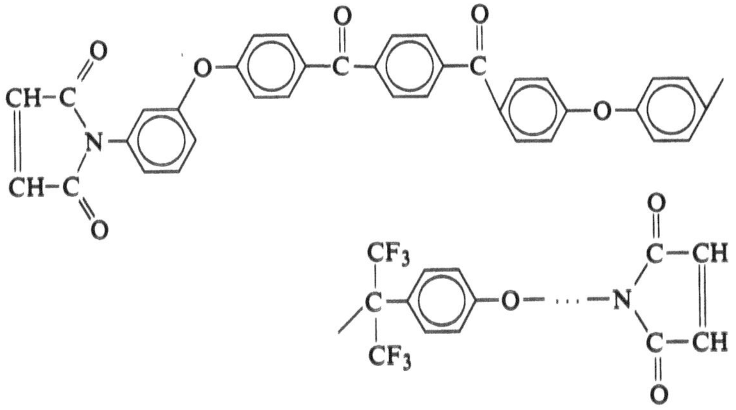

Chain lengthening in article moulding proceeds as a result of thermal polymerization at 270°C or the reaction with curing agent or chain extender (aromatic diamine or dithiol) [226]:

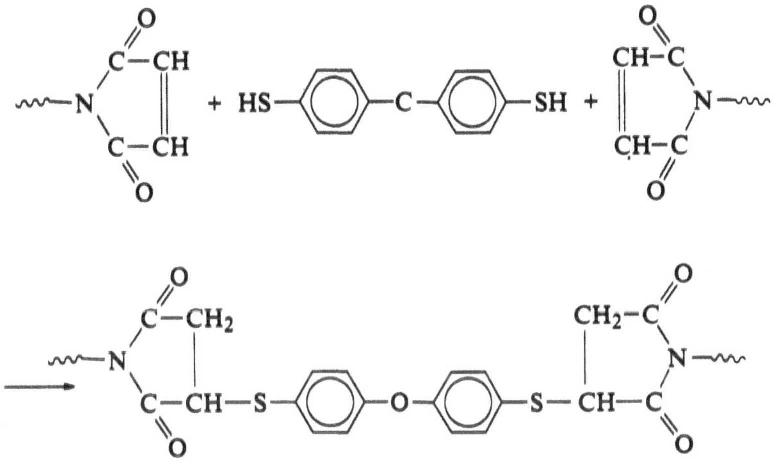

The binder in the initial stage can consist of oligomers of different unit structure but compatible with each other. A particular effect can be achieved by selection of such a composition from low- and high-flexibility oligomers. As a result of their compounding, linear block copolymer with alternating compatible flexible and rigid blocks is formed.

This provides a favourable combination of strength, rigidity and fracture viscosity ('self-reinforced polymers', 'molecular composites'). Polyester block copolymers are related to these polymers [227–230]. Examples of these are the block copolymer from poly(*p*-hydroxybenzoic acid) and poly(ethylene terephthalate), the polyarylate under trademark Xydar or block copolymers consisting of segments of poly(ethylene terephthalate) and poly(*p*-acetamidobenzoic acid) [231]. The melt viscosity is determined by the ratio of rigid-chain segments to flexible-chain ones:

Polymers of such structure are related to thermotropic liquid-crystal structure. The melting temperature of crystalline phase of Xydar polymer equals 370°C; transition to the isotropic state is accompanied by a sharp viscosity decrease (clearing temperature) and begins above 420°C. The polymer is pressed at 450°C in ceramic moulds heated to 200°C. Prepreg coextruded with glass-fibre (30%) above the clearing temperature is reinforced as a result of orientation along the flow.

Polymeric matrix along the flow acquires a bending strength of 152 MPa, perpendicular 76 MPa, and becomes stable to cold flowing under loading. The elasticity modulus of the polymer attains 9650 MPa and retains a value of 2200 MPa at 300°C.

The property indices of some polymers containing rigid-chain segments in macromolecule structure (copolyesters LC-2000, LC-6000, Vectra A950 and Xydar SRT-300) are: breaking stress at extension 110, 190, 210 and 120 MPa, respectively; elongation 4.4, 3.5, 3.0 and 2.9%; elasticity modulus at bending 8500, 10 000, 9000 and 11 000 MPa; and temperature of pressing 240, 330, 300 and 400°C, respectively.

REFERENCES

1. Flory P.F., *Ind. Eng. Chem.*, 1946, **38**, 417.
2. Kwei T.K., *J. Polym. Sci. A*, 1963, **1**, 2977.
3. Flory P.F., *Principles of Polymer Chemistry*, Cornell University Press, New York, 1953.
4. Gordon M., Wasd T.C., Whytney R.S., *Polymer Networks*, New York, 1974, pp. 1–20.
5. Kotova V.V., Mezhikovsky S.M., *et al.*, *High-Molecular Compounds A*, 1987, **8**, 1961.
6. Izjak V.I., Rozenberg B.A., Enikolopyan N.S., *Polymer Networks*, Nauka, Moscow, 1972.
7. Trostyanskaya E.B., Babajewsky P.G., *Successes of Chemistry*, 1971, **1**, 117–41.
8. Andrianova K.A., Emelyanova V.N., *Successes of Chemistry*, 1976, **45**, 1817–41.
9. Babajewsky P.G., Gillham J.K., *J. Appl. Polym. Sci.*, 1973, **17**, 2067–88.
10. Trostyanskaya E.B., Babajewsky P.G., *Mechanics of Polymers*, 1968, **6**, 1033–42.
11. Abenova Z.D., Malkin A.Ya., *et al.*, *High-Molecular Compounds A*, 1989, **11**, 2372.
12. Kulichikhin S.G., *Mechanics of Composite Materials*, 1986, **6**, 1086.
13. Mezhikovsky S.M., *High-Molecular Compounds A*, 1987, **8**, 1571.
14. Vollmert B., Stutz H., Stempes J., *Angew. Makromol. Chem.*, 1972, **25**, 187.
15. Askadsky A.A., *Deformation of Polymers*, Chimia, Moscow, 1973.
16. Rozenberg B.A., Oleinik E.D., Izjak V.I., *Journal of All-Union Chemical Society Named After D.I. Mendeleev*, 1978, **3**, 272–84.

17. Berlin A.A., Kefeli T.Ya., Korolev G.V., *Polyesteracrylates*, Nauka, Moscow, 1967.
18. Nielsen L.E., *J. Macromol. Sci.*, 1969, **3**, 69.
19. Chernikova O.D., Dissertation, Cyolkovsky Moscow Aircraft Technological Institute, 1976.
20. Pludeman E., The phase boundary in polymer composites, in *Composite Materials*, vol. 6, ed. L. Broutman, R.M. Krok, transl. from English, Mir, Moscow, 1978, pp. 181–228.
21. Trostyanskaya E.B., Poimanov V.Ya., Kazansky Yu, N., *Mechanics of Polymers*, 1965, **1**, 26.
22. Trostyanskaya E.B., Shadchina Z.M., Vinogradov V.M., *Plastmassy*, 1979, **7**, 13–15.
23. Trostyanskaya E.B., Grabilnikov A.S., Komarov G.V., *Mechanics of Composite Materials*, 1988, **6**, 991–5.
24. Ponamareva E.A., *et al.*, *Mechanics of Composite Materials*, 1989, **1**, 92–5.
25. Tiefthaler G., Urban M.W., *Composites*, 1989, **2**, 145.
26. Manacha L.M., *Composites*, 1988, **4**, 311.
27. Gillham J.K., *CRC Crit. Rev. Macromol. Sci.*, 1972, **1**, 82–172.
28. Lipatova T.E., Zubko S.V., *High-Molecular Compounds A*, 1970, **7**, 1555.
29. Gillham J.K., *Proc. SAMPE 28th Symp.*, 1983, p. 564.
30. Vinogradov V.M., Yakusevich Yu. I., Trostyanskaya E.B., *Mechanics of Composite Materials*, 1974, **4**, 754.
31. Vinogradov V.M., *Plastmassy*, 1975, **4**, 20–31.
32. Vinogradov V.M., Residual strains in articles from composites, in *Structural Plastics*, ed. E.B. Trostyanskaya, Chimia, Moscow, 1974, pp. 46–75.
33. Maizel N.S., Bazshtein R.S., *et al.*, *High-Molecular Compounds A*, 1987, **9**, 2044.
34. Sultanov R.M., Kiselev V.I., *High-Molecular Compounds A*, 1989, **10**, 2184.
35. Kulik S.G., Flexibilization of the epoxy polymers, Dissertation, Cyolkovsky Moscow Aircraft Technological Institute, 1979.
36. Bott R.H., *et al.*, *Proc. SAMPE 33rd Symp.*, 1988, p. 1177.
37. Levita G., *et al.*, *Polym. Eng. Sci.*, 1989, **1**, 63–73.
38. Butov V.P., Gandelman M.I., *High-Molecular Compounds A*, 1988, **6**, 1139.
39. Trostyanskaya E.B., Shadchina Z.M., Vinogradov V.M., *Int. Polym. Sci. Technol.* 1981, **8/7**, 7/72–3.
40. Malliek P., Broutman L., *Fibre Sci. Technol.*, 1975, **2**, 113–44.
41. Broutman L., Agrawal P., *Polym. Eng. Sci.*, 1974, **7**, 581–8.
42. Peiffer D.S., *Appl. Polym. Sci.*, 1979, **6**, 1451–5.
43. Alpes J.M., Thoman S.J., *Proc. Am. Soc. Compos. 4th Tech. Conf.*, 1989, p. 299.
44. *Handbook of Composites*, vol. 1, ed. J. Lubin, transl. from English, Mashinostroyeniye, Moscow, 1988, pp. 28–219.
45. Babajewsky P.G., in *Structural Plastics*, ed. E.B. Trostyanskaya, Chimia, Moscow, 1974, pp. 75–120.
46. Kulik T.A., Kochergin Yu.S., *et al.*, *Plastmassy*, 1989, **3**, 44.
47. Fisch W., Hofman W., Schmidt R., *J. Appl. Polym. Sci.*, 1969, **13**, 295–308.
48. Smith J., *J. Appl. Polym. Sci.*, 1979, **5**, 1385.
49. Tilsches H., *Angew. Makromol. Chem.*, 1972, **25**, 1.
50. Bellenger V., *et al.*, *J. Mater. Sci.*, 1988, **12**, 4244–50.
51. Trostyanskaya E.B., Kazansky Yu. N., *Plastmassy*, 1975, **12**, 54.
52. Bell J., *J. Polym. Sci. A*, 1970, **3**, 417.
53. Murajama T., Bell J., *J. Polym. Sci. A*, 1970, **3**, 437.
54. Morgan R.J., *et al.*, *Proc SAMPE 28th Symp.* 1983, pp. 596–607.
55. Fischer M., *et al.*, *Makromol. Chem.*, 1980, **6**, 1251–87.
56. Gupta V.B., *et al.*, *J. Mater. Sci.*, 1985, **10**, 34–9.
57. Boofs H.J., Hauschildt K.R., *Angew. Makromol. Chem.*, 1979, **1**, 10.
58. Karakozov V.G., *et al.*, *Plastmassy*, 1989, **4**, 28.
59. Smith J., *J. Appl. Polym. Sci.*, 1979, **5**, 1385.
60. Gaston A., *J. Polym. Sci. A.*, 1986, **22**, 1495–506.

61. Racich J., Koutsky J., *J. Appl. Polym. Sci.*, 1976, **8**, 2111–29.
62. Prudksit P.A., *Plastmassy*, 1988, **12**, 13.
63. Lapicky V.A., *et al.*, *Mechanics of Composite Materials*, 1988, **4**, 585–9.
64. Kovalenko L.G., Stroganov V.T., *Plastmassy*, 1986, **11**, 34–9.
65. Fischer M., *et al.*, *Makromol. Chem.*, 1980, **12**, 1251–87.
66. Lupinovich L.N., *et al.*, *Plastmassy*, 1986, **1**, 8.
67. Vladimirov L.V., *et al.*, *High-Molecular Compounds A*, 1977, **9**, 2104–11.
68. Williams J., Delatycki O., *J. Polym. Sci. A*, 1970, **8**, 295.
69. Phillips D.C., Scott J.M., Jones M., *J. Mater. Sci.*, 1978, **2**, 311–22.
70. Stark E.B., *et al.*, *Proc. SAMPE 28th Symp.* 1983, pp. 581–9.
71. Kinloch A.J., *et al.*, *J. Mater. Sci.*, 1987, **22**, 4111–22.
72. Lee McKaque E., *et al.*, *J. Appl. Polym. Sci.*, 1978, **6**, 1643–54.
73. Boll D.J., *et al.*, *Compos. Sci. Technol.*, 1985, **24**, 253–73.
74. Anikevich A.N., Chromenkov N.E., *Mechanics of Composite Materials*, 1989, **5**, 911–16.
75. Peyses P., Bascow W., *J. Mater. Sci.*, 1981, **16**, 75–83.
76. Bauer R., *Proc. SAMPE 31st Symp.* 1986, p. 1226.
77. Daniely N.D., *et al.*, *J. Polym. Sci. Chem.*, 1981, **19**, 2443.
78. Garton A., *J. Polym. Sci. A*, 1984, **22**, 1495–506.
79. Trostyanskaya E.B., Shadchina Z.M., Vinogradov V.M., *Plastmassy*, 1981, **3**, 16.
80. Wang J.L., Wolcott M., *et al.*, *Proc. Am. Soc. Compos. 4th Tech. Conf.*, 1989, p. 189.
81. Drzal L.T., *Proc. SAMPE 28th Symp.*, 1983, p. 1057.
82. Yurchenko N.P., *et al.*, *Plastmassy*, 1989, **7**, 36.
83. Batzer G., *et al.*, *Angew. Makromol. Chem.*, 1973, **29/30**, 349–411.
84. Vasilewsky V.M., in *Thermostability of the Structural Plastics*, ed. E.B. Trostyanskaya, Chimia, Moscow, 1980, p. 38.
85. Bascom W.D., *et al.*, *J. Appl. Polym. Sci.*, 1975, **9**, 2543–62.
86. Trostyanskaya E.B., Babajewsky P.G., Kulik S.G., *High-Molecular Compounds A*, 1979, **6**, 1328–33.
87. Sultan J., Meggarg F., *Polym. Eng. Sci.*, 1973, **1**, 29–34.
88. Cecere J.A., *et al.*, *Proc. SAMPE 31st Symp.*, 1986, p. 580.
89. Levita G., *et al.*, *Polym. Eng. Sci.*, 1986, **1**, 63–73.
90. Trostyanskaya E.B., Babajewsky P.G., Kulik S.G., *Plastmassy*, 1981, **11**, 31–4.
91. Trostyanskaya E.B., Babajewsky P.G., Kulik S.G., *High-Molecular Compounds*, 1982, **7**, 1456–62.
92. Yurek M., McGrath J.E., *Proc SAMPE 31st Symp.* 1986, p. 913.
93. Bucknall C.B., Partridge T.R., *Polymer*, 1983, **24**(5), 639–44.
94. Trostyanskaya E.B., Babajewsky P.G., *et al.*, in *Collection: Polymer Composite Materials*, AN USSR, Kiev, 1984, p. 21.
95. Bascom W.D., *et al.*, *Compos. Sci. Technol.*, 1980, **1**, 9–18.
96. Korshak V.V., *Thermostable Polymers*, Nauka, Moscow, 1971.
97. Koton M.M., *High-Molecular Compounds*, 1974, **16**, 1199–214.
98. Michailin Yu.A., in *Thermostability of the Structural Plastics*, ed. E.B. Trostyanskaya, Chimia, Moscow, 1980, pp. 33–107.
99. Frazer A.G., *High Temperature Resistant Polymers*, transl. from English, Chimia, Moscow, 1971.
100. Belyakov V.K., *et al.*, *High-Molecular Compounds A*, 1973, **12**, 2635.
101. Kovarskaya B.M., Blumenfeld A.B., Levantovskaya J.J., *Thermostability of the Heterocyclic Polymers*, Chimia, Moscow, 1977, pp. 153, 173.
102. Scola D., *Proc. SAMPE 31st Symp.*, 1986, pp. 1844–55.
103. Trostyanskaya E.B., Michailin Yu.A., *Plastmassy*, 1978, **10**, 18–24.
104. Kumas D.J., *J. Polym. Sci.*, 1980, **4**, 1375–85.
105. Gibbs H.H., *J. Appl. Polym. Sci.*, 1979, **35**, 207–22.
106. Dolmatov S.A., *et al.*, *Plastmassy*, 1985, **1**, 29.

107. Tracesky J., *et al.*, *Proc. SAMPE 33rd Symp.*, 1988, p. 512.

108. Arnold C.A., *et al.*, *Proc. SAMPE 33rd Symp.*, 1988, p. 960.

109. Yurek M., *et al.*, *Am. Chem. Soc. Polym. Prepr.*, 1986, **1**, 315–17.

110. Berger A., *Proc. SAMPE 30th Symp.*, 1985, p. 64.

111. Kwiatkowski G.T., *et al.*, *J. Polym. Sci., Chem. Edn.*, 1975, **4**, 961–72.

112. Serafini T.T., Delvigs P., Lightney G.R., *J. Appl. Polym. Sci.*, 1972, **16**, 905.

113. Trostyanskaya E.B., Michailin Yu.A., Kchochlova L.F., *Plastmassy*, 1977, **2**, 32–6.

114. Perepechko I.I., *High-Molecular Compounds A*, 1974, **9**, 2094.

115. Young P.R., Chang A.C., *Proc. SAMPE 33rd Symp.*, 1988, p. 538.

116. Perepechko I.I., Voloshinov E.B., *High-Molecular Compounds* 1977, **7**, 1620–5.

117. Trostyanskaya E.B., Michailin Yu.A., Kchochlova L.F., *Plastmassy*, 1979, **10**, 29–31.

118. Trostyanskaya E.B., Michailin Yu.A., Mitchenko I.P., in *Collection: Operational Properties of the Structural Materials*, House of Scientific and Technical Propaganda, Moscow, 1984, p. 101.

119. Swapan Dolui, Sukumer Maite, *Angew. Makromol. Chem.*, 1986, **141**, 31–47.

120. Trostyanskaya E.B., Prusakova E.A., Gunyaev G.M., *et al.*, *Mechanics of Composite Materials*, 1990, **5**, 777–82.

121. Trostyanskaya E.B., Michailin Yu.A., Kchochlova L.F., *Plastmassy*, 1977, **3**, 11–14.

122. Bott R.H., *et al.*, *Proc. SAMPE 33rd Symp.*, 1988, p. 1177.

123. Knop A., Scheib W., *Chemistry and Application of Phenolic Resins*, transl. from English, Mir, Moscow, 1983.

124. Hultzsch H., *Chemistry des Phenolhasze*, Springer, Berlin, 1950.

125. Siling M.J., in *Collection: Chemistry and Technology of the High-Molecular Compounds* (*Reviews of the Science and Engineering*, **11**), 1977, p. 119.

126. Ellis K., *The Chemistry of Synthetic Resins*, Reinhold, New York, 1935.

127. Scheiber J., *Chemie und Technologie der Kunstlichen harze*, Stuttgart, 1943.

128. Sprengling G.R., Freeman J.H., *J. Am. Chem. Soc.*, 1950, **72**, 1952.

129. Kammerer H., *Kunststoffe*, 1966, **3**, 154.

130. Kammerer H., *Makromol. Chem.*, 1963, **64**, 181.

131. Imoto M., *et al.*, *Makromol. Chem.*, 1968, **131**, 117.

132. Kammerer H., Grossman H., Umson G., *Makromol. Chem.*, 1960, **1/2**, 39.

133. Safronova A.S., Pak I.K., *et al.*, *Plastmassy*, 1989, **4**, 29.

134. Braun D., Wolde-Giogis K., *Angew. Makromol. Chem.*, 1986, **140**, 153–9.

135. Gupta M.K., *et al.*, *Am. Chem. Soc. Polym. Prepr.*, 1986, **27**, 303.

136. Bhatia J., *et al.*, *J. Mater. Sci.*, 1985, **2**, 1022–8.

137. Mitchell S.J., Pickering B.S., Thomas C.R., *J. Appl. Polym. Sci.*, 1970, **14**, 175.

138. Lukin B.V., Skupov N.K., *Plastmassy*, 1969, **2**, 58.

139. Trostyanskaya E.B., Novikov V.I., Kazansky Yu.N., *Mechanics of Composite Materials*, 1966, **1**, 67–71.

140. Trostyanskaya E.B., Babajewsky P.G., *High-Molecular Compounds A*, 1967, **5**, 1058–65.

141. Trostyanskaya E.B., Pojmanov A.M., Babajewsky P.G., *Mechanics of Composite Materials*, 1975, **5**, 58–65.

142. Tiedeman G.T., Sanclemente M.R., *J. Appl. Polym. Sci.*, 1973, **6**, 1813–18.

143. Ena B., Sluchin V.P., *Plastmassy*, 1989, **3**, 17.

144. Dannels B.F., Shepard A.F. *J. Polym. Sci. A*, 1968, **8**, 2051.

145. Trostyanskaya E.B., Belnik A.R., *et al.*, *High-Molecular Compounds A*, 1972, **2**, 467.

146. Trostyanskaya E.B., Chernikova O.Ya., *Plastmassy*, 1976, **2**, 64.

147. Trostyanskaya E.B., Shadchina Z.M., Reznichenko G.M., *Plastmassy*, 1990, **8**, 81–3.

148. Trostyanskaya E.B., Kazansky Yu.N., Pojmanov A.M., *Mechanics of Polymers*, 1965, **1**, 26.

149. Vinogradov V.M., Yakusevitch Yu.I., Trostyanskaya E.B., *Mechanics of Polymers*, 1974, **4**, 754.

150. Trostyanskaya E.B., Vinogradov V.M., Shadchina Z.M., *Plastmassy*, 1979, **6**, 43.

151. Chang E.P., *et al.*, *J. Appl. Polym. Sci.*, 1982, **12**, 4759–72.

152. Manocha L.M., *Composites*, 1988, **4**, 311.
153. Vinogradov V.M., Yakusevitch Yu.I., *Plastmassy*, 1978, **5**, 46.
154. Bachman A., Muller R., *Plaste und Kautschuk*, 1977, **24**, 158.
155. Haris G.J., *Werkshoffer und Korrosion*, 1971, **22**, 227.
156. Babajewsky P.G., Bucharov S.V., et al., *Physical and Chemical Mechanics of the Materials' Treatment*, 1981, **5**, 32–7.
157. Babajewsky P.G., Bucharov S.V., *High-Molecular Compounds B*, 1982, **24**, 202–5.
158. Trizno M.N., Dissertation, LHTI, Leningrad, 1978.
159. Gorbachev S.C., et al., *Plastmassy*, 1987, **8**, 6–8.
160. Michailova Z.V., Sedov L.N., *Unsaturated Polyesters*, Chimia, Moscow, 1977.
161. Berlin A.A., Korolev G.V., Kefeli T.Ya., Seregin Yu.M., *Acrylic Oligomers and Materials on the Basis of it*, Chimia, Moscow, 1983.
162. Kotova A.V., Mezhikovsky S.M., et al., *High-Molecular Compounds A*, 1987, **8**, 1761.
163. Mezhikovsky S.M., *High-Molecular Compounds A*, 1987, **8**, 1571.
164. Michailova Z.V., Sedov L.N., *Unsaturated Polyesters*, Chimia, Moscow, 1977.
165. Mazhlovich M.M., Dissertation, LHTI, Leningrad, 1974.
166. Molotkov R.V., Levitskaya O.M., et al., *Plastmassy*, 1970, **11**, 18–20.
167. Klosowska Z., *Polimery*, 1968, **11**, 485–7.
168. Hamann R., Funke R., Gilch W., *Angew. Chem.*, 1959, **71**, 596.
169. Boenig H., Walker N., *Modern Plastics*, 1961, **38**, 123.
170. Lee P.Z., Michailova V.V., et al., *Plastmassy*, 1960, **3**, 9.
171. Trostyanskaya E.B., Vinogradov V.M., Kazansky Yu.N., *Plastmassy*, 1962, **7**, 15.
172. Vinogradov V.M., Neverov A.N., et al., *Plastmassy*, 1965, **8**, 38.
173. *Handbook of Composites*, vol. 1, ed J. Lubin, transl. from English, Mashinostroyeniye, Moscow, 1988, p. 51.
174. Vaneso-Szmenebangi J., *Kunststoffe*, 1970, **12**, 1066–71.
175. Fekate F., *Modern Plastics*, 1970, **10**, 154–8.
176. Demmer K., Lawann H., *Kunststoffe*, 1970, **12**, 953–9.
177. Pattison V.A., Hindessinn B., Schwarz W.T., *Appl. Polym. Symp.* 1974, **18**, 2763–71.
178. Trostyanskaya E.B., Babajewsky P.G., Stepanova M.I., Kulik S.G., *High-Molecular Compounds*, 1984, **5**, 1053–60.
179. Dwight D.W., Sabat P.Y., Brinsen H.F., *Proc. Am. Soc. Compos. 4th Tech. Conf.*, 1989, p. 356.
180. Trostyanskaya E.B., Vinogradov V.M., Kazansky Yu.N., *Plastmassy*, 1962, **10**, 14.
181. Trostyanskaya E.B., Stankoj G.G., Kazansky Yu.N., *Plastmassy*, 1966, **9**, 31.
182. Zaicev B.A., et al., *Mechanics of Composite Materials*, 1988, **4**, 579–84.
183. Lukasov S.V., Zaicev B.A., et al., *Mechanics of Composite Materials*, 1988, **5**, 771.
184. Lukasov S.V., et al., *High-Molecular Compounds A*, 1988, **10**, 2196.
185. Volodin V.I., Tarasov A.I., Spassky S.S., *Successes of Chemistry*, 1970, **2**, 276.
186. Pavlova O.V., et al., *High-Molecular Compounds A*, 1987, **8**, 1777.
187. KOichi Ito, et al., *J. Polym. Sci. Chem. Edn*, 1975, **1**, 87–96.
188. Prusinska J., Krolikowski W., *Angew. Makromol. Chem.*, 1977, **64**, 29–42.
189. Alekseev N.N., et al., *Plastmassy*, 1986, **6**, 53.
190. Slysh R., et al., *Polym. Eng. Sci.*, 1974, **4**, 264.
191. Dmitrieva I.V., et al., *Plastmassy*, 1989, **1**, 44.
192. Krajnikova I.G., et al., *Plastmassy*, 1989, **2**, 9.
193. Prokopova T.V., Gritzuk A.N., in *Collection: Technology of the Articles from Polyester Premixes and Prepregs*, House of Scientific and Technical Propaganda, Moscow, 1975, p. 55.
194. Tzvetkov V.N., *High-Molecular Compunds A*, 1974, **5**, 994–64.
195. Tzvetkov V.N., *High-Molecular Compounds A*, 1977, **10**, 2171–89.
196. Papkov S.P., in *Liquid Crystalline Polymers*, ed. N.A. Plate, Chimia, Moscow, 1988, pp. 43–71.
197. Lee H., Stoffey D., Neville K., *New Linear Polymers*, McGraw-Hill, New York, 1970.

198. Frazer A.G., *High Temperature Resistant Polymers*, transl. from English, Chimia, Moscow, 1971.
199. Duthie A.S., *Proc. SAMPE 33rd Symp.*, 1988, p. 296.
200. Shorochov V.M., *et al.*, *Plastmassy*, 1981, **6**, 29.
201. Yudkin B.I., *et al.*, *Plastmassy*, 1987, **8**, 26–28.
202. Golovkin G.S., Shibanov A.K., Ryazancev A.N., *Plastmassy*, 1988, **3**, 40.
203. Bergen W.B., Rigby R.B., *Chem. Eng. Prog.*, 1985, **1**, 36–8.
204. Pfeiffer R.H., Scarle O.B., *Polym. Eng. Sci.*, 1985, **8**, 474–6.
205. Vogel H.A., *J. Polym. Sci. A*, 1970, **8**, 2035.
206. Davis A., *Makromol. Chem.*, 1976, **132**, 23.
207. Winkel J.D., Hurdle J.R., *Proc. SAMPE 33rd Symp.*, 1988, p. 816.
208. Quella F., Schmidt H.F., *Kunststoffe*, 1989, **1**, 52–4.
209. Shue R.S., *Proc. SAMPE 33rd Symp.*, 1988, p. 626.
210. Warshawski L., *J. Mater. Sci.*, 1988, **2**, 497.
211. Vallance M., Tomkinson G., *Proc. SAMPE 31st Symp.*, 1986, p. 410.
212. Cebe P., *J. Mater. Sci.*, 1988, **10**, 3721–31.
213. Chang I.Y., *Proc. SAMPE 33rd Symp.*, 1988, p. 194.
214. Lee J., Portes R., *Polym. Eng. Sci.*, 1986, **9**, 633.
215. Duthie A.C., *Proc. SAMPE 33rd Symp.*, 1988, p. 296.
216. Dorey G., *et al.*, *Compos. Sci. Technol.*, 1985, **33**, 221–37.
217. Volokchina A.V., *Chemical Fibres*, 1990, **3**, 42.
218. Che-Lung Chen, *et al.*, *Proc. SAMPE 33rd Symp.*, 1988, p. 134.
219. Hergenrother P.M., Havens S.J., *Proc. SAMPE 33rd Symp.*, 1988, p. 451.
220. Izyneev A.A., Teplyakov M.M., *et al.*, *Successes of Chemistry*, 1967, **12**, 2090.
221. Koton M.M., *et al.*, *Plastmassy*, 1968, **11**, 18.
222. Benett C., *Proc. SAMPE 33rd Symp.*, 1988, p. 148.
223. Empey R., Wrasidlo W., *J. Polym. Sci. A*, 1967, **5**, 1573.
224. Sheppard C., *Proc. SAMPE 31st Symp.*, 1983, p. 1426.
225. Lyle G.D., *Proc. SAMPE 33rd Symp.*, 1988, p. 1080.
226. Connell J.W., Bass R.G., *Proc. SAMPE 33rd Symp.*, 1988, p. 251.
227. Dieke H.R., Lenz R.W., *Angew. Makromol. Chem.*, 1985, **131**, 95–105.
228. Skorochodov S.S., in *Liquid Crystalline Polymers*, ed. N.A. Plate, Chimia, Moscow, 1988, pp. 161–87.
229. Gordon B., *et al.*, *Am. Chem. Soc. Polym. Prepr.* 1986, **1**, 311–12.
230. Shibaev V.P., *Chemical Fibres*, 1987, **3**, 4.
231. Jackson W.J., Kunfuss H.F., *J. Appl. Polym. Sci.*, 1980, **8**, 1685–94.
232. Shtennikov I.N., in *Liquid Crystalline Polymers*, ed. N.A. Plate, Chimia, Moscow, 1988, pp. 73–120.

2

Some principles for creating fibrous composites with a polymeric matrix

G.M. Gunyaev

2.1 INTRODUCTION

The main feature of polymeric fibrous composites is the possibility of devising (creating) materials and elements with the required properties meeting most fully the operating requirements of the parts and constructions.

The basic principles of and approaches towards creation of polymeric fibrous composites stem from the definition of what a composite is, and seem to be simple enough, but the design of these materials in the course of manufacture has remained up till now more art than science.

This fact is explained in many respects because implementation of the task of optimizing the composition, structure and form of the composite and its construction necessitates the unification of materials technology, technology and design, as well as basic science, i.e. chemistry, physical chemistry, solid-state physics and mechanics of reinforced media. Much success has been achieved thus far in each of the above areas but, on the whole, the combined task awaits a solution. The in-service behaviour of a composite and its production practice are much more complicated than in the case of conventional materials owing to its multicomponent nature and heterogeneous structure, exhibiting appropriate levels and sublevels in the macrovolume and microvolume (layers, bundles of fibres, elementary fibres, binding matrix, boundary fibre and matrix layers, the interface proper, local changes of composition, and structural defects of the composite and components thereof). Like any heterogeneous material comprising integral components, the composite should remain solid through-out the operation period including various types of energy effects, i.e. mechanical, heat, chemical, wave, etc. For this purpose, it is necessary to select and combine the components in the composite so as to ensure their physicochemical, thermodynamic and mechanical (deformation) compatibility, in particular, at the fibre–matrix interface. The approaches to design of polymeric composites also include review of their structure in the macrovolume, in the layers and interface, and in the microvolumes of each component, taking into account their change in the areas of contact.

The complexity of the problem and space limitations have meant that this chapter is confined to disclosure of certain general issues, which, stemming from the author's practical experience, should be taken into account in devising polymeric fibrous composites.

2.2 COMPATIBILITY OF COMPONENTS AND SOLIDITY OF POLYMERIC COMPOSITES

The properties of fibrous composites are influenced primarily by the interface and, first of all, by the strength of fibre adhesion with the polymeric binder. The interface in polymeric composites implies not only the geometric surface between the fibres and binder, but also the areas adjoining the above surface and exposed to the effect of the physical and chemical processes occurring in the case of fibre–matrix interaction at the stage of composite moulding and operation [1, 2].

Moulding and the state of the interface are influenced by the physicochemical and thermomechanical compatibility of the components forming the composite. The first is responsible for the completeness of contact, the nature, number and strength of the physical and chemical bonds that originate as a result of the interaction between the matrix and the fibre surface, the effects produced on forming the structure, the changes of the composition and properties of the matrix and fibres in the boundary layers due to mutual diffusion, selective sorption, the catalytic effect on the binder hardening process, etc.

The second determines the mutual correspondence of components, ensuring their solidity (continuity) in the course of production and operation (heating, swelling, loading, etc.) and the stress level in these conditions of the interface and components, conditioned by the difference in their deformability (change in volume and linear dimensions). The interaction of the binder with the surface of the fibres is initiated at the stage of components production, and starts from wetting whose completeness can be judged in terms of the difference between the surface energies of the fibre and binder. Proper wetting predetermines the high adhesion conditioned by the forces of chemical and physical interaction at the two-phase contact boundary. Quantitatively, it is conventional to express adhesion as the adhesion strength or, to be more exact, the strength of the fibre adhesion with the matrix. Data on the strength of adhesion of some reinforcing fibres with epoxy matrices [3] are presented in Table 2.1. The strength of adhesion is decisive in the joint work of the reinforcer and matrix in the composite and appears in the main design equations. The adhesion strength is judged in terms of the composite's shear and transverse pull test results.

The interphase interaction in the composite material is the result of chemical reactions between the functional groups located on the surface of the filler and binder, as well as the result of physical interaction of these groups in the process of sorption, orientation of the molecules or sections of the binder chains and their mechanical attachment to the surface by filling pores and cracks [4].

Table 2.1 Strength of fibre adhesion with epoxy matrices

Matrix material (trademark)	Fibre	τ_{fm} (MPa)	Nature of breakdown		
			Adhesion	Cohesion affecting matrix	Mixed
EDT	Boron	82	40	12	42
	Carbon	35	66	10	24
	Glass	52	35	14	51
	Thread-like crystals:				
	Aluminium nitride	66	48	44	8
	Silicon carbide	83	–	81	19
	Aluminium dioxide	53	70	–	30
ETPh	Boron	93	45	8	47
	Carbon	32	80	8	12
	Glass	43	44	10	46

The area of the surface of interaction between the components in the composite, accounting for its unit volume, depends on the extent of filling with the reinforcing fibres, their geometric dimensions, the specific surface area, as well as the area between the fibre surface and the matrix.

Table 2.2 presents data on the physical and chemical properties of various fibres, enabling one to judge topology, chemical composition and surface reactivity. The surface energy of the mineral fibres (glass, oxide, carbide, boron) at the instant of their production is high enough, i.e. $(0.25$ to $2) \times 10^{-8}\,J\,m^{-2}$. However, it diminishes rapidly as a result of surface contamination because of adsorption of environmental products, i.e. organic substances, water, etc. This leads to reduction of the interfacial angles, deterioration of the strength of adhesion with the binder or substantial decrease of the adhesion strength under the effect of moisture and other media, in particular, for fibrous composites with hydrophilic fibrous composites. To improve the adhesion strength at the interface and its stability under operating conditions, the manufacture of composites is preceded by treatment aimed at purification and activation or chemical modification (finishing) of their surface [5, 6].

The efficiency of fibre surface treatment can be assessed in terms of the composite's interlayer shear test results using the approximate equation [7]:

$$\tau_{xz} \approx K_v K_s \tau_{xz}^{\circ} \tag{2.1}$$

where K_v is the surface activation coefficient, proportional to the surface energy and adsorption capacity of the fibre surface (adhesion molecular component); K_s is the surface roughness coefficient, proportional to the specific surface of the fibres (adhesion mechanical component); and τ_{xz}° is the ultimate shear stress at the interface.

Table 2.2 Physical and chemical characteristics of fibres[a]

Fibre	d_0 (μm)	S_{sp} ($m^2 g^{-1}$)	S ($m^2 cm^{-3}$) at $v_a = 0.6$	$\sigma_{ult} \times 10^{11}$ (Jm^{-2})	Chemical composition		Possible functional groups
					Bulk	Surface	
Carbon	7	0.350	0.378	26 to 28	C, O, N, H	C, O, H	—C(O)H, —C—OH, C=O
Boron	100	0.015	0.024	28 to 30	B	O, B	—B—OH, B—O—B
Glass	10	0.160	0.240	60 to 130	Si, O, Al, Ca	Si, O, Al	—Si—OH
Silicon carbide	100	0.012	0.252	42	Si, C, O, N	Si, O	—Si—O—Si, —Si—OH

[a] Symbols: d_0, original diameter; S_{sp}, specific surface area; S, surface area; σ_{ult}, ultimate strength; v_a, volume fraction

Those types of surface treatment should be favoured whose effect on the reinforcing fibre is followed by not only increase of its specific surface but also substantial rise in the chemical activity. The potentialities of treatment increasing the shear strength due to mechanical adhesion as the contact area grows are limited by the probability of reaching maximum specific surface (S_{sp}) at which is initiated a noticeable loss of fibre strength (which takes place, for instance, in pickling and whiskerization of carbon fibres).

Chemical activation of boron and carbide fibres is effected by cleaning them so as to dispose of the adsorbed products or remove the thin surface layer. In this case, the specific surface and the degree of fibre roughness remain practically unchanged, whereas the growth of strength in the case of interlaminar shear is determined by the increase of the fibre critical surface tension (Fig. 2.1), i.e. the adhesion molecular component. The thus-cleaned fibre is characterized by substantially higher surface energy and low value of the interfacial angle in wetting with liquids [8] (Table 2.3).

The surface of carbon fibres is activated by pickling and oxidation in liquid and gaseous media with the appropriate increase of the fibres' specific surface and chemical activity. In interaction with an oxidizer in active areas of the carbon-fibre surface, the total content of functional groups increases (Table 2.4). In this case, the shear strength of the carbon-fibre composites increases in proportion to the number of functional groups (Fig. 2.2) or paramagnetic centres on the fibre surface. Pickling of the carbon fibres is followed by the complication of surface microrelief and S_{sp} increase due to opening of pores in pickling of the thin surface layer, which causes a proportionate interlaminar shear strength increase [9, 10]. Presented below are data on the effect of various methods for activation treatment of the surface of carbon-fibre

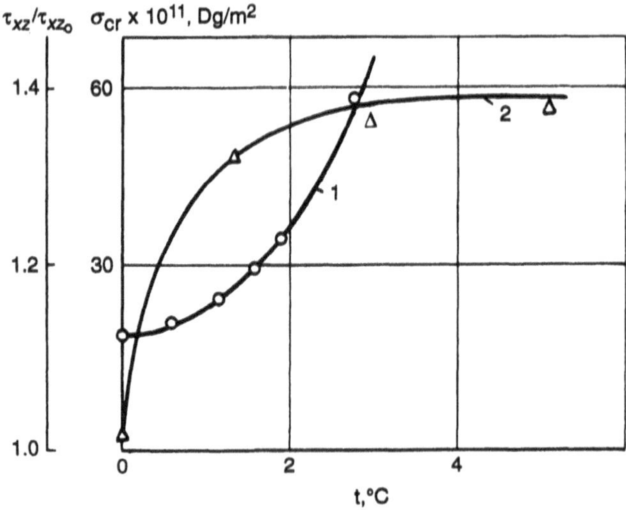

Fig. 2.1 Critical surface tension (1) and relative increase of epoxy boron-fibre composite shear strength (2) versus nitric acid pickling time.

Table 2.3 Effect of boron-fibre surface activation treatment on value of interfacial angle involving wetting with epoxy resin and interlaminar shear strength

| | | τ_{xz} (MPa) | | |
Treatment method	Fibre interfacial angle (deg)	In initial state	After 2 h boiling in water	Preservation of properties (%)
No treatment	30	56	17	31
Boiling in ethanol	7	68	65	95
Oxidation with concentrated acid:				
with subsequent boiling	5	102	97	95
with subsequent washing				
in trichloroethene	5	54.5	41	75

Table 2.4 Effect of high-modulus carbon-fibre oxidation with HNO_3 on surface composition, its wettability with epoxy resin and shear strength of epoxy carbon-fibre composite

| | | Surface content (%) | | | | Interfacial angle (deg) | τ_{xz} (MPa) |
Fibre	S_{sp} ($m^2 g^{-1}$)	Hydrogen	Carboxyl groups	Hydroxyl groups	Oxygen		
Initial	0.8	0.01	0.11	0.13	0.1	62 to 72	23
Oxidized	11.8	0.04	1.00	0.50	1.9	61 to 66	42

strips on the properties of the epoxy carbon-fibre composite. If τ_{xz} of the composite based on carbon fibres without treatment is 18 MPa, τ_{xz} after oxidation of carbon fibres in sodium hypochlorite solution increases up to 55 MPa; after oxidation in nitrogen and air mixture, up to 65 MPa; after oxidation using concentrated nitric acid, up to 75 MPa; and after monomer polymerization on the surface of fibres, up to 70 MPa.

Surface treatment of most mineral fibres (glass, oxide, carbide and other fibres exhibiting a hydrophilic surface) involves the application and attachment to the fibre surface of organosilicon compounds containing different functional groups wherein some are capable of forming chemical bonds with the hydroxyl groups found on the surface of the fibres, whereas others do so with the functional groups of the polymeric binder. As the finishing agent interacts with the mineral fibre, the polycondensation process runs at the same time and involves formation on the fibre surface of a protective film. In doing so, the surface becomes water-repellent. As a rule, finishing insignificantly improves the strength properties of glass-fibre composites but substantially increases their water resistance and atmospheric durability.

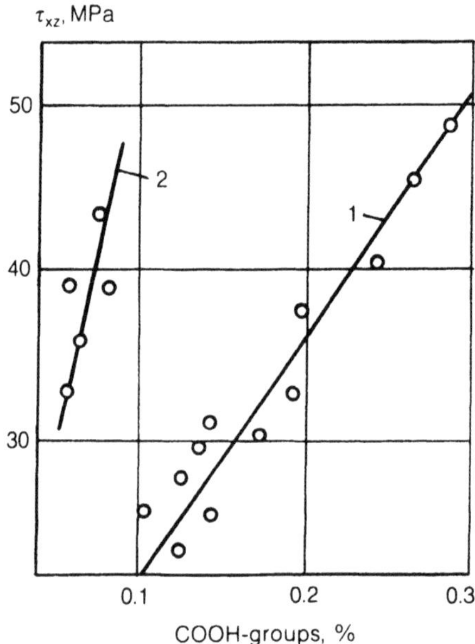

Fig. 2.2 Epoxy carbon-fibre composite shear strength versus content of COOH groups in fibre with specific surface of 0.4 (1) and 1.2 to 1.4 $m^2\,g^{-1}$ (2).

In manufacture of polymeric composites, use is made of multicomponent binders. Hence, the combination of reinforcing fibres with such binders involves complex processes of redistribution of fractions in the bulk and at the interface. As a rule, the boundary layers are enriched in the low-molecular-weight substances, thus leading to a change in the ratio of components in the binder volume and in the boundary layers. This and also the existence on the surface of the fibres of functional groups (exhibiting either acidic or alkaline nature) may produce an inhibitory or catalytic effect on formation of the polymer in the boundary layer, the depth of its hardening and the structure of the polymer network being formed. In some instances, this results in origination of weak binder boundary layers [11]. The application of particular methods for fibre surface modification influences the strength of the binder boundary layers. In this case, it should be borne in mind that an increase in the number of bonds increases the stress level in the boundary layer.

Unlike mineral substances, organic fibres are more susceptible to the effect of the components of the binder with which they are combined in fabrication of composites. In this case, the diffusion of low-molecular-weight products inside the organic fibres is possible, causing their swelling, stress relaxation and, as a consequence, strength reduction. Another specific feature of such systems is their adequate wettability and the possibility of forming chemical bonds between the components; as a result, the interface becomes 'diffuse'.

Along with the physical and chemical compatibility of the components in the composites, of no less importance is their thermomechanical compatibility, i.e. capacity to take up strains in deformation of the entire composite. Even the simplest form of deformation, i.e. tensile strain of a unidirectional composite along the reinforcing fibres, involves the origination of a complex stressed state in the bulk material.

In loading a unidirectional composite along the fibres, the stresses in the fibres are constant over the entire volume, whereas in the polymeric binder they are constant in the reinforcement direction and variable in the plane perpendicular to the reinforcement direction.

The mean values of the longitudinal stresses originating in loading the composite are designated σ_{fx} in the fibres and σ_{mx} in the matrix. As a first approximation, stresses σ_x are determined by [12]:

$$\sigma_{fx} = E_f\sigma_x/(1 - v_f)E_m + v_fE_f \tag{2.2}$$

$$\sigma_{mx} = E_m\sigma_x/(1 - v_f)E_m + v_fE_f \tag{2.3}$$

For highly filled composites in which $E_f \gg E_m$, these expressions are simplified to the following form:

$$\sigma_{fx} = \sigma_x/v_f \tag{2.4}$$

$$\sigma_{mx} = E_m\sigma_x/v_fE_f \tag{2.5}$$

Thus, the longitudinal stresses originating in the matrix as the composite undergoes strain in the fibre orientation direction increase as the modulus of elasticity of the reinforcing fibres employed in the composite decreases.

The volumetric stressed state of the components in axial loading of a unidirectional composite is created due to different values of the Poisson's ratios of the polymeric binder and reinforcing fibres. The existence of a strong bond between the components produces, in the matrix and at the interface, radial and tangential stresses whose values change due to the effect of the adjacent fibres.

Figure 2.3 shows the distribution of radial and tangential stresses in the matrix of a composite with hexagonal arrangement of reinforcing fibres. Radial stresses originating on composite longitudinal loading have maximum value on the fibre–matrix contact surface. Away from this surface, they decrease [13].

Circumferential tangential stresses σ_{fmx} acting on the interface reach maximum value on the 15° line and do not exist on the 0° and 30° lines. They are constant over the entire length of the fibre. The longitudinal tangential stresses act at the ends of the fibres only. Like the radial stresses, the circumferential stresses assume maximum value on the contact surface on the 30° line and decrease as they depart from the latter.

The circumferential and tangential stresses originating on longitudinal loading of the composite increase as the component stiffness coefficient E_f/E_m decreases and the extent of filling increases. Figure 2.4 presents data illustrating the change of the maximum values of the circumferential and radial stresses related to the value of the longitudinal stresses in the matrix for carbon- and glass-fibre composites exhibiting different extents of filling.

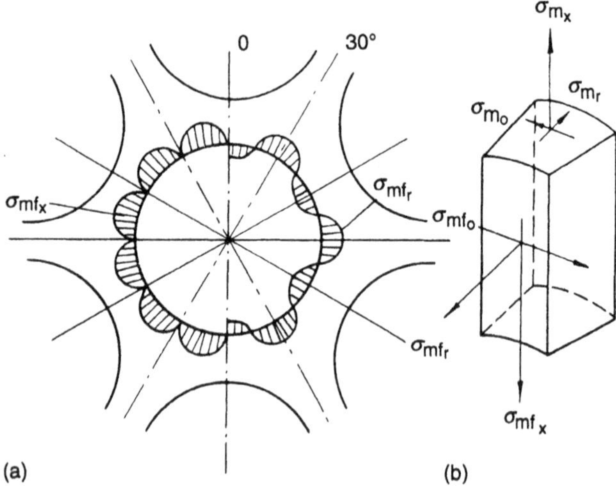

Fig. 2.3 Diagram of matrix element bordering on interface (a) and composite with hexagonal layers lay-up (b); also shown is the distribution of stresses, i.e. radial σ_{fmz} and longitudinal σ_{fmx} at the interface.

Owing to the difference in the linear thermal expansion coefficients of the fibre and matrix ($\alpha_f > \alpha_m$) on cooling below the moulding temperature, an initial stress originates in the composite. In this case, the matrix and the interface witness the origination of longitudinal, tangential (circumferential) and radial stresses whose nature and regularities of distribution are similar to those of the stresses originating in axial loading [13] (Fig. 2.3). Like in the case of longitudinal loading, the stress value increases as the extent of composite filling increases and the ratio E_f/E_m decreases. As the difference in the linear thermal expansion coefficients of the matrix and fibre increases, the stress level of the components increases. As a result of the mutual superposition of the residual thermal stresses and those conditioned by the effect of the external load, the circumferential and tangential stresses become commensurate with the matrix strength and adhesion strength at the interface.

The notion of fibrous composite solidity [14] presupposes the continuity of all components and the absence of disruption of the bonds at the interface in the case of composite deformation till the fibres break due to loss of strength. The solidity conditions obtained in an analysis of the joint deformation of the composite components in various forms of loading are essentially the ratios between the strength and elasticity characteristics of the reinforcing fibres and matrices, and the strength of their adhesion in shear and pull, ensuring their joint work in the composite, taking into account the extent of its filling [15]:

$$\tau_{adh}/\sigma_f \geqslant v_f^{3/2} K\sigma_f(1 - \phi)/(1 + \mu_m)^{0.5}$$

$$\sigma_m/\sigma_f \geqslant (1.4 - 1.6) \times v_f^{3/2} K\sigma_f(1 - \phi)/(1 + \mu_m)^{0.5}$$

$$\varepsilon_m/\varepsilon_f \geqslant 0.5 K\sigma_f[\phi + (3 - 2\phi)v_f(1 - v_f)(E_f/E_m)]$$

$$\sigma_f K\sigma_f h \leqslant [G_m(2 - v_f)/2(1 - v_f)][1 + 0.5v_f(1 - v_f)(E_f/E_m)] \tag{2.6}$$

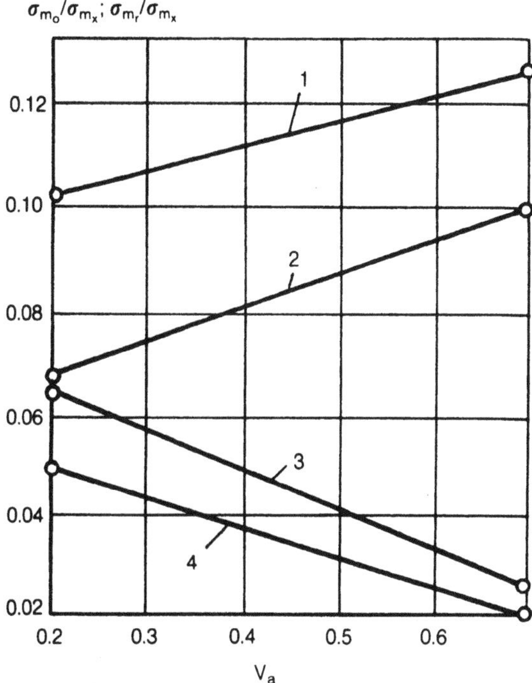

Fig. 2.4 Relation of circumferential (1, 2) and radial (3, 4) stresses and longitudinal stresses in the matrix at an interface versus extent of reinforcement of glass-fibre composite (1, 3) and carbon-fibre composite (2, 4).

Here ϕ is angle of fibre deflection from axis, μ_{m} is Poisson's ratio of matrix E is Young's modulus, v_{f} is volume fraction of fibres, ε is deformation and G is shear modulus.

In this case, the fulfilment of all conditions is mandatory and there is no single condition of solidity. Substitution of the characteristics of the reinforcing fibres in relations (2.6) enables one to determine the elasticity and strength characteristics of the matrix, ensuring production of a solid structure [15, 16]. Table 2.5 also presents data for the glass-fibre composites reinforced with fibres exhibiting different mechanical properties.

As follows from Table 2.5, the requirements for the matrix properties, stemming from the solidity condition, exceed the level of the properties of present-day polymeric binders. For a comparative assessment of the compliance of matrix properties with the requirements of solidity, the following solidity criterion [17] is suggested:

$$M = \varepsilon \eta_i \phi_i \tag{2.7}$$

where η_i is the coefficient of the binder characteristics correspondence to the solidity conditions ($\eta_E = E_{\mathrm{m}}/E_{\mathrm{mp}}$, $\eta_\sigma = \sigma_{\mathrm{m}}/\sigma_{\mathrm{mp}}$, $\eta_{\mathrm{fm}} = \tau_{\mathrm{fm}}/\tau_{\mathrm{fmp}}$, $\eta_\varepsilon = \varepsilon_{\mathrm{m}}/\varepsilon_{\mathrm{mp}}$), and ϕ_i is the

Table 2.5 Binder elasticity and strength properties ensuring glass-fibre composites solidity

Binder properties	Requirements for binder for fibre-reinforced composite	
	$\sigma_f = 2350\,MPa$ $E_f = 75\,GPa$ $\varepsilon_f = 3\%$	$\sigma_f = 4200\,MPa$ $E_f = 95\,GPa$ $\varepsilon_f = 3.5\%$
σ_m (MPa)	140	250
E_m (GPa)	4500	5500
ε_m (%)	4.5	5.25
τ_m (MPa)	94	168
τ_{fm} (MPa)	94	168

coefficient of significance of the binder parameter in various forms of deformation; whereas index 'p' implies the parameter design value determined from the solidity conditions.

As regards glass-fibre composites, the values of the coefficients ϕ_ε, ϕ_E, ϕ_σ and ϕ_{fm} in various forms of deformation are as follows: ϕ_ε, ϕ_σ and ϕ_{fm} in tension reached

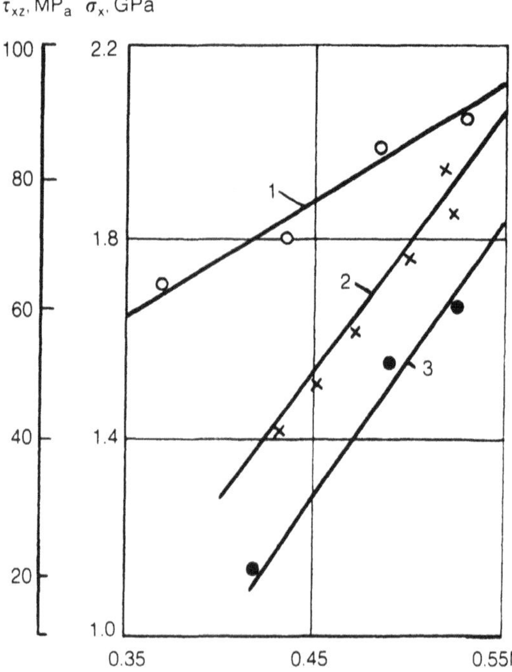

Fig. 2.5 Relation between breaking stress in tension (1), compression (2) and interlaminar shear (3) of glass-fibre composite and solidity criterion M.

0.06, 0.20 and 0.74, respectively; values of coefficients ϕ_E, ϕ_σ and ϕ_{fm} in compression reached 0.17, 0.34 and 0.49, and in shear 0.16, 0.20, and 0.64, respectively.

When the binder properties display full correspondence with the solidity condition, $\eta_i = 1$ and $M = 1$. Data illustrating the relation between the strength properties of the glass-fibre composite and the solidity criterion are presented in Fig. 2.5.

Thus, comparison of the mechanical properties of the reinforcing fibres and polymeric matrices, taking into account their change with temperature and time effect conditions, enables one to assess their thermal and mechanical correspondence to the composites' solidity requirements.

2.3 STRUCTURAL DEFECTS AND MECHANICAL PROPERTIES OF POLYMERIC COMPOSITES

2.3.1 Classification of defects

The methods of calculation and prediction of properties are based on the idealized model of an integral unidirectional composite, where linear elastic reinforcing elements (fibres) with stable geometric dimensions are regularly placed in a solid yielding matrix (binder); and there is good adhesion between them at the interface, providing joint deformation of the fibres and matrix under various types of loading.

Actually, the macrostructure of the fibrous components is quite different from the idealized model. The current engineering level predetermines the presence of regular (distributed in the total volume) and local structure defects in actual composites.

The appearance of local defects, i.e. folds, indents, delaminations, breaks and distortions of layers, scratches, resin extractions from some areas, etc., is, as a rule, the result of deviation from the given parameters or violation of technology, and is of casual nature. On the contrary, regular structure defects, the majority of which, as noted by Tarnopolsky and Rose in 1969 are 'specially embedded' in the material at the current technology level of making components, should be considered and estimated in calculation and prediction of the composite's properties. The carriers of the regular structure defects are the reinforcing filler, polymeric matrix and inter-face in the composite. The main defects [18] are the violation of fibre continuity, misorientation and distortion, differences in fibre size and non-uniformity of their volume distribution in the composite, pores and cracks in the matrix, cracks, delaminations and incompleteness of the phase contact at the fibre–matrix interface. One possible structural defect classification in unidirectional polymeric composites is given in Table 2.6. Some technological factors contributing to defect formation are also listed therein. The effect of the different structural defects in unidirectional composites may be considered by introducing coefficients that represent the ratio of the corresponding indices of the composites properties with the given structure defect type R^* to the composite properties calculated for the idealized model R. In this case:

$$R^* = RK \qquad K = f(K_1, K_\phi, K_v, K_p, K_i, K_s) \qquad (2.8)$$

where K_1, K_ϕ, K_v, K_p, K_i, K_s are coefficients taking into account structure processing

Table 2.6 Processing defects of unidirectional composites structure

Defect carrier	Type of defect	Conditions of defect formation	Designation of structure defectiveness coefficient
Reinforcing filler	Violation of fibre continuity – filaments, threads, bundles, breaks	Filament damage during production and manufacture of reinforcing filler; filaments and threads rupture on tension due to different strengths and lengths of fibres in bundles; the use of discrete fibre yarn for reinforcement	K_1
	Misorientation of fibres in layer plane	Winding by tape with given step; laying-out of layers on tapered, ogival and other double curvature surfaces	K_{ϕ_1}
	Distortion of fibres in layer	Use of fabrics and other reinforcing fillers with controllable interweavings of threads and bundles	K_ϕ
	Space distortion of fibres – twisting	Use of twisted threads or bundles as reinforcing filler; fibre distortion due to stability loss in shrinkage	K_{ϕ_2}
	Non-uniformity of fibre distribution in composite volume	Winding or laying out of fibres with overlapping or discontinuities; fibre agglomeration due to using fibre bundles for reinforcement, in particular, twisted ones; different filament calibre and variation of filament number along thread length	K_v

(continued)

Table 2.6 (*Contd.*)

Defect carrier	Type of defect	Conditions of defect formation	Designation of structure defectiveness coefficient
Matrix and phase boundary	Pores	Extraction of solvents, chemical reaction products and easily sorbed volatile substances	K_p
	Cracks and delaminations	Matrix rupture in volume and along phase boundary due to stressing as a result of composite chemical and thermal shrinkage	K_i
	Imperfections of phase contact	Incomplete wetting of fibre surface by binder pores on phase boundary; local separation of matrix from fibre due to breaks in chemical bonds; physical bonds debonding	K_s

defects such as the violation of fibre continuity, their deviation from linearity, deviation from uniform packing along the cross-section, matrix and interface non-integrity, and incompleteness of the phase contact. The value of R (strength, elasticity modulus) can either be calculated by the known ratios, or it can be determined experimentally, using a model defect-free specimen.

2.3.2 Defects of reinforcing fibres

(a) Violation of fibre continuity

The violation of fibre continuity is the result of damage to the filaments due to abrasion, multiple bends and mechanical damage in the production of the reinforcing fibres, i.e. twisting, weaving, passing through the filament guides of the surface treatment units, combining with the binder, winding or lay-up. There is also the possibility of damage to the filaments and bundles in tensioning as a result of their different stressing due to different strength (θ_σ), modulus (θ_E) and length of the fibres in the thread, bundle, strand or layer.

Integrity violation of continuous fibres results in the decrease of their area in the composite cross-section and in the increase (redistribution) of the stress on adjacent fibres. The danger of damage increases with the increase of the reinforcing element cross-section (diameter d_1). In this case, the coefficient taking into account the violation of the continuous fibres integrity (K_{I1}) at the expense of number of tears (q) in unit cross-section of the composite, the height of which is equal to the non-effective fibre length ($l_{f,cr}$), is determined by the expression suggested by Gunyaev in 1981

$$K_{I1} = q V_{f,max} d^2 f / V_f \tag{2.9}$$

where $V_{f,max}$ is the marginal fibre content in the composite in the given lay-up.

As follows from equation (2.9) and the data given in Fig. 2.6, the danger of this type of defect increases with increase of the diameter and decrease of the fibre content in the composite.

Reinforcement with large-diameter fibres predetermines the higher composite sensitivity to the violation of fibre integrity. As shown by Tumanov *et al.* [19], the number of reinforcing elements per unit area of the composite cross-section is reduced

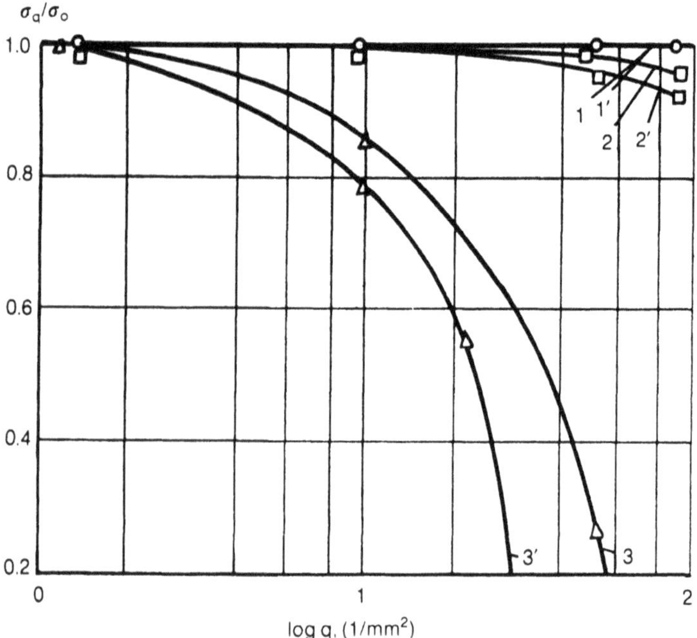

Fig. 2.6 Effect of fibre breaks per unit area of carbon- (1 and 1'), glass- (2 and 2') and boron- (3 and 3') fibre-reinforced plastics cross-section on compression strength: (o) carbon fibres, diameter 9 µm; (□) glass fibres, diameter 18 µm; (△) boron fibres, diameter 95 µm. Reinforcement extent of composites 1, 2 and 3: 0.6; composites 1', 2' and 3': 0.4.

by a factor of more than 100, if the fibre diameter increases from 7–10 µm to 100–150 µm, and is about 100 fibres/mm^2 for boron-fibre-reinforced plastics and about 20 000 fibres/mm^2 for carbon- and glass-fibre-reinforced plastics. On the contrary, the tear strength of the filaments is increased from 0.2 to 0.4 N (carbon and glass fibres) to 25–45 N (boron fibres). The local weakening of the boron-fibre-reinforced plastic in the break of one fibre is identical to breaking 150 to 200 fibres in the glass- or carbon-reinforced plastic.

The use of yarn composed of discrete fibres for composite reinforcement makes it necessary to take into account not only the number of 'breaks' in the cross-section, which may be characterized by the index of the fibres discreteness level, i.e. the ratio of their length (l_f) to the critical length of the fibre in the composite ($l_{f,cr}$), but also the distance between fibre ends Δ.

It is the existence of this gap that leads to the decrease of the fibre cross-section in composites reinforced by discrete fibre yarn. With the increase of the gap between the fibre ends and the fibre discreteness level, the composite's strength and elasticity modulus decrease in the reinforcement direction.

(b) Misorientation and distortion of fibres in layers

The term 'misorientation' signifies the deviation of linear fibres in the composite layers from the given reinforcement direction corresponding to that of load application. Such structural defects occur in the process of layer winding of articles over a mandrel with given step equal, for example, to the tape width, and lay-up of unidirectional layers of reinforcing fibres, tapes, fabrics, etc., on tapered ogival and other complex-shaped surfaces. Thus, for instance, on winding of a tape of single layer width h over a mandrel with diameter D, the adjacent layers turn out to be misoriented through an angle $2\phi_1 = \arctan(h/\pi D)$. While in the first layer the fibres deviate through the angle $+\phi_1$, for the other layer, the deviation is through the angle $-\phi_1$. Such misorientation is called regular antiphase misorientation. If the fibres in each layer deviate through the same angle to one side from the given reinforcement direction, uniform misorientation takes place. In most cases, the latter type of structural defect is not regular and is of random nature.

Another regular type of structure processing defect is fibre distortion, i.e. their deviation from linearity. Regular fibre distortions are inherent in woven fibre-reinforced composites. In this case, depending on the interweaving type in the fabric (satin, serge, linen), regular antiphase distortions are embedded in the composite structure, which are characterized by the extent of fibre bending in the layer. Thus, for the case of sinusoidal distortion $\bar{\phi} = A\pi K/l$, where K is the number of l-based halfwaves, and A is the distortion amplitude. As was shown [20], composites with sinusoidal fibre distortion are similar to a material with broken line-shaped fibre distortion as far as their deformation is concerned, which is why one considers $\phi_i = \phi = \arctan(AK/l_i)$.

Therefore, if the texture parameters, i.e. the diameters and the number of fibres in the weft and warp threads of the fabric d_{fx}, n_{fx} and d_{fy}, n_{fy}, the density of fibre

                    Principles for creating fibrous composites

lay-up in the fabric threads v and l, and the interweaving step, are known, for small angle values, according to [21], when replacing distortion by misorientation, angle $\phi = \phi_1$ is equal to $\phi = [d_{fx}n_{fx}^{0.5} + d_{fy}n_{fy}^{0.5}]v_f^{0.5}/l$. The coefficients, taking into account the effect of regular antiphase distortions and misorientation on composite strength and elasticity modulus, considering that $\phi \leqslant 10°$, $E_x \gg E_y$, may be obtained by using the following equations suggested in [22]:

$$K^E = 1 - 2(1 - v_{xy})[1 - 2(1 + v_{xy})E_x/G_{xz}]\phi^2 \tag{2.10}$$

$$K = \left\{1 + [E_x/G_{xz} - 2(1 - v_{xy})]\phi^2\right\}^{-1} \tag{2.11}$$

With the reduction of the warp and weft threads thickness, the extent of fibre distortion decreases and the composite elasticity modulus, compression and tensile strength increase. As the elasticity modulus and the strength of the reinforcing fibres increase, composite sensitivity to distortion becomes higher (Fig. 2.7). The most dramatic decrease of the mechanical properties values with fibre deviation from the direction of testing or the straight line is inherent in composites with maximum anisotropy, characterized by the ratio of elasticity moduli E_x/G_{xz}, as shown in [23, 24]. The reduction of the anisotropy level due to matrix strengthening (increasing G_{xz}) by introducing single crystals into the interfibre space of the composites, using

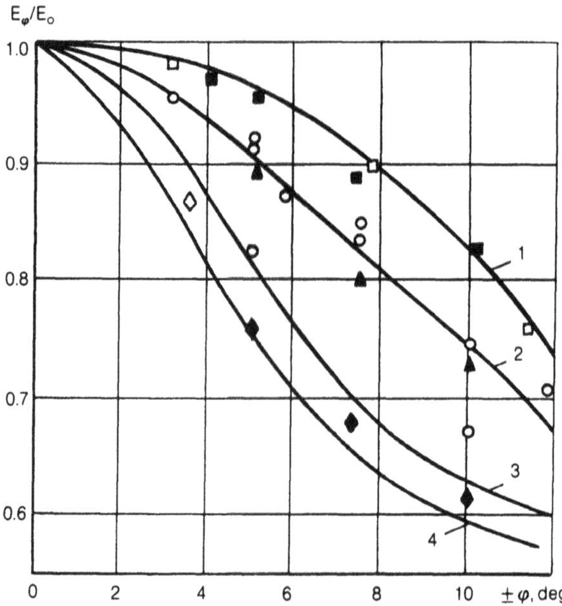

Fig. 2.7 Effect of regular fibre antiphase distortions and misorientation in composite layers with E_x/G_{xz} ratio 10 (1), 20 (2), 40 (3) and 60 (4). Carbon fibres with $E_f = 230\,\text{GPa}$ ($\circ$); whiskerized fibres with $E_f = 450\,\text{GPa}$ ($\diamond$, $\blacklozenge$). Glass fibres with $E_f = 92\,\text{GPa}$ ($\square$, $\blacksquare$). Boron fibres with $E_f = 380\,\text{GPa}$ ($\triangle$, $\blacktriangle$). Open symbols, distortions; full symbols, misorientation.

whiskerization, reduces composite sensitivity to structural defects such as distortion and misorientation.

(c) Spatial distortion of fibres

Structural defects of this type appear in composites when using twisted threads, bundles, etc., reinforcing fibres, and also in the case of fibre distortion due to loss of stability through chemical and thermal shrinkage (the latter are, as a rule, of local nature).

Twisting is widely used as a means of increasing the processability of threads and yarn bundles, and eliminating fluffing and breaking of the filaments. The spatial distortion of the fibres in twisting can be characterized by two parameters, i.e. ϕ_1, the fibre misorientation angle relative to the yarn axis, equal to the pitch of the fibre along the helical line, and the angle describing the distortion of the same fibre due to it encircling the yarn. The relation between the parameters ϕ_1 and ϕ and the yarn texture parameters (bundle, thread), such as twisting K_1 (the number of twists per linear metre), diameter D and density v_f of the yarn, diameter d_1, and number n of filaments in the yarn, is expressed by the equations

$$\phi_1 = \arctan(D)2K_1/1000 = \arctan[d_1(n/v_f)^{0.5}]2K_1 \times 10^{-3} \tag{2.12}$$

$$\phi_2 = \arctan(D)/[D^2 + (1000/2K_1)^2]^{0.5} = \arctan(\sin\phi_1) \tag{2.13}$$

At sparse twistings $\sin\phi \approx \tan\phi \approx \phi$; therefore $\phi = 2\phi_1$. The effect of regular fibre spatial distortions on the strength and elasticity modulus of polymeric composites was studied by Gunyaev in 1981 [31]. It can be taken into account by the corresponding coefficients $K_{\phi_3}^{\sigma}$ and $K_{\phi_3}^{E}$, equal to

$$K_{\phi_3}^{\sigma} = \{1 + [E_x/G_{xz} - 2(1 - v_{xy})]2\phi_1^2\}^{-1} \tag{2.14}$$

$$K_{\phi_3}^{E} = 1 - 2(1 - v_{xy})[1 - 2(1 + v_{xy})E_x/G_{xz}]2\phi^2 \tag{2.15}$$

Figure 2.8 shows data illustrating the effect of twisting on the strength and elasticity modulus of unidirectional epoxy composites reinforced with yarn consisting of 10 000 filaments of 9.5 µm diameter and elasticity modulus of 280 and 440 GPa.

The analysis of the data given shows that the composite strength and elasticity modulus depend on the fibre distortion; the optimum is twisting at an angle ϕ_1 of 1.5 to 2°. At sparse twisting of a yarn consisting of a great number of filaments, the values of the strength and elasticity modulus may exceed the values obtained with the untwisted fibres, as in this case the possibility of fibre misorientation is eliminated, and the differences in the length and stressing are reduced. The extent of the strength and elasticity modulus decrease of the composite with the spatial distortions increases with increase of the fibre elasticity modulus.

(d) Non-uniformity of fibre volume distribution

This type of regular defect can be the result of winding or laying of the layers with regular overlapping or gapping, agglomeration of the fibres when using

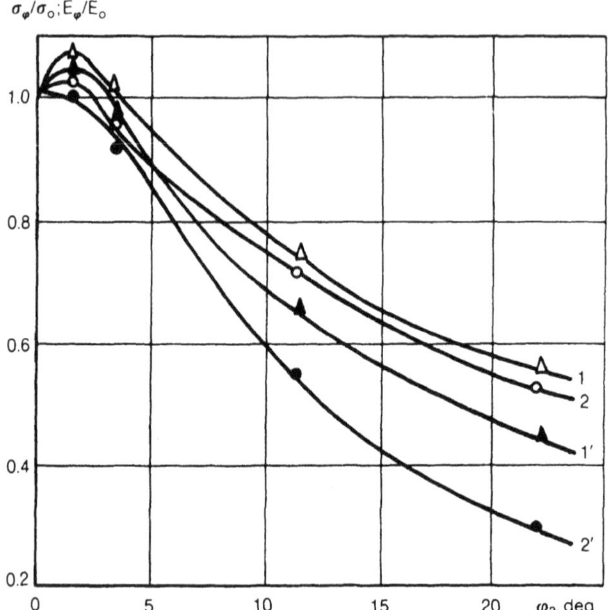

Fig. 2.8 Effect of fibre distortions during yarn twisting on strength (1, 1') and elasticity modulus (2, 2') of epoxy carbon-fibre-reinforced plastics. ($\circ$, $\bullet$) carbon fibres with $E_f = 440\,\mathrm{GPa}$, $\sigma_f = 1.97\,\mathrm{GPa}$; ($\triangle$, $\blacktriangle$) carbon fibres with $E_f = 287\,\mathrm{GPa}$, $\sigma_f = 2.94\,\mathrm{GPa}$.

twisted threads or bundles, woven fibres, the lengthwise variation of the bundle metric number and the dispersion of fibre diameter or cross-sectional area. As a result, the density of fibre packing varies along the composite cross-section, and a difference is observed in the distance between the individual filaments in the structural elements, thread and yarn, and between the structural elements, threads and strands. The local dispersion and the fibre volume fraction in the composite result in different stressing of the matrix volumes and micro-inhomogeneities of its properties, in particular, those sensitive to variations in extent of reinforcing (for example, shear modulus, coefficient of deformation concentration). In the areas of the composite volume where the content of the reinforcing fibres is higher, the concentration coefficient is higher according to Skudra and Bulavs [12], which, combined with technological stresses, due to chemical and thermal shrinkage on curing, leads to the decrease of composite deformation in the transverse direction. The coefficient accounting for the decrease of tear relative elongation value is equal to

$$K_v^\varepsilon = v_f / v_{f,\max} \tag{2.16}$$

It should be taken into consideration that the change in the interface size corresponds to local variation of the reinforcing fibre volume fraction in the composite. In some cases, this results in the change of composition, structure and properties of the polymeric matrix in the boundary layers surrounding the reinforcing fibres as compared to the matrix properties at a definite distance from the fibre.

2.3.3 Defects of matrix and interface

(a) Pores

Pores are formed due to the extraction of volatile products of the chemical reactions of the binder curing, solvents, air and moisture adsorbed by the binder and reinforcing filler. These products, which expand when heated, form a system of open and closed pores varying in shape and size. The effect of the composite porosity is different, depending on the type of the stressed state and strain direction.

It has the greatest effect on the resistance to shear loads and transverse loading, and a lesser effect on the resistance to loading in the reinforcement direction. This is due to the fact that pores reduce the matrix effective cross-section and the area of the fibre–matrix contact, and they are also the sources of crack initiation.

The appearance in the equation of a correction that takes into account the porosity effect, which describes the interlayer shear strength of the unidirectional composite, leads to the following expression [7]:

$$\tau_{xz} = \tau_m S_m K(1 - p) + \tau_{fm} K_s^P S_{fm} \tag{2.17}$$

which in turn makes it possible to obtain the expression for the coefficient taking into account the effect of pores on the interlaminar shear strength:

$$K_p = [A(1 - p) + K_s^P]A + K_s \qquad A = \tau_m S_m / \tau_{fm} S_{fm} \tag{2.18}$$

where τ_m is the interlaminar shear strength of the pore-free matrix; τ_{fm} is the interface shear strength; S_m and S_{fm} are the size of the cross-section accounting for matrix and interface in the composite; p is a coefficient taking into account the matrix cross-section reduction effect on the uniform distribution of spherical voids, proportional to $(v_p/1 - v_f)^{3/2}$; and K_s^P is a coefficient taking into account the reduction of phase contact completeness as a result of pore extension to the interface.

K is the average coefficient of the pore stress concentration in the matrix. With increase of porosity, the axial compression strength decreases, which is the result of the reduction of the filler working cross-section due to the loss of stability of some fibres passing through pores.

The porosity effect on composite tensile strength was accounted for in [18] when studying a unidirectional composite with given ratio of v_f and v_p components, assuming that the fibres in the pore region of length l_p behave like unbonded bundles. In this case

$$K_p = 1 - v_p[1 - (l_p/l_{f,eff})]^{-0.83\theta_o}/(1 - v_f) \tag{2.19}$$

where $l_{f,eff}$ is the effective fibre length in the composite.

Analysis of the properties of composites with different porosity shows that the danger of pores increases with increase of their number and length. Most dangerous are elongated pores whose length exceeds the critical fibre length in the composite. This fact probably accounts for the higher composite sensitivity to the increase of porosity of thinner fibre. Thus, boron-fibre-reinforced plastics are less sensitive to porosity than are glass- and carbon-fibre-reinforced plastics.

(b) Incompleteness of phase contact and cracks

A decisive effect on polymeric matrix–fibre adhesion along the interface is due to incompleteness of phase contact and cracks. The interface is not only a geometric boundary between the fibres and binders, but also a boundary between adjacent areas influenced by the physical and chemical processes taking place during interaction of the fibres and matrix in the composite formulation stage. The interface condition, i.e. its structure and the amount of defects, is affected by the physical, chemical and thermomechanical compatibility of the composite-forming components. The first two define the phase contact completeness, the nature, number and strength of the physical and chemical bonds, generated on interaction of the polymeric binder with the fibre surface, the effects of structure formation, the change of the matrix and fibre composition, and the properties in the boundary layers due to interdiffusion, selective sorption, the catalytic effect on the binder curing process; etc. The third defines the components interrelationship, providing composite integrity and stress level (which can also be related to the structural defects) of the interface and components, determined by the difference in their deformation and thermophysical properties.

The completeness of phase contact in a composite can be characterized by the coefficient K_s, which is the ratio of the area affected by the interaction forces between the functional groups and fractions of the binder and fibre molecules (S_{fm}) to the fibre surface area S_f; thus

$$K_s = S_{fm}/S_f \qquad 0 < K_s < 1$$

The interaction surface area per composite unit volume depends on the extent of reinforcement, fibre geometrical dimensions, their specific surface and topology, fibre wettability by the binder, the mobility of its molecules or their portions facilitating sorption and penetration of molecules into pores and cracks on the fibre surface, and interdiffusion ability. Fibre wettability by the binder and their sorption ability are defined by the fibre surface energy and the presence of polar reactive groups. The surface energy of mineral fibres at the moment of their manufacture is high, but it is quickly reduced as a result of surface contamination due to adsorption of environmental products. This leads to the increase of critical wetting angles and to the loss of adhesion to binders. To prevent this, the reinforcing fibres are subjected to various special treatments that result in cleaning, activation, chemical modification and protection of their surface before making composites.

The effect of phase contact completeness along the interface on composite properties can be estimated by the results of composite interlayer shear tests, using an approximate equation as was suggested in [7]:

$$\tau_{xz} = K_s^p K_s \tau_{xz}^\circ$$

where K_s is the coefficient of contact completeness ($K_s = K_s^p K_s^u$); K_s^p is the coefficient taking into account pore extension to the interface; K_s^u is the coefficient taking into account the surface activity, proportional to the value of the surface energy and adsorbability of the fibre surface; K_s^p is the surface roughness coefficient, proportional to the value of fibre specific surface S_f; and τ_{xz}^o is the ultimate shear stress at the interface, defined by the nature of the components combined.

In the manufacture of polymeric composites, use is made of multicomponent binders; therefore, when combining fibres with these binders, complex processes of fractionation and volume redistribution take place at the interface. The boundary layers are, as a rule, enriched in low-molecular-weight substances, which results in the change of the ratio of components in the binder volume and boundary layers. The above-mentioned factors and the presence of functional groups (of either acidic or basic nature) on the fibre surface can have an inhibitory or catalytic effect on polymer formation in the boundary layer, its depth of cure and the structure of the polymer network being formed. In some cases, this leads to the appearance of 'weak' binder boundary layers characterized by lower density, strength and thermal resistance and higher stress as compared to the binder in the bulk. Contrary to mineral fibres, organic fibres are more sensitive to the effect of the binder with which they are combined in the composite. Diffusion of low-molecular-weight products inside organic fibres may cause their swelling, relaxation and consequently the decrease of the fibre and composite strength. This type of defect can be estimated when comparing the corresponding properties of the actual composite with the corresponding properties of a 'defect-free' composite.

The adhesive interaction of the binder with the fibre prevents the free change of their dimensions in the process of binder chemical shrinkage in curing and subsequent cooling. Owing to the difference in the linear thermal expansion coefficients of the fibre and matrix on cooling to a temperature below the glass transition temperature, some stresses originate in the composite. Longitudinal, tangential (circumferential) and radial stresses are generated in the fibre and matrix whose nature and distribution dependences are similar to those generated in axial loading. The value of the stresses increases with increase in the difference of the fibre and matrix thermal expansion coefficients and elasticity moduli. If the stresses generated are commensurate with the matrix strength or adhesion strength along the interface, cracks are initiated in the composite, aligned in the direction of fibre orientation.

To estimate composite quality, it is reasonable to use the notion of integrity introduced in [14], which assumes the integrity of all components and the absence of binding effects on the composite interface not only in the as-manufactured condition but also on deformation of the composite until the fibres are broken when their strength limit is exceeded. Integrity conditions in the form of a system of inequalities were obtained in [15] when analysing the composite combined deformation under different loadings. The necessity to fulfil all and not only one integrity condition should be taken into consideration. Substitution of the fibre elastic strength properties values into the system of inequalities enables one to determine the desired elastic strength properties of the matrix, providing the integral structure. The complex integrity factor I of the composite is used as the criterion for estimation of composite integrity (the value inverse to defectiveness) and strength:

$$I = K_v \sum \eta_i \lambda_i \qquad 0 < I < 1 \qquad (2.20)$$

where $K = K_1 K_\phi K_v K_p K_i K_s K_{v_\sigma} K_{v_E}$ is a coefficient taking into account the composite microstructural defects K_1, K_ϕ, K_v, K_p, K_i and K_s, and the deviation of the fibre strength and elasticity modulus from the average values (K_{v_σ} and K_{v_E}); η_i is the correspondence coefficient of the binder integrity characteristics, $\eta_E = E/E_T$, $\eta_v = \sigma/\sigma_T$, $\eta_T = \tau/\tau_T$ (index T is the theoretical value of the factor determined from the integrity conditions); and λ_i are the validity coefficients of the binder parameters under various types of deformation.

An example of the coefficient values is presented in Table 2.5, whereas data illustrating the relation of the epoxy glass-fibre-reinforced plastic strength properties to the integrity criterion are given in Fig. 2.6.

In the case of full correspondence of the binder properties to the integrity condition $I = 1$ ($K_v = 1$), the deviations of value I from unity yield an integral value of the extent of composite defectiveness.

2.4 CONTROL OVER PROPERTIES OF COMPOSITES BY CHANGING COMPOSITION AND STRUCTURE

The structure of fibrous composites and the substantial differences in the properties of the combined fibres and matrices are conditioned by the anisotropy of the mechanical, thermal, electrical and other properties of the composites. It is conventional to characterize the extent of anisotropy of properties in terms of the ratio of the indices determined in different directions. Anisotropy manifests itself most distinctly in a comparison of composite properties in the direction of fibre lay-up with those determined at an angle relative to the reinforcement direction or in the plane of fibre lay-up. As follows from Table 2.7, the extent of anisotropy of the physical and mechanical properties of unidirectional glass, carbon and boron fibrous composites, determined in the reinforcement direction and transverse direction, is different (the difference in the properties may amount to two orders). The highest extent of

Table 2.7 Anisotropy of vector properties of unidirectional polymeric composites

Composite	E_x^+/E_y^+	σ_x^+/σ_y^+	σ_x^-/σ_y^-	dy/dx	ρ_x/ρ_y	δ_y/δ_x
Epoxy boron-fibre composite	12 to 15	55 to 75	5 to 7	3 to 4	–	4 to 5
Epoxy carbon-fibre composite	18 to 25	90 to 120	4 to 8	25 to 30	2.5 to 3	4 to 6
Epoxy glass-fibre composite	6 to 8	80 to 150	5 to 7	2.5 to 3.5	1	3 to 4

anisotropy is characteristic of composite tensile strength indices. As the fibre strength and stiffness increase, the differences in the strength and elasticity characteristics of the fibres and matrices increase, and so does the anisotropy of composite properties as indicated by their sensitivity to misorientation, distortions and twisting of the fibres [20, 25, 26].

Fibrous composites with complex structure exhibit an essential difference between the elasticity and strength properties in the reinforcement direction and the indices of shear in the fibre lay-up plane. In shear of such composites, the elasticity modulus and strength differ from the Young's modulus and strength in the reinforcement direction by more than one order.

In some instances, the anisotropy of composite elastic properties increases due to the anisotropy of the properties of the fibres such as carbon and organic. For instance, the axial and transverse elasticity moduli of high-modulus carbon fibres differ 30 to 50 times, and hence the deformability characteristics of organic and carbon fibrous composites should be determined taking into account the anisotropy of the fibres [27]. The use in this case of the known formula $E_y = E_m(1 - v_f)$ obtained according to the Reis model leads to results that differ from the experimental data by 15 to 20%.

Unlike the physical (natural) anisotropy, composites also feature structural anisotropy that is created in the course of their manufacture and is a controlled quantity like composite properties.

Control over the extent of anisotropy and optimization of composite properties indices are achieved by the target-oriented change of their composition or the reinforcement structure. Conventional methods and limits of control over the properties of unidirectional composites by changing the fibre—matrix relation, their properties and the strength of adhesion between the two are of limited nature. The latest methods of control over the properties of unidirectional composites involve the creation of 'hybrid' structures through combination in a single material of fibres or matrices of different nature and different properties, i.e. creation of a heteromatrix [28] and heterofibrous composites. Heteromatrix materials consist of three (or more) solid media separated from one another by interfaces, one of which is the interface between the matrices, whereas another is the traditional interface between matrices and fibres. In this case, the fibres may remain continuous at the matrix interface. These materials enable one to create areas exhibiting different physical and mechanical properties not only in a section of the laminated material due to combination of composites layers with different matrices but also in the reinforcement plane, i.e. it is possible to programme the quantitative and qualitative characteristics of the materials along the axes in compliance with the predetermined loading distribution by combination of the fibres in certain areas with different matrices [29]. Depending on the nature and properties of the matrices, the elasticity, strength, thermal, dielectric and other characteristics of the material change. The example of heteromatrix composites can be furnished by epoxy aluminium-filled boron-fibre composites. Creation of 'hybrid' (heterofibrous) composites by combination in a single material of fibres exhibiting different nature and properties is an effective way to expand the range of target-oriented control over the properties of the composites irrespective of their reinforcement structure. Wide practical acceptance has been gained by three-component composites such as glass carbon plastics, glass organic plastics and organic boron plastics. In analysing the structure of such materials, three levels of hybridization are distinguished, i.e. at the level of monolayers, at the level of the monolayer and at the level of the reinforcing element. In this case, it should be borne in mind that a single material enables one to combine all hybridization levels [30].

The first case provides for alternation of layers composed of different reinforcing fillers. The second alternative involves the use of multicomponent reinforcing filler: cloth, mat or veneer composed of different filaments or bundles. The third alternative provides for distribution of the elementary fibres (filaments) in the bundle, which is the basis for production of the reinforcing filler.

Each of the above alternatives features the appropriate advantages and disadvantages. The third alternative enables one to produce homogeneous filaments with more uniform operative level of the individual fibres. The second and third alternatives make it possible to produce composites exhibiting different properties in terms of the material thickness. The first alternative is most adaptable to manufacture and simple in practice, and most widespread. This alternative, however, only enables one to produce composites exhibiting laminated structure. Spatially reinforced composites can be produced only in the case of hybridization according to the second and third alternatives.

The combination of different-modulus and different-strength fibres enables one to control a number of materials properties; for instance, to increase the modulus of elasticity, fatigue strength and density of glass plastics by adding carbon fibres, to improve the impact strength of carbon and boron plastics by adding organic or glass fibres, and to improve the compressive strength and reduce the cost of organic plastics by adding glass fibres, etc.

As a rule, the limits of control over the properties of hybrid composites are confined to the values characteristic of the two-component combined materials. In this case, the hybrid composite properties can be calculated on the basis of the additivity rule, stemming from the properties and volumes of the two-component materials. This is how matters stand as regards density, modulus of elasticity and some other characteristics [31–33]:

$$R_{\mathrm{HCM}} = \sum_{i=2}^{n} v_{fi} R_{fi} \qquad \text{or} \qquad R_{\mathrm{HCM}} = R_{f1}v_{f1} + R_{f2}v_{f2} + \ldots R_{fn}v_{fn} + R_{m}v_{m}$$

$$\tag{2.21}$$

$$\text{at} \quad \sum_{fi=2}^{n} v_{fi} + \theta_{m} = 1$$

In some instances, however, so-called hybrid effects are observed, both positive and negative. In this case, the 'hybrid' properties not only can be substantially higher or lower than those of the 'hybrid' predicted on the basis of the mixture rule, but also can exceed the limits restricted by the values characteristic of the two-component materials. The effect of synergism (improvement of characteristics) as regards hybrid composites manifests itself, for instance, in tension of boron carbon plastics and compression of organic carbon plastics and boron organic plastics.

The method at issue for the target-oriented control over the extent of anisotropy and the properties of a composite is the change of its reinforcement structure due to cross-sectional or three-dimensional reinforcement.

2.4.1 Cross lay-up

Control over the properties of laminated composites is effected using cross lay-up of the layers along the full vertical extent of the material and changing the angle of fibre orientation in the individual layers (Table 2.8, Fig. 2.9). The simplest and most widespread is cross lay-up, wherein the fibres in the material layers are oriented at an angle $\pm\phi$ relative to the principal axes of symmetry, whose special case is represented by the orthogonal lay-up of the layers at an angle of $0°$ and $90°$ when the relation of the layers arranged in the direction of principal axes changes. As seen from Figs 2.10 to 2.12, the increase of the angle between the directions of loading and orientation of the fibres in the adjacent layers is followed by the monotonic decrease of the strength and elasticity characteristics in the direction of axis x, whereas the above characteristics increase along axis y. The reinforcement plane shear modulus increases and passes through a maximum corresponding to the values of the $45°$ angles and subsequently decreases, thus returning to the initial value. In this case, the elasticity modulus in the case of interlayer shear remains practically constant [34].

In tension of the orthogonally reinforced composite along the direction of fibre lay-up, breakdown occurs in the layers reinforced in the perpendicular direction. On reaching ultimate strain $\varepsilon^+_{x,\mathrm{ult}}$, the material loses solidity as revealed by the characteristic change on curves $\sigma-\varepsilon$ (Fig. 2.13).

This strain figure is determined by the dependence [35]

$$\varepsilon^+_{x,\mathrm{ult}} = \{\sigma_m v_f^{0.5}/[E_f v_f^{0.5} + E_m(1 - v_f^{0.5})]\} + \varepsilon_m(1 - v_f)^{0.5} \qquad (2.22)$$

Table 2.8 Effect of fibre orientation on elasticity modulus in tension and extent of epoxy boron-fibre composite anisotropy

Relative arrangement of fibres	Angle between directions of fibres in adjacent layers (deg)	E (GPa)				Anisotropy indices		
		E_x	E_y	$E_{\pi/4}$	E_z	E_x/E_y	$E_x/E_{\pi/4}$	E_x/E_z
Unidirectional								
1:0	0	162	18.6	15.4	18.6	8.70	10.80	8.70
Cross-planar								
1:1 (0:90)	90	86	85.0	16.2	18.7	1.02	5.30	4.60
1:1:1 (0:60)	60	92	93.0	89.3	18.5	0.99	1.04	4.80

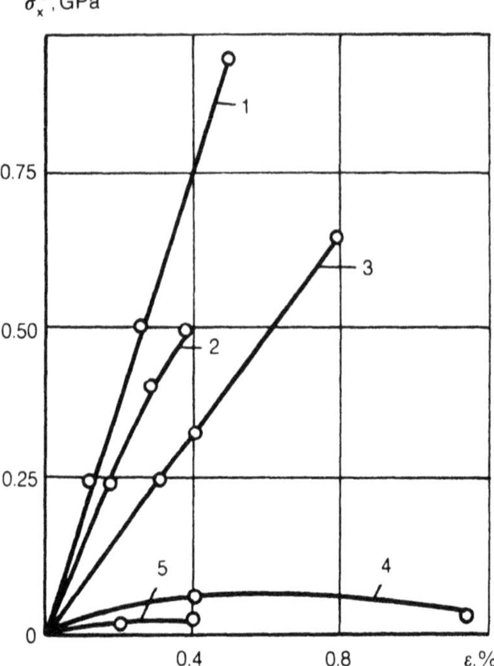

Fig. 2.9 Dependence between stresses and strains in tension of epoxy carbon-fibre composites with different reinforcement structure: 1, $0°$;2, $\pm 30°$; 3, $0°$, $90°$; 4, $\pm 60°$; 5, $90°$.

Hence, the modulus of elasticity and breaking tensile stress of the orthogonally reinforced composites till loss of solidity are calculated by the formulae:

$$E_{x,\text{ult}}^+ = E_x^0 t + E_y^0 (1 - t)$$

$$\sigma_{x,\text{ult}}^+ = E_x^0 t + E_y^0 (1 - t)\varepsilon_{x,\text{ult}}^+$$

(2.23)

where t is fraction of composite unidirectional layers in the loading direction; and E_x^0, E_y^0 are the Young's moduli of the composite unidirectional layer.

On breakdown of the lateral layers, the elasticity modulus and breaking tensile stress of the orthogonally reinforced composite are determined according to the expressions [12]

$$E_x^+ = E_x^0 t \qquad \sigma_x^+ = E_x^0 t \varepsilon_x^0$$

(2.24)

where σ_x^0, ε_x^0 are the breaking stress and relative tensile strain on the composite unidirectional layer.

Control over the orthogonally reinforced composite properties is effected by changing the relation between the unidirectional layers arranged in mutually perpendicalar directions. The compressive strength of the orthogonally reinforced composites

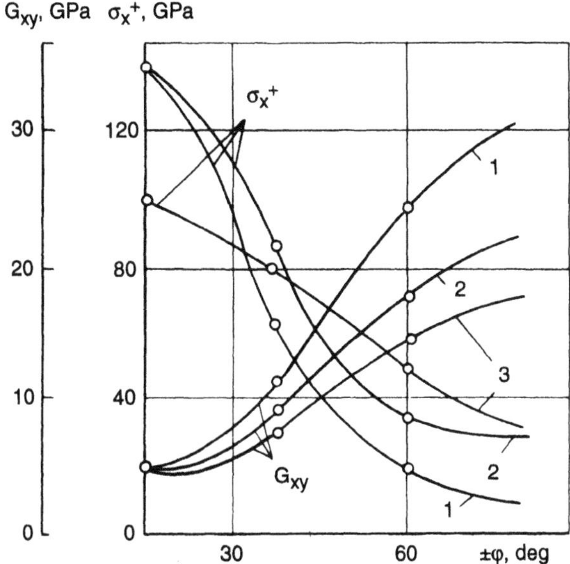

Fig. 2.10 Breaking tensile strength and elasticity modulus in shear in carbon-fibre composite reinforcement plane versus fibre orientation at various angles: 1, $\pm\phi$; 2, 0°, $\pm\phi$; 3, 0°, $\pm\phi$, 90°.

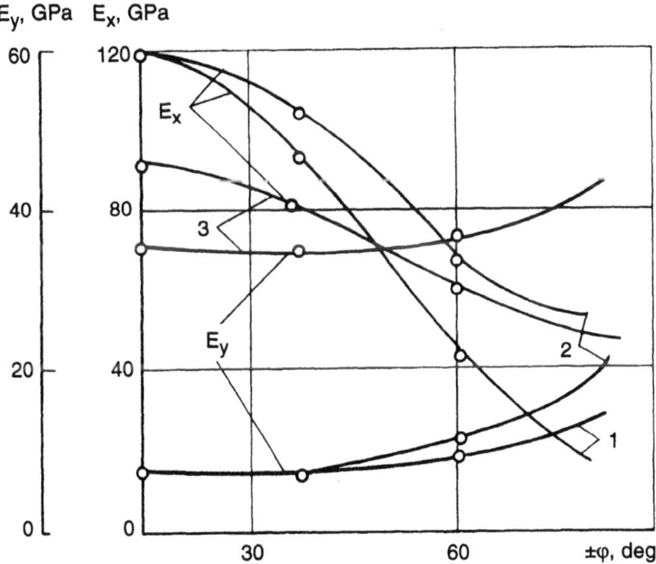

Fig. 2.11 Elasticity modulus in tension in carbon-fibre composite reinforcement plane versus fibre orientation at various angles: 1, $\pm\phi$; 2, 0°, $\pm\phi$; 3, 0°, $\pm\phi$, 90°.

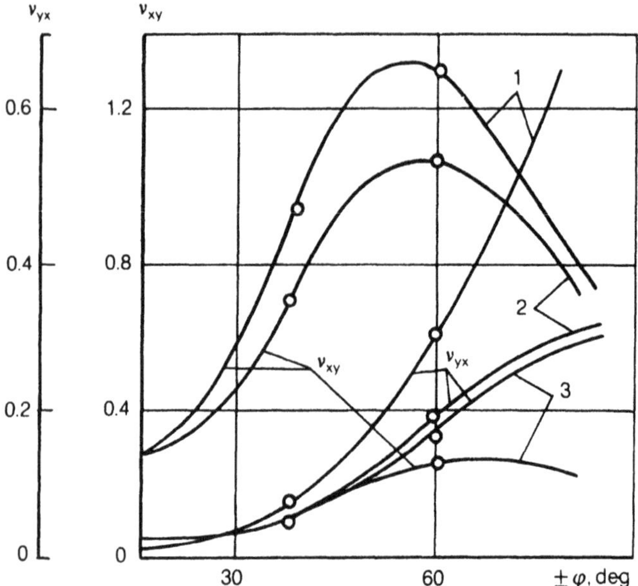

Fig. 2.12 Poisson's ratio in tension in carbon-fibre composite reinforcement plane versus fibre orientation at various angles: 1, $\pm\phi$; 2, $0°$, $\pm\phi$; 3, $0°$, $\pm\phi$, $90°$.

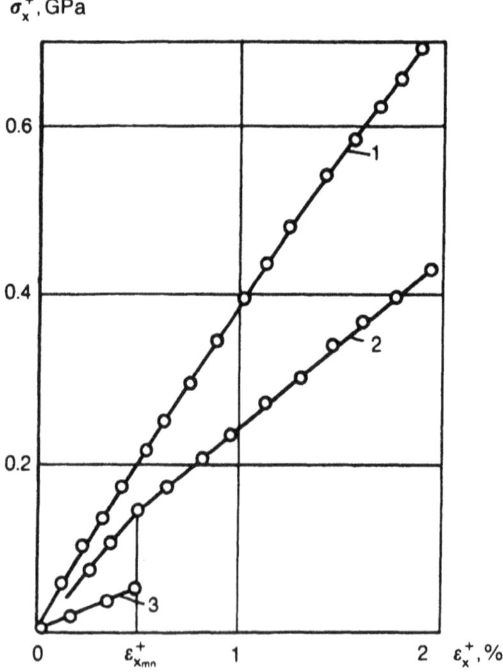

Fig. 2.13 Stress–strain diagrams for unidirectional (1, 3) and orthogonally reinforced (2) glass-fibre composite in tension along (1, 2) and perpendicular to (3) the reinforcement direction.

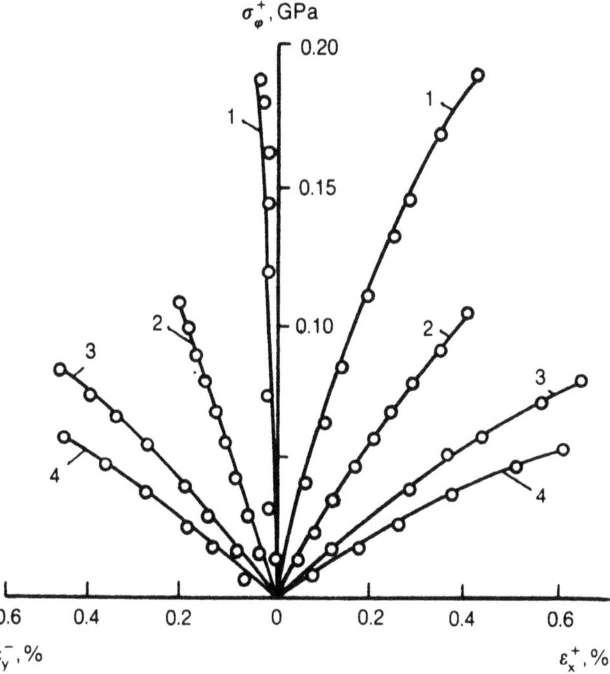

Fig. 2.14 Dependence $\sigma^+ - \varepsilon_x^+$ and $\sigma^+ - \varepsilon_y^-$ of orthogonally reinforced epoxy carbon-fibre composite with lay-up of $0°$, $90°$ ($1:1$) in tension at various angles: $1, 0°$; $2, 15°$; $3, 30°$; $4, 45°$.

decreases somewhat as the thickness of the unidirectional layers increases; hence, it is expedient to alternate the minimum-thickness monolayers.

The change of the number of layers laid along the loading axis in the orthogonally reinforced composites enables one to control the extent of anisotropy; in loading in a direction different from the orientation of the fibres, however, considerable anisotropy of the material properties continues to exist and the stress–strain diagrams become non-linear (Fig. 2.14). To prevent this and also, if required, to match accurately the material resistance field to that of the stresses acting thereon in operation in the construction, use is made of a lay-up that is versatile in terms of complexity.

The strength and elasticity properties of multilayer composites can be calculated on the basis of the experimentally determined characteristics of the unidirectional material or the known properties and relation of the components. The first case enables ones to obtain more accurate results as the extent of implementation of the fibre properties in the composite is taken into account therein. The calculations presuppose that the behaviour of the unidirectional layer exposed to loading is elastic, and the relation between the stress and strain is described by the generalized Hooke's law. A multilayer material composed of differently oriented unidirectional layers is considered as an inhomogeneous solid whose stiffness and strength depend on the properties and arrangement of the individual layers.

The elastic properties of multilayer composites with the orientation of the layers along the full vertical extent in the direction of the arbitrarily chosen axes of elastic symmetry are calculated by formulae [35]:

$$E_x = B_{xx} - B_{xy}^2/B_{yy} \qquad E_y = B_{yy} - B_{xy}^2/B_{xx}$$
$$v_{yx} = B_{xy}/B_{yy} \qquad v_{xy} = B_{xy}/B_{xx} \qquad (2.25)$$
$$G_{xy} = (1/n) \sum_{i=1}^{n} (G_{xyi} + \Delta B \sin^2\phi_i \cos^2\phi_i)$$

In these formulae

$$B_{xx} = (1/n) \sum_{i=1}^{n} (B_{11i} \cos^2\phi_i + B_{22i} \sin^2\phi_i - \Delta B \sin^2\phi_i \cos^2\phi_i)$$

$$B_{yy} = (1/n) \sum_{i=1}^{n} (B_{11i} \sin^2\phi_i + B_{22i} \cos^2\phi_i - \Delta B \sin^2\phi_i \cos^2\phi_i) \qquad (2.26)$$

$$B_{xy} = (1/n) \sum_{i=1}^{n} (B_{12i} + \Delta B \sin^2\phi_i \cos^2\phi_i)$$

$$\Delta B = B_{11i} + B_{22i} - 2B_{12i} - 4G_{xyi}$$

where n is the number of layers laid along the full vertical extent and ϕ_i is the lay-up angle with chosen axis x.

Stiffness matrix coefficients are expressed in terms of the unidirectional layer characteristics according to the equations

$$B_{11} = E_x^0/(1 - v_{xy}^0 v_{yx}^0)$$
$$B_{12} = E_y^0 v_{yx}^0/(1 - v_{xy}^0 v_{yx}^0) \qquad (2.27)$$
$$B_{22} = E_y^0/(1 - v_{xy}^0 v_{yx}^0)$$

The composite properties in directions that do not align with the chosen axes of elastic symmetry are determined taking into account the axes rotation through an angle relative to the earlier chosen axis.

For the strength analysis of cross-reinforced laminated composites, use can be made of the equations obtained stemming from the first strength criterion and taking into account the macrostructural stresses in the reinforcing fibres and matrix, originating in loading each i layer in the reinforcement direction or at an angle relative thereto. In the case when the strength of the reinforcing fibres is decisive, the expression assumes the form

$$\sigma_{xf}^+ \leqslant \sigma_f/\{\{[p_f a_i + (1 + q_f)f_i]/2\} + \{[p_f a_i - (1 - q_f)]^2/4 + C_i^2\}^{0.5}\} \qquad (2.28)$$

If the matrix strength is decisive, the relation takes the form

$$\sigma_{xm} \leqslant \sigma_m/\{\{[p_m a_i + (1 + q_m)b_i]\} + \{[p_m b_i - (1 - q_m)b_i]^2/4 + C_i^2\}^{0.5}\} \qquad (2.29)$$

where σ_f and σ_m are the breaking tensile stresses of the reinforcing fibres and matrix, respectively.

In these formulae

$$p_f = E_f/E^0_{xi} \qquad q_f = (v_f - v_{xy})E_f/E^0_y$$
$$p_m = E_m/E^0_{xi} \qquad q_m = (v_m - v_{xy})E_m/E^0_y \qquad\qquad (2.30)$$
$$a_i = B_{11}T_{1i} + B_{12}T_{2i} \qquad f_i = B_{12}T_{1i} + B_{22}T_{2i} \qquad C_i = 2B_{12}T_{3i}$$

Stiffness coefficients B_{11}, B_{22} and B_{12} are determined from the equations (2.27) as

$$T_{1i} = a_{xx}\cos^2\phi_i + a_{yy}\sin^2\phi_i - 0.5a_{xy}\sin^2\phi_i$$
$$T_{2i} = a_{xx}\sin^2\phi_i + a_{yy}\cos^2\phi_i + 0.5a_{xy}\sin^2\phi_i$$
$$T_{3i} = 0.5(a_{xx} - a_{yy})\sin^2\phi_i + 0.5a_{xy}\cos^2\phi_i$$

$$a_{xx} = (1/n)\sum_{i=1}^{n}(a_{11}\cos^2\phi_i + a_{22}\sin^2\phi_i - \Delta a_i\sin^2\phi_i\cos^2\phi_i) \qquad (2.31)$$

$$a_{yy} = (1/n)\sum_{i=1}^{n}(a_{11}\sin^2\phi_i + a_{22}\cos^2\phi_i - \Delta a_i\sin\phi_i\cos^2\phi_i)$$

$$a_{xy} = (1/n)\sum_{i=1}^{n}(a_{12i} + \Delta a_i\sin^2\phi_i\cos^2\phi_i)$$

$$\Delta a_i = 0.5(a_{11i} + a_{22i} + 2a_{12i} - 1/G_{xyi})$$

Compliance matrix coefficients are expressed in terms of the unidirectional layer characteristics:

$$a_{11} = 1/E^0_x \qquad a_{22} = 1/E^0_y \qquad a_{12} = v_{xy}/E_y$$

The results of calculating the elastic characteristics of the cross-reinforced laminated composites according to the above dependences display good agreement with the experimental data. In calculation of the strength characteristics, in particular, in a direction different from the orientation of the fibres, convergence of results is somewhat worse, which is associated with higher sensitivity of composite strength to structural imperfections and technological factors of material moulding.

The unidirectional and orthogonally reinforced composites exhibit high strength and stiffness in the reinforcement directions but low shear resistance. One method for increasing the shear strength of high-modulus composites is to change the reinforcement patterns, thus enabling one to increase the material shear resistance considerably in the reinforcement plane at a minor reduction of stiffness and strength in the fibre lay-up direction.

In practice, the frequently encountered task of composite structure optimization so as to ensure the required resistance to the action of normal and tangential stresses can be solved in two ways, i.e. change in the misorientation angle of the individual layers relative to axis x, or change in the number of layers exhibiting misorientation at angle $\pm\phi$ relative to the same axis. For instance, the use of reinforcement patterns

 Principles for creating fibrous composites

Table 2.9 Elasticity and strength properties of epoxy carbon-fibre composites with cross misoriented structure

Misorientation angle (deg)	E^+ (GPa)			G (GPa)		σ (MPa)		
	E_x^+	E_{45}^+	E_y^+	G_{xy}	G_{xz}	σ_x^+	σ_{45}^+	σ_y^+
0	125.0	7.1	5.4	3	2.6	1280	66	36
± 15	71.0	15.5	6.2	8	2.9	800	104	50
± 30	24.5	43.7	6.6	16	2.7	400	220	68
± 45	8.4	115.0	7.8	24	2.7	140	840	135

with antiphase orientation at small angles (up to $\pm 15°$) leads to an increase of shear modulus G_{xy} and elasticity modulus E_{45} by about 2.5 times and elasticity modulus E_y by 1.7 times as compared to similar characteristics of unidirectional composites (Table 2.9). In this case, the elasticity modulus reduction amounts to only about 20% [34].

Should the same loads act perpendicularly to the reinforcement plane and the composite work in torsion, one cycle of the lay-up control in the reinforcement plane is not enough. In this case, it is also necessary to control the reinforcement patterns in terms of the material thickness. The torsional stiffness of the height-irregular laminated composites is determined from the equation [36]

$$D = \sum_{i=1}^{n} h_i z_i^2 \{ G_{xyi} + [1/(1 - v_{xyi}v_{yxi})][E_{xi} + E_{yi} - 2v_{xyi}E_{xi}$$

$$- 4G_{xyi}(1 - v_{xyi}v_{yxi})] \sin^2(2\phi_i) \} \tag{2.32}$$

where h_i and z_i are the thickness and distance of the ith layer from the mid-surface; and E_{xi}, E_{yi}, v_{xyi}, and G_{xyi} are the elasticity moduli, Poisson's ratio and shear modulus of the ith layer.

It follows from formula (2.32) that in the case of predetermined fibre lay-up pattern in the reinforcement plane, the composite torsional stiffness depends on the pattern of lay-up along the full vertical extent. For instance, in carbon-fibre composites consisting of 16 layers wherein the external layers are laid up at an angle of $\pm 45°$, modulus $= 220$ GPa, which is much higher than in the composites with the orientation of the external layers along the fibre axis ($G_{xy} = 132$ GPa). And the larger the number of external layers laid up at an angle of $\pm 45°$, the higher is the material torsional stiffness. The lay-up of the layers at the above angle in the centre of the composite is of low efficiency. Thus, the removal of the number of layers laid up at angle relative to the chosen system of coordinate results in the increase of the material torsional strength [36].

In selection of a rational pattern of composite reinforcement along the full vertical extent, it is necessary to take into account the possibility of origination of initial stresses (due to anisotropy of the linear thermal expansion coefficients of the individual layers) and, as a result, warpage. It is necessary to arrange the layers symmetrically about the mid-plane, and the disoriented layers should be located as close thereto as possible.

Control over the properties of heterofibrous composites and their optimization in terms of the required parameters in the reinforcement plane are also effected using the cross lay-up of layers of different-modulus fibres along the full vertical extent of the three-component material lamination [33].

The polar diagrams of the elasticity modulus change (Fig. 2.15) for the cross-reinforced glass carbon-fibre composite have four symmetry axes, two of which correspond to the misorientation angles of the carbon fibres relative to the principal axes along which the glass fibres are located.

The extent of the material anisotropy, characterized by the relations of elastic constants E_x/ε_y and E_y/G_{xy}, decreases from 5.4 and 11 in unidirectional material to 1.6 and 2.1 in a composite with misorientation of the carbon fibres at an angle of $\pm 45°$.

Comparison of the reinforcement plane shear modulus dependences on the misorientation angles of the glass-fibre composite and the glass carbon-fibre composite enable one to conclude that maximum value G_{xy} reached for the glass-fibre composite

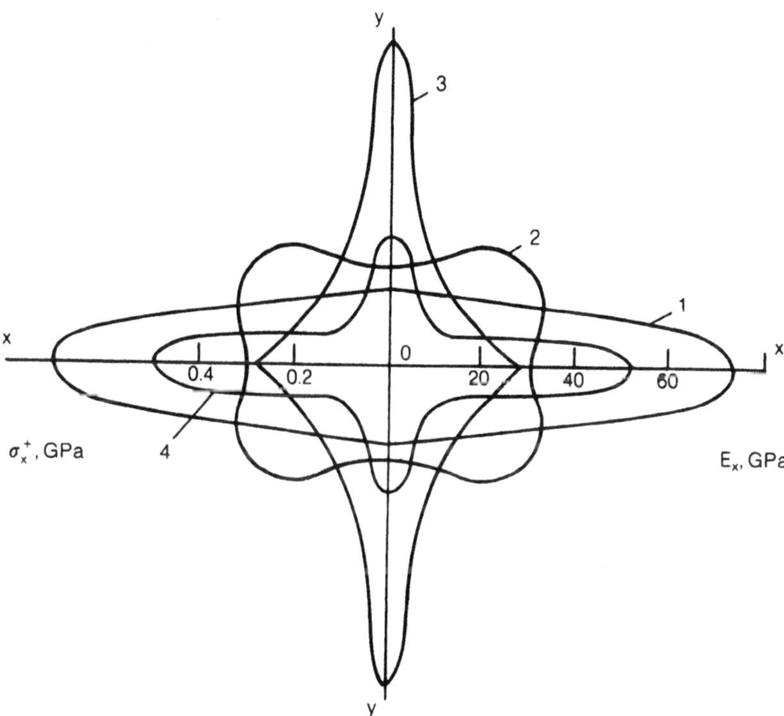

Fig. 2.15 Polar diagram of carbon glass-fibre composite elasticity modulus (1, 2, 3) and strength (4) in carbon-fibre misorientation relative to glass fibres at various angles: 1, 7.5°; 2, ±45°; 3, 4, 90°.

Table 2.10 Comparative properties of glass-fibre composite and carbon glass-fibre composite with layer misorientation ensuring achievement of identical shear moduli in reinforcement plane

Material	$\gamma \times 10^{-3}$ $(kg\,m^{-3})$	σ_x^+ (MPa)	σ_x^- (MPa)	σ_x^b (MPa)	τ_{xz} (MPa)	E_x (GPa)	E_B (GPa)	G_{xy} (GPa)	a_x $(kJ\,m^{-2})$
Glass-fibre composite (15% of layers at an angle of $\pm 45°$)	2.0	530	320	650	120	30	17	12.5	302
Carbon glass-fibre composite (15% of layers at an angle of $\pm 15°$)	1.8	510	390	600	200	70	17	12.5	165
Relation of indices at reduced density	—	1.05	1.35	1.0	1.85	2.6	1.1	1.10	0.6

at a misorientation angle of $\pm 45°$ can be reached in the carbon glass-fibre composite at a carbon-fibre misorientation angle of $\pm 15°$; this leads to a considerable increase of the material strength and elasticity characteristics except for the impact viscosity in the direction of axis x (Table 2.10). Like the carbon glass-fibre composite, its specific indices are higher [31].

2.4.2 Three-dimensional reinforcement

The elasticity moduli and strength in interlayer shear, tension and compression in a direction perpendicular to the plane of the composite layers lay-up exhibit low sensitivity to a change of fibre arrangement in the reinforcement plane. The essential increase of the above composite characteristics is achieved by establishment of interlayer crosslinks, i.e. the three-dimensional arrangement of the fibres. In this case, the characteristics of the composites in the fibre lay-up direction are proportional to the number of fibres in each direction [37].

Depending on the principle of forming the three-dimensional links, the materials are divided into three groups (Table 2.11). The first group includes composites based on multilayer fabrics, wherein the three-dimensional links are formed due to distortion of the warp threads linking the individual layers or passing through the entire thickness of the fabric. The second group includes materials wherein the three-dimensional links of the reinforcing filler are created due to addition of the third-direction fibres. These materials are formed by a system of three threads in the rectangular or cylindrical coordinate system. The third group incorporates composites wherein the three-dimensional links are forced due to the thread-like crystals grown or applied to the surface of the fibres, strips or fabrics and introduced between the fibres of the main filler.

As regards the glass, organic and carbon fibrous composites (composites based

Table 2.11 Mechanical properties of composites with various reinforcement structures

Reinforcement structure	Glass-fibre composite			Carbon fibre composite		
	E_x (GPa)	G_{xz} (GPa)	τ_{xz} (MPa)	E_x (GPa)	G_{xz} (GPa)	τ_{xz} (MPa)
Layer	–	–	–	20[a]	1.8[a]	12[a]
	70	2.0	25	180	3.5	30
Three-dimensional, crosslinked	22	3.0	103	15[a]	2.5[a]	30
Isotropic, reinforced	22	4.5	120	50[a]	–	45[a]
Layer, whiskerized	72	5.2	45	180	5.4	50

[a] Based on carbon fibres produced from viscose rayon ($E_f = 60$ GPa).

Table 2.12 Properties of glass plastics based on three-dimensional glass fabrics

	Composite characteristic, ϕ (deg)		
Properties (MPa)	$\phi = 10°$, $v_f = 65\%$	$\phi = 19°$, $v_f = 55\%$	$\phi = 32°$, $v_f = 55\%$
$E_x \times 10^{-3}$	32.5	25.0	13.0
$E_y \times 10^{-3}$	23.8	19.4	19.8
$G_{xz} \times 10^{-3}$	2.6	2.9	33.4
$G_{xy} \times 10^{-3}$	11.5	–	40.5
σ_x^+	440	370	189
σ_x^-	290	260	120
σ_y^+	390	350	340
σ_y^-	350	330	270
τ_{xz}	42	50	63

on low-modulus fibres), use is made of multilayer fabrics. The properties of these materials are determined by the extent of distortion of the warp threads as seen from the data of Table 2.12, wherein are specified the characteristics of the glass–cloth–base laminates made from such fabrics.

As regards composites based on high-modulus carbon and boron fibres, most acceptable is the three-dimensional reinforcement pattern that enables one to preserve the straightness of the reinforcing fibres. In this case, various directions admit the lay-up of various fibres, thus enabling one to form a multicomponent material. Table 2.13 presents the properties of such a material reinforced in direction z with silica

Table 2.13 Properties of composites with orthogonal three-dimensional reinforcement system

Elastic properties (GPa)	Fibres		Strength properties (MPa)	Fibres	
	Silica	Quartz + silica		Silica	Quartz + silica
E_x^+	9.0	14.7	σ_x^+	160	138
E_y	8.4	15.6	σ_y^+	180	240
E_z	9.1	10.8	σ_x^-	183	218
G_{xz}	2.0	2.47	σ_y^-	168	197
G_{xy}	1.72	2.4	σ_z^-	178	161
			τ_{xz}	70	84
			τ_{xy}	70	111

fibres and, for the purpose of comparison, the properties of a material reinforced exclusively with silica fibres [37].

The manufacture of such materials involves certain difficulties in selection of the binder, which should exhibit low viscosity and should not contain a solvent, as only the hydrovacuum or autoclave impregnation methods are used for manufacture of these materials.

Whiskerization, and reinforcement with discrete fibres of the polymeric matrix filling the interfibre space in composites with continuous fibres [31, 38], enable one not only to increase the strength and elasticity characteristics of composites in shear without deterioration of their properties in the reinforcement direction but also to change some physical properties of the composite. The introduction into the composite of discrete fibres with physical properties different from those in the main (continuous) fibres enables one to influence the thermal conductivity and conductivity, dielectric constant and dielectric loss factor of the composites.

The introduction of discrete fibres into the matrix contributes to the increase of its elasticity and strength indices and heat resistance of the composite is proportion to the number and mechanical characteristics of the discrete fibres [39]. This, in turn, produces a favourable effect on the increase of composite resistance to deformation in loading in a direction different from the continuous fibres orientation direction. As the extent of filling the polymeric matrices increases, their strength and elasticity properties improve. With the content of discrete fibres amounting to 1%, the elasticity and strength properties of the matrices can be determined with an accuracy sufficient for engineering calculations using the equations [38]:

$$E_{Mx} = E_{My} \approx 0.3\, E_f v_f \qquad \sigma_{Mx} = \sigma_{My} \approx (0.2 \text{ to } 0.3)\sigma_f v_f \qquad (2.33)$$

$$\sigma_{Mxy} \approx 0.13\, E_f v_f \qquad \tau_{Mxy} \approx 0.5\, \sigma_{Mx} \qquad v_{Mxy} \approx 0.3 \text{ to } 0.35$$

The particularly significant increase of strength and stiffness of polymeric matrices is achieved by strengthening them with thread-like crystals. The introduction of thread-like crystals into the polymeric matrix in the amount of 11–13% (by volume) leads to a drastic increase of the indices of its properties. For instance, the elasticity modulus in tension increases 5–7 times, in shear 5–6 times, and the breaking tensile stress, 3–4 times. At the same time, the similar characteristics for the same matrix reinforced with discrete glass fibres increase accordingly only 3, 2.5 and 2 times [39].

REFERENCES

1. Lipatov Yu.S., *Physical Chemistry of Filled Polymers*, Khimia, Moscow, 1977.
2. Pludeman E., in *Composite Materials*, vol. 6, *Interfaces in Polymeric Composites*, Mir, Moscow, 1978, pp. 181–228.
3. Andreevskaya G.D., Gorbatkina Yu.A., Ivanova-Mumzhieva V.G., Epifanova S.S., *Polymer Mechanics*, 1974, **1**, 37–42.
4. Skola D., in *Composite Materials*, vol. 6, *Interfaces in Polymeric Composites*, Mir, Moscow, 1978, pp. 228–92.
5. Kobets L.P., Prigorodov V.N., Gunyaev G.M., *et al.*, *Fibrous and Dispersion-Strengthened Composite Materials*, Nauka, Moscow, 1976, pp. 159–65.

6. Aslanova M.S. (ed.), *Chemical Treatment of Glass Fibre Surface*, Khimia, Moscow, 1966, p. 111.
7. Kobets L.P., Gunyaev G.M., *Polymer Mechanics*, 1977, **3**, 445–51.
8. Trostyanskaya E.B. (ed.), *Structural Plastics*, Khimia, Moscow, 1974.
9. Varshavsky V.Ya., Results of Science and Engineering, Ser.: *Chemistry and Technology of High-Molecular Compounds*, vol. 9, VINITI under the USSR Academy of Sciences, Moscow, 1977.
10. Kaverov A.T., Ermilov N.K., Chernenko L.A., *et al.*, *Carbon-Based Structural Materials*, vol. 13, Moscow, 1978.
11. Lipatov Yu.S., *Journal of All-Union Chemical Society Named After D.I. Mendeleev*, 1978, **23**(3), 305–9.
12. Skudra A.M., Bulavs F.Ya., *Structural Theory of Reinforced Plastics*, Zinatne, Riga, 1978.
13. Shami K., in *Composite Materials*, vol. 6, *Interfaces in Polymeric Composites*, Mir, Moscow, 1978, pp. 42–87.
14. Rabinovitch A.L., *Introduction to Reinforced Polymer Mechanics*, Nauka, Moscow, 1970.
15. Roginsky S.L., Kanovitch M.Z., Koltunov M.A., *High-Strength Glass Plastics*, Khimia, Moscow, 1978.
16. Roginsky S.L., *Journal of All-Union Chemical Society Named After D.I. Mendeleev*, 1978, **23**(3), 261–4.
17. Roginsky S.L., Koltunov M.A., Natrusov V.I., *et al.*, *Polymer Mechanics*, 1978, **4**, 743–6.
18. Tumanov A.T., Gunyaev G.M., Lutsau V.G., Stepanichev E.I., *Mekhanika Polimerov*, 1975, **2**, 248.
19. Tumanov A.T., Gunyaev G.M., Yartsev V.A., Svoystva kompositsiy, uprochnennykh voloknami bolshogo diametra, in *Voloknistie i Dispersnouprochennye Kompositsionnye Materialy*, ed. N.V. Ageev, A.T. Tumanov, M.Kh. Shorshorov, K.I. Portnov, V.I. Bakarinova, Nauka, Moscow, 1976, p. 156.
20. Tarnopolsky Yu.M., Rose A.V., *Osobennosti Rascheta Detaley iz Armirovannykh Plastikov*, Zinatne, Riga, 1969.
21. Gunyaev G.M., Optimizatsiya sostava i makrostrukturni pri konstruirovanii polymernikh materialov, in *Voloknistie i Dispersnouprochnennye Kompositsionnye Materialy*, ed. N.V. Ageev, A.T. Tumanov, M.Kh. Shorshorov, K.I. Portnov, V.I. Bakarinova, Nauka, Moscow, 1976, p. 141.
22. Tarnopolsky Yu.M., Rose A.V., Zhigun I.G., Gunayev G.M., *Mekhanika Polimerov*, 1971, **3**, 676.
23. Gunyaev G.M., Zhigun I.G., Sorina T.G., Yakushin V.A., *Mekhanika Polimerov*, 1973, **3**, 492.
24. Gunyaev G.M., Sorina T.G., Termoustoichivost' napolnennikh plastikov na osnove phenolo-formal 'degidnikh i Epoksidnykh Svyazuyutshikh, in *Termoustoichivost' Plastikov Konstruktsionnogo Naznacheniya*, ed. E. Trostyanskaya, Khimiya, Moscow, 1980, p. 161.
25. Gunyaev G.M., *Polymer Mechanics*, 1972, **6**, 1123–5.
26. Tarnopolsky Yu.M., Rose A.V., Zhigun I.G., Gunyaev G.M., *Polymer Mechanics*, 1971, **4**, 676–85.
27. Bulavs F.Ya., Auzukalis Ya.V., Skudra A.M., *Polymer Mechanics*, 1972, **4**, 631–9.
28. Belozerov L.G., Tumanov A.T., Gunyaev G.M., *et al.*, in *Proc. 3rd All-Union Conf. on Composite Materials*, ed. A.A. Baikov, IMET under the USSR Academy of Sciences, Moscow, 1974, p. 122.
29. Belozerov L.G., *Polymer Mechanics*, 1978, **6**, 963–9.
30. Gunyaev G.M., Rabotnov Yu.N., Rumyantsev A.F., *et al.*, *Plasticheskiye Massy*, 1976, **9**, 31–3.
31. Gunyaev G.M., *Struktura i Svoistva Polimernikh Voloknistikh Kompositov*, Khimiya, Moscow, 1981.
32. Gunyaev G.M., in *Composite Materials*, Moscow University Press, Moscow, 1979, pp. 309–25.
33. Gunyaev G.M., *Polymer Mechanics*, 1977, **5**, 819–26.

34. Gunyaev G.M., Zhigun I.G., Dushin M.I., *et al.*, *Polymer Mechanics*, 1974, **6**, 1019–27.
35. Skudra A.M., Bulavs F.Ya., Rotsens K.A., *Creep and Static Fatigue of Reinforced Plastics*, Zinatne, Riga, 1971.
36. Zhigun I.G., Gunyaev G.M., Dushin M.I., *et al.*, in *Fibrous and Dispersion-Strengthened Composite Materials*, Nauka, Moscow, 1976, pp. 91–3.
37. Zhigun I.G., Polyakov V.A., *Properties of Three-Dimensional Reinforced Plastics*, Zinatne, Riga, 1978, p. 215.
38. Gunyaev G.M., Zhigun I.G., Sorina T.G., Yakushin V.A., *Polymer Mechanics*, 1974, **3**, 492–501.
39. Gunyaev G.M., Sorina T.G., *Plasticheskiye Massy*, 1971, **11**, 13–15.

3

Structural carbon-fibre-reinforced plastics and their properties

T.G. Sorina and G.M. Gunyaev

3.1 INTRODUCTION

Carbon-fibre-reinforced plastics (CFRP) are composites that are based on a polymeric matrix (thermosetting and thermoplastic binders) containing carbon fibres (tows, tapes, fabrics, webs) as reinforcing filler.

CFRP strength and stiffness are mainly defined by the properties of the carbon fibre and service temperature by the thermal stability of the matrix.

CFRP possess a unique complex of properties and exceed the specific stiffness, fatigue strength, chemical and radiation resistance of traditional structural materials (metals, glass-fibre-reinforced plastics). They have high thermal conductivity, low linear thermal expansion coefficient and allow one to control mechanical and electrical characteristics in a wide range.

Basic CFRP types can be used instead of metals during production of medium- and high-load aerospace structures, automobile body elements, transmission gears, cases, machine and instrument operating elements, electrical engine cases and drives, bridge framework elements, building heating panels, etc.

CFRP containing continuous fibres in combination with epoxy matrices possess good mechanical properties, low shrinkage, high adhesion to carbon fibres, resistance to agressive media and relatively high thermal resistance to 180°C, and are widespread.

Currently, structural CFRP based on thermoplastic matrices, such as poly(ether ether ketone), polysulphone, poly(ether sulphone), poly(ether imide), etc., are being developed and used.

High resistance to impact loading (low brittleness), moisture and chemical resistance, fitness for recycling and weldability are the positive features of these CFRP, compared to materials based on thermosetting binders.

CFRP use allows one to produce structures with high weight perfection and ensures weight reduction by 20–25%, service life increase by 1.5–2.0 times, labour consumption decrease by 40–60% and material utilization coefficient increase up to 85%.

Carbon fillers and matrices used for CFRP production, namely structural ones, are briefly considered in this chapter; the properties and their variation under environmental conditions of the most widespread materials are also given. The mutual effect of components on CFRP thermal stability and the ability to withstand long-term operation at temperatures of 250–300°C are described. Particular questions of structure-simulating modelling during development of CFRP with given properties are discussed.

3.2 COMPONENTS FOR STRUCTURAL CARBON-FIBRE-REINFORCED PLASTICS

3.2.1 Carbon fillers

Carbon fibres, obtained by oxidation and carbonization of polyacrylonitrile (PAN) organic fibres, are used as structural CFRP fillers [1]. Carbon fibres are divided into high-strength and high-modulus ones according to their mechanical properties. Both are produced in the form of carbonized thin cord tapes with weak technological carbon weft under trademarks such as LU, ELUR and LGU, and as untwisted tows with different filament contents (UKN, Kulon, Granit).

The creation of an assortment of different textiles and carbon fillers with different elasticity and strength properties allows carbon-fibre application fields to be extended into aerospace and other branches of industry. As reinforcing fillers determine a number of basic composites properties (tensile, interlaminar shear, fatigue strength, fracture toughness, etc.), the study of their mechanical properties, structure, surface state, geometrical and energy parameters is of paramount importance [2]. Elastic and strength properties of fillers are given in Table 3.1 and surface physical and chemical properties for some of them are given in Table 3.2.

UKN and ELUR high-strength carbon fibres, having similar final treatment temperature, but produced from variously moulded PAN fibre, differ significantly in orientation degree, carbon stack sizes, fibril cross-sectional area and have more friable structure compared to high-modulus Kulon, Granit and LU-24 fibres.

Filler thermal oxidation resistance reflects the difference in their structural organization [3]. Activation energy ε_a of surface oxidation and bulk oxidation, total weight losses at elevated temperatures, T_m, thermal effects of temperature ranges and other parameters (defined from thermograms, obtained under both dynamic and isothermal heating conditions of samples) are significantly different. So, activation energy ε_a of surface oxidation of different fibres varies between 25.0 and 60.5 kcal mol^{-1}; in this case it is considerably higher than for bulk oxidation. This factor confirms the ideas about radial anisotropy of carbon-fibre structure. High level of activation energy, the highest oxidation temperatures and minimum weight losses are naturally typical for Kulon, Granit and LU-24P high-modulus fibres. The large weight losses of two other types of fibres are due to a deep oxidation effect on fibre friable structure.

Surface inertness of high-modulus fibres and large wetting angles by epoxy binder present difficulties for interphase boundary formation in a composite, which defines low interlayer shear and compression strength values.

Table 3.1 Physicomechanical characteristics of carbon fibres

Binder grade	Fibre diameter, d (μm)	Diameter coefficient of variation, v (%)	Density, ρ ($kg\,m^{-1}$)	Strength by microplastic, σ_b (MPa)	Strength coefficient of variation, V (%)	Elasticity modulus, E (GPa)	Textile configuration
UKN-P	7	7	1750	2800	10	230	Tow 2.5k; 5k
UKN-P/0.1	5.2	5.5	1760	3500	10	240	Tow 5k; 8k
ELUR-P	6.4	6	1740	2500	8	230	Tape
Kulon N-24P	6	7.5	1880	1900	12	450	Tow 5k
LU-24P	7.0	7.0	1800	2700	8	350	Tape

Table 3.2 Structure and physicochemical properties of carbon-fibre surface

Fibre grade	Parameters of thermal oxidation resistance[a]				Specific surface, S (g/m)	COOH content (mg-eq/g)	OH content (mg-eq/g)	Wetting angle of epoxy binder, Θ (deg)
	T_o (°C)	T_q/T_m (°C)	E_a ($kcal\,mol^{-1}$)	$\Delta m/m$ (%)				
UKN-P	405	565/645	25.6	21	0.66	0.00168	1.6–2.2	19
ELUR-P	480	705/800	35.2	3.5	0.53	0.0056	–	16
Kulon N-24P	520	820/992	60.5	0.7	0.88	–	–	30

[a] T_o, oxidation onset temperature; T_q, maximum heat emission temperature in DTA curve; T_m, maximum oxidation rate temperature in DTA curve; E_a, activation energy; m, mass; Δm, mass change.

High-modulus and high-strength carbon fibres, designated for use as structural CFRP fillers, are subjected to obligatory surface treatment; sizing composition in the amount of 1.5–3.5 wt% is additionally applied. These operations are carried out for compatibility improvement of carbon fibre with binders and in order to ensure the possibility of processing tows into different textile forms. Surface treatment allows one to increase CFRP shear properties by 2–3 times [4]. Absence of size on the surface, good wetting by various types of binders and shape stability with monolayer thickness of 0.08–0.1 mm are obvious advantages, as is the universality of LU-P and ELUR-P (P-surface treatment) carbon types.

Tape fillers are widely used for production of thin skins and allow one to realize complex lay-ups with the retention of equiaxial, low-strained structure with a low material thickness up to 1 mm.

Equal-strength UT-900 fabrics of serge weave and unidirectional UOL-300 cord tapes with the properties given in Table 3.3 have been developed on the base of UKN-P carbon tows with different numbers of filaments. Thin glass or organic thread is used as weft in UOL-300 tapes.

The physical and mechanical properties of epoxy CFRP based on tested fillers are given in Table 3.4. According to CFRP properties, given in Table 3.4, a number of Russian carbon fibres are at the level of traditional fillers widely used in world engineering.

Table 3.3 Characteristics of woven carbon fillers

	Filler grade			
Parameter	*UT-900-2.5*	*UOL-300-2.5*	*ELUR-P-0.08*	*LU-24P*
Weave	Serge 2/2	Cord	Cord	Cord
Warp thread	UKN-P-2.5k	UKN-P-2.5k	Twisted thread	Twisted thread
Weft thread	UKN-P-2.5k	Glass fabric	–	–
Width (mm)	900	300	255	85
Weight of linear metre (linear density) $(g\,m^{-1})$	218	62	15	11
Weight of one cubic metre (volumetric density) $(g\,m^{-3})$	240	–	–	–
Thread density				
Warp	6	10	57	57
Weft	6	1	–	–
Monolayer thickness in plastic at pressure[a] $P_{sp} = 6$ atm (mm)	0.2	0.18	0.09	0.13

[a] Specific moulding pressure.

Table 3.4 Physicomechanical characteristics of unidirectional carbon-fibre-reinforced epoxy plastics

			Strength (MPa)		Tensile elasticity modulus, E(GPa)
Fibre grade	Tensile strength, σ_b	Compressive strength, σ_{comp}	Bending strength, σ_{bend}	Shear strength, τ_{shear}	
UKN-P	1500	1300	1800	80	140
UKN-P0.08	1000	1000	1200	80	–
UKN-P0.1	1800	1500	2000	85	140
LU-24P	1200	950	1500	58	220
Kulon N-24P	980	600	980	30	250
Granit	1500	800	1400	45	200

3.2.2 Polymeric matrices

Epoxy compositions, possessing the highest mechanical properties in the cured state in the temperature range from -60 to $+180°C$, good adhesion to carbon fibres and curing at comparatively low temperatures with minimum shrinkage, are particularly widely used now as binders for CFRP in structural applications.

Components of the composites are produced based on such binders at low pressures, which is particularly important in the case of using carbon fibres with high stiffness, and consequently brittleness [5].

Not only CFRP strength properties but also their thermal resistance (which changes according to starting polyphenol structure and selected hardener type) depend on chosen binder composition. The chemical structure of epoxy resin and hardener greatly affect prepreg pot-life and composite technological properties.

Epoxy–novolac and polyfunctional epoxy resins, cured by amines or latent hardeners (for example, ion-type ones), are used to create heat-resistant epoxy matrices (ENFB, VS-2526) with service temperatures of 150–160°C. The use of such a catalyst allowed one to obtain epoxy–novolac binder with prepreg pot-life equal to more than one year at room temperature. Other advantages of an ionic catalyst are: the absence of volatile products during curing; accurately fixed curing temperature during moulding; and resulting absence of the need to carry out technologically complicated staged curing conditions.

It should be noted that increase of binder and prepreg pot-life not only is aimed at easy technological process organization but also can be the source of improvement of material mechanical properties.

Self-ordering processes, resulting in formation of optimum three-dimensional structures and finally creation of polymers with higher strength properties, and taking place in binders in long-term contact with carbon-fibre filler, are theoretically possible and experimentally shown by the example of EFNB and EDT-69 compositions.

Bisphenol A-type epoxy aniline phenol–formaldehyde resins, cured by oligomers, etc., are used in the less thermally resistant 5-211B and EDT-69 compositions (with T_g up to 130°C).

The most technologically efficient heat-resistant polymers, which have found application in the aerospace industry and machine building, are PAIS (poly(amine imide)s) and OMI (oligo(ether maleimide)s) [6, 7]. Poly(amine imide) binder PAIS-104, an oligomer mixture of diphenylmethane bismaleimide and diaminodiphenylmethane, cured by polymerization reaction without low-molecular-weight product emission, ensures CFRP solid structure (porosity 1–3%) and high physical and mechanical properties in the temperature range from −120 to 250°C.

Basic chemical and technological properties of OMI binder consist in the fact that introduction of heat-resistant structures into molecular chains was transferred from binder synthesis stage to CFRP formation stage. Heat-resistant structure formation begins at a temperature of 180°C and finishes at 250°C by high-density network copolymer formation.

Prepregs with long-term pot-life, which are processed into products by pressing, vacuum and autoclave moulding, winding and contact moulding in closed moulds, are made with the combination of these binders and different carbon fillers.

A typical representative of thermoplastic binders, used as a matrix for structural CFRP, is polysulphone, which possesses thermal stability up to 170°C. It is produced in the form of film and powder; and can be dissolved in available solvents and processed into composites at comparatively low pressure of 10–15 atm and temperatures of 260–280°C.

Mechanical properties of some cured binders are given in Table 3.5, and typical σ–ε stress–strain diagram is shown in Fig. 3.1 under tension for VS-2526 m, OMI thermosetting and thermoplastic matrices.

Figure 3.2 shows the change of resin viscosity in the process of heating and curing. VS-2526 and OMI binders are related to low-viscosity systems; CFRP based on them require complex curing cycles, including exposure at high temperature during which resin viscosity in increased, and moulding pressure is applied only after this stage.

Viscosity characteristics of UNDF-4a, ENFB and 5-211B allow one to perform simple curing cycles for CFRP with the application of maximum pressure already at the beginning of the cycle.

The duration of the moulding cycle for CFRP with polysulphone matrix does not exceed 30 min and envisages placing the stack into the hot mould with simultaneous application of maximum pressure and short-term exposure for ∼10 min at the temperature of 280°C with subsequent rapid cooling.

The aggregate state of non-cured binders predetermines the technology of semiproducts manufacturing. Thus, for example, PAIS-104 binder-based prepregs are produced by electron beam spraying of powder with definite fractional composition on carbon filler surface with subsequent melting in a thermal furnace; UNDF-4a, OMI and VS-2526 binder-based semiproducts are produced by melting technology.

Film-type thermoplastic PSN binder is combined with carbon tape by means of lamination. Table 3.6 shows technological characteristics of prepregs based on different

Table 3.5 Properties of cured binders

Properties[a]	Epoxy resins					Oligo(ether maleimide)	Polysulphone	Bismaleimide
	UNDF-4a	EDT-69	VS-2526m	ENFB	VS-2561	OMI	PSN	PAIS-104
$\rho\,(\text{kg m}^{-3})$	1340.0	1270.0	1240.0	1230.0	1230.0	1230.0	1240.0	1250.0
σ_b (MPa)	62.0	55.0	78.0	65.0		65.0	90.0	80.0
ε (%)	2.0	2.5	2.5	2.0	3.8	2.2	9.0	2.5
E (MPa)	3750.0	2900.0	4500.0	3600.0	3800.0	4050.0	2600.0	–
σ_{comp} (MPa)	200.0	250.0	280.0	170.0	–	250.0	–	–
τ_{shear} (MPa)	80.0	70.0	90.0	70.0	85.0	60.0	60.0	60.0
T_g (°C)	140.0	100.0	190.0	160.0	160.0	260.0	190.0	260.0
Ultimate moisture sorption (%)	4.0	4.8	3.8	2.7	–	3.0	1.1	3.5

[a] Density ρ, ultimate bending strength σ_b, ultimate tensile strain ε, tensile elasticity modulus E, ultimate compressive strength σ_{comp}, ultimate interlaminar shear strength τ_{shear}, and glass transition temperature T_g.

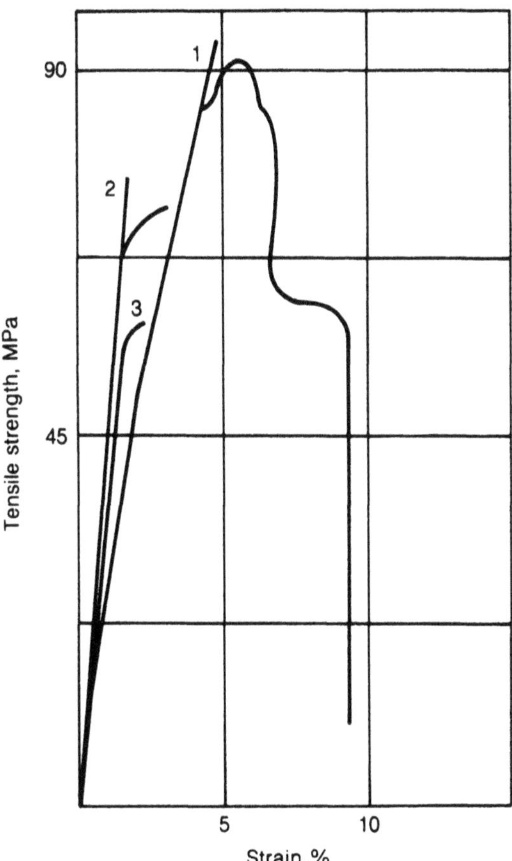

Fig. 3.1 Typical stress–strain (σ_b–ε) curves of cured binders: 1, polysulphone PSN; 2, epoxy VS-2526; 3, oligo(ether maleimide) OMI.

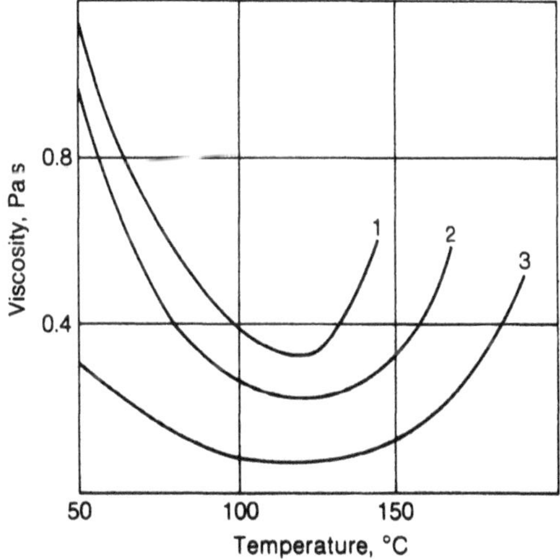

Fig. 3.2 Viscosity of binder melts as a function of temperature: 1, UNDF-4; 2, VS-2526; 3, OMI.

Table 3.6 Technological properties of binders

Binder	Prepreg pot-life at $T = 20°C$ (months)	Moulding temperature and time (°C/h)	Heat treatment temperature and time (°C/h)	Specific pressure (atm)
UNDF-4a	2	170/3	Not required	3–5
EDT-69	4	120/3	Not required	1–5
VS-2526m	1	170/3	190/2	5–7
ENFB	12	160/4	Not required	3–5
OMI	6	250/4	270/2	5–7
PAIS-104	12	250/5	Not required	7–10
	24	180/0.5	250/2	7–10
PSN	24	280/0.5	Not required	15–20

binders. The specific role of filler, associated with fibre surface condition, should be noted while discussing the problems of composite processibility and economic efficiency, which are considered to be fully dependent on matrix composition and its physicomechanical and physicochemical characteristics.

Filler surface treatment is known to allow considerable increase of CFRP compression and interlayer shear strength; the mechanism of this phenomenon is described in detail. The effect of filler activation treatment on matrix curing process is also of great interest. Figure 3.3 shows as an example the reactivity change of VS-2526 m epoxy binder, applied on UKN-P carbon-fibre surface.

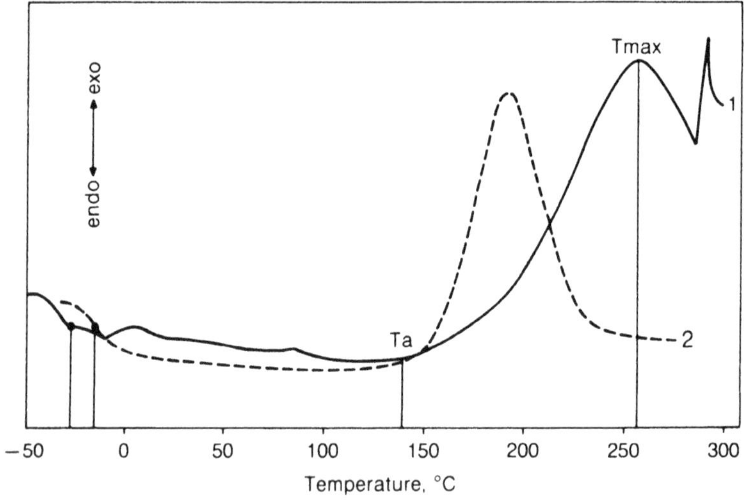

Fig. 3.3 DSC curves for curing of epoxy VS-2526 binder (1) and UKN-P carbon tow-based prepreg (2).

This fact should be taken into consideration when selecting optimum conditions for composite moulding.

3.3 EPOXY CARBON-FIBRE-REINFORCED PLASTICS

3.3.1 Production technology and basic physical and mechanical properties

The physicomechanical properties of some medium- and high-modulus epoxy CFRP are listed in Tables 3.7 and 3.8.

CFRP are made of prepregs, which comprise unidirectional tapes and fabrics, impregnated with the binder melt or solution.

The texture form of carbon tow (separate filaments, wound on bobbins), its higher brittleness compared to organic and glass fibres and high property anisotropy predetermined the necessity of developing a new technological process compared to woven fillers and special equipment for prepreg production.

The essence of the method for producing rolled-out tapes—prepregs of carbon tow involves the combination of reinforcing filler with thermosetting binder, free of volatile organic solvents, on heated calender rolls with application of definite pressure.

The technology of prepreg production by this method consists of the following: a thin binder layer, the thickness of which is defined by the gap between the spinneret and paper substrate with anti-adhesion coating, is applied on moving paper substrate having anti-adhesion coating. The stack obtained by this method is placed into the heated calender, and here filler impregnation with the required filler content is performed, i.e. prepreg formation. Impregnation quality is defined by the size of the gap between the hot calender rolls.

The formed prepreg, coated with anti-adhesion substrates on both sides, is removed from the heated calender to the drying chamber for postcure, defining its subsequent processability (tackiness).

Final prepreg formation by thickness is carried out on a cooled calender. The prepreg formed on paper anti-adhesion substrate is covered from the top side with polyethylene film and is advanced onto a receiving bobbin. Technological parameters (temperature in the main regions, carrying substrate movement rate) of producing prepreg based on carbon thread and binder melts have been worked out in order to ensure optimum prepreg technological characteristics: binder build-up, monolayer thickness, soluble fraction percentage and tackiness.

Conditions for filler and binder combination require that melt viscosity should be 3–10 Pa s. Only in this case can uniform, high-quality filler impregnation over the whole volume, and its good rolling out to required monolayer thickness with strictly regulated binder build-up (38–42 wt%), be achieved.

Table 3.9 shows properties of prepregs containing 39–40% resin (tackiness, yielding) as a function of storage time. Some decrease of conditional yielding and tackiness, and monolayer thickness increase do not exclude the possibility of using UNDF-4a binder-based prepreg after 40 days of storage at room temperature, while the pot-life of VS-2526 m binder-based prepreg is 25 days. Pot-life of both semiproducts

Table 3.7 Mechanical properties of carbon-fibre-reinforced plastics based on UNDF-4a epoxy binder

Properties	Load application direction (deg)	Filler ($V_f = 62 \pm 2\%$)					
		UKN-P		UOL-300		UT-900	
		22°C	100°C	22°C	100°C	22°C	100°C
Ultimate tensile	0	1500	1300	1350	1100	650	620
strength (MPa)	90	34	30	28	21	630	–
Elasticity modulus	0	140	135	125	115	68	59
(GPa)	90	7.1	5.3	7.7	4.74	63	–
Ultimate compressive	0	1200	950	1000	770	600	475
strength (MPa)	90	170	141	160	110	600	–
Compressive elasticity	0	132	115	98	90.4	65	55
modulus (GPa)	90	8.0	6.5	9.0	8.6	4.5	–
Ultimate bending	0	2000	1700	1770	1170	790	620
strength (MPa)	90	63	57	61	41	820	–
Bending elasticity	0	140	130	118	102	62	49.5
modulus (GPa)	90	9	8.2	8.7	7.4	60.5	–
Ultimate interlaminar shear strength (short beam method) (MPa)	0	85	60	78	55	62	58
Shear modulus in-plane (GPa)		5.2	4.5	5.2	3.2	6.9	5.8
Ultimate relative	0	1.1	1.05	1.08	0.96	0.9	0.9
elongation (%)	90	0.48	0.57	0.37	0.54	0.9	0.9
Poisson's ratio		0.42	0.36	0.33	0.28	0.07	–

Table 3.8 Mechanical properties of VS-2526 epoxy binder-based CFRP

| | | Filler ($V_f = 62 \pm 2\%$) | | | | | | | | |
| | | LU-24P | | UOL-300 | | UT-900 | | UKN-P | | UKN-P/0.1 | |
Properties	Load application direction (deg)	22°C	150°C	22°C	150°C	22°C	150°C	22°C	150°C	22°C	150°C
Ultimate tensile	0	970	850	1540	1360	640	580	1650	1340	2050	1600
strength σ_b (MPa)	90	25	16	34	28	630	520	34	26	36	288
Elasticity modulus,	0	215	195	133	118	68	63	146	132	156	140
E (GPa)	90	6.4	4.6	8.8	7.0	67	59	9.2	5.3	–	–
Ultimate compressive	0	750	650	1210	860	640	460	1350	1140	1500	1150
strength, σ_{comp} (MPa)	90	190	170	200	160	630	430	230	160	235	160
Compressive elasticity	0	170	155	118	104	61	55	125	115	145	115
modulus, E_{comp} (GPa)	90	6.8	5.5	9.5	7.4	60	52	10	7.0	10	7
Ultimate interlaminar shear strength (short beam method), τ_{shear} (MPa)	0	62	41	78	58	52	42	89	58	85	60
Shear modulus in-plane, G (GPa)		–	–	5.9	4.4	8.0	7.0	5.2	5.6	4.0	4.0
Ultimate relative	0	0.44	–	1.12	–	1.05	–	1.1	1.25	1.1	1.1
elongation, ε (%)	90	0.45	–	0.43	–	0.95	–	0.34	0.38	0.52	0.52
Poisson's ratio, μ		0.26	–	0.36	–	0.07	–	0.38	0.27	–	–

Table 3.9 Technological properties variation for UKN-P/5000 tow based prepreg as a function of storage time

	Storage time at $T = °C$ *(days)*				
Properties	Initial	10	25	30	40
VS-2526 binder ($V_f = 39\%$)					
Tackiness (mm)	200	200	205	220	300
Condition of yield resin (%)	16	16	14	11	8
Monolayer thickness (mm)	0.15	0.15	0.151	0.155	0.17
UNDF-4a ($V_f = 40\%$)					
Tackiness (mm)	250	250	255	265	310
Condition of yield resin (%)	14	14	14	13	12
Monolayer thickness (mm)	0.15	0.15	0.15	0.151	0.16

is increased to 4–6 months on storing them in sealed packages at a temperature of $-18°C$. Both CFRP are cured at a temperature of $170°C$ and a pressure of 5–7 atm for 3 h. However, owing to the low viscosity of VS-2526m binder melt, the material requires a complex moulding cycle (Fig. 3.4), including preliminary curing, and its serviceability at $150°C$ is ensured by additional heat treatment at $190°C$ outside the autoclave for 2 h. The viscosity of UNDF-4a binder allows one to apply maximum pressure at the beginning of the curing cycle (Fig. 3.5). UNDF-4a binder-based CFRP

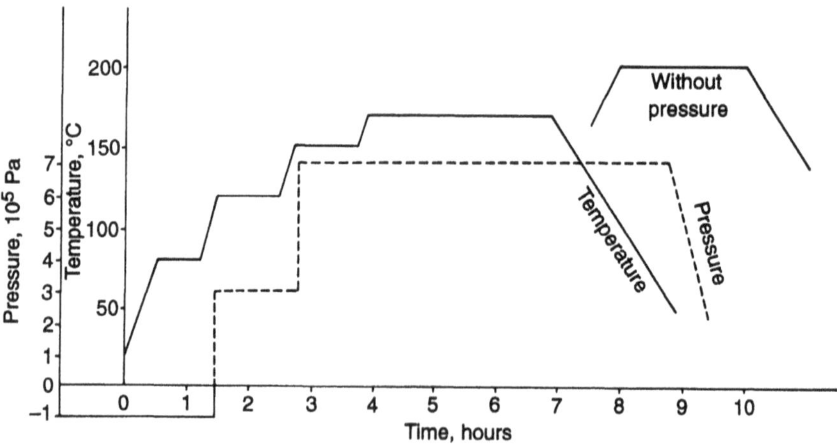

Fig. 3.4 Curing conditions of VS-2526 binder-based CFRP.

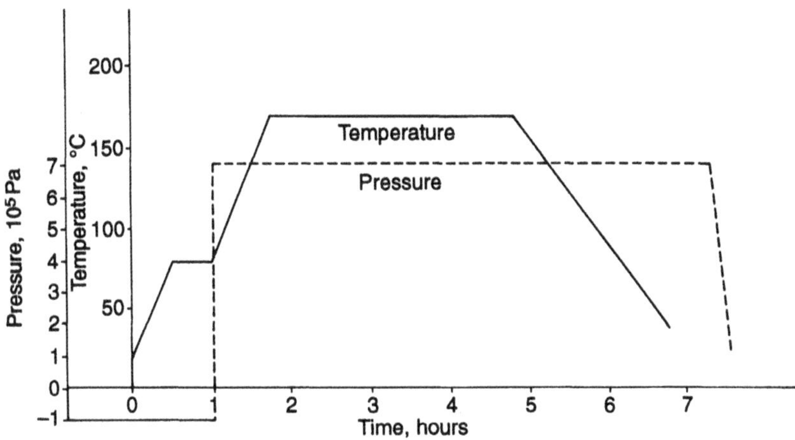

Fig. 3.5 Curing conditions of UNDF binder-based CFRP.

has lower glass transition temperature T_g compared to VS-2526m binder-based composition (160–190°C, respectively). Above 120°C the strength characteristics of UNDF-4a matrix-based CFRP decrease much more rapidly (Figs 3.6 to 3.9).

Table 3.10 compares the properties of two unidirectional composites based on UKN-P carbon tow in combination with epoxy VS-2526 and oligo(ether maleimide) OMI-P binders, produced by continuous drawing of the filler, impregnated with the binder melt through heated shape-forming spinneret-pultrusion. This method is used for making shapes of wide variety. Strict fibre orientation and composite curing under

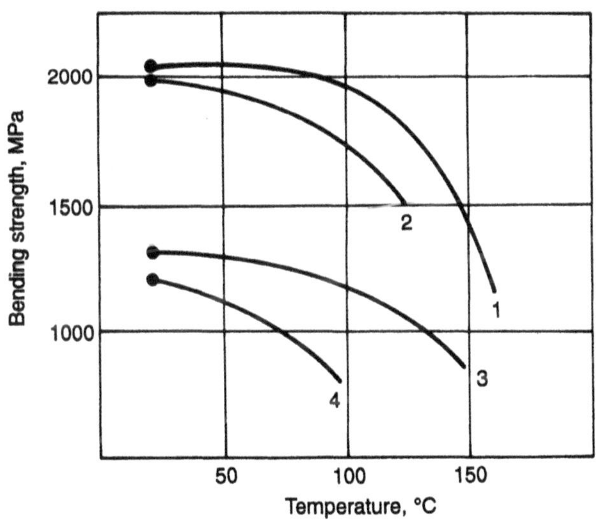

Fig. 3.6 Bending strength σ_{bend} of unidirectional epoxy CFRP as a function of temperature: 1, UKN-P0,1/VS-2526; 2, UKN-P/UNDF-4a; 3, ELUR-P/ENFB; 4, ELUR-P/EDT-69.

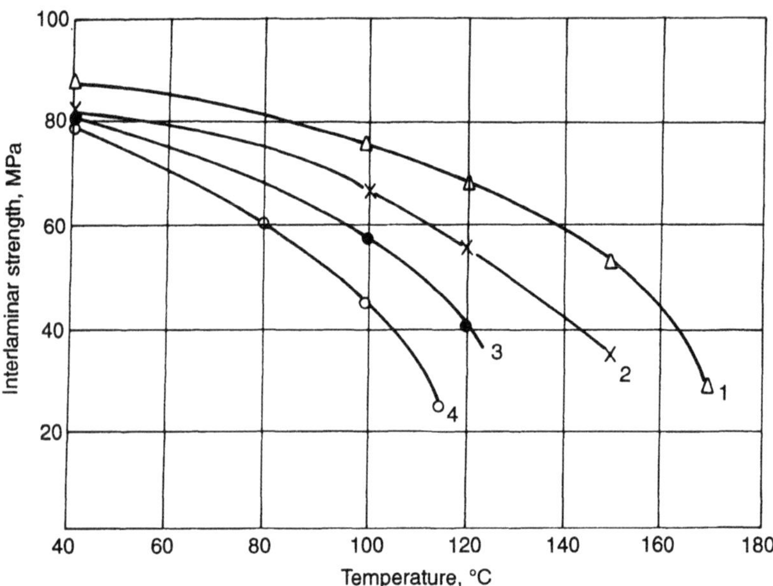

Fig. 3.7 Interlaminar shear strength τ of unidirectional epoxy CFRP as a function of temperature (short beam method): 1, UKN-P0,1/VS-2526; 2, ELUR-P/ENFB; 3, UKN-P/UNDF-4a; 4, ELUR-P/EDT-69.

tension allow fuller realization of carbon-fibre strength and stiffness characteristics, increase of CFRP property stability, and reduction of tensile strength coefficient of variation by almost two times, compared to the characteristics of composites processed by other methods. Resin formulation is optimized from the viewpoint of curing rate, which exceeds by several tens of times the curing rates when using traditional production methods, and shrinkage; the necessity of introducing modifying additions into the binder, reducing material friction over spinneret walls, is a complicated task in this process.

CFRP based on thin ELUR carbon tapes and modified ENFB and EDT-69 epoxy binders are now widely used for producing medium-loaded structural components. Both materials are highly technologically efficient, and they are manufactured and processed by prepreg technology.

Prepregs for ENFB-based CFRP exceed all this class of materials known in terms of pot-life, which is equal to 12 months at the ambient temperature in production premises of 15–25°C. Pot-life of prepreg for EDT-69 binder-based CFRP is 6 months, when storing them under similar conditions. Texture characteristics of 0.08–0.1 mm thick carbon tapes guarantee good shape stability of prepregs during manual handling and automated lay-up. Controlled tackiness of prepregs ensures rapid and easy lay-up of large-sized and complex-shaped components. Sheet prepregs first compacted at elevated pressure and temperature with different reinforcement schemes suitable for subsequent final moulding can be manufactured of ENFB binder-based prepregs.

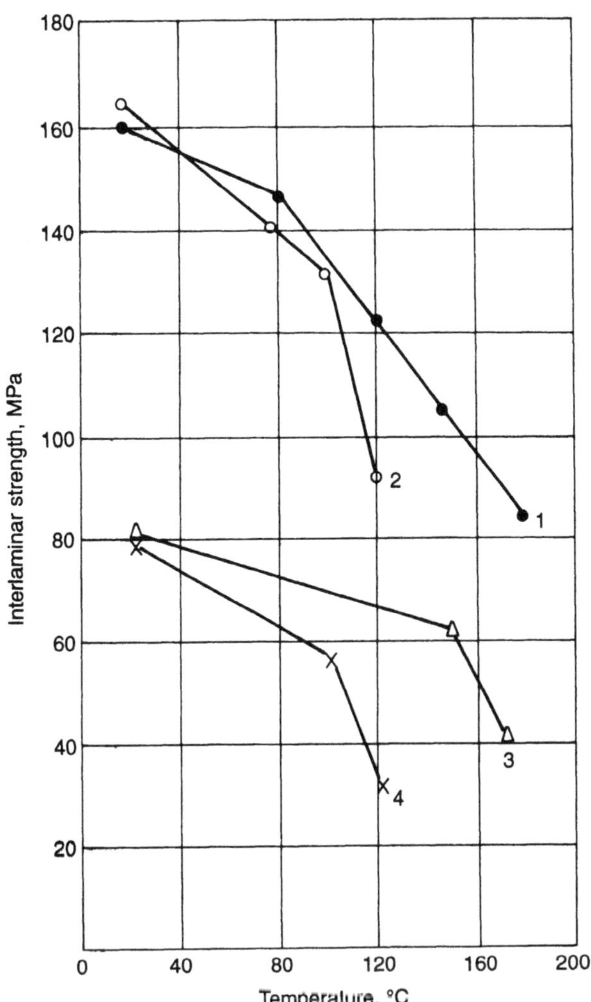

Fig. 3.8 Shear strength in-plane of epoxy CFRP as a function of temperature: 1, UKN-P0,1/VS-2526 (lay-up ± 45°, 0°, 90°); 2, UKN-P/UNDF-4a (lay-up ± 45°, 0°, 90°); 3, UKN-P0,1/VS-2526 (lay-up 0°); 4, UKN-P/UNDF-4a (lay-up 0°).

Materials are processed by autoclave and vacuum moulding, direct pressing winding, etc.

Final curing temperature and time for ELUR + EDT-69 CFRP are 160 and 125°C, 4 and 2 h, respectively, and service temperature is −60 to +150°C and −60 to +80°C, respectively.

Properties of tape filler and ENFB binder-based CFRP are listed in Tables 3.11 to 3.13.

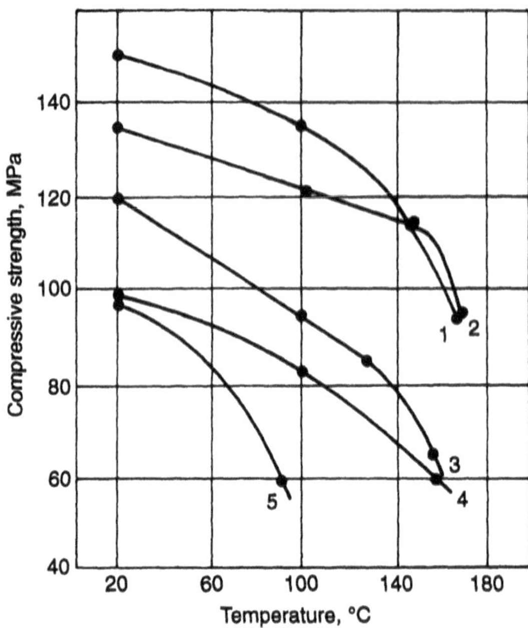

Fig. 3.9 Compressive strength of unidirectional epoxy CFRP as a function of temperature: 1, UKN-P0,1/VS-2526; 2, UKN-P/VS-2526; 3, UKN-P/UNDF-4a; 4, ELUR-P/ENFB; 5, ELUR-P/EDT-69.

Table 3.10 Mechanical properties of CFRP, produced by pultrusion method, for two binders

| | | Filler: UKN-P (V_f = 60%) | | | |
| | Load appli-cation di-rection (deg) | Oligo(ether bismaleimide) OMI | | Epoxy VS-2526 | |
Properties		22°C	250°C	22°C	150°C
Ultimate tensile	0	1550	1130	1650	1300
strength, σ_b (MPa)	90	26	10	30.5	19.5
Tensile elasticity	0	140	131	143	135
modulus, E (GPa)	90	7.6	6.7	9.0	4.7
Ultimate compressive	0	1000	770	1200	780
strength, σ_{comp} (MPa)	90	99	68	166	110
Compressive elasticity	0	110	90	135	115
modulus, E_{comp} (GPa)	90	7.6	6.7	7.8	6.3
Bending elasticity	0	–	–	130	107
modulus, E_b (GPa)	90	–	–	7.9	6.0
Ultimate interlaminar shear strength (short beam method), G (MPa)	0	60	32	82	47
Shear modulus in-plane (Gpa)		5.2	4.2	6.4	3.8
Ultimate relative elon-gation, ε (%)	0	0.93	0.77	1.2	1.1
	90	0.41	0.20	0.43	0.46
Poisson's ratio		0.36	–	0.36	0.32

Table 3.11 Properties of unidirectional CFRP based on ELUR-P tape filler and ENFB binder

Properties	Load application direction (deg)	$T_{test}(°C)$	
		20	*150*
Physical			
Density ($kg\,m^{-3}$)		1530	–
Moisture absorption for 24 h (%)		0.2	–
Mechanical			
σ_b (MPa)	0	1050	920
	90	24	19
E (GPa)	0	120	118
	90	9.35	8.25
σ_{comp} (MPa)	0	1000	870
	90	95	72
E_{comp} (GPa)	0	120	90
τ (short beam) (MPa)	0	80	62
G (GPa)	0	6.5	2.3
σ_{bend} (MPa)	0	1350	1050
E_b (GPa)	0	110	900
Poisson's ratio		0.265	–
Thermophysical			
Thermal conductivity ($W\,m^{-1}\,K^{-1}$)	0	0.63–0.82	
Thermal diffusivity ($10^{-7}\,m^2\,s^{-1}$)	0	5.1	
Heat capacity		0.85	
Monolayer thickness (mm)		0.09	

Table 3.12 Properties of ELUR-P/ENFB CFRP with different reinforcement schemes

Reinforcement structure	Tensile strength, σ_b (MPa)	Characteristics	
		Tensile elasticity modulus, E (GPa)	Compressive strength, σ_{comp} (MPa)
$(0°, 90°, 0°)_4$	620	90	540
$(\pm 45°, 0°, \pm 45°)_2$	280	39	360
$(0°, +30°, -30°, 0°)_2$	530	76	570
$(0°, +45°, -45°, 0°)_2$	510	71	530
$(\pm 45°)_4$	125	14	180

Table 3.13 Elastic strength properties of ELUR-P/ENFB CFRP with typical lay-up ($\pm 45°$, $0°$, $90°$)

Properties	$T_{test}(°C)$	
	20	150
Ultimate tensile strength, σ_b (MPa)	310	300
Elasticity tensile modulus, E (GPa)	47	46
Ultimate compressive strength, σ_{comp} (MPa)	440	380
Elasticity compressive modulus, E_{comp} (GPa)	48	35
Ultimate shear strength in-plane (MPa)	180	120
Shear modulus in-plane, G (Pa)	16	13
Poisson's ratio	0.26	
Crack resistance ($N\,mm^{-3/2}$)		
tensile	800	820
compression	850	440
shear	900	570

3.3.2 Change of properties due to effects of moisture, loading and temperature

Study of CFRP behaviour in different media, under the action of temperature and mechanical loading, is necessary for rational usage of available CFRP and development of new structural CFRP with stable physicomechanical properties.

It is known that the effects of different media on composite properties depend on matrix and filler nature, interface condition and material quality. In particular, CFRP moisture absorption is defined by three simultaneous processes: moisture absorption by the binder, accumulation in the material pores and absorption by the matrix–fibre interface (the latter factor being of prime importance).

Moisture in CFRP does not affect carbon-fibre properties, which is why composite mechanical characteristics, defined mainly by fibre properties, do not change considerably with moisture concentration variation in the material. However, parameters that depend on polymeric matrix properties and interface condition (tensile strength in transverse direction, interlaminar shear and compression strength) are considerably decreased.

Of interest are the results of studying the dependences of epoxy CFRP mechanical characteristics at different temperatures on water and kerosene content after long-term exposure in heated media and in the process of preliminary and simultaneous effects of loading and medium.

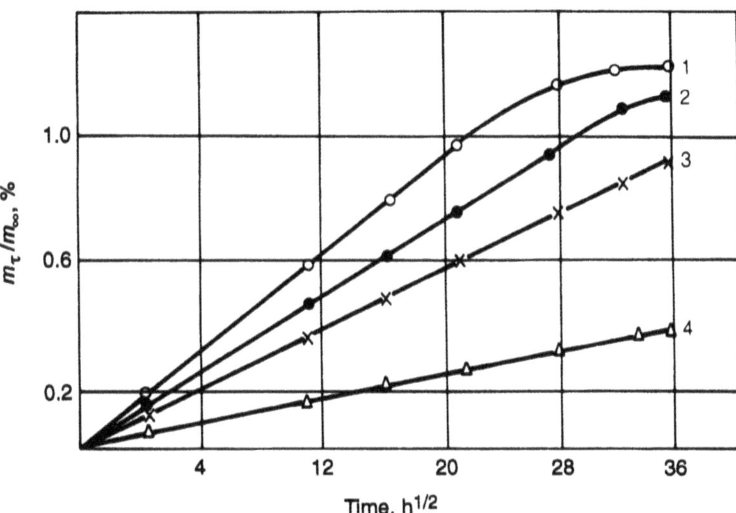

Fig. 3.10 Water sorption kinetics of ELUR/ENFB CFRP (specimen weight increase m_τ/m_∞) as a function of time and water temperature: 1, $T = 80°C$; 2, $T = 60°C$; 3, $T = 40°C$; 4, $T = 20°C$.

Bidirectional (0°/90°), ($\pm$ 45°), unidirectional and complex reinforced (0°/90°/$\pm$ 45°) CFRP of 1.6–2.0 mm thickness were used for testing suspended specimens with unprotected edges immersed into distilled water or kerosene, heated to the desired temperature and exposed for 1000 h. The effect of these media was estimated by variation of mechanical characteristics: shear, compression and bending strength, maximum liquid absorption and its diffusion coefficient into the material.

Figure 3.10 shows typical water sorption kinetic curves of CFRP, the linear character of which is the range $m_\tau/m_\infty = 0.5–0.7$ (m_τ and m_∞ are the amount of substance at the present moment and in the saturation state) indicates that up to this value water sorption by CFRP does not depend on its concentration in the material. Sorption isotherms asymptotically approach the equilibrium value of water absorption. Maximum water absorption is independent of temperature, but the time required to achieve it is closely connected with the temperature.

Analysis of Table 3.14 shows that, with temperature increase from 20 to 80°C, the diffusion coefficient of CFRP reinforced with UKN-P carbon fibres with activated surface is increased by 10–11 times. For CFRP reinforced with untreated UKN fibres, this coefficient is weakly dependent on temperature. Porosity is the key factor in the absorption process.

Table 3.15 and Figs 3.11 to 3.14 show the change of epoxy CFRP mechanical characteristics as functions of exposure time and water temperature. The shear modulus of moisture-saturated specimens decreases over the whole temperature range studied (Fig. 3.11). Plasticizing effect of water is manifested in the fact that the glass transition

Table 3.14 Coefficients of water and kerosene diffusion ($D_x \times 10^9\,\mathrm{cm^2\,s^{-1}}$) into epoxy CFRP as a function of temperature

	Water				Kerosene	
Material[a]	*20°C*	*40°C*	*60°C*	*80°C*	*20°C*	*80°C*
ELUR/ENFB, $V_f = 57\%$, porosity $= 1.2\%$	1.47	4.51	9.02	16.12	0.65	1.51
UKN-P/VS-2526, $V_f = 64\%$, porosity $= 1.5\%$	2.3	6.1	14	20.1	0.7	1.9
UKN/VS-2526, $V_f = 62\%$, porosity $= 3.2\%$	3.6	6.4	15	26	0.81	2.0

[a] V_f = volume fraction of fibre.

Table 3.15 Water exposure effect on elastic strength characteristics of unidirectional ELUR/ENFB CFRP (tests at 20°C)

	Property retention (%) after exposure in water for 1000 h			
Properties	*20°C*	*40°C*	*60°C*	*80°C*
σ_{comp} (MPa)	96.5	92.3	90	85
σ_b (MPa)	100	98	95	92
σ_{bend} (MPa)	95	93	90	87
E_{bend}	100	100	95	90
	98	92	89	87

temperature range is shifted to lower values, from 160 to 115°C. After drying, T_g and mechanical characteristics return to their initial state.

Irreversible property losses (8–10%) were observed for specimens exposed in water, heated to 80°C.

Isotherms of compression (Fig. 3.14) and shear (Fig. 3.12) strength indicate that, with water temperature increase, CFRP heat resistance is decreased to a greater extent

Fig. 3.12 Moisture effect on ultimate shear strength of unidirectional UKN-P/VS-2526 CFRP at different temperatures: 1, dry specimen; 2, wet specimen exposed for 1000 h at relative humidity $\phi = 98\%$ and temperature $T = 70°C$, moisture content $W = 1.56\%$; 3, wet specimen exposed for 72 h in boiling water, moisture content $W = 1.54\%$.

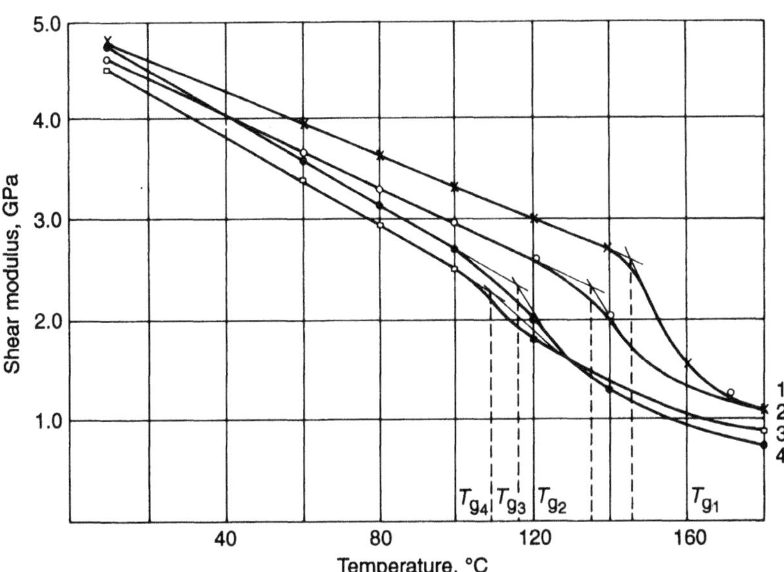

Fig. 3.11 Shear modulus as a function of temperature after specimen exposure in water for ELUR/ENFB CFRP: 1, dry; 2, wet (water temperature $T = 40°C$, exposure time $t = 1600\,h$, moisture content $W = 1.34\%$); 3, wet ($T = 60°C$, $t = 1200\,h$, $W = 1.35\%$);4, wet ($T = 80°C$, $t = 1080\,h$, $W = 1.4\%$).

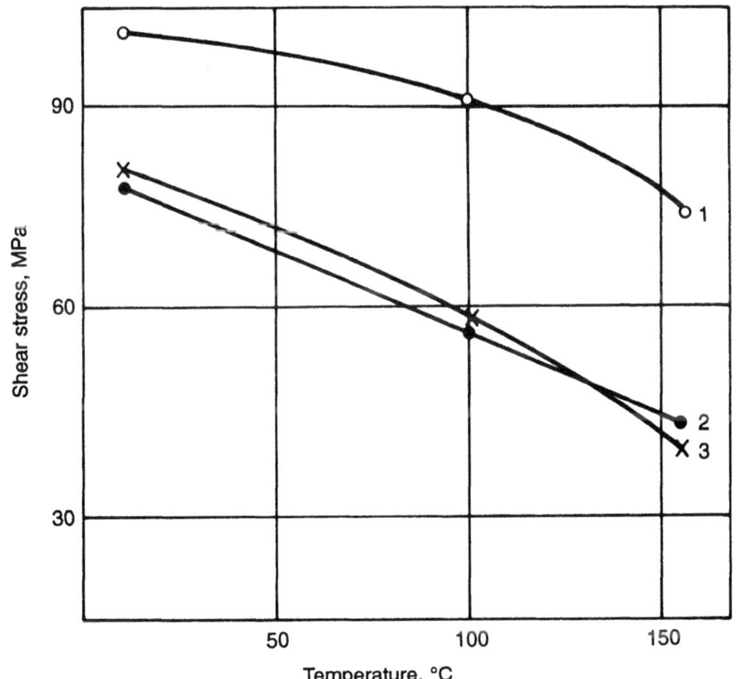

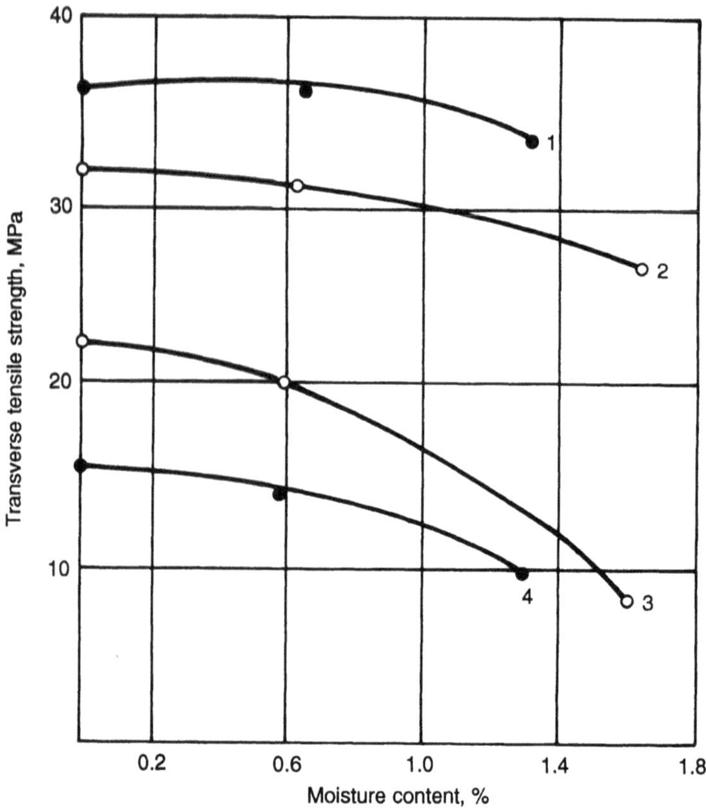

Fig. 3.13 Variation of ultimate tensile strength in the 90° direction for unidirectional CFRP as a function of moisture content (specimens were exposed at $\phi = 98\%$, temperature = 70°C for 1000 h): 1, ELUR/ENFB (test temperature = 20°C); 2, UKN-P/VS-2526; 3, UKN-P/VS-2526 (test temperature = 150°C); 4, ELUR/ENFB.

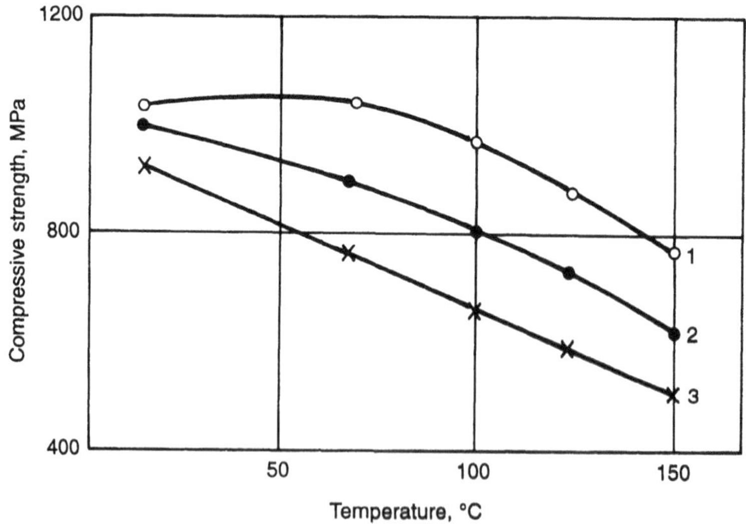

Table 3.16 Elastic strength characteristics of unidirectional CFRP as a function of exposure time in kerosene heated to 120°C

Material	Properties	Test temperature (°C)	Property change (%) after exposure in kerosene at $T = 120°C$			
			50 h	100 h	500 h	1000 h
ELUR/ENFB,	σ_{bend}	20	105	118	110	108
$V_f = 57\%$		150	102	106	108	108
	E_{bend}	20	98	98	97	96
		150	96	93	91	91
UKN-P/VS-2526,	σ_{bend}	20	102	105	105	100
$V_f = 64\%$		150	103	104	103	100
	E_{bend}	20	96	96	100	100
		150	93	91	90	90
UKN-P/UNDF-4a,	σ_{bend}	20	111	115	115	110
$V_f = 62\%$		100	107	109	103	105
	E_{bend}	20	100	100	100	96
		100	98	98	94	92

at similar levels of its content in specimens, which is likely to be associated not only with epoxy binder plasticization but also with fracture processes in the material.

Safe moisture content is 0.5–0.6%.

The effect of long-term exposure in heated kerosene on CFRP mechanical characteristics is shown in Table 3.16. Diffusion coefficients, calculated for kerosene, vary from 0.65×10^{-9} to $1 \times 10^{-9}\,cm^2\,s^{-1}$ at 20°C and are increased by 1.5–5 times with temperature increase to 80°C. Expected strength reduction, proportional to the increase of specimen mass, was not observed. Subsequent kerosene analysis confirmed the absence of chemical interaction between the material and medium in the studied temperature range. Long-term exposure of CFRP in kerosene heated to 120°C even contributed to some increase of strength characteristics by 10–15%. A strength increment occurs during the first 50–100 h, after which material properties become stabilized. Elasticity bending modulus remains practically constant or it is slightly (by 5–7%) decreased. It can be supposed, by comparing material behaviour in the process of ageing in air (Table 3.17) and kerosene, that an inert liquid medium protects the material from oxidative degradation, by promoting residual stress release, and the process of polymer additional structurization proceeds under mild conditions.

Fig. 3.14 Compressive strength isotherms for ELUR/ENFB CFRP, exposed in water at 80 and 20°C: 1, dry specimen; 2, wet specimen (moisture content $W = 1.2\%$, water temperature $T = 20°C$, exposure time $t = 2800\,h$); 3, wet specimen ($W = 1.2\%$, $T = 80°C$, $t = 1000\,h$).

Table 3.17 Effect of long-term thermal cycling in air on epoxy CFRP properties

Material	Properties	Temperature(°C) Ageing	Test	Initial values (MPa)	Property retention (%) after the effect of T 100 h	500 h	1000 h
ELUR/ENFB,	σ_{comp}		20	1000	100	99	93
$V_f = 57\%$,		150	150	870	100	100	100
	τ (short		20	80	100	94	92
lay-up $(0°)_n$	beam)		150	67	100	100	92
	σ_{bend}		20	140	100	96	92
			150	100	100	100	92
UKN-P/VS-2526,	σ_{comp}		20	520	100	102	102
$V_f = 64\%$,		150	150	420	100	100	100
	τ		20	85	100	97	95
lay-up $(\pm 45°,$			150	65	100	98	98
$0°, 90°)_n$							
UKN-P/UNDF-4a,	σ_{comp}		20	450	100	100	100
$V_f = 62\%$,		120	100	380	100	100	100
	τ		20	80	100	97	94
lay-up $(\pm 45°,$			100	65	100	99	95
$0°, 90°)_n$	σ_{bend}		20	480	100	100	95
			100	400	100	100	98

Table 3.18 lists the results of studying the effect of thermal cycling, simulating seasonal ($\pm 60°C$) and daily ($\pm 25°C$) temperature gradients, conditionally equivalent to 5–20 years storage of CFRP in open sites, on their properties [8]. A regime of one cycle of $\pm 60°C$ ($-60°C$, 1 h; $+60°C$, 1 h) and 47 cycles of $\pm 25°C$ ($-25°C$, 1 h; $+25°C$, 1 h) at relative humidity of 95–98% simulated variation of seasonal and daily temperature gradients for one year [9, 10]. As CFRP structures in the process of service and storage can be affected by thermal–humidity complexes, resulting in moisture accumulation in the material, both dry and wet specimens have been subjected to thermal cycling. Material resistance to the effect of the factor studied was estimated by the change of Rx service characteristic, representing the ratio of its current value x at moment i to initial value x, expressed as a percentage:

$$Rx = (x_i/x_0) \times 100 \tag{3.1}$$

Decrease of mechanical characteristics for preliminarily wetted specimens, subjected to thermal cycling, conditionally equal to 20 years of storage, is 20% at the test temperature of 20°C, which is practically twice as much as for dry specimens.

In order to estimate the combined effect of water and preliminary loading, the latter was performed on dry specimens, which then were moisture-saturated and tested for tension [11].

Table 3.18 Effect of thermal cycling, simulating daily and seasonal temperature gradients, on epoxy CFRP properties

Material	Specimen condition	Conditionally equivalent storage life (year)	Affecting factors	T_{test} (°C)	Property retention (%)	
					$R\sigma_{bend}$	$R\sigma_{comp}$
ELUR/ENFB: Lay-up $(0°, 90°)_n$		5	$\pm 25°C$ $\phi = 95-98\%$	20 150	100 98	98 92
$\sigma_{bend}^{20} = 560\,MPa$	Dry	20	$\pm 60°C$ $\phi = 95-98\%$	20 150	100 91	91 90
$\sigma_{bend}^{150} = 460\,MPa$ $\sigma_{comp}^{20} = 570\,MPa$ $\sigma_{comp}^{150} = 450\,MPa$	Wet, $W = 1.45\%$	5	$\pm 25°C$ $\phi = 95-98\%$	20 150	97 93	88 80
$V_f = 56$ vol %		20	$\pm 60°C$ $\phi = 95-98\%$	20 150	93 90	80 80
UKN/UNDF-4a: Lay-up $(\pm 45°, 0°, 90°)_n$		5	$\pm 25°C$ $\phi = 95-98\%$	20 100	92 84	93 82
$\sigma_{bend}^{20} = 586\,MPa$ $\sigma_{bend}^{100} = 553\,MPa$ $\sigma_{comp}^{20} = 470\,MPa$ $\sigma_{comp}^{100} = 360\,MPa$ $V_f = 62$ vol %	Dry	10	$\pm 60°C$ $\phi = 95-98\%$	20 100	90 85	92 84
UKN-P/VS-2526: Lay-up $(\pm 45°, 0°, 90°)_n$ $\sigma_{bend}^{20} = 630\,MPa$ $\sigma_{bend}^{150} = 450\,MPa$ $\sigma_{comp}^{20} = 520\,MPa$ $\sigma_{comp}^{150} = 420\,MPa$ $V_f = 64$ vol %	dry	5 10	$\pm 25°C$ $\phi = 95-98\%$ $\pm 60°C$ $\phi = 95-98\%$	20 150 20 150	92 84 90 83	93 83 92 84

The stress level chosen for preliminary loading did not disturb the integrity of ELUR/ENFB CFRP with the layer orientation $\pm 45°$ and introduced no defects into its structure, which is shown by the dependence $D_x = f(\delta)$. On the contrary, with stress increase from 0.3 to $0.7\sigma_b$ monotonic decrease of diffusion coefficient from 1.34×10^{-9} to $0.8 \times 10^{-9}\,cm^2\,s^{-1}$ was observed, which is likely to be associated with the free-volume reduction.

The following points have been noted as a result of conducted experiments:

1. No effect of preliminary static loading on the material properties was observed on dry specimens at any temperatures.
2. No moisture effect at 20°C was observed in the absence of preliminary loading.
3. At high temperature presence of moisture results in drastic lowering of strain diagrams.
4. On wet specimens, preliminary loading leads to increase of strength values, maximum increase being obtained at the level of $0.5\sigma_b$ without the change of maximum strains.
5. When passing through the glass transition temperature, the effect of extent of moisture is increased and results in sharp strength reduction, preliminary loading effect being decreased in this case.

Figure 3.15 shows maximum stress values as a function of preliminary loading for different test temperatures. It is seen from the figure that the higher the temperature, the lower is the preliminary loading effect.

With temperature increase, plasticizing moisture effect on CFRP is observed, which

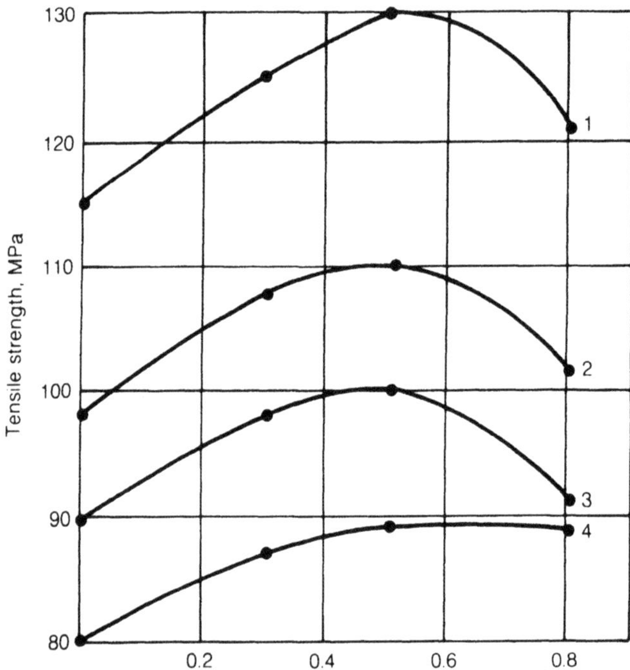

Fig. 3.15 Strength of moisture-saturated ELUR/ENFB CFRP specimens as a function of preliminary loading level at different temperatures (specimens with lay-up $(\pm 45°)_n$ were exposed in water at 80°C for 1000 h, moisture content $W = 1.20\%$): 1, test temperature $T = 20°$C; 2, $T = 100°$C; 3, $T = 125°$C; 4, $T = 150°$C.

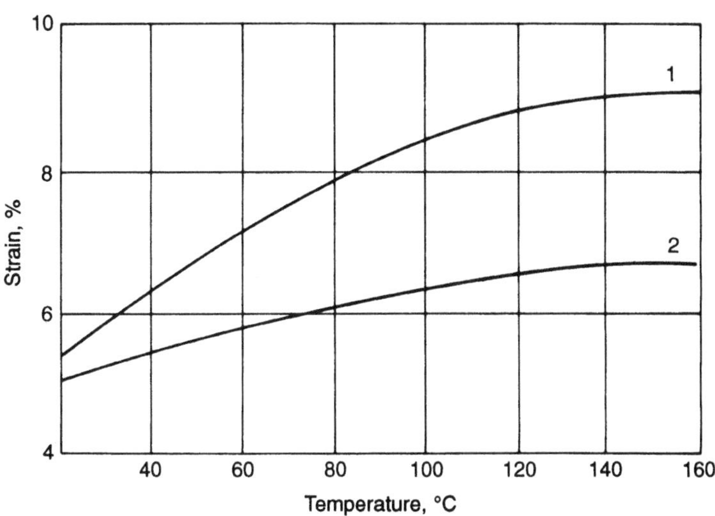

Fig. 3.16 Ultimate strain as a function of temperature for ELUR/ENFB CFRP with $(\pm 45°)_n$ lay-up for wet specimens $W = 1.20\%$ (curve 1) and for dry specimens (curve 2).

results in glass transition temperature decrease by 20–25°C and suppresses the positive effect of preliminary loading at elevated temperatures. The behaviour of epoxy–novolac CFRP, containing 1.2% moisture, at temperatures exceeding T_g is similar to that of polymer–plasticizer system in a high-elasticity state, when the presence of polar plasticizer results in a sharp decrease of molecular interaction energy and reduction of high-temperature strength and stiffness. As seen from Fig. 3.15, strain diagrams of wet specimens at 125 and 150°C are significantly lower compared to those corresponding to dry specimens.

Figure 3.16 shows the values of maximum strains as a function of temperature in the presence and absence of moisture.

The second interesting observation consists of the fact that, in the case of moisture saturation, preliminary loading makes strain diagrams higher. (Figures 3.17 and 3.18, in order not to make the figures too overloaded, show only diagrams for $\delta = 0.5$.) It can be noted that the diagrams at $T = 20°C$ are identical for cases $W = 0$ (at any level of preliminary loading) and $W = 1.2\%$ at $\delta = 0$. This means that at this temperature moisture affects the material properties only when it is first loaded. The strength increase in the temperature range from 20 to 80°C (lower than T_g) of preliminarily loaded moisture-saturated CFRP (Fig. 3.17) can be explained as follows: On applying mechanical loading, ordering of polymeric matrix chains takes place, owing to their contacts sliding relative to each other with simultaneous breakage of weak physical and chemical bonds. Moisture penetrating into the material by activated diffusion mechanism acts as the ingredient necessary for regeneration of broken bonds and formation of new non-chemical ones and also enhancement of molecular

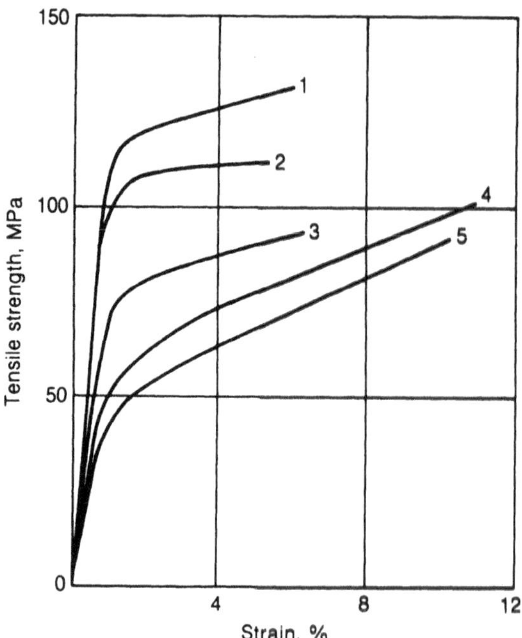

Fig. 3.17 Strain diagrams of ELUR/ENFB CFRP with $(\pm 45°)_n$ under the following conditions: 1, $T = 20°C$, $W = 1.2\%$, $\delta = 0.5\sigma_b$; 2, $T = 20°C$, dry specimen ($W = 0$) with any level of preliminary loading; 3, $T = 100°C$, $W = 0\%$, $\delta = 0.5\sigma_b$ (or any); 4, $T = 100°C$, $W = 1.2\%$, $\delta = 0.5\sigma_b$; 5, $T = 20°C$, $W = 1.2\%$, $\delta = 0$.

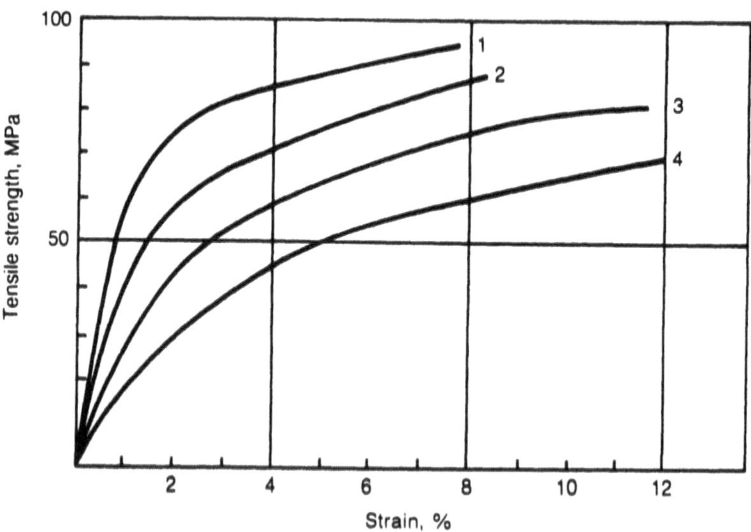

Fig. 3.18 Strain diagrams of ELUR/ENFB CFRP with $(\pm 45°)_n$ lay-up under the following conditions: 1, $T = 125°C$, $W = 0\%$, $\delta = 0.5$ (or any); 2, $T = 150°C$, $W = 0\%$, $\delta = 0.5$ (or any); 3, $T = 125°C$, $W = 1.2\%$, $\delta = 0$; 4, $T = 150°C$, $W = 1.2\%$, $\delta = 0$.

interaction efficiency. In the absence of moisture, all strain diagrams for different δ levels practically coincide.

Thus, it is seen that the presence of moisture significantly changes the material's mechanical behaviour. In this case, a positive effect is clearly manifested at a temperature lower than the glass transition temperature: the strength is being increased, especially in the case of preliminary loading to the level of $\delta = 0.5\sigma_b$. The glass transition temperature itself is shifted to the area of lower temperatures, and, when passing through it, the strength of moisture-saturated specimens is sharply decreased (Fig. 3.19).

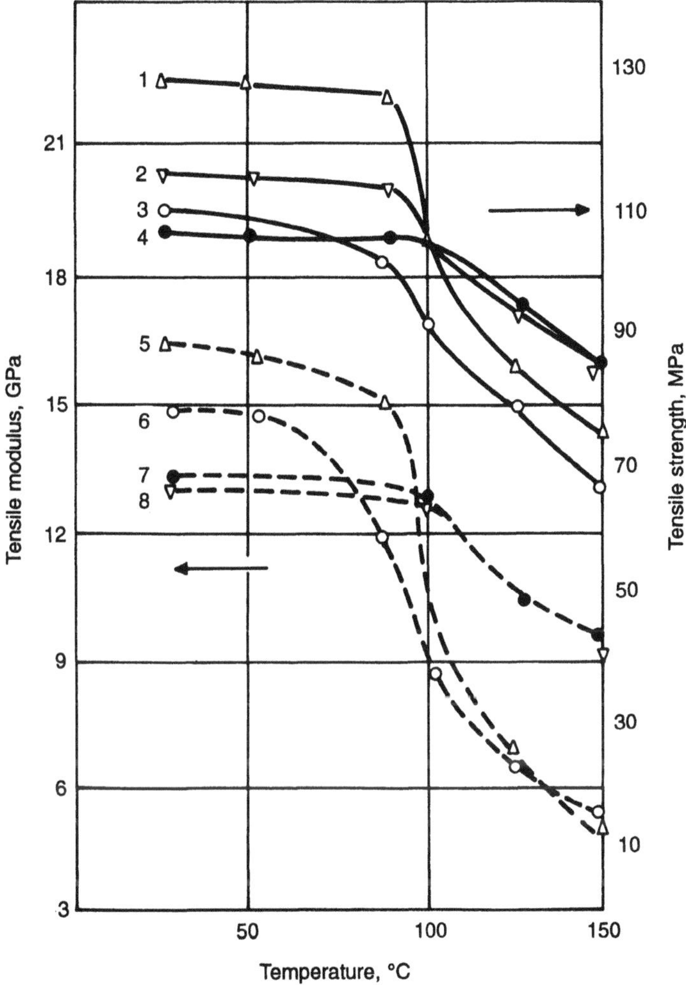

Fig. 3.19 Strength σ_b (curves 1–4) and elasticity modulus E (curves 5–8) for dry and moisture-saturated specimens of ELUR/ENFB CFRP with $(\pm 45°)_n$ lay-up as a function of temperature: 1 and 5, $\delta = 0.5\sigma_b$, $W = 1.2\%$; 2 and 8, $\delta = 0.5\sigma_b$, $W = 0\%$ (dry specimen); 3 and 6, $\delta = 0.5\sigma_b$, $W = 1.2\%$; 4 and 7, $\delta = 0$, $W = 0\%$ (dry specimen).

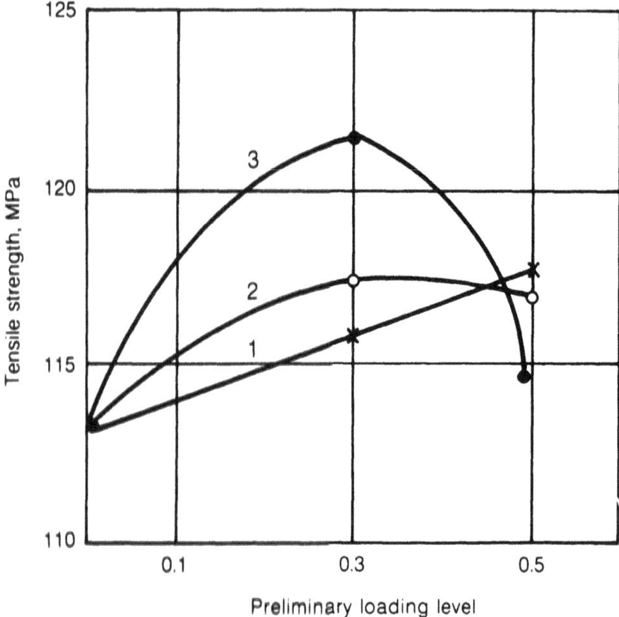

Fig. 3.20 Strength of wet specimens of ELUR/ENFB CFRP with $(\pm 45^{\circ})_n$ lay-up as a function of preliminary loading level on cyclic exposure with a frequency of 0.1 Hz: 1, $W = 1.2\%$; 2, $W = 0.8\%$; 3, $W = 0.6\%$.

The effect of moisture and preliminary loading on specimens of epoxy CFRP with layers lay-up of $\pm$ 45° was studied in [12]. Preliminary load levels δ at cyclic loading performed with the frequency of 0.1 Hz for 5 min were 0.3 and 0.5 σ_{max}. With moisture in the material, cyclic loading in all considered cases did not give a negative result. It should be noted that the strength of wet specimens was increased (Fig. 3.20), though the increase was less significant compared to single preliminary loading. Maximum strength value was shifted to the area of lower cyclic load amplitudes, owing to the additional process of damage accumulation, compared to single definition level loading, the effect of which was mainly reduced due to partial decrease of residual stresses and reorganization of wetted material structure.

The complex effect of medium and loading was studied in [13]. Tests for stress rupture and low-cycle fatigue of ELUR/ENFB CFRP under exposure to static and pulsed tensile load with frequency of 0.7 cycles per minute were performed in air and water.

Stresses sustained by specimens in different media for 1000 h were determined during these tests. The average strength value for specimens (Table 3.19) tested in water with $T = 20°C$ was equal to $0.75\sigma_b$. Scatter of values was not greater than 6.6%. Test results at normal temperature confirmed the absence of an effect due to water on tensile stress rupture of epoxy CFRP. Similar to short-term strength, water

Table 3.19 Water effect on tensile rupture stress of ELUR/ ENFB CFRP (lay-up $\pm$ 45°, 0°, 90°, 0°, $\pm$ 45°), volume fraction of fibres, $V_f = 56\%$

Test conditions	Rupture stress (MPa) after various times		
	100 h	*500 h*	*1000 h*
Air, $T = 20°C$	320	317	310
Water, $T = 20°C$	325	320	315
Air, $T = 80°C$	310	307	305
Water, $T = 80°C$	295	285	273

heating up to 80°C promoted the enhancement of its activation effect and decreased stress rupture level after 1000 h by 8–10%. Comparative low-cycle fatigue test results in water and air at 20 and 60°C on the basis of 50 000–70 000 cycles with subsequent determination of residual strength are tabulated in Table 3.20.

Water temperature increase to 80°C results in the reduction of fatigue and residual strength by 15–18%, compared to those for specimens tested in air (durability of

Table 3.20 Water effect on tensile low-cycle fatigue of ELUR/ENFB CFRP (lay-up $\pm$ 45°, 0°, 90°, 0°, $\pm$ 45°; $V_f = 56$ vol %)

Number of specimens	Stress (MPa)	Cycles (number)	Residual strength (MPa)	Number of specimens	Stress (MPa)	Cycles (number)	Residual strength (MPa)
	Air, $T = 20°C$, 312(0.75σ_b)				Air, $T = 80°C$, 312(0.75σ_b)		
1		63 000	437	15		58 000	441
2		58 810	421	16		62 000	408
3		65 000	415	17		50 000	412
4		61 000	420	18		54 000	403
5		54 000	446	19		65 000	415
6		60 000	440	20		57 000	398
7		53 000	425	21		50 000	410
	Air, $T = 20°C$, 312(0.75σ_b)				Air, $T = 80°C$, 312(0.75σ_b)		
8		61 000	436	22		728	fracture
9		55 000	424	23		979	fracture
10		50 000	425	24		650	fracture
11		58 000	418		260(0.60σ_b)		
12		61 000	421	25		50 000	340
13		51 000	413	26		68 000	315
14		50 000	420	27		72 000	318

CFRP tested in heated water at stress level of $0.75\sigma_b$ is decreased by 8–10 times); specimen fracture with delamination was observed. The latter indicates the reduction of adhesion strength between carbon fibre and matrix.

3.3.3 Fatigue and stress rupture

An obvious advantage of CFRP from the viewpoint of fatigue strength is their high endurance limit, exceeding by 2–2.5 times that of aluminium alloys on the basis of 10^7 cycles.

At the level of $0.6\sigma_b$ the majority of CFRP at normal temperature sustain the basis of 10^7 cycles without loss of initial static strength. For steel and aluminium the endurance limit on the basis of 10^7 cycles is $(0.3–0.4)\sigma_b$. Evidently, service life of CFRP structures can be calculated by endurance curves, i.e. calculation can be made for infinite service life, as is done in general machine building.

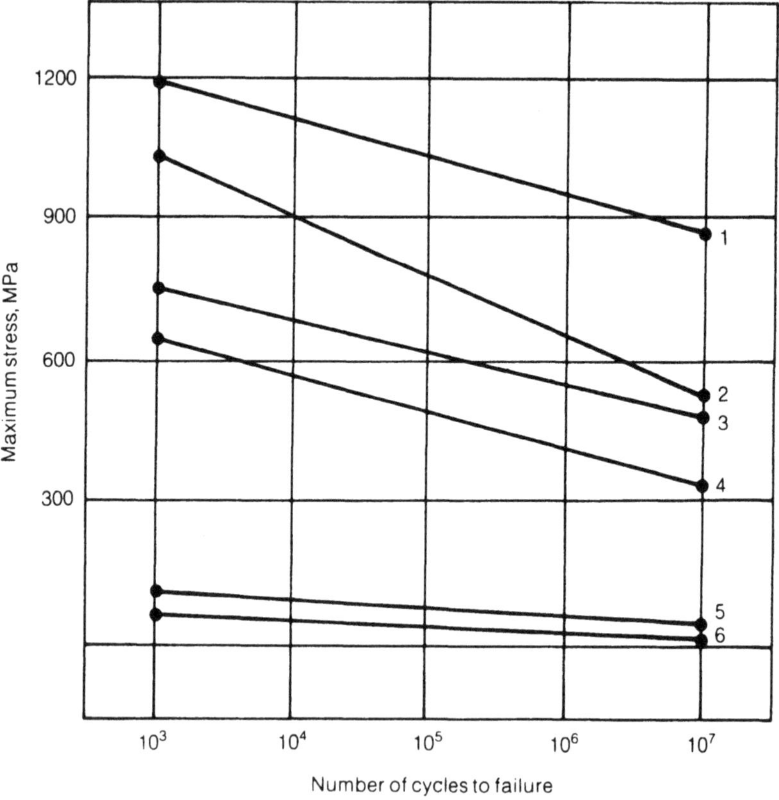

Fig. 3.21 Effect of reinforcement scheme on fatigue strength of UKN-P/VS-2526 (frequency 50 Hz) at $T = 20°C$: 1, $(0°)_n$, $R = 0–0.1$; 2, $(0°)_n$, $R = -1$; 3, $(0°, 90°)_n$, $R = 0–0.1$; 4, $(0°, 90°)_n$, $R = -1$; 5, $(\pm 45°)_n$, $R = 0–0.1$; 6, $(\pm 45°)_n$, $R = -1$.

CFRP fatigue strength is defined, to a great extent, by fibre orientation (Fig. 3.21). Unidirectional materials possess considerably higher fatigue properties, compared to cross-reinforced ones [14].

The latter endurance is greatly dependent upon CFRP transverse characteristics. The increase of transverse and tear strength, normal to the specimen plane, by 8–10 units results in a double increase of endurance limit (Fig. 3.22).

Fatigue tests of epoxy CFRP (Figs 3.23 and 3.24) at elevated ($T = 150°C$) and low ($T = -60°C$) temperatures have shown that negative temperatures do not lead to endurance and residual strength reduction. It is only the change in the failure mode of wet specimens in the process of further static tests that can be observed. Cooling of CFRP with moisture resulted in the increase of the material fracture zone, and made the fracture more loose.

Elevated temperature ($T = 150°C$) noticeably (by 15–20 times) reduces specimen durability at a loading level of $(0.6-0.7)\sigma_b$ on the basis of 10^7 cycles (Fig. 3.23). Such significant reduction of parameters at elevated temperatures is, first of all, associated with the change of specimen stiffness. The most dangerous is cyclic loading under conditions of 'compression' and 'tension–compression'. Fatigue curves for these loading modes are practically parallel to each other. Residual strength in this case is decreased to $0.5\sigma_{-1}$.

Epoxy CFRP are fairly insensitive to variation of loading frequency (Table 3.21). Frequency change from 200 to 2400 cycles per minute revealed no drastic reduction of endurance. One can only mention the certain tendency to endurance decrease at low loading frequencies.

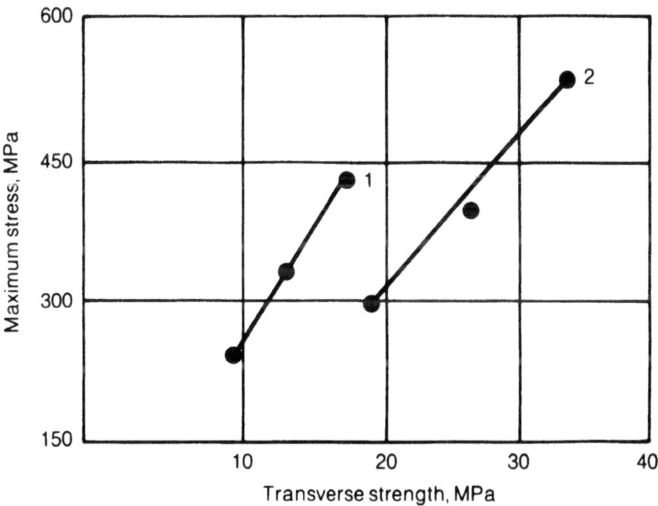

Fig. 3.22 Fatigue strength of epoxy CFRP with quasi-isotropic lay-up (with the same stiffness and strength) as a function of tear strength normal to specimen plane (curve 1) and transverse strength σ_{90} for unidirectional layer (curve 2).

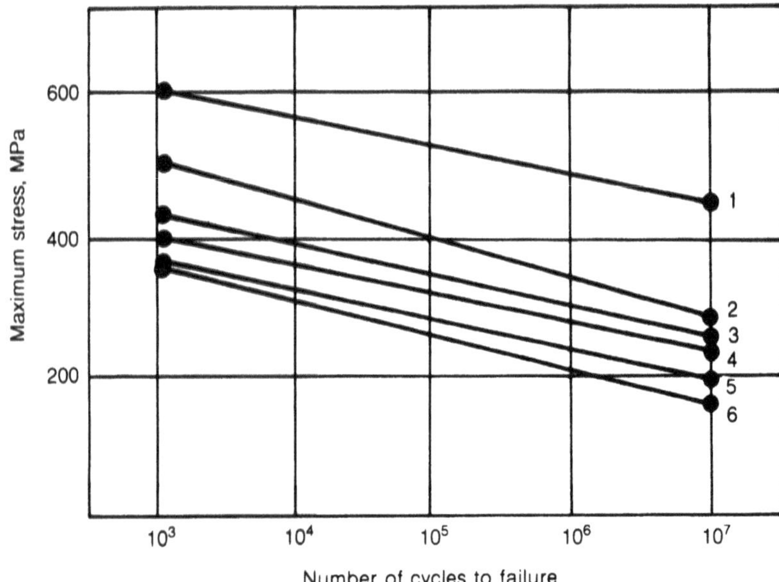

Fig. 3.23 Temperature effect on fatigue strength of UKN-P/VS-2526 CFRP with ($\pm$ 45°, 0°, 90°)$_n$ lay-up (frequency 50 Hz): 1, $T = 20°C$, $R = 0$–0.1; 2, $T = 20°C$, $R = \infty$; 3, $T = 20°C$, $R = -1$; 4, $T = 150°C$, $R = 0$–0.1; 5, $T = 150°C$, $R = \infty$; 6, $T = 150°C$, $R = -1$.

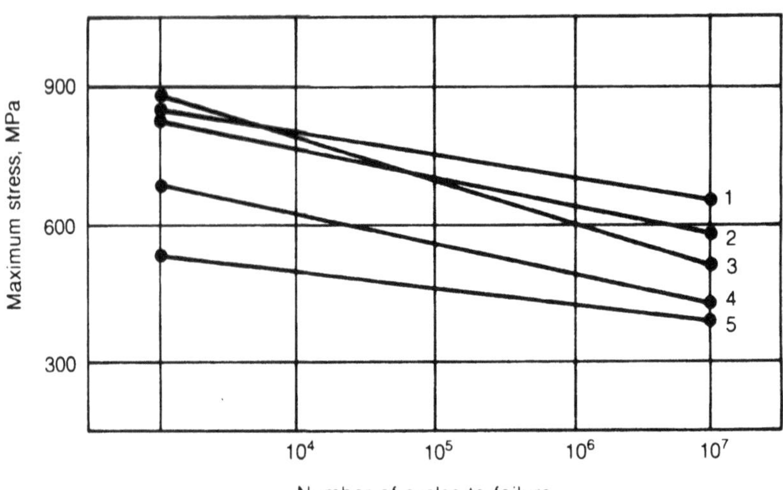

Fig. 3.24 Effect of temperature and moisture on bending fatigue strength of unidirectional ELUR/ENFB (frequency 50 Hz): 1, dry specimen, test temperature $T = 20°C$; 2, dry specimen, $T = -60°C$; 3, wet specimen, $W = 1.2\%$, $T = -60°C$; 4, dry specimen, $T = 150°C$; 5, wet specimen, $W = 1.2\%$, $T = 150°C$.

Table 3.21 Fatigue properties of ELUR/ENFB epoxy with lay-up ($0°$, $\pm 45°$, $90°$) as a function of frequency (cycle: tension–compression)

Cycles (number)	Fatigue strength (MPa)		
	$f = 2400\,c/min$	$f = 600\,c/min$	$f = 200\,c/min$
10^3	310	310	310
10^5	260	260	250
10^6	240	230	230
10^7	220	210	200

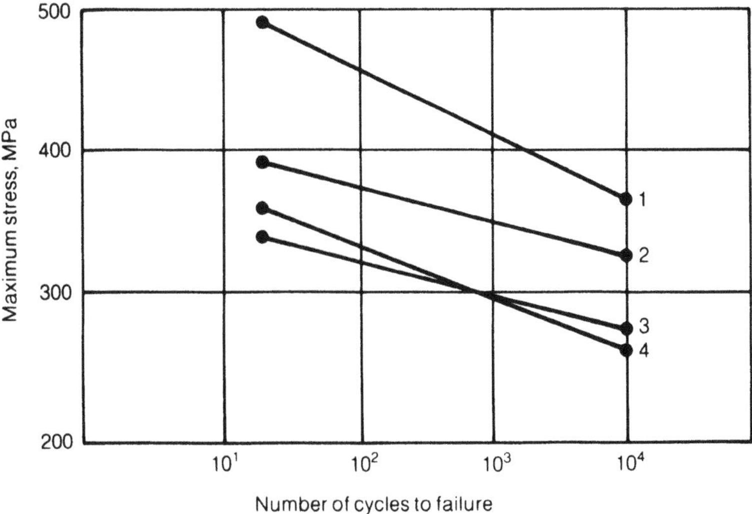

Fig. 3.25 Fatigue collapse ($R = 0–0.1$) of UT-900/VS-2526 carbon-fabric-reinforced plastic (curves 1 and 4) and UKN-P/VS-2526 (curves 2 and 3) with quasi-isotropic lay-up as a function of cycle number (frequency 0.2 Hz) at 20°C (curves 1 and 2) and at 150°C (curves 3 and 4).

Figure 3.25 shows the change of fatigue strain for CFRP with $(0°, 90°)_n$ lay-up and CFRP based on rolled out tape prepregs with quasi-isotropic structure. With the increase of test base to 10^5 cycles, the difference in the materials' behaviour is practically eliminated.

Tables 3.22 to 3.24 and Figs 3.26 and 3.27 present data on long-term static loading effect on CFRP behaviour. It is found that creep rate at the same temperature and stress depend, to a great extent, on the material reinforcement scheme: the presence of layers oriented parallel to the applied stress considerably retards the material creep rate. Hence, creep rate with the reinforcement scheme ($\pm 45°$) is more than two orders of magnitude higher than creep rate with the reinforcement scheme ($0°$, $0°$,

Table 3.22 Stress rupture of ELUR-P/ENFB unidirectional epoxy CFRP for different modes of deformation

Deformation	Test temperature (°C)	Stress (MPa) after various times				
		0.1 h	*1 h*	*10 h*	*100 h*	*500 h*
Tension	20	960	940	930	910	900
	150	780	770	760	750	740
Compression	20	780	700	680	660	650
	150	690	620	600	540	500
Bending	20	1000	990	980	970	960
	150	900	800	750	650	600
Shear	20	55	51	50	49	48
	150	38	30	28	26.5	25

Table 3.23 Stress rupture of ELUR/ENFB epoxy CFRP with cross lay-up (tested in 0° orientation)

Deformation	Layer orientation	Temp. (°C)	Stress (MPa) after various times					
			0.1 h	*1 h*	*10 h*	*100 h*	*500 h*	*1000 h*
Tension	$(\pm 45°)$	20	120	115	110	108	105	100
		150	75	70	65	62	61	60
	$(0°, 0°, \pm 45°,$	20	470	465	435	425	400	395
	$\pm 45°, 0°, 0°)$	150	390	380	370	360	350	340
Compression	$(\pm 45)_6$	20	120	100	90	80	70	60
		150	100	80	70	6	5.5	5
	$(0°, 0°, \pm 45°,$	20	58	56	54	52	50	48
	$\pm 45°, 0°, 0°)$	150	36	34	32	30	28	26

Table 3.24 UKN-P-0.1/VS-2526 CFRP stress rupture, lay-up $(\pm 45°, 0°, 90°)$

Deformation	Test temperature (°C)	Stress (MPa) after various times					
		0.1 h	*1 h*	*10 h*	*100 h*	*500 h*	*1000 h*
Tension	20	580	570	560	550	520	500
	150	540	520	515	500	480	470
Compression	20	480	460	450	440	400	380
	150	320	300	280	260	240	220

$\pm 45°$, 0°, 0°), which is equal to 1.06×10^{-5} and $5.3 \times 10^{-5}\% \, h^{-1}$ at stresses of 10 and 20 kgf mm^{-2}, respectively. It is practically impossible to plot a trustworthy creep curve at such creep rates, as the strain accumulated during this period between two measurements lies outside the limits of measurement accuracy. That is why only total creep strain during test duration can be recordered sufficiently accurately. Figure 3.26 shows creep strain for CFRP with $\pm 45°$ reinforcement scheme at different loading levels, and Table 3.23 lists stress rupture data at tension and compression. Figure 3.27 shows the results of studying accumulated creep strain effect on the material residual strength estimated by $\bar{\sigma}_b/\sigma_b$ ratio, where $\bar{\sigma}_b$ is material strength at

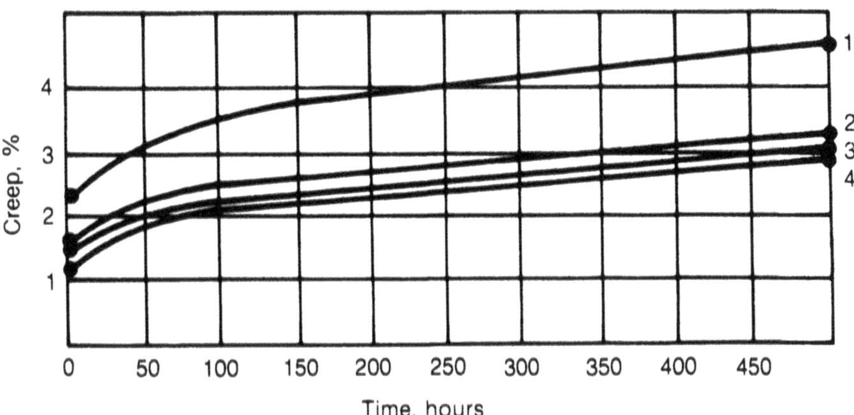

Fig. 3.26 Creep curves of epoxy CFRP with reinforcement scheme ($\pm 45°$) at the acting stress values: 800 MPa (1), 60 MPa (2), 50 MPa (3) and 40 MPa (4).

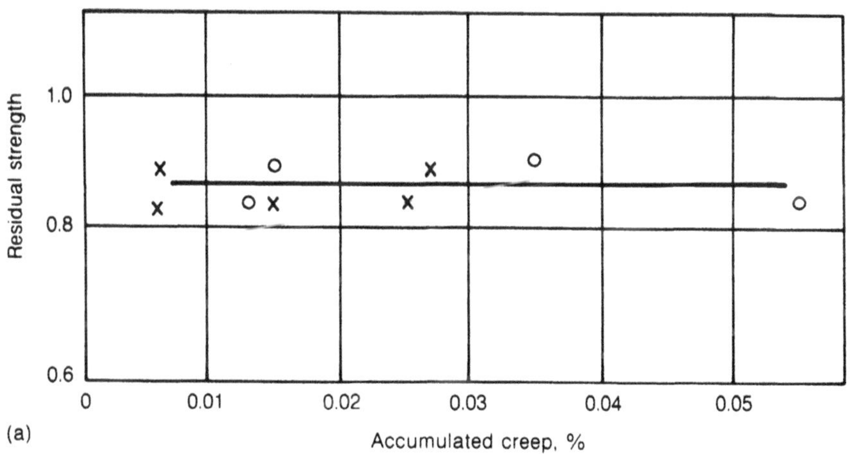

Fig. 3.27 Creep strain effect on residual strength of CFRP with different reinforcement schemes: (a) (○) (0°, 90°) and (×) (0°, 0°, $\pm 45°$, $\pm 45°$, 0°, 0°); (b) (▲) ($\pm 30°$) and (●) ($\pm 45°$).

(Continued overleaf)

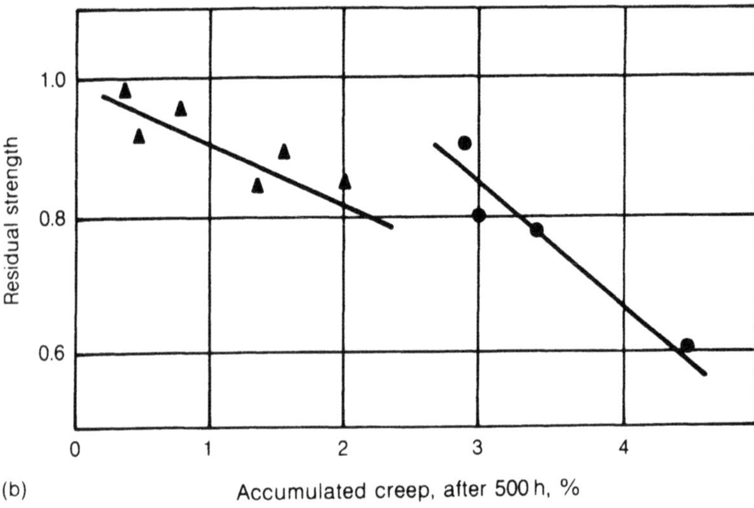

Fig. 3.27 (*continued*)

temperature $T = 20°C$ after creep tests for 500 h, and σ_b is material initial strength. The strengh of CFRP with $(0°, 0°, \pm 45°, \pm 45°, 0°, 0°)$ reinforcement scheme is decreased by 15–16%, independent of creep strain degree. This variation is characteristic of any CFRP, having layers oriented along the loading axis in their structure. With $\pm 45°$ reinforcement scheme, residual strength is decreased with preliminary creep strain increase. In this case the contribution of polymeric matrix to creep strain is more significant, compared to materials containing unidirectional layers.

Tables 3.22 to 3.24 present the values of stress rupture on the basis of 1000 h for unidirectional and cross-reinforced epoxy CFRP. With the increase of test base, stress rupture in compression and shear is decreased at normal and elevated temperature much more significantly, compared to tensile strain.

3.4 CARBON-FIBRE-REINFORCED PLASTICS WITH THERMOPLASTIC MATRIX

A typical representative of thermoplastics used for CFRP production is polysulphone. Filler surface laminating by polysulphone film with given thickness, which ensures the required ratio of components in the material, is used for production of semiproducts based on continuous fibres.

CFRP based on polysulphone, the physical and mechanical properties of which are given in Table 3.25, possess high properties of ultimate tensile strength and elongation in transverse orientation, residual strength after impact, high moisture and acid resistance, reduced moulding time, capability for recycling and also weldability compared to epoxy CFRP.

CFRP physical and mechanical properties significantly depend not only on fibre and matrix properties but also on wetting conditions and adhesion strength at the

Table 3.25 ELUR-P/PSN CFRP mechanical properties

Properties	Load application direction (deg)	Test temperature (°C)	
		20	150
UNIDIRECTIONAL CFRP			
Tension			
Ultimate strength (MPa)	0	1000	950
	90	45	40
Elasticity modulus (GPa)	0	130	120
	90	6.9	5.8
Relative elongation at break (%)	0	0.75	0.74
	90	0.7	0.7
Poisson's ratio	0	0.32	0.30
Compression			
Ultimate strengh (MPa)	0	850	800
	90	110	90
Elasticity modulus (GPa)	0	112	–
	90	6.5	–
Shear			
Ultimate interlaminar strength (MPa)	0	60	45
Shear modulus (GPa)	0	3.12	–
LAY-UP (0°, + 45°, 90°)			
Tension			
Ultimate strength (MPa)	0	370	370
Elasticity modulus (GPa)	0	41.5	–
Relative elongation at break (%)	0	0.78	–
Poisson's ratio	0	0.31	–
Compression			
Ultimate strength (MPa)	0	450	380
Elasticity modulus (GPa)	0	42.2	38
Shear			
Ultimate strength in-plane (MPa)	0	120	90
Shear modulus (GPa)	0	12	–

interface, as bonding strength defines stress transfer efficiency via the interface. The difficulty of producing qualitative materials based on tape and tow carbon fillers is determined by high viscosity and polysulphone melt surface tension at the processing temperature of 270–290°C (Fig. 3.28). In order to achieve the maximum contact area between polymer and fibre surface, plastics solidity and effective realization of the advantages of the polysulphone matrix, interface formation is carried out by a barrier

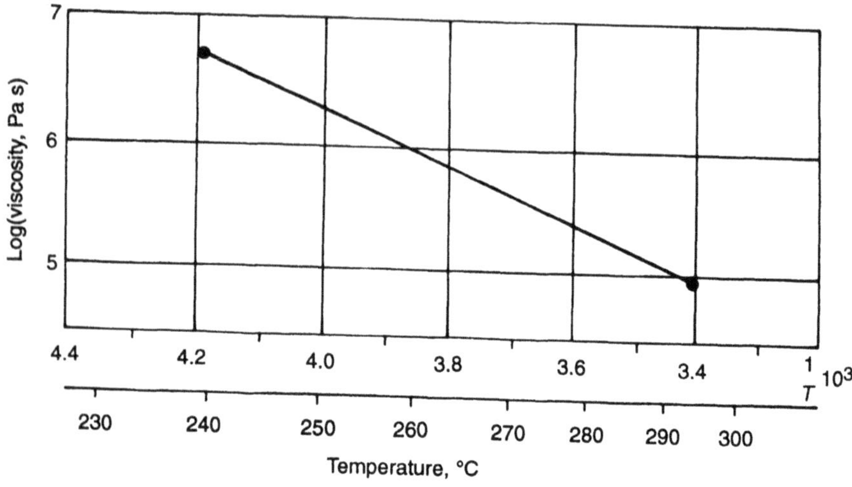

Fig. 3.28 Toughness of PSN polysulphone as a function of temperature.

coating application, ensuring edge wetting angle decrease from 60 to 30° on the carbon fibre [15]. Edge wetting angles of carbon filler surface by polysulphone and poly(ether ether ketone) melts are given in Table 3.26. The results presented in Table 3.26 show that thermoplastics do not moisten carbon fibres equally. So, polysulphone possesses lower carbon fibre wetting ability ($\theta = 50$–60°) compared to poly(ether ether ketone) ($\theta = 29°$). Data on edge wetting angle measurements are in good agreement with mechanical test results of CFRP based on carbon fillers with different barrier coatings (Table 3.27). Strength properties of ELUR-P/PSN CFRP dry and wet specimens versus temperature are given in Figs 3.29 and 3.30. Unlike epoxy

Table 3.26 Edge wetting angle of carbon-fibre surface by thermoplastic melt

Fibre	Binder	Barrier layer (tradename)	Edge wetting angle, θ (deg)	Coefficient of variation, v(%)
ELUR-P	Polysulphone	Untreated	60	10.1
		RTA	35	9.06
	Poly(ether ether ketone)	Untreated	29	11.38
UKN-P	Polysulphone	Untreated	50	11.5
		RTA	35	9.9
		SP	30	12.0
	Poly(ether ether ketone)	Untreated	29	10.96
		RTA	50	10.2
		AEF	24	12.3
		SP	25	10.9

Table 3.27 Mechanical characteristics of unidirectional CFRP based on unsized and sized carbon fillers and polysulphone

Properties	UKN-P	UKN-P + RTA (2.5 wt%)	UKN-P + SP (1.5 wt%)	ELUR-P	ELUR-P + RTA (2.5 wt%)
Along fibres (0°)					
Ultimate tensile strength (MPa)	1300	1380	1380	900	1100
Ultimate compressive strength (MPa)	700	760	860	780	850
Ultimate shear strength (MPa) (short beam)	50	55	60	50	60
Across fibres (90°)					
Ultimate tensile strength (MPa)	40	52	51	40	45
Relative elongation at break (%)	0.5	–	0.7	0.52	0.68

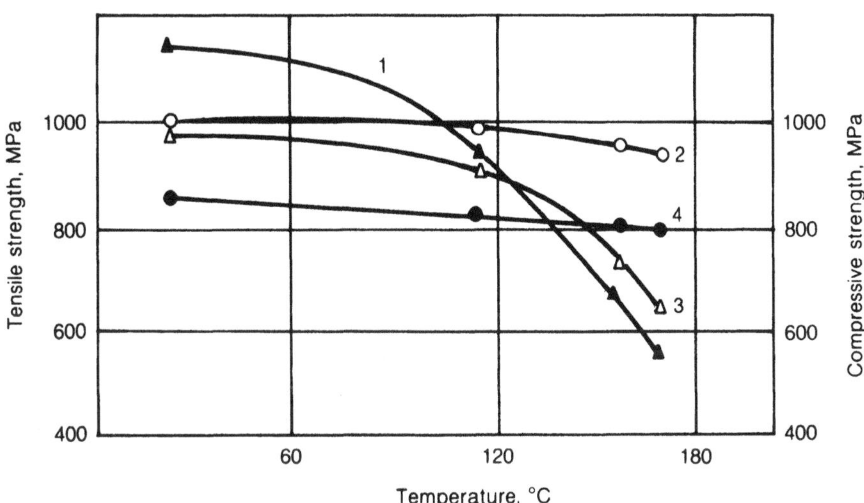

Fig. 3.29 Ultimate compressive (1, 4) and tensile (2, 3) strength for ELUR-P carbon-tape-based CFRP with epoxy ENFB (1, 3) and polysulphone PSN (2, 4) and epoxy (2, 3) binders as a function of temperature.

CFRP, the material based on polysulphone hardly changes its properties after exposure to increased humidity and temperature ($\phi = 98\%$, $T = 70°C$) for 30 days.

Thermoplastic matrix deformation properties positively effect CFRP fracture mechanism at impact. ELUR-P/PSN CFRP residual compression strength after impact of 68 J exceeds that of ELUR-P/ENFB epoxy CFRP by 30–40%; a through-hole appeared on epoxy CFRP specimens after impact (Table 3.28) [16].

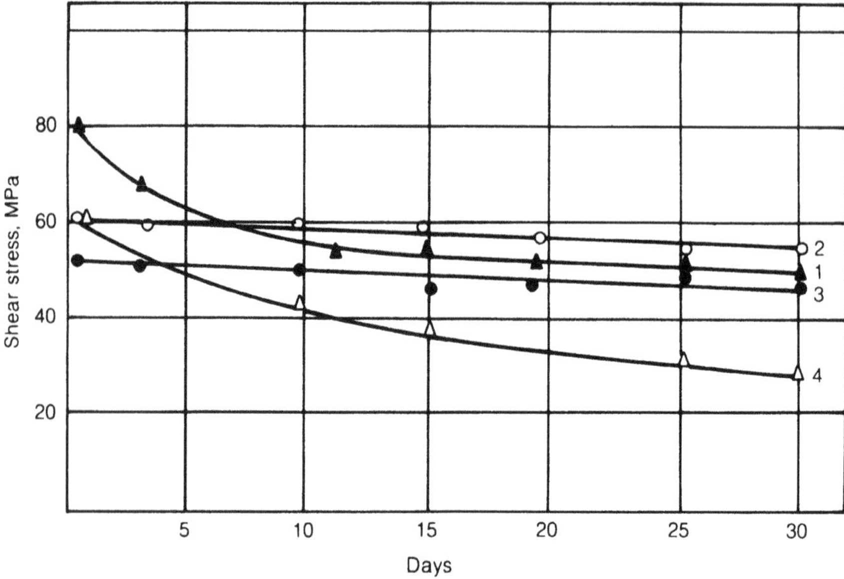

Fig. 3.30 The change of ultimate shear strength τ_{shear} for CFRP based on UKN carbon fibres and (2, 3) polysulphone and (1, 4) epoxy binders as a function of exposure time in a wet atmosphere at $\phi = 98\%$ and temperature $T = 70°C$: 1, 2, test temperature 20°C; 3, 4, test temperature 150°C.

Table 3.28 Impact resistance of CFRP based on epoxy and polysulphone matrices; Lay-up ($\pm 45°, 0°, 90°$), sample thickness $\delta = 2$ mm

	Residual compression strength (MPa)	
Impact energy $(J\,mm^{-2})$	*ELUR-P + ENFB,* $V_f = 57\ vol\%$	*ELUR-P + PSN* $V_f = 56.5 \cdot vol\%$
2	175	280
4	144	190
6	120	175
8	110	150

Fatigue properties of CFRP with polysulphone and epoxy matrices and quasi-isotropic lay-up under tension on the base of 10^7 cycles are close in value. ELUR-P/PSN CFRP ultimate endurance under sign-variable loading (cycle of tension–compression) at 20 and 150°C is lower by 30–40% compared to epoxy CFRP of similar structure (Fig. 3.31). The difference in fatigue strength values is increased with frequency rise from 25 to 50 Hz. The reason for the observed phenomenon is probably connected with ELUR-P/PSN material heating and additional loss of stiffness under testing.

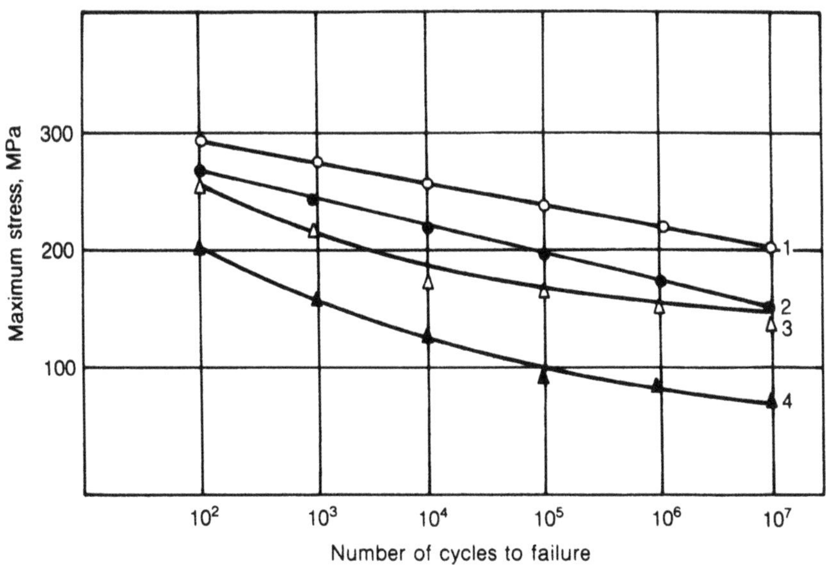

Fig. 3.31 Endurance limit of CFRP with quasi-isotropic lay-up, based on ELUR-P carbon tape with polysulphone (3, 4) and epoxy (1, 2) matrices as a function of tension–compression cycle number N (frequency 25 Hz at 20°C (1, 3) and 150°C (2, 4).

3.5 COMPONENT PROPERTIES AND HIGH-TEMPERATURE STRENGTH OF HEAT-RESISTANT CARBON-FIBRE-REINFORCED PLASTICS

Different component effects on high-strength heat-resistant CFRP properties in the process of testing were found, as follows. If material short-term strength is defined by the initial strength of the reinforcing fibres and matrix, then so are filler thermo-oxidation resistance and its surface preparation. Testing different compositions in combination with high-temperature matrices showed that the realization of maximum component mechanical properties is achieved through use of compounds that ensure an elastic uniform film on the fibre surface, good textile properties of filler and capability to form heat-resistant rings on matrix curing, as boundary layers.

Testing new heat-resistant binder modifications of different nature (phenolic, bismaleimide, complex unsaturated oligoether, polyimide) allowed high-strength CFRP to be obtained for operating temperatures from 210 to 300°C. The optimum ratio of cyclic oligoethers and bismaleimides, needed to create rapidly cured compositions for material production with long-term serviceability at 250°C by pultrusion, has been found.

The following technological methods greatly affect heat-resistant CFRP solidity and serivce properties: use and optimization of melt and solution technology for component compatibility, and treatment according to thermomechanical and micro-structural analysis results of curing conditions. Placing moulded semiproducts based on polycondensation binders into a vacuum increases compression strength by 1.5–2 times. Fibres tension in the process of curing improves structure ordering and decreases residual stresses, as a result of which CFRP strength is increased by 20–30%.

One of the reasons for decrease of CFRP strength properties at high temperature can be oxidation of sodium and potassium ions and also other impurities present on the carbon-fibre surface. The presence of alkali metals is probably due to their significant content in the initial PAN fibres, and also deposition from the furnace atmosphere in the process of fibres carbonization. The material oxidation process usually begins on part open surfaces and edges, and passes from fibre to matrix along interfaces, causing material delamination.

Weight loss thermograms of UKN-P carbon fibres with a content of 0.015% (curve 1) and 0.2% ions (curve 2) are given in Fig. 3.32. As seen from the diagrams, the onset temperature of fibre thermo-oxidation destruction increases from 400 to 500°C with decrease of metallic impurities.

Curves of compression strength losses along fibres of CFRP based on heat-resistant OMI binder during air ageing at 250°C are shown in Fig. 3.33. Curves 1 and 2 characterize the thermal resistance of plastic reinforced with fibres of high oxidation resistance; curves 3 and 4 are for plastic reinforced with traditional UKN-P fibres. Decrease of metallic impurities content in carbon fibre considerably increases composite thermal resistance.

The following requirements have been taken into account when developing binders for production of high-strength CFRP to operating temperatures of 200–300°C:

1. Compatibility with carbon fibre.
2. Long-term thermal oxidative stability of matrix at the material operating temperature.

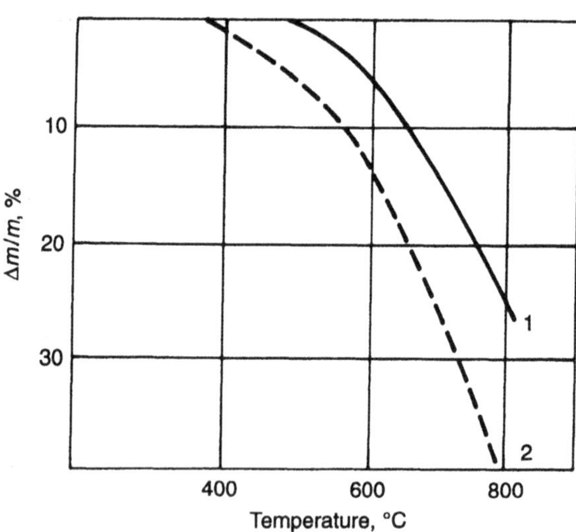

Fig. 3.32 Weight losses of carbon fibres in the process of heating as a function of sodium impurities content: 1, 0.02% Na; 2, 0.1% Na.

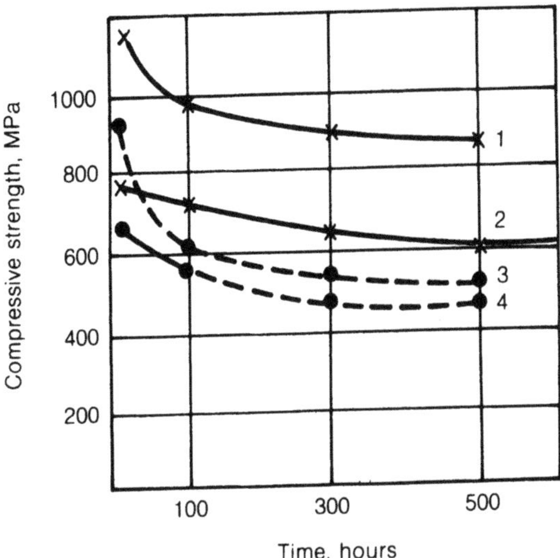

Fig. 3.33 Compressive strength of unidirectional CFRP based on carbon fibres with different content of Na and OMI binder as a function of exposure time at 250°C: 1, 2, UKN-P filler with sodium content 0.02%, at CFRP test temperature of 20°C (1) and 250°C (2); 3, 4, UKN-P filler with sodium content 0.1%, at CFRP test temperature of 20°C (3) and 250°C (4).

3. Acceptable composite moulding conditions.
4. Prepreg pot-life not less than one month.
5. Low toxicity.

Individual results of studying CFRP based on bismaleimide, non-saturated, oligoether and polyimide aromatic binders and their modifications are presented in [17].

Binder thermal stability was estimated by weight losses of cured non-saturated specimens in the course of heating (Fig. 3.34(a)) and long-term thermal ageing at 250°C in air (Fig. 3.34(b)). Based on the presented relationships, temperatures have been established for long-term serviceability of composites based on:

1. Bismaleimide (curve 3) – 200 to 230°C.
2. Oligo(ether maleimide) (curve 2) – 250°C.
3. Aromatic polyimide (curve 1) – 300°C.

Special properties are characteristic of the OMI binder based on non-saturated oligoethers and oligoesters, which is in the liquid-phase state under normal conditions, and can be combined with fibrous fillers, using melt technology, for example, on rolls. Binder viscosity is decreased by an order of magnitude when heated to 50–70°C, which allows its use in different technological processes.

The binder is non-toxic. It is cured by a polymerization–polycondensation mechanism with the evolution of a small quantity of low-molecular-weight products and the formation of crosslinked and cyclic polymeric structures. Cured polymer has mechanical

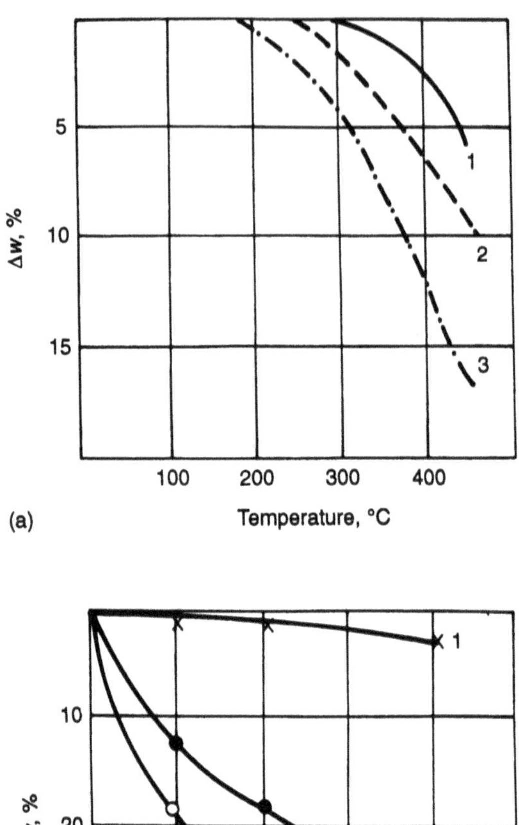

Fig. 3.34 Weight change of binder cured specimens in the process of heating (a) and long-term exposure at a temperature of 250°C (b): 1, SPTsM; 2, OMI; 3, PAIS.

characteristics on the level of epoxies, and its heat resistance is defined by curing temperature and reached 300°C.

Pot-life of OMI binder-based prepregs is 6 months. CFRP items are produced by pultrusion, pressing, autoclave moulding and winding methods.

The high-temperature strength of CFRP depends, to a great extent, on the nature of the boundary layer between the components. The possibility was studied for developing barrier layers for carbon tows used with OMI binder. Sizing compositions of the following types were investigated:

1. Epoxy sizes.
2. Silicone sizes.
3. Solutions of imide-forming monomers.
4. Mixtures of non-saturated oligoethers with bismaleimides.
5. Furan sizes.

Figure 3.35 shows the curves of weight loss on heating UKN-P carbon fibre with different sizes. Sizes having imide groups in their polymeric chain possess the highest heat resistance of all those studied.

A microstructural analysis of heat-resistant CFRP reinforced with fibres having various sizes was performed. Epoxy sizes were found to degrade already on high-temperature composite curing, and their bearing capacity deteriorated.

Good technological characteristics of the filler are provided by a size based on a mixture of non-saturated bismaleimide oligoethers. It forms a uniform and thermostable (up to 300°C) boundary layer in CFRP.

Thermomechanical analysis of cured CFRP revealed a boundary-layer effect on the material glass transition temperature. Substitution of oligo(ether maleimide) size for epoxy one increased T_g for OMI binder-based CFRP from 200 to 260°C. As seen from Table 3.29, CFRP initial compression strength was increased approximately by 2 times at 250°C, and interlaminar shear strength was increased by 1.5 times.

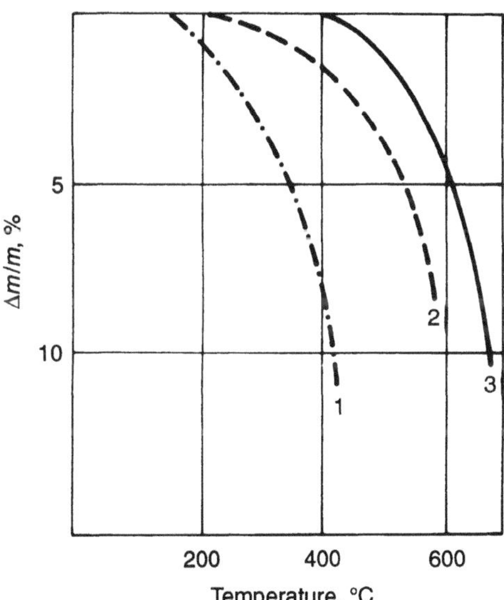

Fig. 3.35 The change of weight losses for specimens of CFRP with different sizes as a function of temperature: 1, epoxy; 2, unsized; 3, OMI.

Table 3.29 Effect of carbon fillers with different sizes on thermal stability of plastics based on oligo(ether maleimide) binder

Size	Compression (MPa)		Interlaminar shear (MPa)	
	20°C	250°C	20°C	250°C
Without size	1100	750	50	30
Epoxy	790	430	36	25
Oligo(ether maleimide)	1100	800	60	35

After ageing for 500 h at 250°C, the shear strength of OMI binder-based CFRP reinforced with oligo(ether imide)-sized fibre is 50% higher than that of CFRP reinforced with epoxy-sized fibre.

It should be noted that, with the growth of requirements for heat resistance, the temperature of final curing reaction stages is also increased and usually it exceeds the CFRP service temperature in structure elements.

In the process of polyimide binder testing, acceptable conditions for autoclave moulding of evacuated prepreg multilayer stacks were found, which made possible the production of CFRP with porosity less than 3%. The material compression strength was enhanced from 800 to 1200 MPa at normal temperature and from 500 to 1000 MPa at 250°C, compared to pressing technology.

The obtained levels of tensile, compressive and shear (along the fibres) strength at temperatures of 20 and 250°C for unidirectional CFRP are demonstrated in Table 3.30.

Bismaleimide CFRP possess good stability of physicomechanical characteristics to 230°C. At this temperature they retain 70% of initial strength values; at a temperature of 250°C, this value amounts to 50–60%.

The use of carbon filaments with microplastic strength of 3500 MPa and containing no more than 0.015% alkali-metal impurities, sized with heat-resistant size, in combination with OMI binder melt allowed the manufacture of CFRP specimens

Table 3.30 Strength properties of CFRP based on thermostable matrix and UKN-P-0.1 carbon fibre

Matrix (tradename)	Ultimate tensile strength, σ_b (MPa)		Ultimate compressive strength, σ_{comp} (MPa)		Interlaminar shear (short beam) (MPa)	
	20°C	250°C	20°C	250°C	20°C	250°C
PAIS-104	1770	1000	1100	550	63	30
OMI	2000	1530	1000	800	60	35
SPTsM	1800	1680	1200	1000	65	48

Table 3.31 Mechanical properties of CFRP based on OMI oligo(ether bismaleimide) binder

Properties	Load application direction (deg)	Filler					
		UKN-P		UOL-300		UT-900	
		$22°C$	$250°C$	$22°C$	$250°C$	$22°C$	$250°C$
Ultimate tensile strength, σ_b (MPa)	0	1500	1300	1250	1100	500	340
	90	25	12	22	11	500	300
Tensile elasticity modulus, E (GPa)	0	141	125	110	90	60	49
	90	6.5	4.1	5.1	3.2	60	49
Ultimate compressive strength, σ_{comp} (MPa)	0	1000	770	800	600	380	260
	90	97	–	94	62	375	250
Compressive elasticity modulus, E_{comp} (GPa)	0	110	77	90	70	52	35
	90	6.5	–	5.0	3.9	50	37
Ultimate bending strength, σ_{bend} (MPa)	0	1600	1040	1500	750	600	–
Bending elasticity modulus, E_{bend} (GPa)	0	123	107	120	100	50	–
Interlaminar ultimate shear strength, τ_{shear} (short beam) (MPa)	0	50	30	53	30	35	20
Shear modulus in-plane, G (GPa)		5.6	4.8	5.0	4.3	6.0	4.8
Relative elongation at break, ε (%)	0	0.9	0.85	0.9	0.7	0.75	0.8
	90	0.26	0.18	0.3	0.2	0.9	0.87
Poisson's ratio, μ		0.42	0.32	0.37	–	–	–

Table 3.32 UKN-P/OMI CFRP fatigue strength and stress rupture

Fatigue strength at coefficient of asymmetry	Reinforcement scheme	T (°C)	Fatigue strength (MPa) after number of cycles shown			
			10^4	10^5	10^6	10^7
$R = -1$	$0°$	20	510	470	430	400
		250	410	370	340	300
$R = 0–0.1$	$0°, \pm 45°, 90°$	20	330	310	295	280
		250	240	210	191	173
$R = -1$	$0°, \pm 45°, 90°$	20	320	310	290	260
		250	235	213	190	165

Stress rupture in tension or compression	Reinforcement scheme	T (°C)	Stress (MPa) based on time (h) shown				
			0.1	1	10	100	500
Tension	$0°$	20	1200	1100	1100	1000	1100
		250	850	830	810	800	790
Compression	$0°$	20	730	710	680	660	650
		250	360	350	340	330	320
Tension	$0°, \pm 45°, 90°$	20	380	360	350	340	335
		250	300	280	270	260	256
Compression	$0°, \pm 45°, 90°$	20	230	210	200	190	180
		250	220	200	180	175	170

whose tensile strength amounts to 2000 MPa. The developed material retains 70–90% of initial strength characteristics at a temperature of 250°C. Tables 3.31 to 3.33 present the physicomechanical characteristics of heat-resistant CFRP, based on oligo(ether maleimide) OMI binder and various carbon fillers, and Fig. 3.36 shows the material curing conditions.

In conclusion, it can be stated that serviceability of high-strength CFRP at elevated temperatures is defined by carbon-fibre oxidation resistance, polymeric matrix heat resistance, and size thermal resistance and compatibility with matrix.

3.6 STRUCTURE-SIMULATING MODELLING OF CARBON-FIBRE-REINFORCED PLASTICS WITH GIVEN TENSILE AND COMPRESSIVE PROPERTIES

3.6.1 Tension

Production of a composite possessing given properties is associated with the problem of optimum selection of its structural components, i.e. fibres and matrix, and also reinforcement orientation. Appropriate binder selection plays an important part in ensuring composite maximum strength properties.

Table 3.33 Thermophysical properties of UKN-P/OMI CFRP

Properties	$T(^\circ C)$	Property values
Thermal conductivity	-100	0.58
$(W\,m^{-1}K^{-1})$, 0° lay-up	-50	0.59
	0	0.60
	25	0.60
	100	0.61
	150	0.61
	200	0.61
	250	0.62
	300	0.63
Thermal diffusivity	-100	9.0
$(10^7\,m^2\,s^{-1})$, 0° lay-up	-50	5.5
	0	4.7
	25	4.0
	100	3.4
	150	3.15
	200	3.0
	250	2.55
	300	2.70
Specific heat capacity	-100	0.45
$(kJ\,kg^{-1}K^{-1})$	-50	0.75
	0	0.90
	25	1.05
	100	1.25
	150	1.35
	200	1.45
	250	1.70
	300	1.70
Linear expansion coefficient	Along fibres	
$(10^{-6}K^{-1})$	-80 to -60	0.4
	-60 to 40	0.6
	40 to 100	0.7
	100 to 200	0.8
	200 to 300	0.8
	Across fibres	
	20 to 40	22.0
	40 to 100	28.0
	100 to 120	31.0
	120 to 180	36.0
	180 to 220	38.0
	220 to 240	34.0
	240 to 260	32.0
	260 to 300	30.0

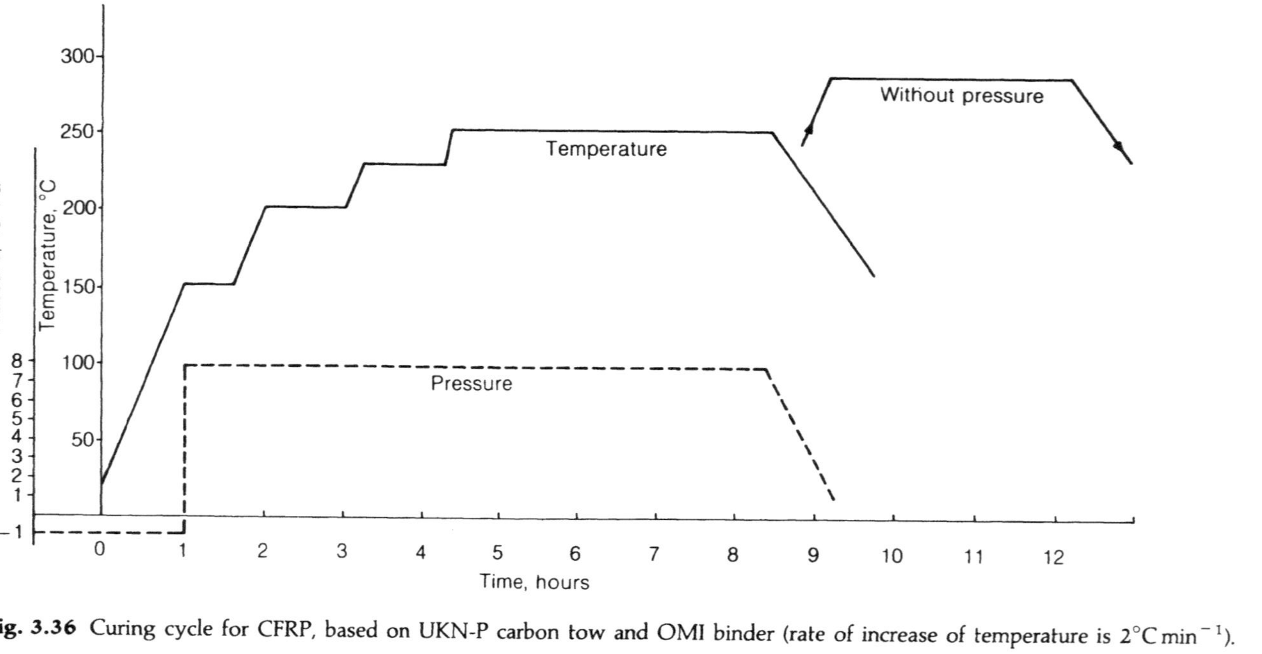

Fig. 3.36 Curing cycle for CFRP, based on UKN-P carbon tow and OMI binder (rate of increase of temperature is $2°C\,min^{-1}$).

Table 3.34 Initial data for plotting calculated dependences of CFRP tensile strength on components elastic strength properties

	Fibre		Matrix	
Properties	B_1	B_2	M_1	M_2
Ultimate tensile strength (MPa)	3520	5050	65	90
Coefficient of strength variation (%)	25.0	18.3	10.0	8.0
Elasticity modulus (MPa)	286 000	270 000	3600	3300
Ultimate elongation (%)	1.3	1.8	1.5–1.7	3.8–4.4
Ultimate shear strength (MPa)	–	–	60	80

The fracture of unidirectional CFRP, differing in structural components properties, was studied under tension in [18, 19]. Matrix deformation properties were shown to affect significantly the realization of carbon fibre strength in the composite. Hence a material having lower-strength fibres can be stronger than one with stronger fibres and matrix possessing low ultimate elongation.

Structure-simulating modelling was used for analysis of the effect of CFRP component properties on the process of deformation and fracture [20].

The fracture of four types of CFRP with different combination of reinforcing fibres B_1 and B_2 and matrices M_1 and M_2, the properties of which are given in Table 3.34, was simulated on a computer. The studied composite fracture depends on loading redistribution nature from fractured fibres onto unfractured fibres. A fibre ceases completely to bear a loading at breaking, and loading redistribution onto adjacent fibres with given K_{f1} coefficient occurs. If previously fractured fibres are within the adjacent ones, then the loading is not redistributed onto them, but is transferred onto the next fibres with lesser K_{f2}, K_{f3} coefficients, etc. Loading increment on these fibres is proportional to the product $K_{f1}K_{f2}K_{f3}$. Hence, K_f loading redistribution coefficients define stress concentration level on fibres adjacent to fractured ones and characterize matrix capability to decrease this concentration. Composite cross-sectional damage, determined by relative number of fractured fibres, was calculated as $w = N_p/N$, where N_p is the number of fractured fibres at average stress on them and N is total number of fibres in modelling section.

Curves of damage accumulation versus average stress on fibres for four studied CFRP types are given in Fig. 3.37.

An analysis of CFRP fracture processes showed that a low-elongation matrix (M_1) is fractured after fibre breakage, increasing the stress concentration on adjacent fibres. This defines large values of loading redistribution coefficient, and breaking first the weakest fibres leads to instantaneous composite fracture due to primary crack propagation. A matrix with high elongation (M_2) allows fibre deformation ability to be realized and the fracture mechanism is changed – uniform accumulation of damage

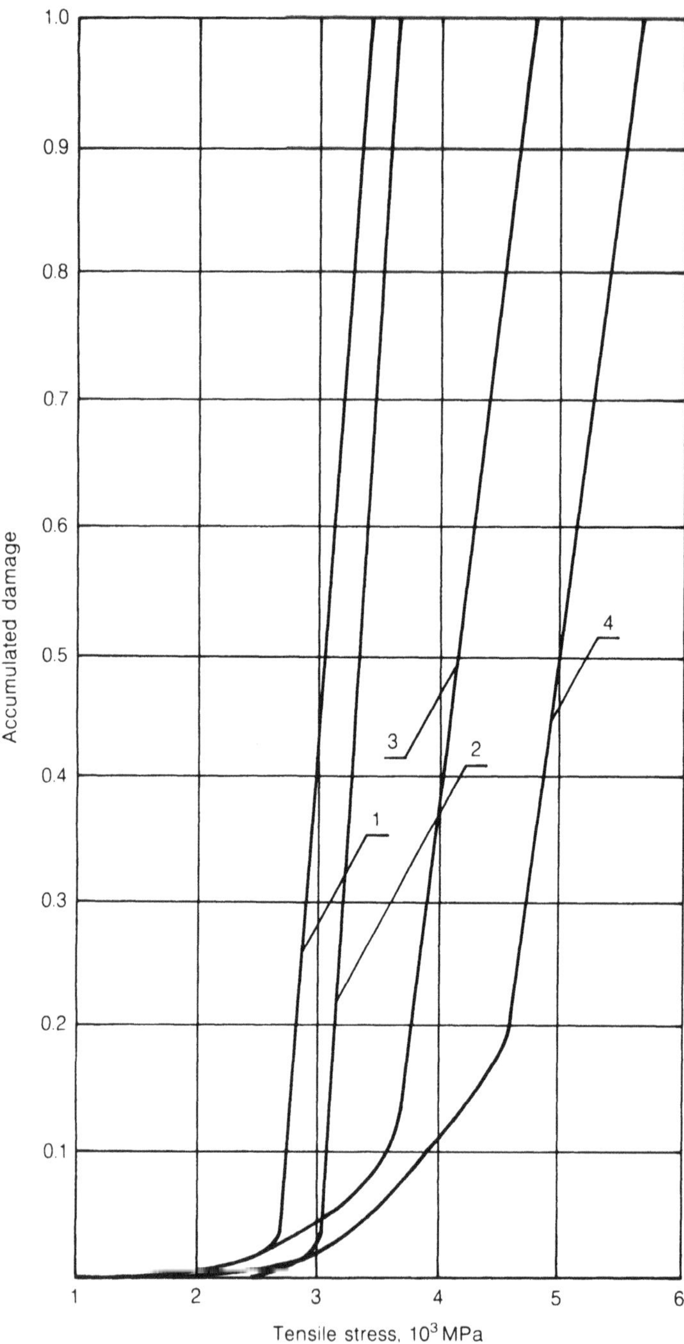

Fig. 3.37 Damage accumulation curves as a function of mean stress on fibres: 1, B_1M_1 material; 2, B_2M_1 material; 3, B_1M_2 material; 4, B_2M_2 material.

Table 3.35 Characteristics of CFRP

Material	Experimental σ_b (MPa)	Calculated			
		σ_b (MPa)	ε (%)	K_{f_1}	K_{f_2}
B_1M_1	1500	1650	1.1	0.15	0.12
B_1M_2	1450	1950	1.3	0.13	0.10
B_2M_1	1800	1750	1.15	0.23	0.20
B_2M_2	2420	2400	1.52	0.18	0.16

in material volume takes place, loading redistribution coefficients are decreased and composite strength is found to be higher. This factor is proved by relationships given in Fig. 3.37, where curves 1 and 3, corresponding to materials based on B_1M_1 and B_2M_1, show the onset of avalanche fracture at considerably lower levels of applied stress compared to curves 2 and 4 for materials based on M_2 matrix.

Design data were proved experimentally on CFRP specimens based on B_1 and B_2 carbon fibres and two types of binders, having elasticity and strength properties close to those in Table 3.35. Fibre volume fraction in specimens was 62–64%.

A comparison of design and experimental mechanical properties values of CFRP with different component combinations, shown in Table 3.35, showed their good similarity. As can be seen, experimental data mainly confirm design ones. A discrepancy with design data is observed only for the case of B_1M_2. This is probably caused by the fact that the fibre selected for CFRP production has lower strength compared to that put into the model calculation.

Using the same approach [21], the high-deformation properties effect of polysulphone matrix ($\varepsilon_p = 9\%$) on the realization of carbon fibre strength in a unidirectional material was analysed. For comparison, two materials were studied: thermoplastic CFRP based on ELUR-0.08P carbon tape and polysulphone (material I) and epoxy CFRP based on ELUR-0.08P carbon tape and ENFB binder (material II).

Materials mechanical properties values and fracture diagrams are given in Table 3.36 and in Fig. 3.38. Figure 3.38 shows that deformation levels from which avalanche fracture begins are close for both materials. In this case the significantly higher ultimate elongation value ($\varepsilon = 9\%$) of thermoplastic matrix compared to that of epoxy matrix ($\varepsilon = 2\%$) does not affect the load redistribution process in unidirectional CFRP structure under tension.

Data analysis on rate relationships of the strength of the two studied materials, given in Tables 3.36 and 3.37, allowed one to make definite conclusions about matrix deformation properties effect on material behaviour. Diagrams of material I deformation and curves of damage accumulation in it at different rates are given in Fig. 3.39. Figure 3.39 shows that material damage is not constant and depends on loading rate. Polysulphone matrix-based CFRP strength increase (see Table 3.36) is explained by lower levels of damage at the moment prior to material macrofracture. In contrast

Table 3.36 Strength as a function of rate at equiaxial tension for material I

Loading rate, σ (mm min^{-1})	Ultimate strength, $\bar{\sigma}$	Coefficient of strength variation, V_σ (%)	Elasticity modulus, E (MPa)	Elongation, ε (%)	Poisson's ratio, μ
Reinforcement orientation 0°					
0.5	1010	6.51	1.20×10^5	0.84	–
5.0	1075	5.08	1.34×10^5	0.80	0.32
50	1150	3.31	1.44×10^5	0.76	–
Reinforcement orientation (0°, ±45°, 90°)					
0.5	348	8.20	4.55×10^4	0.77	–
5.0	360	9.2	4.49×10^4	0.86	0.31
50	365	5.3	4.50×10^4	0.78	–
Reinforcement orientation (±45°)					
0.5	175	6.17	1.23×10^4	12.5	–
5.0	182	5.2	1.20×10^4	12.5	0.73
50	184	6.1	1.19×10^4	12.9	–
Reinforcement orientation 90°					
5.0	43	10.5	6.90×10^4	0.7	0.026

Table 3.37 Strength as a function of rate at equiaxial tension for material II

Loading rate, σ (mm min^{-1})	Ultimate strength, $\bar{\sigma}$	Coefficient of strength variation, V_σ (%)	Elasticity modulus, E (MPa)	Elongation, ε (%)	Poisson's ratio, μ
Reinforcement orientation 0°					
0.5	1150	5.0	1.25×10^5	0.90	–
5.0	940	3.5	1.20×10^5	0.78	0.265
50	1050	4.7	1.35×10^5	0.85	–
Reinforcement orientation (0°, $\pm 45°$, 90°)					
0.5	310	2.0	4.70×10^4	0.7	–
5.0	300	2.4	4.85×10^4	0.64	0.32
50	290	3.7	4.93×10^4	0.61	–
Reinforcement orientation ($\pm 45°$)					
0.5	110	6.2	1.40×10^4	5.1	–
5.0	109	5.4	1.38×10^4	5.1	0.67
50	113	5.9	1.48×10^4	5.3	–
Reinforcement orientation 90°					
5.0	28	11.0	0.90×10^4	0.28	0.021

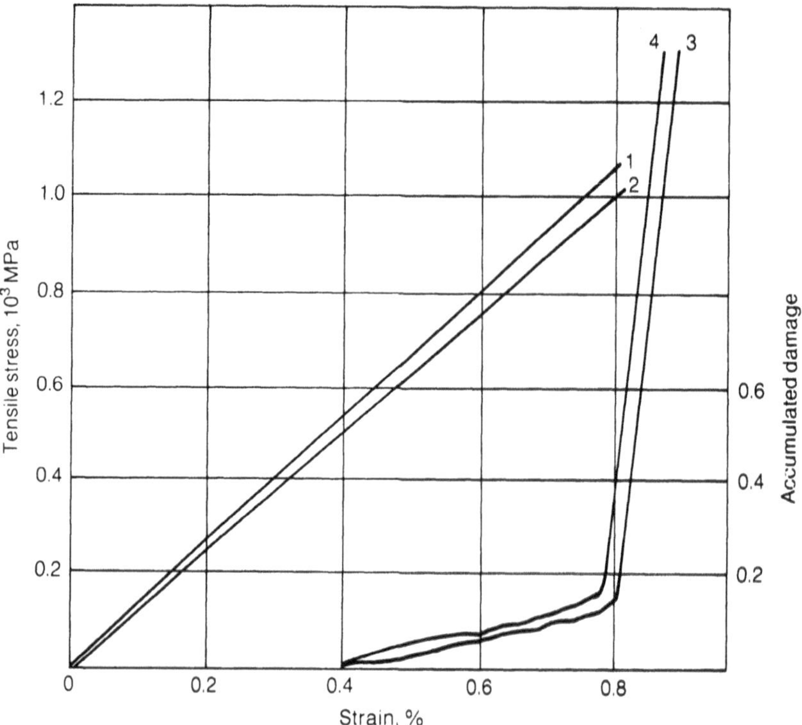

Fig. 3.38 The $\sigma-\varepsilon$ diagrams and damage accumulation curves for two CFRP (loading rate, 5 mm min^{-1}): 1, 3, material I; 2, 4, material II.

to epoxy CFRP [22] material I strength drop was not revealed in the limits of the studied rates. This means that fracture mechanism corresponding to damage accumulation is carried out and thermoplastic matrix performs stress redistribution in its structure.

Effect of CFRP reinforcement scheme on its mechanical properties is shown in Fig. 3.40. Diagrams of material I deformation are given as an example. Deformation relationships versus reinforcement schemes for material II are similar. Deviation from a linear relationship at a loading level of 0.8–0.9% of the critical one is observed in materials reinforced in transverse directions. The comparison of the properties of CFRP reinforced in the transverse direction shows that matrix deformation properties with such lay-up affect significantly material cracking process. Use of polysulphone with ultimate elongation of 9% as binder for material I determines CFRP strength increase by 1.5 times and ultimate elongation by 2.1 times compared to those of material II.

Calculations and experimental results showed that matrix deformation properties affect realization coefficient value of unidirectional layer strength in cross-reinforced

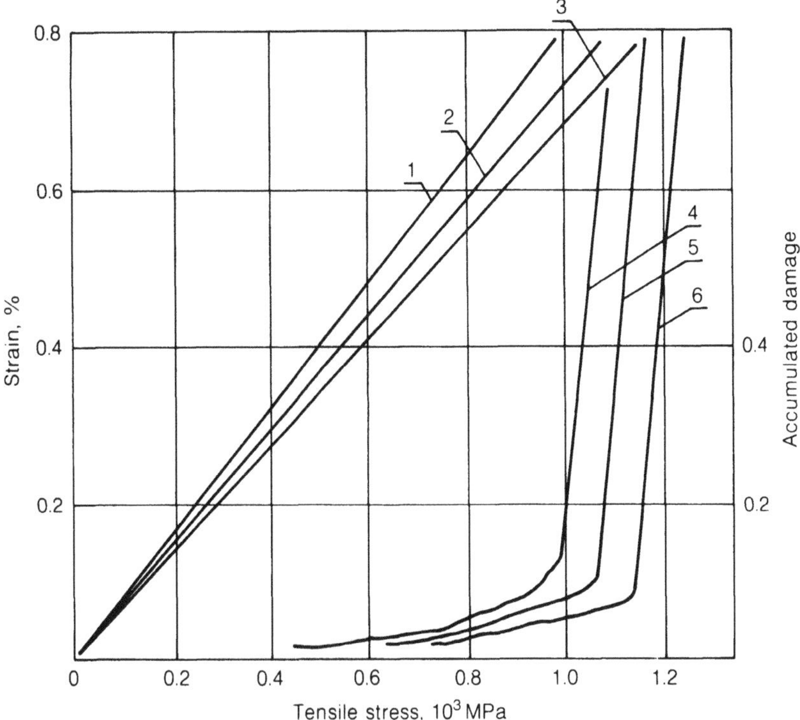

Fig. 3.39 The $\sigma-\varepsilon$ diagrams and damage accumulation curves as a function of loading rate for material I: 1, 4, $\dot\sigma = 20\,\mathrm{MPa\,s^{-1}}$; 2, 5, $\dot\sigma = 2.15 \times 10^2\,\mathrm{MPa\,s^{-1}}$; 3, 6, $\dot\sigma = 2.25 \times 10^3\,\mathrm{MPa\,s^{-1}}$.

CFRP. In material I, the matrix of which possesses higher ultimate elongation compared to that of material II matrix, realization coefficient value of unidirectional layer strength R in a stack (0°, $\pm45°$, 90°) is higher than 100%, while in material II unidirectional layer strength in a stack is not fully realized. The revealed feature of cross-reinforced CFRP behaviour can be explained by the fact that, owing to good matrix deformation ability in material I, the layer reinforced in transverse orientation is fractured under larger loadings than in material II, the matrix deformation ability of which is not sufficient for ensuring joint deformation of layers in a stack (0°, $\pm45°$, 90°) down to macrofracture.

A thermoplastic matrix allows fracture type to be changed and CFRP residual strength to be increased after impact loadings effect. Comparative data on impact resistance of ELUR/ENFB and ELUR/PSN materials are given in Table 3.38. Residual tensile strength value of ELUR/PSN CFRP is increased by 1.25–1.3 times depending on impact energy compared to ELUR/ENFB CFRP. In this case the material delamination zone in relation to the point of concentrated impact loading application is practically absent on specimens of CFRP with thermoplastic matrix.

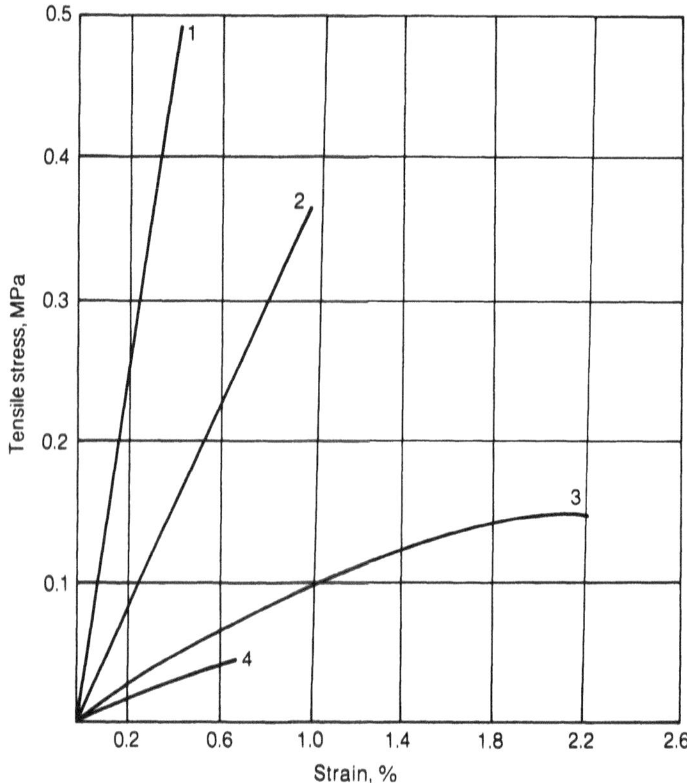

Fig. 3.40 The $\sigma-\varepsilon$ diagrams for material I under uniaxial tension with different reinforcement schemes: 1, lay-up $0°$; 2, $(0°, \pm45°, 90°)$; 3, $(\pm45°)$; 4, $90°$ (loading application direction $0°$).

Table 3.38 Residual strength comparison for CFRP based on polysulphone and epoxy matrices

Impact energy, $(J\,mm^{-2})$	Residual tensile strength, σ_b (MPa)	
	ELUR/PSN $(V_f = 0.52)$	ELUR/ENFB $(V_f = 0.55)$
2	340	250
4	200	160
6	170	140[a]

[a] Through-hole with delamination zone.

3.6.2 Compression

Composite compression properties are defined only partially by the mechanical properties of the individual components. The key factor in improving compression bearing capacity of composites is material structure optimization both at the level of reinforcing fibres and at the level of blocks and layers.

Let us specify from the great number of observed fracture mechanisms of composites under compression along the fibres the most frequent ones, at which local stability loss of particular fibres, blocks or layers is accompanied by delamination and flicking out of this layer sideways.

Microstructural analysis of studied unidirectional CFRP specimens has revealed the presence of a block-layered structure extremely sensitive to CFRP production technology. In composite cross-sectional areas with dense lay-up of carbon fibres, separated by polymeric matrix interlayers, can one consider certain blocks to be shown with cross-sectional shape close to rectangular with thickness h_e. Separate blocks, in their turn, are combined into layers.

Hence, it is possible to suppose that one of the basic mechanisms responsible for initiation and propagation of fracture under compression is stability loss of structure elements – namely blocks (or group of blocks with their delamination and swelling sideways (Fig. 3.41)).

The condition of stability loss with delamination of a certain rectangular-shaped element (layer) with thickness h_e, width B_e and length l_e can be obtained from the energy relation [23]:

$$U_{\text{comp}} = U_{\text{bend}} + P \tag{3.2}$$

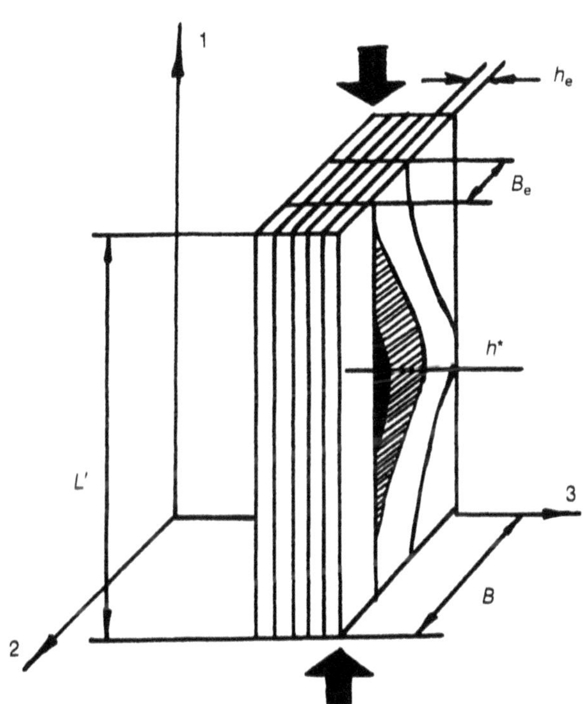

Fig. 3.41 CFRP compression fracture scheme.

where U_{comp} is compression energy, accumulated before stability loss; U_{bend} is block bending energy after stability loss; and P is energy of new surface formation.

Blocks in CFRP are formed in the process of composite production out of tows, containing 2500, 5000, etc., densely laid up carbon fibres, impregnated with matrix material; parameter h^* shows not only block thickness, which can be varied with the change of technological regime, but also their shape, as block area remains practically unchanged.

Taking into consideration experimental data concerning the significant effect of transverse strength, i.e. bonding strength between blocks in the direction of possible breaking or bonding strength σ_{b3} of edge layer with other layers in direction 3 of critical fracture stress, and assuming that interlayer thickness h^*, in which tensile strains are generated under delamination of a certain element, is proportional to this element thickness:

$$h^* = K_i h_e \tag{3.3}$$

where K_i reflects the energy transitions under discussion, the criterion for stability loss of a certain element with delamination and swelling can be obtained as follows:

$$(\sigma_{comp}/E_e) \geqslant (\sigma_{comp}/E_e)(2/3\pi)(h_e)^{2/3} + (\sigma_{b3}/E_m E_e)K_y \tag{3.4}$$

where

$$E_e = E_f V_f + E_m (1 - V_f) \tag{3.5}$$

E_e is elasticity modulus of structure element, V_f is fibre volume fraction, h_e is structure element thickness, E_m is matrix elasticity modulus, σ_{b3} is bonding strength between blocks in direction 3, and E_f is fibre elasticity modulus.

The effect of bonding strength σ_{b3}, matrix elasticity modulus E_m and element thickness h_e on the value of critical compression strength was studied, using the above expression.

Taking into account the complexity of this problem, structure-simulating modelling of the processes was used for its solution, similar to the tension case [24, 25]. Strength scatter between the elasticity moduli of the individual blocks, which were present by statistical distributions, have been taken into account while plotting the structural model.

Materials loading was simulated on the computer by stepped increase of compression stress levels. The possibility of excluding edge elements, groups of elements, layers and the specimen as a whole out of the work was controlled at each step. Compression stresses were released from the elements, which lost stability, and redistributed to remaining elements. Initial data for plotting the design dependences are given in Table 3.39.

Relatively small change of fibre volume fractions ($V_f = 0.53-0.65$) results in noticeable stiffness increase of composite and its structural elements (blocks) and, as shown both by experiments and calculations (in particular, for the material with M_2 matrix), in the composite average compression increase from 1270 MPa to 1510 MPa (Fig. 3.42).

Table 3.39 Initial data for plotting calculated dependences of σ_{comp} on structural elements' elastic strength properties[a]

Matrix	E_f *(GPa)*	E_m *(MPa)*	σ_{b3} *(MPa)*	σ_{b2} *(MPa)*	h_e *(mm)*	σ_{comp} *(MPa)*
M_1 (UNDF-4a)	240	3750	12	35	0.125	1270
M_2 (VS-2526M)	240	4500	15	28	0.125	1500
M_3 (OMI)	240	4050	7	16	0.125	930

[a] $V_f = 0.65$.

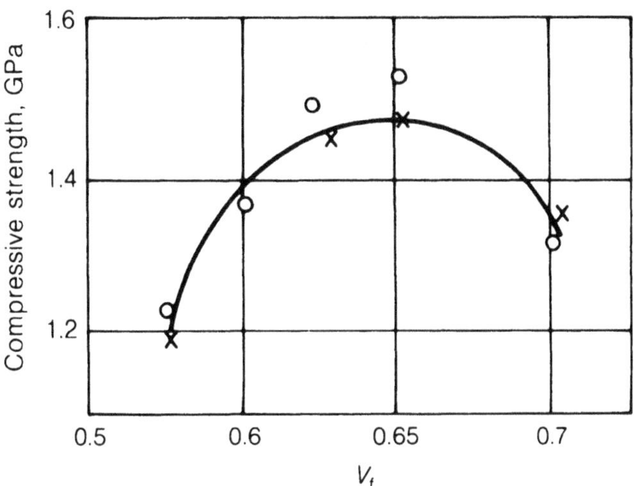

Fig. 3.42 CFRP compressive strength σ_{comp} as a function of fibre volume fraction.

Non-traditional results were obtained during the analysis of stiffness effect on composite compression strength. Matrix elasticity modulus variation (E_m) in the range of $(300-450) \times 10^{-1}$ MPa showed that composite strength (σ_{comp}) is monotonically decreased with E_m increase. This result is unexpected, as it contradicts the 'microswelling' traditional models, according to which the composite compression strength should be increased with modulus raise (in particular, G_m matrix shear modulus).

Experimental points for M_1, M_2, M_3 basic materials, differing in not only E_m values, but also σ_2 and σ_3 (see Table 3.39), effectively provided a set of 'pure' single-factor relationships between σ_{comp} and E_m derived from simulating experiments with various types of materials (Fig. 3.43(a)).

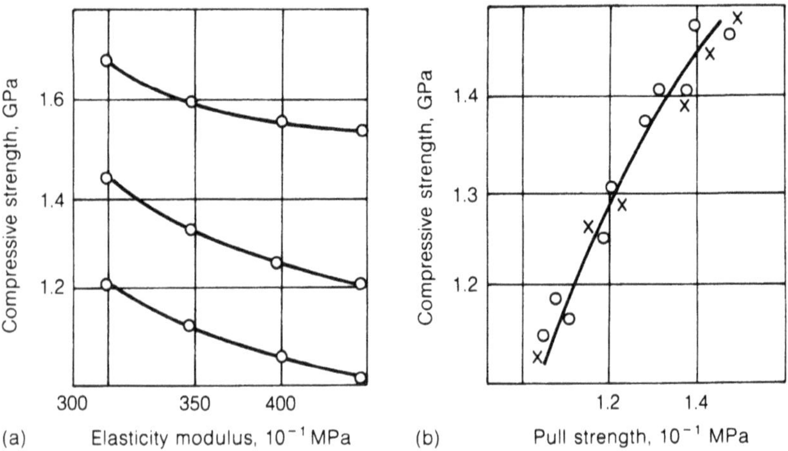

Fig. 3.43 Plots of σ_{comp} as a function of matrix elasticity modulus in the interlayer (a) and B_3 blocks bonding strength (b): ($\times$) experimental points; ($\circ$) calculated values.

This circumstance states that the assumptions made are correct and in accordance with them the matrix stiffness decrease causes energy consumption increase, which is necessary for the realization of losing stability peeling.

However, matrix stiffness effect on composite strength is not as considerable as the effect of bonding strength between blocks. Calculations showed considerable effect of transverse strength σ_{b3} on composite strength (Fig. 3.43(b)). Not only experimental points for M_1 material with some different values of σ_{b3}, but also experimental strength values for M_2 and M_3 materials, substantially differing by values of σ_{b3}, are similarly placed close to the calculated curve (Fig. 3.43(b)). This circumstance indicates that the level of composite transverse strength in the direction of element possible flicking out has the greatest effect on compression strength. Having increased the transverse strength by 2–4 MPa, it is possible to get composite compression strength increase by 300–500 MPa.

The greatest increase and practically most important results were obtained with the structure elements (block shapes) varying, i.e. with their thickness b_e and width h_e changing under the condition $H_e\, B_e = \text{const}$. The simulation experiments show that composite compression strength can be increased both through element thickness decrease and also its increase (Fig. 3.44).

Strength increase with element thickness increase is explained by individual stability increase (the right branch, Fig. 3.44(a)). At the same time, with element thickness decrease the composite compression strength is also increased (the left branch, Fig. 3.44(a)), but already through increasing bonding surface of elements with adjacent ones.

Studying materials with different shapes of structural elements confirmed the strength increase effect with element thickness decrease. The thinnest elements (points 1, 2, 3 in Fig. 3.44(b)) are in line with prepreg produced by rolling-out, thicker elements

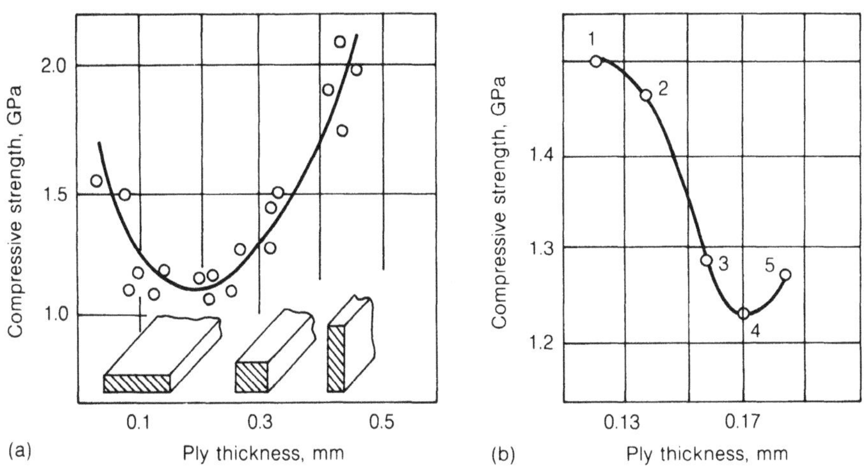

Fig. 3.44 Compressive strength as a function of strucutral element shape: (a) calculated values; (b) experimental dependence.

(point 4) were produced by tow winding on materials, while element shapes close to square (point 5) were produced by pultrusion method (impregnated tows were passed through spinnerets with simultaneous polymerization of matrix).

As a whole, the studies performed showed that compression strength can be controlled in wide ranges by varying structure element block shapes, and also other structure parameters closely connected with composite production technology.

REFERENCES

1. Konkin L.A., *Carbon and Other Heat-Resistant Fibrous Materials*, Khimiya, Moscow, 1974.
2. Tamuzh V.P., Azarova, M.G., Fracture of unidirectional CFRP and realization of fibre strength properties in them, *Mekhanika Kompozitnykh Materialov*, 1982, **1**.
3. Bezruk L.I., Khoreva G.B., Carbon fiber as an example of a self-reinforced composite, *Mekhanika Kompozitnykh Materialov*, 1982, **3**, 387–9.
4. Kobetz L.P., Effect of surface treatment of high-modulus fibers on their compatability with polymeric binders, in *Composites*, Nauka, Moscow, 1981, pp. 201–6.
5. Gunyaev G.M., Khoroshilova I.P., The effect of epoxy matrix composition on CFRP properties and technological efficiency, in *Composites*, Nauka, Moscow, 1981, pp. 214–18.
6. Dovgyalo V.A., Yurkevich O.R., Zinovyev S.N., Pomerantseva K.P., Technological properties of composites, based on the dispersed polymers and reinforcing fillers, *Izvestiya, BAS Series of Physico-Technological Sciences*, 1985, **6**, 3–6.
7. Zaitsev B.A., Kisilyova R.F., On the formation mechanism of oligomers, based on biatomic secondary fatty aromatic alcohols, *Vysokomolekulyarnye Soedininiya, series A*, 1981, **23**(8), 1783.
8. Startsev O.V., Vapirov J.M., Yartsev V.A., Mitrofanova E.A., Krivonos V.V., Effect of long-term atmospheric ageing on CFRP properties and structure, *Mekhanika Kompozitnykh Materialov*, 1986, **4**, 637–42.
9. Mitrofanova E.A., Yartsev V.A., Lukashina L.A., Tkachenko V.N., Atmospheric resistance of epoxy CFRP, *Proc. TGU*, Dushanbe, 1983, pp. 10–14.

10. Bulmanis V.N., Yartsev V.A., Krivonos V.V., Serviceability of polymeric composites structures under exposure to static loads and climatic factors, *Mekhanika Kompositnykh Materialov*, 1987, **5**, 915–20.

11. Makhmutov I.M., Sorina T.G., Suvorova J.V., Surgucheva A.I., Composites fracture with regard to temperature and moisture effect, *Mekhanika Kompositnykh Materialov*, 1983, **2**, 245–50.

12. Makhmutov I.M., Sokolovsky S.V., Sorina T.G., Suvorova J.V., Effect of moisture and preliminary loading on the strength of polymeric matrix composites under uniaxial tension, *Mashinovedenie*, 1985, **5**, 62–6.

13. Sorina T.G., Surguccheva A.I., Buyanov G.I., Finogenov G.N., Yartsev V.A., CFRP behaviour under complex effect of environment and loading, in *Composites*, Nauka, Moscow, 1981, pp. 218–23.

14. Gunyaev G.M., *Structure and Properties of Polymeric Fibrous Composites*, Khimiya, Moscow, 1981.

15. Sorina T.G., Filippova E.J., Determination of wetting edge angle of carbon fibers surface by thermoplastics melt, *Zavodskaya Laboratoriya*, 1991.

16. Gunyaev G.M., Sorina T.G., Beider E.Ya., Filippova E.J., Interfaces formation in CFRP with thermoplastic matrix, Abstracts, *Moscow Int. Conf. on Composites*, 1990, p. 109.

17. Gunyaev G.M., Sorina T.G., Strebkova T.S., Pavlova T.M., CFRP components properties and temperature strength, Abstracts, *Moscow Int. Conf. on Composites*, 1990, p. 204.

18. Suvorova J.V., Sorina T.G., Gunyaev G.M., Effect of matrix deformation properties on the realization of fibers strength in composites, *Mekhanika Kompositnykh Materialov*, 1987, 630–34.

19. Suvorova J.V., Sorina T.G., Gunyaev G.M., Rate dependencies of CFRP strength, *Mekhanika Kompositnykh Materialov*, 1990, **4**, 654–6.

20. Ovchinsky A.S., Computer aided structure-simulation modelling and its application to composites, *Mekhanika Kompositnykh Materialov*, 1987, **3**, 433–9.

21. Dobrynin V.S., Filippova E.J., Khairetdinov A.Kh., Effect of structural parameters on the mechanical properties of cross-reinforced composites, *Mekhanika Kompositnykh Materialov*, 1990, **5**, 831–5.

22. Suvorova J.V., Sorina T.G., Viktorova I.V., Mikhailov, V.V., Loading rate effect on composites fracture mode, *Mekhanika Kompositnykh Materialov*, 1980, **5**, 847–51.

23. Kachanov L.M., Composites fracture by delamination, *Mekhanika Kompositnykh Materialov*, 1976, **5**, 918–22.

24. Ovchinsky A.S., *Composites Fracture Processes: Computer-Aided Simulation of Micro- and Macromechanisms*, Nauka, Moscow, 1988.

25. Sorina T.G., Gunyaev G.M., Ovchinsky A.S., CFRP structure and strength, Abstracts, *Moscow Int. Conf. on Composites*, 1990, p. 203.

4

Organoplastics

V.D. Protassov

4.1 REINFORCING MATERIALS

Modern technology has led to the necessity to develop and use new types of reinforcing materials that are superior – in their specific strength and stiffness – to existing materials. They also make it possible to upgrade considerably present-day highly loaded structures.

Organic fibres can be considered as advanced reinforcing materials, some types of which have tensile strength more than 30 MPa, elastic modulus more than 1000 MPa and relatively low density of less than $1.5 \, \text{g cm}^{-3}$. The first superstrong high-modulus fibres were made from reinforced polyamides in two countries: the USSR [1] and the USA (by Du Pont). A little later, Moncinto, an American firm, produced high-strength X-500 fibres on the basis of poly(amidohydrazides).

Later studies in this field have led to the feasibility of producing superstrong fibres of very different chemical structures. Superstrong and high-modulus synthetic fibres from poly(Schiff bases) [2], aromatic copolyesters [3], poly(oxadiazole hydrazides) [4] and others have been developed.

Compared with the other fibre types existing today, fibres based on aromatic polyamides possess the highest mechanical parameters. This has determined their wide use as reinforcing materials for highly loaded structures.

The process of producing high-strength fibres is complicated and has particular features for fibres of various chemical compositions. The major conditions for obtaining superstrong fibres are [5–7]:

1. Presence of fibre-forming polymers with a sufficiently high molecular mass.
2. Building-up, in the fibre, of a maximum straightened conformation of macro-molecules with a stable elemental structure.
3. Maximum homogeneity of the fibre's macro- and microstructure.

These conditions are met by polymers based on a number of chloroanhydrides of aromatic diacids and aromatic diamines. They have a very high backbone stiffness of their chains while in solution their macromolecules are stretched into a rod form.

The thermodynamic equilibrium state of these polymers is matched by a certain ordered system. When flowing from the spinner openings, their solutions maintain a sufficiently high degree of molecular orientation, which can be enhanced and stabilized in the process of additional drawing and thermal treatment. The high degree of orientation reached in forming and the high structural homogeneity accompanying the former ensure production of superstrong high-modulus fibres.

It was considered before [8–10] that combined conditions needed to ensure high strength of fibres are possible only for utterly rigid-chain polymers, in particular, for aromatic polyamides of *para* structure. However, studies of the principles for high-strength polyamide fibre formation have allowed one to obtain high mechanical properties for other types of fibres, too; and not only on the basis of rigid-chain polymers but also on the basis of such flexible-chain ones as polyolefins and aliphatic polyamides.

However, the most developed and most used in production are methods that obtain fibres from utterly rigid-chain polymers. Among the latter, aromatic polyamides take the leading place. Fibre production from rigid-chain aromatic polyamides consists of two processes [5]: the first one being synthesis of fibre-forming polyamides and the second one being fibre forming.

Synthesis of fibre-forming polyamides is a low-temperature polycondensation of chloroanhydrides of aromatic bicarbonates of aromatic diamino acids [11–13] in a homogeneous or heterogeneous system of non-mixing solvents. The methods of obtaining these systems from the melt cannot be used owing to the non-meltability of aromatic polyamides and low thermal stability of initial monomers. In principle, any types of polycondensation in solution can be applied: high-temperature, interphase, emulsion and others. However, the plastics industry has used low-temperature polycondensation widely because it features low energy consumption and simple equipment. The advantage of this method is also the possibility of obtaining – in the process of synthesis – a spinning solution that can be used directly to form fibres. In this case, fibres are formed by the so-called wet method. Also known is a dry method of forming high-strength fibres. By this method, the polymer obtained in the synthesis is precipitated in water, then dried and dissolved in 98–100% sulphuric or methylsulphuric acid. To bring down the content of gel particles, the polymer is first mixed with a solvent while being cooled down to 5–15°C and then heated up to 50–80°C. From the solution thus obtained, fibres are formed by drawing through precipitation baths containing water solutions of lithium, sodium, potassium, ammonium, copper, zinc, aluminium and magnesium salts. The concentration of spinning solutions must correspond to the critical concentration of transition into an anisotropic state (11.5–12%). After bath precipitation, the newly formed fibres are thoroughly washed, additionally stretched in hot water or a vapour medium, dried and thermally treated at 500°C in a nitrogen flow.

On production and pilot-plant scales, high-strength high-modulus reinforcing fibres are made in the USA, the USSR (also tradenamed SVM at present) and Germany (Arenka). According to the literature [14], Kevlar and SVM fibres are quite near to each other in their properties (Table 4.1).

Table 4.1 Properties of reinforcing organic fibres

Fibres	Density $(g\,cm^{-3})$	Equilibrium humidity at RH 65% and 20°C (%)	Failure stress in tension (MPa)	Elongation at break (%)	Relative strength in wet state (%)	Specific strength	Elastic modulus in tension (MPa)	Specific elastic modulus $\times 10^{-2}$	Destruction temperature (°C)
Poly(p-benzamide) (Kevlar 49)	1.45	2.0	37–40	1–4	94–96	255–275	1200–1400	83–97	300–400
Poly(amidohydrazide) (X-500)	1.46	2.0	19.85–22.5	3–4	–	136–154	860–1058	59–72	525
Aramid (Terlon)	1.46	2.0	11.7–14.6	1–1.5	80–82	80–100	850–1200	58–82	500
Aramid (SVM)	1.45	3.7	35–40	2–4	86–96	241–275	1200–1300	83–90	350
High-strength carbon	1.7–2.0	–	20–35	0.5–0.8	–	118–175	2000–6000	118–300	–
Aluminoborosilicate glass	2.6	1–2	14–22	2–3	75	54–85	600	23	700 melting

Table 4.2 Physicomechanical properties of Kevlar fibres

Properties	Kevlar	Kevlar 29	Kevlar 49
Density ($g\,cm^{-3}$)	1.44	1.44	1.45
Linear density (tex)	1.70	1.70	–
Elongation (%)	4.0	3.0–4.0	2.8
Strength ($cN\,tex^{-1}$)	185	194	254
Elastic modulus ($cN\,tex^{-1}$)	4190	4240	8950
Shrinkage at $160°C$ (%)	0.2	0	–

Having been developed quite recently, aramid fibres are expanding into various fields of modern engineering in place of other materials. The current production of aramid fibres makes up about 4000 tonnes per year, while in the next five years or so it will probably increase almost 10-fold.

Under the general trademark of Kevlar, the USA produces a group of fibres with somewhat different properties: Kevlar is designed for production of mine cord, Kevlar 29 for rubber engineering parts and ropes, and Kevlar 49 for reinforcement of structural materials [15]. The properties of the three types of Kevlar fibres are given in Table 4.2.

Aramid fibres have very valuable properties. In their tensile strength and elastic modulus, they are superior to all organic and glass fibres (Table 4.1). Their strengths are twice and moduli four times as high as those of nylon and 50% higher than those of glass fibres.

The combination of both high strengths and moduli of aramid fibres with low densities leads to considerably higher values of their specific strengths as compared with all other fibres.

The fibres possess good thermal stability: in particular, they maintain the initial level of properties after being heated up to $260°C$. These fibres neither burn nor melt, they do not have a clearly expressed glass transition temperature and when heated above $350–400°C$ they get continuously graphitized, which causes fibre destruction at higher temperatures. As the destruction process is time-dependent, short-time temperature exposures can be withstood by the fibres themselves, even at rather high values of temperature. The limit here is usually determined not by the fibres but by the matrix material that is reinforced by these fibres.

Aramid fibres can be considered practically as non-shrinking when compared with other types of chemical and synthetic fibres (viscose, acetate, polyamide, Lavsan, polyolefin, polyacrylonitrile and others). SVM fibres have a small (0.04%) shrinkage at a temperature of $350°C$. Aramid fibres maintain their properties when impregnated by a binder in the process of making a composite material.

They do not require any additional surface treatment for better wettability by binders and higher adhesion. The adhesion strength of SVM fibres to epoxy binders (0.4–0.5 MPa) is at the same level as it is for mineral fibres.

Aramid fibres feature high fatigue strength and inconsiderable creep at both ambient and elevated temperatures.

Absolute deformation of these fibres in long-time loading does not exceed 1%.

The fibres possess higher resistance to cutting and abrasion, which facilitates their processing into goods. Thus, for example, in weaving they lose 10–20% of their initial strength while glass fibres lose up to 50% [16].

Stability of strength in machining is due to the fact that polymer fibres have little sensitivity to damage while their strengths in tension depend on the fibres much less in comparison with other types of fibres such as glass, carbon, boron and others.

Aramid fibres maintain relatively high strength at break in a loop and in a knot. For instance, a thread of PRD-49 fibres (280 filaments, diameter of fibres 11.6 μm) has loop and knot strengths of 7.7 and 13.2 MPa respectively, and the thread's strength after textile processing is 24.8 MPa (initial strength of the thread being 27.6 MPa) [17, 18].

Aramid fibres are resistant to water, solvents, fuels and lubricants. Acids and alkalis attack them only at higher concentrations and temperatures.

Aramid fibres have low heat conductivity ($0.12\,\text{kcal}\,\text{m}^{-1}\,^{\circ}\text{C}^{-1}$), their heat capacity is $0.30\,\text{kcal}\,\text{kg}^{-1}\,^{\circ}\text{C}^{-1}$ and their coefficient of thermal expansion is $2.5 \times 10^{-6}\,^{\circ}\text{C}^{-1}$ (values are given at 20°C).

Among the disadvantages of these fibres, one should note their somewhat lower compression strength and relatively large coefficient of variation of strengths and elastic moduli as well as a higher build-up of static electric charges.

Aramid fibre strength is influenced by ultraviolet (UV) radiation, which requires corresponding protective measures for the material against its effects.

All the positive features mentioned above for organic fibres (SVM, Kevlar 49) have determined their wide use in making pressure vessels by filament-winding methods.

The winding machines, tape-guiding systems, thread-laying heads, impregnation baths, winding software and other machinery developed earlier to produce goods from glass fibres can also be used successfully to make items from organic fibres. Organoplastic parts are wound by a tape formed from twisted threads or tows according to the part's dimensions.

As for any other reinforcing fillers, realization of the organic fibres' strength in a plastic depends to a considerable degree on their textile structure: twist, fibre diameter, their number and linear density.

Investigations carried out to determine how the structure of organic fillers influences the strength properties of a composite material have basically confirmed the mechanisms influencing glass fillers' structure. They have also shown some structural peculiarities of threads made of aramid fibres.

Among the textile structures of twisted threads developed recently, the most effective one – considering the technological capabilities of their processing into parts – is an SVM thread of linear density 29.4 tex, twisted 100 turns/metre and consisting of 200 fibres with 12 μm diameter. Among the tows, the most effective one is an SVM roving of linear density 1000 tex, from fibres with 12 μm diameter.

Threads are subjected to textile processing. The most widely used is a satin-weave cloth. Table 4.5 shows the characteristics for some types of cloths made of organic fibres.

Along with aramid fibres, recent times have seen the use of high-strength polyethylene fibres. They have strengths up to $450\,kg\,mm^{-2}$ at specific weight of $0.97\,g\,cm^{-3}$. This gives specific strengths at the level of 460 km.

Aramid fibres are known to be sensitive to atmospheric effects: moisture and UV radiation. High-strength polyethylene (PE) fibres do not have these disadvantages. Practically, 100% retention of properties in both ordinary and sea water gives these fibres a substantial advantage. PE fibres show characteristic low reduction (20%) as compared with aramide fibres (15%) for strength parameters after exposure to UV radiation for one year.

The advantage of polyethylene fibres in specific moduli can be as high as 50% compared with aramid fibres.

Other advantages of polyethylene fibres are also worth mentioning: chemical and wear resistance, impact strength and radiotransparency.

It should be noted that serious attention has been paid recently to the physical processes responsible for thread and tow failure. This makes it possible to take certain measures to gain more complete use of the fibres' initial strength [19, 20].

4.2 BINDERS

When producing parts from composite materials, an important role is played by the polymer matrix (binder). In a cured state, the latter ensures a monolythic composite material, fixes the form of the part and relative position of the filler's particles, distributes the stresses acting within the material's volume, provides uniform loading of the reinforcing fillers, redistributes loads when the filler's particles fail and so on.

The choice of a binder to make a composite material with the necessary operating properties is determined by the possibility to realize these properties when combining this binder with a reinforcing filler, by the possibility to set up an effective technology to produce parts from composite materials and by economic considerations in some cases [19, 20].

As a rule, requirements for a composite material, and thus for its binder, are set not only for their mechanical characteristics but also – depending on operating conditions – for a number of special performances: for example, stability of mechanical characteristics in a given a temperature range, water and chemical resistance, radiation stability, non-flammability, non-toxicity and the like.

In their structural arrangements, the binders can be polymers of either network or linear structure. The former are known as thermosetting polymers while the latter are thermoplastics.

The most widely used to make plastics for engineering purposes are polymers of space-network structure. The basic component of most network polymers is a mixture of reactive oligomers, usually called a resin, and more seldom an individual oligomer of a given structure.

The nature of oligomers, i.e. type, number and arrangement of reactive groups in them as well as their molecular weight, determine the resin's softening point and solubility, melt or solution viscosity, the impregnating and wetting capabilities, curing conditions and processes taking place, and the structure and properties of the cured binder. If the oligomeric functional groups are capable of reacting among themselves, forming a network polymer, binder curing is accomplished at elevated temperature (in the case of thermosetting resins) or by using a catalyst (or initiator). Curing catalysts or initiators are usually introduced in small quantities. Catalysts do not form part of the composition of the obtained polymers.

To cure resins whose functional groups are not capable of reacting with each other, low-molecular-weight polyfunctional compounds are used -- hardeners, which become links in a network polymer when reacting with oligomers. The reaction of a low-molecular-weight hardener with oligomers can be accompanied by smaller steric and diffusion restrictions as compared with the reaction of oligomers among themselves. This enhances a higher degree of conversion for the oligomers' reactive groups and thus a higher degree of cure. Most curing binders contain resins of hard-to-reproduce compositions, which makes it difficult to obtain binders of a strictly controllable composition, and thus thermosets and parts made of them, with reproducible properties. The necessity to solve this problem has brought about a desire to replace the mixture of oligomers by an individual oligomer of a given composition, with fixed number and arrangement of reactive groups [21, 22].

Beside an oligomer or resin, hardener, catalyst or initiator of cure, inert solvents are introduced into the binder's composition when needed. Solvents lower the binder's viscosity, thus facilitating its matching with the filler. Before moulding parts, the solvent must be completely removed from the thermoset. In the cases when moulding conditions require the solvent's presence in the binder to ensure its low viscosity at the moulding stage, so-called active solvents are used – monomers or low-viscosity oligomers capable of taking part in the curing process or of entering the network polymer's structure.

If the binder's viscosity is so low that the binder – under certain conditions – can drip out from the reinforcing filler, some thickeners or thixotropic additives are introduced into its composition. Thickeners are usually polymers or high-viscosity oligomers that dissolve in the binder, while thixotropic additives are finely dispersed powders with high surface energy.

Production of composite plastics widely uses curing binders based on esters, epoxy binders and binders that cure by a polycondensation mechanism (phenolic, amino-aldehyde and silicone resins) as well as binders based on cyclic oligomers with functional end-groups.

When organic fibres are used as a reinforcing material, the requirements are more fully – in comparison with other known binders – met in the case of using, as a matrix, binders based on epoxies.

Much more seldom, phenolic and acetal binders are used. Table 4.6 compares data for organoplastics based on epoxy and polyester matrices.

The major component of epoxy binders [23–25] is a mixture of oligomer products

Table 4.6 Mechanical properties of laminated composite materials based on epoxy and polyester matrices and cloths (epoxy/polyester values in this order) from aramide (Kevlar 49) fibres [6]

Filler	Density $(g\,cm^{-3})$	Binder content (vol%)	Failure stress (MPa)			Elastic modulus (GPa)	
			In tension	In flexure	In interlaminar shear	In flexure	In tension
Aramide cloth, trademarks							
243	1.31/−	44.60/−	561/−	312/−	36.20/−	29.99/−	40.82/−
281	1.29/1.26	36.0/37.0	499/445	311/239	35.30/30.48	27.10/18.89	25.93/24.06
285	1.28/1.24	32.5/29.8	500/454	349/292	30.82/25.37	23.86/21.93	27.30/25.44
328	1.34/1.29	54.4/46.4	370/378	140/139	20.34/19.92	16.00/15.38	20.34/18.41
1050X	1.28/1.23	32.1/25.6	512/481	303/290	31.79/19.65	22.68/21.65	25.58/31.17
1033X	1.30/1.27	38.6/39.9	374/350	239/156	28.96/22.27	19.10/15.31	24.41/21.65

with epoxy groups in the end links, epoxy resins. The latter are obtained in two ways. The first one is the reaction of epichlorohydrin with diatomic (more seldom, polyatomic) alcohols or phenols, forming diglycidyloxy ethers

$$CH_2-CH-CH_2-O-R\!\left(\!O-CH_2-CHOH-CH_2-O-R\right)_{\!n}$$
$$\underset{O}{\diagdown\diagup}$$
$$O-CH_2-CH-CH_2$$
$$\underset{O}{\diagdown\diagup}$$

(R is diatomic phenol or alcohol radical; $n = 0-10$) or with aromatic amines, forming diglycidyloxyamines

$$CH_2-CH-CH_2-NR\left(NR-CH_2-CHOH-CH_2-NR\right)_{\!n}$$
$$\underset{O}{\diagdown\diagup}$$
$$CH_2-CH-CH_2$$
$$\underset{O}{\diagdown\diagup}$$

(R is aromatic amine radical; $n = 0-3$). The second method is epoxidation of compounds that have double linkages (cycloaliphatic epoxies).

Wider use is made of epoxies obtained from epichlorohydrin and diphenylolpropane (bisphenol A), called diane resins (ED-type resins), or from epichlorohydrin and polycondensation products of methylolphenols, called polyepoxy or epoxyphenol resins (EPh, EM and other resins). Recent times have seen the use of resins from epichlorohydrin and aniline (EA resin), diaminodiphenylmethane (EMDA resin), p-aminophenol (UP-610 resin), cyanurates (EC resin), cycloaliphatic epoxies (UP-612, UP-632), dioxydicyclopentadiene (DDCPD resin) and others. Low-viscosity reaction products of epichlorohydrin with di- and triethylene glycol (DEG and TEG resins) are used as active thinners or modifiers of epoxy binders based on ED and EPh resins.

The high reactivity of epoxy groups as well as the presence of hydroxyl groups in epoxy oligomers cause various processes to cure epoxy resins. As a rule, cure takes place with little volume shrinkage and without producing low-molecular-weight substances.

As catalysts in curing of epoxy resins, widest use is made of triamines and Lewis acids (triethanolaminotitanate, BF_3 monoethylamine complex and others), while their most frequent hardeners are polyfunctional amines (aliphatic, aromatic and low-molecular-weight polyamides and others) and acid anhydrides (maleic, phthalic, tetrahydrophthalic, etc.).

The curing agent is mixed with an epoxy resin in the melt. If the melt viscosity is great or the melting point of one of the components is too high, mixing is done with the use of an inert solvent, for example, acetone, which must later be thoroughly removed, or using an active thinner, a low-viscosity epoxy resin (DEG, TEG and the like).

Among all the binders considered, epoxy binders have the greatest variability of chemical structure [26]. The choice of epoxy oligomer composition and molecular weight, type and amount of curing agent and catalyst gives the possibility to change – in wide ranges – the nature of links and population of chemical groups in a network polymer. When epoxy binders are cured in the presence of catalysts, polymers of an open network are formed, in which oligomer links are connected with each other by flexible simple oxyether links. Network polymers from epoxy resins cured by amines are dense-network polymers containing oxyamine and oxyether groups. When epoxy resins are cured by anhydrides of acids, the oligomer links are connected by oxyester groups. In the presence of curing catalysts, the reaction of epoxy and hydroxyl groups brings about oxyether and oxyester links simultaneously. The ratio of oxyether to oxyester links depends on the amount and activity of the catalyst.

Cured binders are three-dimensional polymers whose links and chains are connected to each other at polyfunctional points (chemical groups), thus forming a common spatial structure – a polymer network.

The network density in cured polymers is determined by the chain length between the oligomers' reactive groups, by the functionality of the oligomers and curing agents, and by the conversion degree of reactive groups (curing ratio).

The overwhelming majority of cured binders are amorphous polymers. Owing to the different reactivities of functional groups and to the effects of phase separation when binders pass from the liquid state into the solid one, the curing process of multicomponent binders is accompanied by the appearance of microregions with different densities in the network polymer. The cured binder usually has a heterogeneous structure – a microgel, micrograin or globular one.

After moulding, parts made of thermosets are cooled down to room temperature, which leads to the polymer's glass transition and fixation of the cured binder's non-equilibrium packing of network chains. A more equilibrium packing of chains in the network polymer can sometimes be obtained by additional heat treatment at a temperature above T_g of the cured polymer with a slow cooling to follow. If T_g of the network polymer exceeds its thermal destruction temperature, the chain packing of the polymer network that results during cure will be irreversible, while thermal treatment at a temperature below T_g can only bring about some additional cure of the binder or its cracking.

Binders based on epoxy resins and their modifications show characteristic universal properties. They possess the high mechanical and adhesion properties necessary to produce a monolithic composite material and to realize the reinforcing fillers' properties as much as possible; a higher thermal stability, characterizing their resistance to temperature/time factors when properties are reversible; high processibility, which allows the processes of impregnating the reinforcing filler and forming the part to be combined into one technological cycle or, when needed, to be separated; a relatively low cure temperature; and a number of other useful properties.

Production of load-bearing structures from composite materials by the method of wet winding makes greatest use of binders EDT-10, K-35-90, KU-5-75, K-69M, K-10-56 and others, which have high processibility and provide a quite high level of physicomechanical characteristics for composite materials. Basically, these are binders based on diane resin ED-20 with various thinners and modifiers. The common disadvantage of these binders is their great loss of strength at temperatures of 150–200°C. For example, it amounts to 79–94% when the binders are exposed to flexure loads.

As a number of structures are to be used at elevated temperatures, thermal stability of polymer matrices is required. This makes it a must to find thermally stable binders.

Optimization of binders has the purpose of producing new types of binders and modifying existing ones, which can have higher thermal stability and whose application will allow one to bring down structure weight and enhance processibility in their production as well as performance of parts.

Higher thermal stability is provided by binders based on cycloaliphatic resin UP-632, epoxidized chloroamine EHD, epoxycyanurate and epoxyaniline resins EC and EA, and dioxydicyclopentadiene DDCPD. Table 4.7 gives compositions, technological and physicomechanical characteristics of binders based on the resins mentioned above and, for comparison, those of a binder based on diane resin ED-20, which is not thermally stable.

The values of physicomechanical properties for the given binders are approximately at the same level except for their values of ultimate strains and the lower – as compared with the other – tensile strength of binder ECA-EM. However, high thermal stability of binders is connected, as a rule, with their lower elasticity and tendency to brittle failure and cracking. This leads to deteriorating properties of parts made on the basis of such binders, in particular, after prestressing. In this case, thermally stable binders are sometimes modified to enhance their deformability via chemical addition of low-molecular-weight elastomer chains to an epoxy resin having functional end-groups. Here, the elastomer provides conditions necessary to transfer deformation energy to rubber particles formed due to segregation in the curing process.

To raise the elasticity of thermally stable binders, one can use modifiers such as carboxyl-terminated butyl acrylate rubber, methylvinylpyridine rubber, butadiene–acrylonitrile rubber SKN-20 KTPA and others.

Investigations have shown that binders modified by synthetic rubbers improve their mechanical characteristics while lowering thermal stability a little.

For example, an EHD-resin binder modified by butadiene–styrene rubber SKN-20 KTRA and a DDCPD-resin binder modified by methylvinylpyridine rubber enhance their strength by 15–20% and elongation by 1.5–2 times while lowering thermal stability by as little as 10%.

The physicomechanical characteristics of a composite material are known to be determined by the properties of fibres and polymer matrix, their structure, amount and distribution, and also they depend – to a considerable degree – on the connections between them. The major condition for obtaining effective connections is the wetting, as fully as possible, of fibres by liquid binder in the impregnation process. The better the wettability of fibres, the better is the binder's flow over their surface. This results in fewer voids, which are stress concentrators in a plastic and a cause of premature failure of a structure.

Table 4.7 Composition, technological and physicomechanical characteristics of epoxy binders

| Binder[a] | | | Thermal stability, | Processing | Physicomechanical parameters | | |
Trademark	Formulation (weight parts)	Cure cycle (°C/h)	Martens (°C)	temperature (°C)	σ (MPa)	ε (%)	E (MPa)
UP-632M	UP-632 (100) i-MTGPhA (109) UP-606/2 (0.5)	80/6, 120/4, 160/10	165	20–25	45–55	1.0–1.5	140
EHD-M	EHD (100) i-MTGPhA (10)	80/6, 100/2, 120/2, 160/10	165	30–35	45–50	1.5–2.0	160
ECA-EM	EC (50) EA (50) i-MTGPhA (112)	80/6, 100/2, 120/2, 160/10	145	30–35	35–40	3.0–3.5	150
UP-278L	DDCPD (100) UP-63 (50) MA (150) ET (4) BF_3 aniline (0.5)	120/6, 140/4, 160/8	195	45–50	45–50	1.0–1.5	170
EDT-10	ED-20 (100) DEG-1 (10) TEAT (10)	100/1	95	60–65	45–50	3.0–4.0	150

[a] UP-606/2 is 2,4,6-tris(dimethylaminomethyl)phenol.
i-MTGPhA is isomethyltetrahydrophthalic anhydride.
DEG-1 is epoxidized ethylene glycol.
TEAT-10 is triethanolaminotitanate.

UP-63 is diglycidyl ether of resorcinfurfuryl alcohol.
MA is maleic anhydride.
ET is ethylene glycol.
BF_3 is a complex catalyst.

Wettability problems have been discussed quite fully in the literature [3]. Therefore, this study shows, as an example, the wettability kinetics of an SVM synthetic fibre with some binders as determined by a standard method: capillary rise of a liquid in cathetometer KM-6, the temperature of each binder corresponding to its processing temperature.

The wettability kinematics for an SVM fibre indicates that this process is slow for all binders and considerably exceeds the time when the binder is in contact with the fibre directly in the impregnating bath and the time when it moves along the thread-feeding trajectory of the winding machine to the mandrel, which is usually not more than 30 s. Besides, the wetting rates of fibres with various binders are different, the highest one being for binder EDT-10.

High impregnability (wettability) is a necessary but by far not a sufficient condition to make a composite material with high mechanical characteristics. The polymer matrix must also possess a complex of properties to ensure maximum strength of a reinforced material. Studies of a plastic based on an SVM fibre with $\sigma = 27–30\,\mathrm{MPa}$, $E = 1100\,\mathrm{MPa}$, $\varepsilon = 30\%$ and $\gamma = 1.45\,\mathrm{g\,cm}^{-3}$ have shown the composite's strength to depend – in a certain form – on the wettability of fibres with binders (Table 4.8). Comparison and analysis of data allow one to see the dependence of an organoplastic's strength on wettability of fibres with a binder.

As capillary action of a binder on fibres increases (better wettability), an organoplastic's strength substantially improves, which implies – without considering effects of the binder's physicomechanical and deformation properties – a direct dependence of the organoplastic's strength on the value of adhesion between the fibre and polymer matrix at their interface. This is confirmed when data in Tables 4.8 and 4.9 are compared, where adhesion strengths of the binders under investigation to an SVM fibre are shown.

The burning problem of further optimizing epoxy resins is not only improvement of their performance but also enhancement of their technological properties, in particular, improving their life or, more generally, the possibility to regulate the cure rate. The latter allows one to make – on their basis – preimpregnated and partially cured tapes and tows (prepregs). Long life of the binder in a prepreg opens up the possibility

Table 4.8 Physicomechanical characteristics of organoplastics when tested in tension

Binder trademark	Tensile strength, σ (MPa)	Elastic modulus, E (MPa)
UP-632M	14.8	697
ECA-EM	15.1	600
EHD-M	16.4	747
UP-278L	11.4	660
EDT-10	17.6	789

Table 4.9 Adhesion strength of binders to an SVM fibre

Binder trademark	Average adhesion strength (MPa)	Number of specimens tested
UP-632M	0.375	61
ECA-EM	0.384	54
EHD-M	0.386	75
EDT-10	0.424	56

to separate and specialize technological processes of obtaining winding semiproducts and for winding products proper. In production of heavy parts, this can substantially raise economic efficiency of prepregs being used.

Binder life generally depends on both chemical composition and structure of oligomers, and on the curing system applied. It is essential that when storing a prepreg the binder does not change its properties for as long as possible, which would preserve the material's processibility during the entire production cycle and thus provide stability of cycles in prepreg processing and stable characteristics of the parts obtained. Usually prepregs are stored and processed at a temperature of $25 \pm 5°C$, at which epoxy oligomer gradually reacts with the curing agent to produce later non-linear polymers of higher molecular weight. As the molecular weight of a linear polymer grows, its glass transition temperature rises and, at the point where it exceeds ambient temperature, the polymer becomes brittle while the prepreg on which it is based becomes non-processible.

Most epoxy resins considered above in combination with matching curing systems have limited life.

Epoxy oligomers whose molecular weight is above 500 (for example, epoxy-novolacs EN-6, EM-18, UP-643 and diane resins ED-16, ED-8, E-49) cannot be used for making prepregs because the glass transition temperature of a linear polymer — which results in the reaction of an epoxy oligomer with a curing agent — rises above 20°C already in the first 10 days while the material losses its processibility properties.

Aminoepoxy oligomers (triglycidyl ether of *p*-aminophenol) contain tertiary nitrogen in their structures, which catalyses the resin/curing agent system, and thus lowers this system's life. Besides, these products can self-cure at elevated temperatures.

Epoxy resins based on polyatomic phenols (diglycidyl ether) have high strength parameters in the cured state but, considering their high reactivity and increased toxicity, their use is extremely inadvisable.

For cycloaliphatic resins UP-612, UP-632 and others, the major curing agents are acid anhydrides, which when stored for a long time react quickly with moisture from the air forming diacids. This leads to quick formation of linear polymers and thus to worsening of prepreg processibility.

Of greatest interest — from the point of view of prepreg life and obtaining utmost mechanical characteristics in a plastic — are diane epoxy resin ED-20 based on bisphenol A, whose molecular weight does not exceed 500, as well as diane resins modified by aliphatic glycidyl ethers.

Binder EDT-10 based on a KDA product (liquid epoxy compound) and the curing agent triethanolaminotitanate (TEAT) has already found use in producing preimpregnated materials. Composite materials produced using this binder have high strength parameters; however, the life of prepregs based on them is only 10–15 days. When TEAT is replaced by another curing system, their life can be considerably prolonged.

Among the great number of various curing agents for binders with long storage life described in the literature, one should note chlorinated aromatic amines and urea derivatives. The investigations mentioned have shown that a binder with a chloro-aminobenzylaniline curing agent ensures a prepreg's life for 45 days, while a binder with an ethyleneurea does so for 90 days.

One of the important technological operations in production of composites is binder cure, which takes place from the moment the binder is prepared and is over in the finished part.

It is at this stage that the part's shape is fixed and the composite material's final structure is created. Therefore, the processes going on in the binder when cured determine – to a decisive degree – the technology to make a reinforced plastic and parts from it, and the material's performances. The formation of any network polymer includes two successive stages: the initial one, till the moment when a polymer network appears, i.e. till the so-called gel point; and the final one, the formation of a polymer network after gelatinization. At the moment a polymer network appears, the binder loses its ability to pass into a viscous flow state and to dissolve. At the final stage, the depth and rate of cure are characterized by conversion of the binder's reactive groups in the curing process or by population of chemical junctions generated in the polymer network.

Kinetic features when binders are cured at both the initial and final stages are determined by the mechanisms of reactions going on during cure.

Cure of epoxy binders proceeds by a mechanism of step polymerization, which – at the initial stage – is characterized by gradual growth of viscosity until flowability and solubility are lost at the gel point, with reactive groups reaching a certain degree of conversion. The latter is mainly defined by the functionality of oligomers and curing agents. The gel time decreases as the number of functional groups and their reactivities increase. The rate of curing reactions is decisively affected by temperature and catalysts. Cure by the step mechanism usually slows down after the gel point due to steric and diffusional restrictions. The cure process terminates either due to exhaustion of reactive groups or as a result of sharp growth of steric and diffusional restrictions, which hinder the reacting functional groups. Thus, the rate and depth of a binder's cure in the final stage after gelatinization are determined mainly by the mobility and flexibility of chains in the polymer that is being formed.

The binder's cure process resulting in a network polymer is accompanied by transition of the binder from a viscous liquid into the solid state. In the cure process at constant temperature, the binder's viscosity at the initial stage or its stiffness at the final stage can increase drastically not only as a result of gelatinization but also because of glass transition of the binder due to increasing molecular weight, formation of new chemical and physical ties or increasing density of the network. The polymer's physical state at any stage of its formation and, thus, the flexibility of its chains is determined by the relationship between the polymer's glass transition temperature and its curing temperature. If the curing temperature becomes lower than the glass transition temperature of a binder that has reached a certain cure stage, further cure at this temperature stops due to diffusional restrictions. Thus, if the cure temperature is lower than the polymer's glass transition temperature at the gel point, the sharp rise of the binder's viscosity will not be the result of gelatinization but that of the binder's glass transition.

As the cure temperature rises above the binder's T_g at a given stage, the curing reaction will continue and, having reached a stage where the polymer's T_g again exceeds the cure temperature, the reaction will slow down anew. Thus, curing reactions can take place till absolute exhaustion of reactive groups only at temperatures above the glass transition temperature of a completely cured binder. At temperatures below T_g, complete cure is unattainable. The polymer still has reactive groups but their reactions with each other or with other groups are hindered due to loss in segmental mobility of the network polymer as a result of its glass transition. It is as if the cured binder is in a 'non-equilibrium' state, and when parts based on it operate at elevated temperatures, the cure process will continue accompanied by changing dimensions, shape and properties of the part.

The cure cycles for binders UP-632M, EHD-M, ECA-EM, UP-278L and EDT-10 given in Table 4.8 have a step-growth nature and are presented on the basis of the above information.

Major characteristics of the cure process also include volume shrinkage and amount of low-molecular-weight substances (volatiles) produced. Volume shrinkage during cure is due to the binder's growing density in the cure process as a result of the great number of new chemical ties that come into being. It should be distinguished from temperature shrinkage, which is connected with cooling. Volume shrinkage decreases as the oligomer's molecular weight increases and the number of functional groups in it comes down. In the process of moulding parts, volume shrinkage can be lowered by precuring the binder. For epoxy binders, it is 3–6%.

Volatile products resulting in cure are low-molecular-weight products of reactions, remaining solvent, easily evaporating components of a thermoset or partial destruction products of a thermoset's components under cure conditions. Volume shrinkage and the pressure caused by volatiles lead to residual stresses, changes in dimensions and shape of the parts, pores and cracks in the material.

The new class of high-strength fibres based on high-molecular-weight polyethylene has raised the problem of choosing polymer matrices for polyethylene composites. In this case, the problem is worsened by low adhesion of binders to polyethylene fibres by low temperatures of thermal treatment, which are limited by the heat stability of fibres (not more than 100–120°C). Investigation shows that the best wettability is ensured by binders based on polyethers but their adhesion capability is lower than that of epoxy binders. Wettability of polyethylene fibres can be enhanced by introducing small amounts of adhesion promoters of vinyltriethoxysilane type. The best results have been obtained for a composition based on a diane resin, with a cure promoter. But binder adhesion to polyethylene fibres is not sufficient without activating the latter's surface, for example, by plasma treatment.

4.3 COMPOSITES

The major direction in making new materials for load-bearing assemblies and structures is connected with obtaining as great a specific strength as possible and its optimal distribution according to the field of acting loads. This problem can be successfully solved when using composite materials with tailored anisotropy of properties.

Beside the high level of specific mechanical characteristics, the material must meet a number of other technical requirements that are essential for normal operation in the process of its operation. These requirements are: resistance of the plastic to heat, creep, radiation, chemical agents, climatic effects, biological damage and others.

Composite materials are heterogeneous systems whose stiffness and strength are determined by reinforcing elements, while joint work of these elements is ensured via an isotropic binder and certain technological cycles of processing.

Table 4.10 Processing cycles for organic fibres (method of filament winding)

Technological parameters of processing	SVM fibre		ARMOS fibre		Polyethylene fibre
	Thread	Tow	Thread	Tow	Thread
Nominal density (tex)	29.4	1000	100.0	600	40
Optimum width of tape (mm)	10–30	30–50	10–30	30–50	20
Amount of fibre in tape (filament, tow)	70	9	70	15	80
Type of binder used	EHD-MK	EXD-MK	EHD-MK	EHD-MK, EHD-MD	Epoxy
Content of binder in tape for pressure vessels (%)	41 ± 2	42 ± 2	33 ± 2	33 ± 2	35 ± 2
Tension (kgf) filament	0.6	16–20	2.0	10–15	70
tape	40–200	150–250	50–250	180–300	–
Realized strength (max.) in unidirectional ring (MPa)	310	295	340	315	170
Realized strength (max.) in structure ($kgf\,mm^{-2}$)	280	250	300	280	–
Specific time to wind a pressure vessel ($h\,m^{-3}$)					
from filament	10.0	8.0	10.0	8.0	10.0
from tow	3.0	2.8	2.0	2.8	–

Table 4.10 gives some processing cycles for organic fibres by the method of filament winding.

Other methods of technological processing (moulding, contact spraying of moulds, etc.) are easily chosen by the usual method [19, 20].

As noted before, the most promising reinforcing fillers for structural plastic that have maximum specific strength for each type of fibres can be named:

1. VM-1 and VMP fibres for glass-fibre plastics.
2. SVM fibre for organic plastics.
3. VMN-5 fibre for carbon plastics.
4. BN-1 boron filament for boron plastics.

Physicomechanical characteristics of fibres and trademarks of fillers based on them used to produce unidirectional plastics by the method of filament winding are given in Table 4.11.

Comparing specific strengths of the cited fillers, one can note considerable advantages of organic fibres over all other fibres. Of all the polymer binders developed today and providing the best mechanical properties and performance of parts, as well as having high processibility, we should distinguish binders based on epoxy oligomers: ED-20, EHD and UP-643.

Unidirectional epoxy composite materials based on advanced fillers have a high level of physicomechanical characteristics (Table 4.11). Their stress–strain relationship in tension is practically linear up to material failure.

Comparison of characteristics for unidirectional composites shows that plastics based on SVM fibre have greatest specific strengths, of the order of 136 km, which

Table 4.11 Physicomechanical properties of unidirectional epoxy plastics (filler volume fraction 62–65%)

Type	Density $(g\,cm^{-3})$	Failure stress in tension (MPa)	Failure stress in tension (km)	Modulus of elasticity (MPa)	Modulus of elasticity (km)	Coefficient of fibre strength realized in plastics
Glass-fibre plastic (VM-1 + EDT-10)	2.08	17.5	84.1	580	2.8	0.70
Glass-fibre plastic (VMP + EHD-M)	2.12	18.5	87.3	610	2.9	0.68
Organoplastic (SVM + EDT-10)	1.32	18.0	136.4	700	5.3	0.83
Carboplastic (VMN-5 + UP-632)	1.40–1.50	10.0–11.0	69–76	1500	10.3	0.62
Boroplastic (BN + EDT-10)	2.0–2.1	11.0–12.0	54–59	2200	10.7	0.67

is 36% more than those of glass-fibre plastics. Specific strength characteristics for carbon and boron plastics are lower than those of glass-fibre plastics by 13 and 20%, respectively.

An important factor characterizing a composition's efficiency is the realization coefficient of the reinforcing filler's strength, which is the ratio of the fibre's strength in a plastic to the fibre's initial strength. Organoplastics have the highest realization coefficient of the fibre's strength (0.83), which exceeds similar coefficients for glass-fibre plastics by 14–15% and for carbon and boron plastics by 25–15%.

The major requirement of parts is the highest possible specific strength of the part's material with a certain level of deformability. Hence, of all the materials considered above, organoplastics can be the most promising for production of large parts because they have the highest specific strength and relatively high specific stiffness, about two times greater than glass-fibre plastics.

Organoplastics' behaviour under mechanical loading features a relatively small scatter of strength characteristics when testing material of the same batch. If this scatter reaches 20–30% for glass-fibre plastics, changes in strength parameters for organoplastics do not exceed 10%. Two main reasons for high reproducibility of a materials' strength parameters can be noted. First of all, organoplastics have few pores – concentrators of stress, which weaken the material's cross-section and promote cracks developing in it. Besides, the phase separation boundary in a plastic is blurred owing to the binder's diffusion into the fibre, which lessens the probability of a crack initiating and spreading along the fibre's surface.

Failure mechanisms of organoplastics under external loading have not been studied much. Taking into account the above peculiarities of composition and structure in these materials, as well as the anisotropy of properties in the fibre proper, one can assume that stress distribution between the components, their deformation, and initiation and growth of microdefects leading to failure differ substantially from the ones in known composite materials. The filler's polymer nature and complicated physicomechanical relationships of the binder with fibres in organoplastics make it necessary to evaluate the components' mechanical properties directly in a composition.

In organoplastics based on aramide fibres, elongations at break of fibres and epoxy matrices are close in value ($\varepsilon_f = 1.5$–3%, $\varepsilon_m = 3$–4%), which ensures a monolithic composition that is practically linear up to failure when a unidirectional plastic based on SVM fibres is subjected to tension.

The above considerations make one expect that the weakest link in organofibre plastics is the border between structural elements of the anisotropic polymer fibre (fibrils or individual macromolecules). In this connection, failure of high-strength organofibre plastics appears to be accompanied by fibre splitting, the crack growing in that part of a fibre where the binder strengthening the fibre has not reached. Fibre splitting also seems to explain the relatively low strength of unidirectional organoplastics in tension in the direction normal to the fibres (Table 4.12).

Failure stress in tension across fibres in organoplastics is more than twice as low as that of glass-fibre plastics.

The major disadvantage of organofibre composites is their low strength in

Table 4.12 Mechanical characteristics of epoxy plastics based on organic and glass fibres (volume fraction of fillers 60–63%)

Characteristics	Organoplastics		Glass-fibre plastics	
	PRD-49-III	*SVM*	*VM-1*	*Prepreg 800R9HTS2/80*
Density (g cm^{-3})	1.30	1.32	2.1	2.16
Failure stress in tension (MPa)				
along fibres	0.17	18.1	17.6	16.7
across fibres	0.28	0.11	0.27	–
Failure stress in compression (MPa)				
along fibres	2.8	3.0	6.2	6.4
across fibres	1.3	0.76	0.8	1.5
Failure stress in flexure (MPa)	–	6.6	9.7	10.5
Bearing failure stress (MPa)	–	1.0	1.4	2.6
Failure stress in cutting (MPa)	–	2.1	2.2	2.5
Failure stress in shear (MPa)	0.45	0.33	0.48	0.55
Modulus of elasticity in tension (MPa)				
along fibres	843	720	570	635
across fibres	56	31.4	104	–
Shear modulus (MPa)	21	17.2	52.3	–
Poisson's ratio	0.34	0.28	0.27	–

compression, which is more than twice as low as that of glass-fibre plastics. When a unidirectional organoplastic is compressed, stresses depend on strains – at relatively low strains – linearly, then distinct plastic deformations develop in the material. This is the reason why some investigators have compared organoplastics' behaviour in compression with that of metals.

This failure mechanism can be explained in the following way: First, at a certain level of loading, the reinforcing fibres and polymer matrix experience elastic deformation. Then, in further loading, the fibres swell, forming bundles of fibrils with small diameters, which leads to lower compression strengths of organofibre composites.

In transverse compression of plastics, aramide fibres exhibit elastic-plastic deformation. Owing to this, even high compression loads leave filaments intact. Under the same conditions, carbon fibres fail completely.

The major disadvantage of organofibre composites – low compression strengths – can be avoided by making combined composites with boron, carbon and other fibres. The most successful combination is that of aramide and carbon fibres, owing to their longitudinal coefficients of thermal expansion being close and the problem of internal thermal stresses being not so great as, for example, when combining aramide and glass fibres. Combining such fibres, one can obtain composites that possess higher compression strengths than organofibre composites and higher impact strengths than carbon-fibre composites.

Low compression strengths are closely linked with organofibre composites' behaviour in flexure. Organoplastics having lower strengths in flexure – about 20% less as compared with glass-fibre plastics – can operate for a long time without brittle failure at a certain level of flexure loads. In flexure, organoplastics are deformed but they neither break nor delaminate, though the maximum tension zone sometimes develops fibre fracture.

For load-bearing structures, very important are materials capable of withstanding long-term static loads. Organofibre composites can withstand 1000 hours exposure to loads that are 90% of failure stresses in tension.

Changes of mechanical characteristics with higher temperatures depend on thermal stability of fibres and binders. Table 4.13 shows temperature effects in the range of 20–150°C on the properties of an epoxy organoplastic based on SVM fibres and epoxy binder EDT-10. As can be seen, all mechanical characteristics of composites depend on temperature to a considerable degree. That is why these losses should be taken into account when designing load-bearing structures.

Use of a more thermally stable binder, for example EHD-M instead of EDT-10, allows one to bring down the losses of mechanical characteristics by 10–15% at elevated temperatures. Effects of negative temperatures on organoplastic strength are not considerable. Thus, at a temperature of $-160°C$, tensile failure stress of epoxy PRD plastic decreases by 4%.

Thermal properties of structural organoplastics in the operating temperature range (up to 100–150°C) are characterized by heat conductivity, heat capacity and coefficient of linear thermal expansion (CLTE). Average (in the operating temperature range)

Table 4.13 Properties of epoxy organoplastics (SVM fibres; EDT-10 binder) as a function of temperature

| *Material characteristics*[a] | *Test temperature (°C)* | | | |
$(kgf\,mm^{-2})$	*20*	*50*	*100*	*150*
Tensile strength in reinforcement direction	20.6 (5.6)	19.8 (5.1)	14.5 (3.1)	12.0 (3.2)
Tensile strength across fibres	0.22 (9.7)	0.2 (12.5)	0.12 (14)	0.02 (23)
Compression strength in reinforcement direction	0.29 (2.5)	0.24 (4.9)	0.77 (14)	0.6 (17)
Compression strength across fibres	0.67 (3.4)	0.58 (3.6)	0.19 (6.2)	0.09 (7.2)
Shear strength along fibres	0.41 (3.7)	0.4 (4.0)	0.12 (10)	0.005 (22)
Modulus of elasticity in reinforcement direction	742 (3.7)	690 (3.5)	665 (3.6)	645.5 (3.3)
Modulus of elasticity across fibres	33 (7.6)	27 (8.4)	5.6 (10)	3.3 (13)
Shear modulus	21.6 (6.2)	15.1 (11)	2.2 (14)	1.5 (13)

[a] In parentheses are given the variations, V (%).

Organoplastics

Table 4.14 Thermophysical properties of epoxy organoplastics

Filler type of organoplastics	Heat conductivity, across fibres $(W\,m^{-1}\,K^{-1})$	Heat capacity $(kJ\,kg^{-1}\,K^{-1})$	CLTE, along fibres $(\times\,10^{-6},\,K^{-1})$
Kevlar	0.21	1.17	-4.1
SVM	0.24	1.70	-4.3
ARMOS	0.20	1.50	-6.3
High-strength polyethylene	0.29	2.20	-9.1 to -16.0

thermophysical characteristics for typical structural organoplastics (based on an epoxy binder) are shown in Table 4.14.

The characteristics given in the table for domestic materials are determined by standard methods with laboratory thermophysical equipment produced in batches. Analysis of the table's data yields that the most thermally conductive is the polyethylene-based organoplastic, but at the same time it has the highest heat capacity. This results in structural elements from this material heating up (in one-side heating) to somewhat smaller depths than do structures made from other materials. It should also be noted that the heat conductivities of various organoplastics are close to each other.

Heat conductivities of organoplastics (in the above temperature range) depend rather weakly on temperature: the changes are well within 5–7%. This gives a reason to consider this dependence approximately constant.

Temperature dependence of the heat capacities is stronger: the growth in the range 20 to 150°C reaches 50%, which is explained by 'defreezing' (in heating) various degrees of freedom in polymer macromolecules.

Negative values of CLTE for the considered materials show that, when heated in the temperature range of 20 to 150°C, they shrink. As seen from the table, the greatest shrinkage belongs to the polyethylene-fibre organoplastic.

At higher temperatures, organoplastics begin to decompose chemically, which is accompanied by loss of strength and elasticity, gas emission, formation of porous coke and more intensive shrinkage.

In those structural elements where the organoplastic carries out not only load-bearing' functions but also heat-protective ones (for example, external layers of a casing exposed to aerodynamic heating), the above physicochemical processes are taken into consideration when designing the heat-up by additional components in the heat conductivity equation.

To determine the parameters of these processes, specially developed methods are used based on measuring the dynamics of gas emission and heat absorption when small-sized specimens of the material under investigation are heated.

Epoxy organoplastics possess the high fatigue strength needed for materials of large-sized load-bearing parts. Fatigue characteristics (Table 4.15) obtained under

Table 4.15 Fatigue characteristics of epoxy organoplastic

		Characteristics of fatigue diagram	
		Limited fatigue	Fatigue ratio,
Types of deformation	Cycle characteristics	range, $\sigma_{R,max}$ (MPa)	$K = \dfrac{\sigma_{R,max}}{\sigma_b} \times 100$
Tension (ring speci-mens)	$N = 10^6$, $f = 12.5\,\mathrm{Hz}$	5.75	33.0
	$N = 10^6$, $f = 20\,\mathrm{Hz}$	5.35	31
Compression (coupons)	$N = 10^6$, $f = 12.5\,\mathrm{Hz}$	0.8	34.8
Flexure (coupons)	$N = 10^6$, $f = 20\,\mathrm{Hz}$	1.4	32.6
	$N = 2 \times 10^6$, $f = 0.16\,\mathrm{Hz}$	2.5	52.6

loading conditions, i.e. asymmetry coefficient of the cycle $\zeta = 0.1$, loading frequency $f = 0.16\,\mathrm{Hz}$ (basic number of cycles $N = 2 \times 10^4$) and $f = 10\text{--}20\,\mathrm{Hz}$ (basic number of cycles $N = 2 \times 10^4$), have shown that values of limited fatigue ranges for these materials in cyclic loading based on up to 10^6 cycles are 30–55% of short-term static strength, depending on the deformation type. Multiple cyclic loading of organoplastics at frequencies up to 20 Hz does not lead to lower static strengths if the level of effective stresses does not exceed the fatigue limits. In fatigue tests on the basis of 10^6 cycles, strength parameters of unidirectional organoplastics are 25% higher than those of glass-fibre plastics.

Organoplastics have inconsiderable creep at room temperature, which rises somewhat at elevated temperatures.

The most objective evaluation of how the temperature/time factors affect the material under static loads can be obtained in investigations carried out on scaled-down shells that have no effect of cut fibres. In this case, fatigue tests have been done on organoplastic models of diameter 400 mm, which were loaded with internal pressure $p = 0.3p_0$ (where p_0 is the model's burst pressure in the initial state) and temperatures of 20, 60 and 80°C for 20 days. After that, the positive pressure was relieved while deformation was measured to find the values of reverse creep (Table 4.16). To evaluate residual strengths, these models were tested under positive pressures at 20°C (Table 4.17). Analysis of test results has shown that creep of organoplastics at room temperature does not exceed 0.2%, thus being at the level of creep for glass-fibre plastics. As temperature rises to 80°C, organoplastic creep goes up to 0.8%, which can be attributed to the synthetic nature of reinforcing fibres and particularities of their structure.

Test results of shells by internal pressure allow one to conclude that long-term loading of organoplastic models at loads that are 30% of the bursting strength, and at temperatures up to 80°C, does not only maintain the initial level of strength but also enhances it somewhat.

Table 4.16 Characteristic deformations of model shells made of an epoxy organoplastic

Type of deformation	Characteristic values of deformation	Test temperatures (°C)		
		20	60	80
Axial	Elastic deformations at loading, ε'_x (%)	0.396	0.491	0.550
	Elastic deformations at unloading, ε''_x (%)	0.392	0.462	0.490
	Total deformations, ε_x (%)	0.427	0.637	0.741
	Creep deformations, $\varepsilon_{xc} = \varepsilon_x - \varepsilon''_x$ (%)	0.095	0.175	0.251
	Relative creep deformations, $\varepsilon_{xc}/\varepsilon'_x$	0.242	0.380	0.512
Tangential	Elastic deformations at loading, ε'_y (%)	0.627	0.767	0.828
	Elastic deformations at unloading, ε''_y (%)	0.602	0.649	0.700
	Total deformations, ε_y (%)	0.770	1.209	1.474
	Creep deformations, $\varepsilon_{yc} = \varepsilon_y - \varepsilon''_y$ (%)	0.168	0.560	0.774
	Relative creep deformations, $\varepsilon_{yc}/\varepsilon'_y$	0.278	0.863	1.11

Table 4.17 Test results for model organoplastic shells in initial state and after long-term loading

Test results	Without pre-loading at $T = 20°C$	After exposure for 20 days to $p = 0.3p_0$ at temperatures (°C)		
		20	60	80
Burst pressures (atm)	110, 116 116, 118	125, 125 113	122, 124	125, 128
Average burst pressures (atm)	115	127.7	123	126.5
Changes of average burst pressures compared with the initial one (%)	–	+ 11	+ 7	+ 10

Permeability of composite material depends on the permeability of each component, the mechanism of reaction between them and presence of micro- and macrodefects inherent to composite materials. As organoplastics based on high-strength aramide fibres have fibres of fibrillar structure and considering the plastic's high reinforcement ratio, one can expect the permeability of composites to be high. Unidirectional plastics most sharply exhibit anisotropy of strength and stiffness properties, which cause various defects (cracks, delaminations and others) appearing under high test and operation strains. Studies of water permeability have shown unidirectional organoplastics in the initial state to be impermeable to water in the transverse direction and permeable in the end-face direction. The material from a structure element made by helical winding and preloaded is permeable both in the end-face and transverse directions.

Absence of penetrating water in the initial material in the transverse direction proves the material's continuity provided by a film of the polymer matrix covering the composite's facing layer. Water penetration in the end-face direction proves the presence of defects and channels communicating with each other and opening a path for water to penetrate along fibres. Leakage appearing in the transverse direction in specimens of a structure shows the material's broken continuity and through-defects as a result of testing stresses. Therefore, organoplastics based on aramide fibres always have communicating channels and defects in the end-face direction while high deformability of the material also causes defects initiating and growing in the transverse direction after working and test stressing of parts. To provide impermeability of load-bearing unidirectional shells in large parts, one can use rubbers, thermoplastics materials, foils and others.

Organoplastics based on SVM fibres are not characterized by gas emission at either ambient or elevated temperatures. Organoplastics investigated at temperatures of 20, 50, 60 and 80°C and exposure times of 60 days, 100 hours, 20 and 15 days, respectively, have shown that all volatiles (alcohol, acetone, toluene, epichlorohydrin, hydrogen chloride) assumed to be possible in this material are absent. Epoxy organoplastics based on SVM fibres are chemically stable to individual effects of glycerine, ethanol and water as well as to their successive action.

A number of structural materials designed for use in space engineering are required to be resistant to radiation. Impulse γ radiation of epoxy organoplastics and glass-fibre plastics, with intensity 10^{-3} rad s^{-1} and absorption dose of 10^5 and 10^6 rad, does not change their mechanical characteristics in tension. When studying these effects on SVM organic fibres and glass fibres VMP, VM-1 and on polymer binders, fibers SVM, VMP and VM-1 were found to maintain (in the dose range investigated) their initial mechanical properties while the strengths of the epoxy binders decreases.

As the absorbed dose increases, the strength decrease rises monotonically and for EDT-10 binder this decrease ratio is about 37% at absorption dose of 10^3 rad.

Strength changes of epoxy binders under γ radiation are connected with the changing electron relief of molecules and weakening intermolecular ties.

The above strength changes of binders under penetrating radiation must lead to lower mechanical characteristics of a composite material, which are determined by the matrix properties and ties at the fibre—matrix interface.

The ability of an epoxy organoplastic based on SVM fibres to absorb penetrating radiation is very small; thus, a material 3—10 mm thick practically will not (when radiated by neutrons of a 0.1 MeT nuclear blast) reduce the incident neutron flow.

Reliability and lifetime of structural elements made of composite materials depend to a considerable degree on the intensity of ageing processes taking place within these materials under atmospheric conditions with simultaneous effects of sunlight, temperature, air moisture, dust, wind, precipitation (rain and snow), as well as smoke and gases specific for given climatic zones.

Ultraviolet light changes the binder's supermolecular structure, and enhances the cracking of the surface layer and its mechanical spalling. Besides, it actively affects the synthetic fibres in organoplastics, thus lowering their strengths. Sharp temperature changes intensify the processes of microcracking in the binder. When penetrating into microcracks, moisture builds up wedging pressure, enhancing gradual growth of surface defects. The physical conversions taking place in composites exposed to moisture are accompanied by chemical changes (hydrolysis, generation of reactive groups, promotion of free radicals formed, changes of the binder's polymerization rate and others).

Finally, wind and precipitation carry away individual particles of the binder off the surface, considerably accelerating the erosion processes of the resin film in the composite parts.

Natural ageing of polymer materials is most fully characterized by changes in their performances. However, the data needed can be obtained only after a considerable period of time in this case. Therefore, to characterize climatic stability of polymer materials, the method of accelerated climatic testing is most often used. Generally, it is based on a model of thermal ageing at elevated moisture, taking into account the results of statistical analysis for weather conditions in various climatic zones. The

Table 4.18 Properties of structural materials

		Properties of structural material			
		Properties of reinforcing filler			
		Tensile strength		*Tensile modulus*	
Structural material	*Density* $(g\,cm^{-3})$	*Absolute (MPa)*	*Specific (km)*	*Absolute (GPa)*	*Specific (km)*
Organoplastic	1.34	2200	164	70	5224
(SVM)	*1.45*	*4200*	*290*	*140*	*9655*
Organoplastic	1.36	2800	209	97	7239
(ARMOS)	*1.45*	*5000*	*345*	*145*	*10000*
Organoplastic	1.36	2156	158	93	6838
(Kevlar)	*1.45*	*4200*	*290*	*145*	*10000*
Organoplastic (polyethylene, first	1.05	1800	170	75	7143
generation)	*0.97*	*3500*	*351*	*110*	*11340*
Organoplastic (polyethylene, second	1.05	2250	215	100	9534
generation)	*0.97*	*4000*	*412*	*140*	*14432*
Steel 30HGSA	7.85	1200	15	210	2600
Titan VT-4	4.5	1800	40	110	2500
Aluminium AMG-6	2.7	320	11	70	2600

investigations carried out on epoxy organo- and glass-fibre plastics based on SVM, VM-1 and VMP fibres, with the use of standard procedures, guarantee the mechanical properties of these materials to be maintained for 10 years. Organo- and glass-fibre plastics are resistant to biological vermin. They withstand effects of rodents, damage of termites and mould fungi, and therefore parts made of these plastics do not require additional protection.

Organoplastics based on high-strength flexible-chain polyethylene fibre have distinguished features as compared with aramide organoplastics based on stiff-chain fibres. This is exhibited by low compression strengths of polyethylene plastics; when loaded in flexure, they do not fail. Negative effects of these features can be eliminated by hybridization with glass and carbon fibres.

In structures loaded in tension, polyethylene organoplastics have highest specific characteristics among structural materials. Table 4.18 shows comparative data for composites obtained recently.

REFERENCES

1. Superstrength synthetic fiber Vniivlon N. Information of VNIIV, *Chemical Fibres*, 1971, **1**, 76.
2. *Second Int. Symp. on Chemical Fibres*, Kalinin, Offset duplicator of VNIIV project, 1977, issue 4.
3. Kudryavtsev G.I., Shein T.I., Advances in making and using of high-strength synthetic fibres (Review), *Chemical Fibres*, 1978, **2**, 5–15.
4. Sokolov L.B., *et al.*, *Thermally Stable Aromatic Polyamides*, Chimia, Moscow, 1975.
5. Papkov S.P., Kulichikhin V.G., *Liquid-Crystal State of Polymers*, Chimia, Moscow, 1977.
6. Papkov S.P., in *Theory of Forming Chemical Fibres*, ed. G.I. Kudryavtsev, O.P. Papkov, Mytishchi, 1975, p. 3.
7. *Int. Symp. on Chemical Fibres*, Kalinin, Preprints, 1974, issue 7.
8. Sokolova T.C., Efimova S.G., *et al.*, Investigation of poly(N-phenyleneterephthalamide) in concentrated sulphuric acid, *Chemical Fibres*, 1974, **1**, 26–9.
9. Table properties of thermal stable and heat resistant fibres, *Chemical Fibres*, 1975, **3**, 36.
10. Ermolayev, B.I., Major directions of work abroad in the field of composite materials, in *Int. Conf. on Composite Materials*, Switzerland–USA, 1975, pp 15–17.
11. *Composites*, **1** (4), 31–8.
12. *Structural Plastics*, ed. E.B. Trostyanskaya, Chimia, Moscow, 1974.
13. Chou T., Ko F., *Textile Strucural Composites*, Amsterdam, 1989.
14. *Composite Materials, A Guide-Book*, Mashinostroyeniye, 1990.
15. Rosenberg B.A., *et al.*, Binders for composite materials, *Journal of D.I. Mendeleyev Chemical Society*, 1978, **23** (3), 273.
16. Trostyanskaya E.B., Babajewsky P.G., *Advances in Chemistry*, 1971, **40** (1), 117–41.
17. *Handbook of Plastics*, ed. M.I. Garbar, V.M. Kataev, M.S. Akutin, Chimia, Moscow, 1967, p. 1.
18. Lee H., Nevill K., *Handbook on Epoxy Resins*, ed. N.V. Alexandrov transl. from English, Energy, Moscow, 1973.
19. *Structural Plastics (Thermosets)*, ed. E.B. Trostyanskaya, Chimia, Moscow, 1974, pp. 101–3.
20. Lipatov Yo.S., *et al.*, *Physical Chemistry of Polymer Compositions*, Naukova Dumka, Kiev, 1974.
21. Voyutsky S.S., *Autohesion and Adhesion of High Polymers*, Rostechizdat, Moscow, 1960.
22. Lipatov Yo.S., *Physico-Chemistry of Filled Polymers*, Naukova Dumka, Kiev, 1967.
23. *Structural Plastics*, ed. E.B. Trostyanskaya, Chimia, Moscow, 1974.
24. *J. Compos. Mater.* 1972, **4**.
25. *Handbook of Composites*, ed. G. Lubin, New York, 1988.
26. *Composite Materials*, vol. 6, Academic Press, New York, 1974.

5

Glass plastics

B.A. Kiselev

5.1 INTRODUCTION

Glass plastics were essentially the first high-strength polymeric composite materials widely used in various branches of engineering and for domestic purposes. In spite of the ever-growing application of novel high-strength and high-modulus polymeric composite materials based on carbon, organic and other fibres, the volume of consumption of the various types of glass plastics continues to increase.

Glass plastics remain the basic material used in the production of consumer goods, transport, building and other fields. Even in the aerospace industry and military engineering the output and application of glass plastics exceeds 50% of the total range of polymeric composite materials application. The situation is similar in both Western Europe and the former Soviet Union.

The broad application of glass plastics is associated primarily with the relatively low cost and accessibility of the raw materials, the comparatively low power input of glass-fibre production, versatility and the possibility of control over the physical and mechanical properties of glass plastics over a broad range. Glass plastics can be a component of hybrid and combined laminated materials.

Like all composite materials, glass plastics are essentially multiphase heterogeneous systems whose properties depend on those of its components, i.e. the glass-fibre filler and the polymeric matrix, and the nature of their interaction. Achievements in the field of creating composite materials in general, and glass plastics in particular, are substantially the result of basic studies in the fields of development of reinforcing fillers, synthesis of polymeric binders, study of interfaces and the possibility of control over the nature of the phase interaction.

The basic trends of studies in the field of glass plastics in recent years have been the creation of materials exhibiting high physical and mechanical characteristics (strength in static conditions, in extended and repeated loading, crack resistance). Attention has also been concentrated on creation of heat-resistant glass plastics. Technological studies aimed at improvement of the ecological conditions in production of items from glass plastics are gaining in importance. These studies are determined largely by the growing use of glass plastics in the production of highly loaded domestic items.

5.2 GLASS-FIBRE FILLERS

Currently developed and used in the production of glass plastics are a large number of fibres, i.e. high-strength and high-modulus ones, those with changed configuration (hollow) or high heat resistance, as well as those with different dielectric characteristics. Table 5.1 presents the physical and mechanical properties of some high-strength and high-modulus glass fibres. Quartz fibre exhibits high strength. The fibre strength amounts to 6000 MPa [1].

Further increase in the stiffness of glass fibres is accompanied by increase of density and complications in the moulding process, thus making the production of these fibres practically inexpedient.

The fibre strength drops due to an effect produced by adsorbed moisture from the air and development of microcracks. Degradation of fibre strength is also observed in the process of textile processing.

The chemical composition of the glass most widely employed in fibre production is presented in Table 5.2. It should be noted that the glass compositions in the bulk and on the surface of the fibre are different. Fibres made of magnesium aluminosilicate

Table 5.1 Physical and mechanical properties of high-strength glass fibres [1]

			Glass grades			
Indices	*'E'*	*USA:* *'S', 'S-2'*	*Japan:* *'T'*	*France*	*USSR:* *UP*	*USSR:* *VMP*
Density ($kg\,m^{-3}$)	2.540	2.490	2.490	2.550	2.470	2.560
Tensile strength (MPa)						
at 22°C	3.500	4.700	4.400	3.600	5.000	4.500
at 540°C	1.750	2.450	2.460	–	2.500	2.400
Elasticity modulus (MPa)	73 500	86 800	86 000	83 000	85 000	95 000
Elongation at rupture (%)	4.8	5.4	5.5	–	5.0	5.0
Linear thermal expansion coefficient over range of 20 to 300°C ($10^{-7}\,°C^{-1}$)	51	28.8	28	–	26.3	34.9
Refractive index	1.5250	1.5294	1.5490	–	1.5186	1.5460
Dielectric constant at 10^{10} Hz at 23°C	6.23	5.21	5.20	–	5.18	5.93
Dielectric loss tangent at 10^{10} Hz at 23°C	0.011	0.0068	0.0260	–	0.0080	0.0100
Softening temperature at viscosity of 10^{6} Pa s	840	970	975	–	1.156	945

Table 5.2 Chemical composition of glass [2]

Type of glass	Elements (%)							Content
	Si	*Al*	*Mg*	*Ca*	*B*	*Fe*	*O*	
'E'	18.6	6.1	2.2	6.3	4.1	0.4	61.8	In
'S'	21.8	9.7	5.1	–	–	–	63.5	bulk
'E'	24.1	8.4	0.7	1.8	3.0	1.8	61.1	On
'S'	18.1	12.7	17.2	–	–	–	52.0	surface

glass and quartz can be used at high temperatures. The quartz fibres, however, exhibit rapid loss of strength when exposed to heat due to the relatively fast recrystallization processes.

A silica fibre close to the quartz fibre in terms of composition is used in heat-shield materials.

A broad range of fillers intended for the various types of glass plastics has been created on the basis of glass fibres exhibiting different chemical composition, i.e. threads and bundles, canvasses, fabrics, knitted goods, three-dimensional multilayers and sewing cloths.

Used on a particularly broad scale in the production of glass plastics for structural and radio purposes are satin-textured glass fabrics; more seldom used are card-woven fabrics made of fibres with diameter ranging from 6 to 12 μm. As regards glass plastic items bearing compressive loads, it is expedient to use fibres with a diameter of 16 μm and over.

As regards glass-cloth-based laminated items wherein the major form of loading exhibits one prevailing direction, use is made of cord glass fabrics.

In insulating glass-cloth-based laminates, of particular importance are thin-rating fabrics with thickness ranging from 0.04 to 0.1 mm, which enable the production of glass-cloth-based laminates featuring a smooth surface and high dielectric strength.

Table 5.3 Properties of some glass fabrics used in production of glass-cloth-based laminates

Fabric grade	Thickness (mm)	Density of threads per 1 cm		Mass of 1 m² (g)	Mean breaking load, min. (kgf)		Texture
		Warp	*Weft*		*Warp*	*Weft*	
T-10-80	0.24	36	20	290	320	180	Satin
TS 8/3-K-TO	0.28 ± 0.03	36 ± 1	20 ± 1	290 ± 20	100	55	Quartz satin
KT 11-8/3	0.58 ± 0.06	19 ± 1	13 ± 1	575 ± 60	140	100	Silica satin
T-45P-76	0.26 ± 0.03	–	–	215 ± 20	120	80	Hollow satin
T-25(VM)-78	0.30	10	6	365	450	35	Cord

The properties of some types of glass fabrics for structural glass-cloth-based laminates are specified in Table 5.3. The names of the fabrics shown in Table 5.3 fall far short of covering their grades used as plastics fillers.

Multilayer and sewing fabrics enable the production of materials with high shear characteristics in the glass plastic plane. The application of glass-fibre fillers of various types enables one to change the properties of glass plastics over a broad range, thus enabling the production of materials exhibiting different physical, mechanical, dielectric and process characteristics.

5.3 BINDERS OF GLASS PLASTICS

5.3.1 Introduction

The past years have witnessed growth in the number of types and grades employed in the production of glass plastics binders. To ensure the high characteristics of glass plastics, the binders should exhibit an adequate wetting power and adhesion to the glass fibre, the setting shrinkage should result in the minimum appearance of cracks formed as a result of binder setting, and the matrix should display high cohesion strength. Hence, the synthesis of new binders suitable for the production of high-strength and high-modulus composite materials, including glass plastics, is a complicated task because of the need to take into account all the requirements for the binder, the absence of a direct relation between the chemical structure of the matrix and its mechanical properties, in particular, as well as the need to take into account the specific features of the crosslinking process in the polymer boundary layer.

The binders should exhibit certain processing properties, i.e. sufficiently long working life in a comparatively fast curing process. The possibility of using one or other method for moulding glass plastics and items therefrom depends largely on the rheological properties of the binder.

In reviewing the problems associated with the creation of poorly combustible and heat-resistant glass plastics, it should be noted that these properties of the material depend primarily on the type of matrix. Naturally, it does not mean that the filler properties do not affect the material behaviour in heating. It was noted above that the glass filler composition determines its behaviour in heating, in particular at high temperatures.

The phenol–formaldehyde, polyester, polyepoxy, polysiloxane and other binders used on a broad scale for glass plastics do not always enable the production of materials with the required package of properties.

For a long time, extensive use has been made of chemical modification of the above-specified types of polymers so as to satisfy the major requirements imposed on the binders. In this case, as a rule, the matrix was essentially a copolymer of mainly homogeneous structure.

The widespread introduction of composite materials in structural members exposed to high mechanical and vibration loads caused the development of binders whose curing resulted in the formation of hybrid or polymatrix systems. The hybrid matrices are essentially a microheterogeneous system whose properties are determined by the composition and shape of the micro-areas [3]. The reinforcing filler influences the forming structure of the hybrid matrix in a composite material.

To produce such matrices, use is made of multicomponent binders (solutions, melts) whose curing process involves the formation of two or more interpenetrating covalently unbonded networks or covalently unbonded network or linear polymer [3].

A hybrid matrix or polymatrix system can also be produced by processing as a result of the double impregnation of the glass-fibre filler in a closed mould under vacuum and then under pressure. In this case, a fairly porous structure is formed during the first impregnation of the filler and the binder curing, for instance, a binder of the condensation type. During the second impregnation, the pores are filled with a binder of the polymerization type.

Heat-resistant matrices are based on polymers containing hetero-organic fragments, linear or network carbon chains and heterocyclic aromatic systems.

5.3.2 Polyesters

The first period of broad use of glass plastics in various branches of engineering is associated largely with the synthesis of unsaturated polyesters. Binders based thereon are essentially solutions of unsaturated oligoesters in a monomer containing the initiator and accelerator. Depending on the type of initiator and accelerator, the binder curing process can be effected in heating and normal conditions. The production of glass plastics employs two types of unsaturated oligoesters, i.e. oligoester maleinates containing in addition to the double bonds —OH and —COOH functional groups and oligoester acrylates. In both cases, styrene, monoester acrylates and other compounds can be used as active copolymerization solvents.

The cured polyester matrix exhibits multiheterogeneous structure, which is determined by both the binder composition and the glass plastic moulding process parameters.

Glass plastics and structural elements made therefrom and based on a specific type of binder can be manufactured using various processing techniques. Polyester binders

Table 5.4 Properties of glass-cloth-based laminates based on polyester binders

Glass-cloth-based laminate grades	Properties		
	Tensile strength (MPa)	*Compressive strength (MPa)*	*Static flexural strength (MPa)*
VPS-21[a]	360	234	400
VPS-24[b]	450	240	400

[a]T-11-GVS-9 glass fabric.
[b]T-10-80 glass fabric.

are used in the production of relatively low-loaded and experimental parts using the simplest auxiliaries. The contact method of moulding at room temperature has gained the widest acceptance.

Table 5.4 presents the properties of VPS-24 glass-cloth-based laminates based on MKT oligoester acrylate binder and of VPS-21 glass-cloth-based laminates based on PNT polyester maleinate binder, manufactured at a specific pressure of 0.3 MPa. At the contact moulding pressure, the glass-cloth-based laminate properties are somewhat lower.

In addition to direct extrusion and contact moulding, the production of items from glass plastics based on polyester binders may involve the use of autoclave moulding and pultrusion.

5.3.3 Epoxy binders

Most widely used in the production of glass plastics are epoxy binders [4]. The high wetting power and adhesion to glass fibre, low setting shrinkage, considerable cohesion strength and adequate dielectric characteristics have resulted in the employment of epoxy binders in fabrication of glass plastics for structural, radio and insulating purposes.

The heat resistance of epoxy matrices is within the range 60 to 225°C. Many works have dealt with the study of epoxy binders exhibiting lower-than-usual combustibility [5]. The reduction of combustibility is achieved by adding halogen, mainly bromine, or phosphorus atoms to the epoxy oligomer or curing agent. Domestic specimens of binders based on halogen-containing epoxy oligomer were produced using the N,N'-tetraglycidyl derivative of 3,3'-dichloro-4,4'-diaminodiphenylmethane (EXD) and other compounds. Reduction of combustibility of glass plastics based on halide- and phosphorus-containing epoxy matrices brings about a drastic increase of smoke formation owing to the existence of the phosphorus and halide atoms, which is a disadvantage of these materials.

Also used in the production of reinforced plastics with higher-than-usual heat resistance are cycloaliphatic epoxy oligomers and oligomers formed by epoxidation of novolacs, triphenol and other compounds. For the purpose of improving the heat resistance of epoxy binders and glass plastics based on them, it is expedient to employ epoxy oligomers whose structure incorporates imide rings. Items based on epoxy binders are manufactured by direct, autoclave, vacuum moulding, pressure impregnation and pultrusion methods. The properties of some glass-cloth-based laminates are presented in Table 5.5.

5.3.4 Phenol–formaldehyde

Phenol–formaldehyde binders started to gain acceptance in the production of glass plastics before other oligomers owing to the availability of the raw materials, experience of their use in the production of cloth laminates and fibrous composites, and relatively low cost.

Table 5.5 Properties of glass-cloth-based laminates based on epoxy binders (filler:T-10-80 glass fabric)

	Glass-cloth-based laminate grade			
Properties	*VPS-7*	*ETF-T*[a]	*ST-2216FK*	*VPS-25*
Type of binder	Epoxy 4,4'-isopropylide-nediphenol	Epoxy triphenol	Epoxy imide	Epoxy phosphazene
Glass transition temperature ($^{\circ}$C)[b]	80	150	210	170
Tensile strength (MPa)	560	605	614	590
Flexural strength (MPa)	573	730	650	760

[a] Filler: TS 8/3-T glass fabric containing titanium.
[b] Determined by the electrophysical method.

The present-day employment of phenol–formaldehyde binders is associated largely with the adequate ablation properties of the glass plastics based thereon. This determined their broad application in rocketry and the production of heat-shield materials.

The relatively high refractoriness, low smoke generation and low toxicity in burning conditioned the application of glass plastics based on these binders in the production of passenger airplane interiors.

To improve the physical and mechanical properties of glass-cloth-based laminates, heat resistance, ablation and other characteristics, use is made of the modification of phenol–formaldehyde resol or novolac oligomers.

From the viewpoint of producing materials exhibiting high heat resistance and high carbon residue, which largely determines the ablation properties, of undoubted interest are the carborane-containing phenol–formaldehyde polymers [6]. To produce the carborane-containing resols, use was made of phenolcarborane (1,2-bis(4-oxy-phenyl)carborane) and formaldehyde in a molar ratio of 1:2.5 (Table 5.6). Slow heating up to 200°C yields resites with a yield of insoluble products up to 91%, and, at a temperature of 900°C, high carbon yield.

Table 5.6 Flexural strength of glass-cloth-based laminate based on carborane-containing phenol–formaldehyde polymer

Ageing temperature and time	*20°C, 0.5 h*	*300°C, 0.5 h*	*350°C*		*400°C*	
			0.5 h	*500 h*	*0.5 h*	*500 h*
Flexural strength (MPa)	260	370	267	20	250	60

The best processing properties are displayed by the mixed carborane containing resols produced from phenolcarborane and phenol. These resols exhibit higher-than-usual flow and are cured under the ordinary conditions for phenol–formaldehyde resols. Compared to the phenol–formaldehyde resites, the carbon yield in the carborane-containing resites is also noticeably higher [7].

When exposed to high temperatures, the phenol–formaldehyde resins are subjected to destruction, primarily, in terms of the methylene groups. Hence, in synthesis of more heat-resistant phenol oligomers for glass plastics, attempts were made to exclude or partially substitute the CH_2- groups with those more resistant to heat. Owing to the partial substitution of the methylene group by lactone ring, the phenolphthalein–formaldehyde polymers exhibit higher-than-usual heat resistance and a large yield of carbonized polymer [8]. The use of this polymer enabled one to improve also the mechanical characteristics of the glass-cloth-based laminate. For instance, the glass plastic based on phenol–formaldehyde resin has an impact viscosity of $18\,kgf\,m\,cm^{-2}$ and static bending of 64 MPa, whereas the glass plastic based on phenolphthalein–formaldehyde polymer has values of $70\,kgf\,m\,cm^{-2}$ and 207 MPa, respectively.

The processing characteristics of phenol binders and the possibility of their use in the process of glass-cloth-based laminate and item fabrication using vacuum and pressure impregnation methods were improved on adding reactive multifunctional solvents to their composition.

The properties of glass-cloth-based laminates based on phenol binders containing active solvents are presented in Table 5.7.

5.3.5 Organosilicon binders

Interest in organosilicon binders is conditioned by their higher thermal oxidation stability, adequate electrophysical properties and a number of specific characteristics. The disadvantages inherent in this class of binders include the relatively low mechanical strength of glass plastics based thereon and high processing temperatures. Hence, the major investigations were aimed at the study of the causes of the poor mechanical strength, and the creation and modification of organosilicon binders that allow the production of glass plastics with higher mechanical strength that can be processed at lower temperatures.

Table 5.7 Properties of glass-cloth-based laminates based on phenol binders containing active solvents

Glass-cloth-based laminate grade	Tensile strength (MPa)	Compressive strength (MPa)	Static bending strength (MPa)	$\tan\delta$ at $10^6\,Hz^a$	ε at $10^6\,Hz^a$
FN	288–400	119–197	229–353	0.0135	4.3
FFA	400–420	200–300	550–600	0.0064	4.26

[a] Dielectric loss tangent ($\tan\delta$) and dielectric constant (ε).

Unlike the polymers with organic composition, the thermal oxidation of organo-silicon matrices, though not affecting the main chain of the polymer, leads to cleavage of —Si—C— bonds and formation of siloxane bonds stable to oxidation, thus causing additional structurization of the polymer. In this case, the overall thermal oxidation stability of the system increases owing to a reduction of the relative content of —Si—C— bonds therein and formation of higher polymer crosslink density. In this case, the mechanical and some other properties of the material may improve till a certain time.

The study of the mechanical properties of organosilicon matrices as a function of the structure, as well as the properties that govern the nature of their interaction with glass fibres (Table 5.8), demonstrated that, probably, the major cause of the relatively low mechanical properties of glass plastics is represented by the large values of the linear expansion coefficients compared to organic polymers (2 to 10 times) and the extremely large difference with change of glass-fibre dimensions. The latter causes the appearance of residual stresses in the plastic. This can be clearly traced in a comparison of glass plastic ultimate strength with the linear expansion coefficient as a function of the heating temperature [9]. The minimum strength of the glass plastic coincides with the maximum value of the linear expansion coefficient.

The adhesion strength values of polyorganosiloxanes are quite comparable to the adhesion values of polyester and phenol–formaldehyde matrices [10].

For the purpose of improving the mechanical strength of composite materials based on polyorganosiloxanes while preserving adequate thermal stability, use is made of their modification. In this respect, a binder containing in the initial state 50% by weight of polymethylphenylsiloxane and 50% by weight of phenol–formaldehyde oligomer is of interest. As the binder gets cured, a hybrid matrix is formed in the glass plastic. Such a matrix is formed due to the difference in binder components' curing rates, as is quite clear from the change of elementary composition (Table 5.9) of the matrix. The properties of a glass-cloth-based laminate based on the given binder are shown in Table 5.10. A material with higher mechanical properties was

Table 5.8 Properties of organosilicon matrices

	Matrix		
Properties	*Polymethyl-siloxane*	*Polymethyl-phenylsiloxane*	*Polyphenyl-siloxane*
Compressive strength (MPa)	48	52	12
Elasticity modulus at static bending (MPa)	955	1.320	1.330
Shrinkage (%)	4.3	3.0	2.7
Linear expansion coefficient from -10 to $240\,^\circ\mathrm{C}$ $(10^{-6}\,^\circ\mathrm{C}^{-1})$	80–200	80–140	6–121
Adhesion strength (MPa)	27.5	27.5	10.7

Table 5.9 Elementary composition and distribution of components in insoluble polymer

	Content of elements (%)			Ratio of components of organosilicon and phenol–
Curing condition	Si	C	H	formaldehyde oligomers
In initial state	11.84	61.45	6.24	50/50
80°C, 3 h	0	72.90	8.38	0/100
120°C, 3 h	1.52	73.10	7.29	6.4/93.6
160°C, 3 h	1.93	73.10	6.90	8.1/91.9
200°C, 3 h	7.54	67.85	6.27	31.8/68.2
200°C, 24 h	12.43	65.00	5.60	52.5/47.5

Table 5.10 Properties of glass-cloth-based laminate based on 50 wt% polymethylphenylsiloxane and 50 wt% phenol–formaldehyde binder

Tensile strength (MPa)	
at 20°C	370
at 250°C after 250 h ageing	260
Static bending strength (MPa)	
at 20°C	261
at 250°C after 200 h ageing	75
Compressive strength (MPa)	
at 20°C	160
at 250°C after 200 h ageing	61.5
Dielectric constant at 10^{10} Hz	4.1
Dielectric loss tangent at 10^{10} Hz	0.02

produced by modification of the polymethylphenylsiloxane using phenolphthalein–formldehyde oligomer.

Of particular interest are binders composed of organosilicon oligomers exhibiting spatial branched structure and unsaturated monomers. Divinylbenzene, diallyl isophthalate, triallyl cyanurate, etc., were investigated as such monomers. Peroxide initiators were added to the binder. The binder with the monomer as active solvent is suitable for the manufacture of items using the pressure impregnation method. In this case, the matrix is essentially a hybrid exhibiting heterogeneous structure.

Production of glass plastics based on organosilicon binders usually employs glass fillers presubjected to heat treatment for removal of the lubricant.

Both the mechanical characteristics of materials based on polyorganosiloxanes and the thermal oxidation stability depend to a certain extent on the relation between the phenyl and methyl radicals in the polymer.

The employed organosilicon binders and composite materials are processed as a rule at temperatures of 200°C and above, which, in some instances, limits their application. The glass-cloth-based laminate based on organosilicon binders composed of a mixture of low-molecular-weight liquid organosilanes, Si—H groups and oligo(organosilanes) with vinyl radicals on the silicon atom is moulded at lower temperatures (100 to 150°C). Glass plastics based on similar binders exhibit lower thermal stability compared to binders cured on the basis of the polycondensation mechanism.

The use of special curing agents enabled the production of K-9X organosilicon binder, which provides the basis for the manufacture of glass-cloth-based laminate at a temperature of 100 to 150°C. In terms of their mechanical characteristics, the glass-cloth-based laminates based on this binder are highly competitive with materials processed at high temperatures (Table 5.11).

As a rule, the binders are used in the production of glass plastics as solutions in alcohol, acetone, ethyl acetate and other organic solvents. The application of film binders contributes to an improvement in the working conditions. Usually, thermoplastic films are used on quite a broad scale as film binders. These binders should exhibit certain rheological characteristics ensuring adequate impregnation of the glass-fibre filler. The PK-2a film thermosetting organosilicon binder exhibits a thickness of 0.3 ± 0.05 mm at a mass of 320 to $340 \, \mathrm{g \, m^{-2}}$, a content of solubles of 90 to 100% and adequate drapability [11].

The tensile strength of such a film is 0.02 to 0.03 MPa, and elongation at rupture is 12%. Depending on the type of glass fabric, the film thickness and mass can be selected. Figures 5.1 and 5.2 present the binder viscosity and gel forming time chracteristics. The PK-2a binder was used in the manufacture of glass-cloth-based

Table 5.11 Properties of glass-cloth-based laminates based on organosilicon matrices (heat-treated T-10 fabric)

	Grades of glass-cloth-based laminates				
Properties	SK-9A[a]	SK-9FA	SK-9X	SK-10S	SK-9P[b]
Strength (MPa)					
in tension	288	387	380	421	400
in compression	126.5	115	100	130	190
along layers in bending	212	200	205	191	450
Elasticity modulus in tension (MPa)	21 300	26 200	31 600	38 200	–
Dielectric constant at 10^6 Hz	3.55	3.58	4.39–4.72	4.04–4.35	3.5
Dielectric loss tangent at 10^6 Hz	0.003	0.002	0.002	0.03	0.0029

[a] Filler: KT-11 fabric.
[b] T-10-80 fabric.

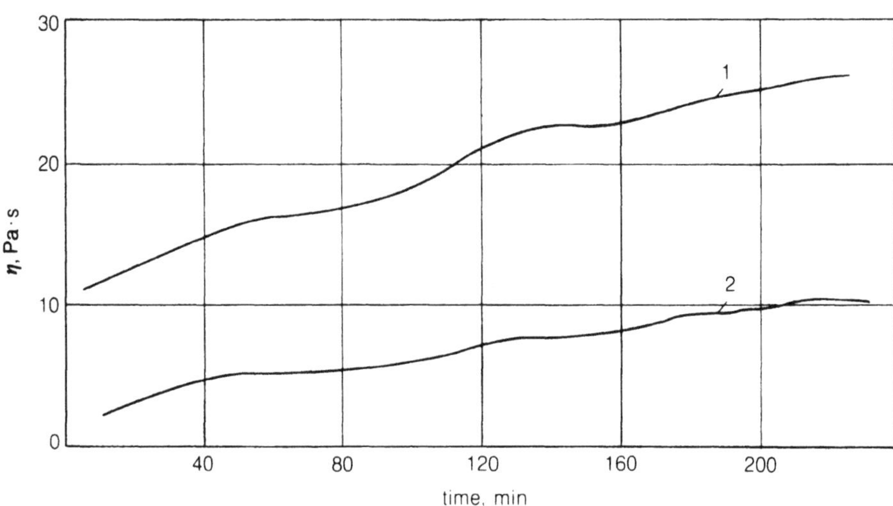

Fig. 5.1 PK-2a binder viscosity versus time at 100°C (1) and 120°C (2).

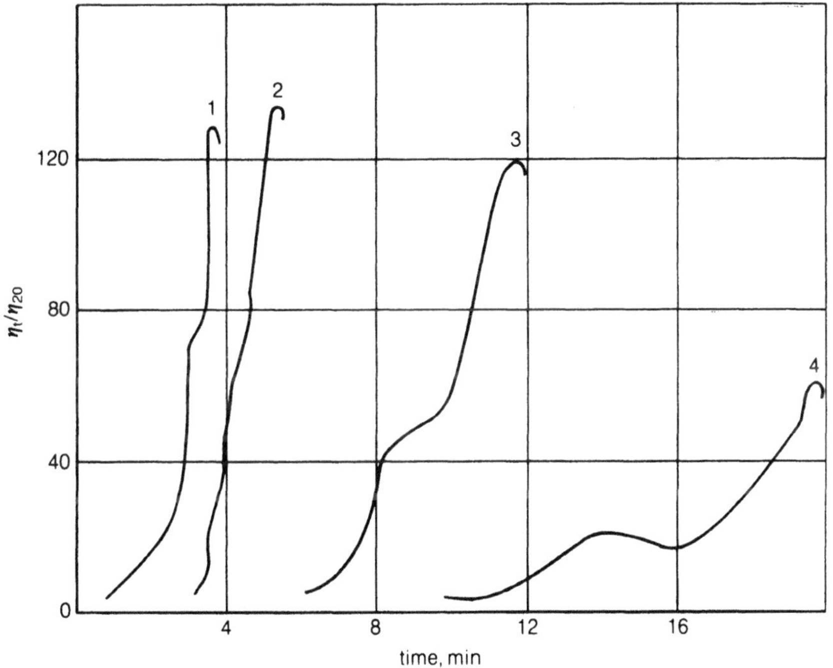

Fig. 5.2 PK-2a binder gel formation time versus time at 210°C (1), 200°C (2), 190°C (3) and 180°C (4).

laminate, grade SK-9P, on T-10-80 fabric and with other glass-fibre fillers produced on the basis of lubricants containing aminosilanes.

The past years have witnessed work on the creation of composite materials, i.e. ceramoplastics, designed for operating temperatures above 1000°C. Specimens of such materials were produced by the high-temperature processing of plastics based on polyorganosiloxanes in the absence of oxygen. The relatively low sintering temperature and low shrinkage (1 to 2%) make these materials rather promising [12].

5.3.6 Polyimides and related compounds

Particular importance in the production of composite materials including glass plastics has been gained by polyimide binders. The production of glass plastics employs mainly polyimides cured via the condensation or polyaddition reactions. The polyimides exhibit high thermal stability and radiation resistance, and retain their physical and mechanical characteristics at a sufficiently high level at elevated temperatures. The excellent thermal stability of such polymers is the result of the stability of the aromatic and heterocyclic chain fragments, as well as the intermolecular interaction.

Most acceptance in the production of glass-cloth-based laminate has been gained by binders representing the solution of the SP-6, SP-97 and SP—CM polyimide-forming components. The disadvantages of the above binders include the high extrusion temperature of 300 to 350°C. Another specific feature of moulding glass plastic items is determined by the rheological properties of the binder, i.e. the abrupt drop of their viscosity in heating up to 200°C, which requires the use of special techniques for moulding the items.

The condensation nature of polyimide binder curing and the existence of a large amount of volatiles in the prepreg cause the higher-than-usual porosity of glass plastics made therefrom.

The basic properties of glass-cloth-based laminates based on polyimide binders are presented in Table 5.12.

Studies associated with modification of polyimide binders have stimulated the search for better processing properties and reduction of porosity in polyimide glass plastics without a substantial decrease of the mechanical strength, heat resistance and refractoriness. The existence of —NH—, and —COOH functional groups in the polyimide-forming components enabled selection of special curing agents that ensure curing temperature reduction; for instance, for the STP-97K glass-cloth-based laminate, down to 170°C.

Addition of curing agent did not affect the working life of the binder and the prepregs based thereon.

The study of the kinetics of the SP-97K binder curing process demonstrated that the structural-group changes taking place in polyimide terminate much earlier than in the case of the pure polyimide. The properties of the STP-97K glass-cloth-based laminate cured at a temperature of 170°C are presented in Table 5.12. The extrusion temperature reduction was followed by glass plastic heat resistance decrease to some

Table 5.12 Properties of glass-cloth-based laminates based on polyimide matrices (T-10-80 fabric)

| | Glass-cloth-based laminate grades | | | | |
Properties	STP-6	STP-97S	STP-97K	STP-CM	STP-CMK[a]
Breaking stress (MPa)					
in tension	482	500	495	650	777
in compression	340	353	407	–	600
along layers in					
flexure	544	645	400	900	760
Elasticity modulus	30 500	34 200	30 000	–	38 600
in tension (MPa)					
Dielectric constant	4.39	4.71	4.57	4.1	4.15
at 10^6 Hz					
Dielectric loss tangent	0.0155	0.0116	0.0081	0.01	0.01
at 10^6 Hz					

[a] T-44 (UP)-76 glass fabric.

extent, but the indices of the mechanical strength, combustibility and smoke generation did not undergo any practical changes.

The study of the temperature–humidity effects on polyimide glass-cloth based laminates also demonstrated that modification of the binder does not cause a noticeable change in the indices of the material properties in various operating conditions (Fig. 5.3).

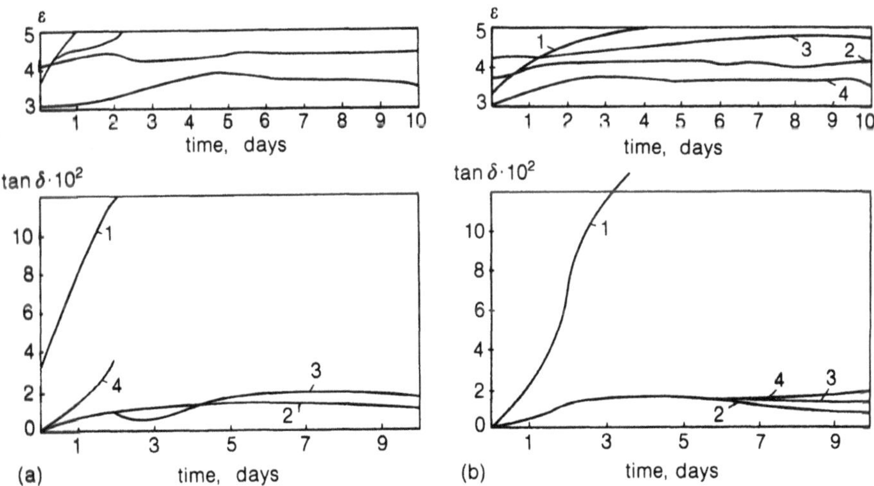

Fig. 5.3 Change in dielectric constant (ε) and dielectric loss tangent (tan δ) of polyimide glass-cloth-based laminate when held in water (a) and at relative humidity $\phi = 98\%$ (b).

Addition of carborane groups (the STP-CMK glass-cloth-based laminate) to the polyimide structure contributes to higher heat resistance of glass plastics based on the polyimide matrices.

Glass plastics exhibiting lower porosity and higher mechanical characteristics were produced on the basis of oligoimides cured without liberation of by-products due to opening of double bonds. Some of these binders are represented by oligoimides SPEN, PAIS-104 [13] and PIK-250 [14].

Polyimide cured as a result of the polyaddition reaction was produced from 4,4'-diaminodiphenylmethane, benzophenonetetracarboxylic dianhydride and nadic anhydride.

The SPEN oligoimide based on the aromatic anhydrides enables production of glass-cloth-based laminates with operating temperature of 300 to 350°C, and glass-cloth-based laminates with operating temperature up to 250°C are produced on the basis of the PAIS-104 and PIK-250 oligoimides, which employ maleic anhydride as the imide-forming component. Their synthesis is effected through copolymerization of excess bismaleinimides with diamines.

Curing results in the formation of network polyaminoimides. Industry employs most widely the polyaminoimide oligomer that is essentially the product of copolymerization of N,N'-4,4'-diphenylmethanebismaleinimide with excess 4,4'-diaminodiphenylmethane.

Stemming from process considerations, the process of moulding the STM-F glass-cloth-based laminate based on polyaminoimide oligomer is run in two stages with heating up to 180°C and subsequent heat treatment at a temperature of 200°C. The properties of the STM-F glass-cloth-based laminate based on the T-10-80 glass fabric are as follows: breaking tensile strength 476 MPa, elasticity modulus 3350 MPa and porosity 1%.

The binder represented by the PIK-250 oligoimide solution in the diallyl ester of isophthalic acid is suitable for manufacture of glass-cloth-based laminated items using the pressure impregnation method [14].

The polymaleinimide binder is used successfully for production of items using the winding method.

A low-porosity, hermetic, crack- and water-resistant glass plastic is produced on the basis of a hybrid polyimide epoxy matrix. To manufacture items based on such a matrix, use is made of a production process comprising two stages: the first stage deals with impregnation of the glass filler with oligoimide and subsequent curing of the binder; the second stage is associated with the vacuum and pressure impregnation of the moulded glass plastic with epoxy oligomer and curing.

The properties of the glass plastic based on the hybrid heterogeneous matrix depend on the physical and chemical processes not only at the heterogeneous matrix–filler interface but also between the matrices. The study of the kinetics of curing of the epoxy binder based on the polyimide matrix using the initial rate studies (IRS) methods demonstrated that the structural changes in the epoxy binder run more intensively. In doing so, an interaction between the remaining unreacted carboxyl and amide groups of polyimide is observed.

Some properties of the STP-97VE glass-cloth-based laminate based on the hybrid polyimide–epoxy matrix are as follows: breaking tensile strength 547 MPa, static bending strength 750 MPa and porosity 0.5%.

5.3.7 Other binders

One promising trend in the synthesis of heat-resistant polymeric matrices is the polycyclotrimerization reaction [15]. The polycyclotrimerization process runs without liberation of volatiles and with formation of network polymeric structures. Among the most frequently used binders are isocyanates, carbodiimides, cyanic esters and substituted and unsubstituted polyfunctional cyanamides.

The Institute of Hetero-Organic Compounds in Moscow have developed binders of the TSNCP type (thermosetting nitrogen-containing polymers), which are essentially liquids exhibiting different viscosity with extended times of storage, processed without using solvents and cured at a temperature of 170°C.

Properties of TSNCP binders are: glass transition temperature 320–380°C, density $1.28\,g\,cm^{-3}$, bending strength 105 MPa, flexural modulus 2800 MPa and dielectric loss tangent 0.005.

Glass plastics can be produced using the method of wet winding, extrusion and pressure impregnation. The basic properties of a glass plastic based on the TSNCP binders are as follows: flexural strength 1250 MPa, tensile strength 1620 MPa and compressive strength 330 MPa. As regards heat resistance, these binders outperform epoxides and polyaminoimides.

Oligocarbodiimides are multipurpose and used as binders, adhesives and for some other purposes. The physical and mechanical properties of glass plastics whose production involves the use of these binders are high enough.

It should be noted that the carbodiimide-containing polymers, because of the high reactivity of the —N=C=N— groups, can easily enter into reaction with monomers and oligomers with different functional groups, thus enabling the processing and physical and mechanical properties of the materials based thereon to be changed over a wide range.

As shown below, the temperature of extended operation of glass plastics based on organic and hetero-organic matrices does not exceed 400°C. In various branches of engineering, however, there is a need for more heat-resistant materials. Hence, use is made in the glass-cloth-based laminates of matrices based on phosphate bundles. These materials exhibit comparatively low mechanical strength due to the filler loss of strength when exposed to heat. Moreover, the strength degrades as a result of the matrix corroding effect on the glass fibre.

To protect the composite against breakdown, a thin film of the hetero-organic or inorganic coating is usually applied to the glass fabric. The mechanical properties of a glass-cloth-based laminate based on an aluminophosphate matrix are presented in Table 5.13.

Further improvement of matrials based on the phosphate matrix is possible provided that more thermally stable and high-strength reinforcing fillers are used.

Table 5.13 Mechanical properties of glass-cloth-based laminates based on aluminophosphate binder at 600°C

Grade of glass-cloth-based laminate	Heating time (h)	Flexural strength (MPa)	Tensile strength (MPa)	Compressive strength (MPa)
STAF	0.5	30	5	–
	200	18	Inop.[a]	30
STAF-1	0.5	77.5	50	50
	200	80	42.5	45
STAF-2	0.5	102	60.0	91.0
	200	90	45.0	81.0

[a]Inoperative.

5.3.8 Interface effects and stabilization

It was already noted above that the properties of glass plastics are determined largely by the state of the glass—binder interface.

The properties of the boundary layer and the condition of interaction between the matrix and the reinforcing glass fibres can be controlled using chemically active compounds. In this case, it is possible to increase not only the extent of employing the glass-fibre strength but also the stability of the glass plastic physical and mechanical properties in the higher-than-usual humidity conditions and water.

Several methods of stabilizing the properties of glass plastics are known, i.e. finishing or thermochemical treatment of the glass filler, application of direct lubricants to the glass fibre and addition of active compounds to the binder. Organofunctional silanes or titanates, which are capable of interaction with the filler surface and binder, are used as finish or active compounds. The effectiveness and possibility of using one or other method depend on the type of binder and chemically active compound.

It was stated earlier that modification of glass plastics takes place due to interaction of the chemically active compound at different ends with the polymer and filler, respectively. Recent years have witnessed the appearance of data that enable one to assert that modification of glass plastics with organofunctional silanes results in more complex transformations [16]. Along with the possible graft of the active compound in terms of the glass —OH groups, polyorganosiloxane is likely to form. The above review notes that, under the effect of water, the finishing agent can be subjected to hydrolysis and removed partially from the filler surface. Owing to the reversible nature of the hydrolysis process, however, the filler—finishing agent—polymeric matrix interface is preserved. This is also contributed by the existence of functional groups, the binder interacting with the molecules. In doing so, the extent of curing in the boundary layers increases and the polymer structure undergoes compaction, thus preventing to a certain extent water ingress. The binder adhesion to the glass fibre increases.

The results of the physical and chemical studies enabled the identification of some mechanisms of the effect of active compounds when added to the binder. Part of the active compounds (about 25 to 30% of the added binder) migrates towards the glass filler surface and is not removed therefrom after the solvent-assisted extraction or after heating in vacuum.

The compound formed on the surface differs from the finishing agent in terms of the composition and is essentially a partially crosslinked polyorganosiloxane. The latter shows high heat and water resistance, and also contains functional groups (for instance, amine groups in the case of using γ-aminopropyltriethoxysilane), which interact with the binder in the course of their curing, influencing not only adhesion but also cohesion properties of the matrix. Addition of active compounds to the binder somewhat decreases the internal stresses at the filler–matrix interface.

γ-aminopropyltriethoxysilane (the AGM-9 product) was found to be an effective stabilizer of the properties of glass plastics based on phenol–formaldehyde, epoxy, polyimide and some other binders.

The examples of polyimide and modified phenol–formaldehyde binders were used for the purpose of comparative analysis of the properties (Table 5.14) of glass plastics stabilized in various ways (using the AGM-9 product as a finishing agent, as part of active lubricant No. 80 and addition to the binder).

The results of the mechanical tests indicated that finishing improves the plastic strength as compared to specimens produced on the basis of the heat-treated glass fabric. In this case, the absolute strength values, however, were lower than in the case of adding the AGM-9 product to the binder or direct lubricant. Moreover, glass plastic specimens produced on the basis of the finished glass fabrics displayed, after immersion in water, inadequate reproducibility of strength values.

Table 5.14 Properties of glass plastics stabilized using various methods

| | | Glass plastic flexural strength (MPa) | | |
| | | In initial state | After 2 h stay in boiling water | Strength preservation (%) |
Type of binder	Stabilization method			
Polyimide	Heat treatment	205	44	20
	Finishing	332	328	99
	Direct lubricant	455	395	87
	Addition to binder	434	407	94
Modified phenol–formaldehyde	Heat tratment	374	127	34
	Finishing	400	226	56
	Direct lubricant	534	477	90
	Addition to binder	480	464	96

Thermal ageing results indicated that, on adding the active compounds to the lubricant or binder, the glass plastic strength is maintained at rather high level. In particular, this effect manifests itself in the combined exposure to heating and water. (In this case, the strength of the stabilized polyimide glass plastic is 10 to 12 times higher than that of the unstabilized plastic.)

The nature of the change in the dielectric characteristics of glass plastics depending on the stabilization method is different. Application of active lubricant to the glass fibre, in contrast to glass fabric finishing and addition of the active additive to the binder, does not lead to the same extent to stabilization of the material dielectric characteristics when held in water and at $\phi = 98\%$ (Fig. 5.3). This should be taken into account in manufacture of radio items.

In reviewing the methods of stabilizing the properties of glass plastics, the type of binder should be noted, e.g. for epoxy binders added to active compounds with amino and alkoxy groups, the working life of the binder and prepreg noticeably decreases.

5.4 PROPERTIES OF GLASS PLASTICS

As noted above, the physical and mechanical properties of glass plastics are determined by the structure of the composite material, which depends on the type of glass-fibre filler, the matrix and their interaction. This enables the material characteristics to be controlled over a broad range. The filler and its location largely govern the anisotropy of the properties of glass plastics and items made from them. The selected method for production of parts from fibrous composite materials influences the possible nature of defects in the source material. These defects may be caused by the existence in the filler (i.e. fabric, bundle, threads) of disintegrated fibres, distortions and shears of the fibres and threads, which results in a high spread of fibrous filler characteristics with respect to the average value. For this reason only, the mechanical strength variation factor of sheet glass-cloth-based laminates amounts to 10% and over. In conditions of exposure to elevated temperature and humidity, the glass plastics properties variation factor increases.

Depending on design and manufacturing processes, non-uniform impregnation of the filler in the course of moulding a part is possible, which causes the formation of pores and, in some instances, the non-uniform distribution of binder over the material volume. All these factors affect the entire package of mechanical, thermal, dielectric and other characteristics of glass plastics.

When in service, the properties of a composite material and items made therefrom depend on the medium's physical and chemical conditions and the loading conditions.

5.4.1 Mechanical properties

For a long time, mechanical characteristics have been determined mainly under static conditions. However, their application in vital structural parts operating in complex loading conditions required the determination of fatigue and long-term strength of composite materials.

Table 5.15 Mechanical properties of glass-cloth-based laminates

Properties		EDT-69N epoxy binder			SP-97K polyimide binder		MF polyaminomaleinimide binder	
		T-10-80[a]	T-25-78[b]	T-15[c]	T-10-80	T-15	T-10-80	T-25-78
Tensile strength (MPa)	Warp	630	1.178	460	495	200	476	1.190
	Weft	325	91	300	220	125	270	85
Compressive strength (MPa)	Warp	551	579	310	407	220	644	836
	Weft	358	228	260	210	132	418	300
Flexural strength (MPa)	Warp	898	1.191	500	400	310	697	1.380
	Weft	525	185	310	265	213	232	190
Elasticity modulus in tension (MPa)	Warp	29 800	49 900	21 000	30 000	26 000	33 500	65 000
	Weft	19 000	18 700	18 000	13 500	18 100	21 000	22 000
Poisson's ratio	Warp	0.144	0.238	0.187	0.92	0.139	0.155	0.258
	Weft	0.125	0.103	0.154	0.06	0.076	0.145	0.085
	45°	0.554	0.503	0.530	—	0.512	0.75	0.416

[a] Satin-weave fabric.
[b] Cord glass fabric.
[c] Hollow-fibre fabric.

Table 5.16 Mechanical properties of unidirectional glass plastics produced using winding method

Properties	Glass-cloth-based laminate grade				
	VPS-9	VPS-18	VPS-27	STPN-97SK	STNM-F
Type of binder	Epoxy	Epoxy	Epoxy	Polyimide	Polyaminomale-inimide
Tensile strength (MPa)	1.700	2.140	1.840	1.100	1.500
Compressive strength (MPa)	690	710	567	410	–
Static bending strength (MPa)	900	1.000	1.180	830	–
Elasticity modulus in tension (MPa)	50 700	64 000	58 700	40 200	70 000
Poisson's ratio	0.31	0.26	0.33	0.22	–
Fatigue tensile strength based on 10^7 cycles (MPa)	–	250	550	–	–

The basic mechanical properties of glass plastics based on various binders have been specified above. The general conclusion that can be drawn from these data and on the basis of the known publications is that higher mechanical properties in normal conditions are exhibited by glass plastics based on epoxy matrices; the properties of glass plastics based on polyimides and polyesters are somewhat lower.

Tables 5.15 and 5.16 present the basic properties of glass-cloth-based laminates based on various glass fabrics and unidirectional wound glass plastics. In analysing the mechanical characteristics of glass-cloth-based laminates based on various fabrics, it should be noted that lower anisotropy of properties is a feature of glass-cloth-based laminates based on satin-woven glass fabric; the highest anisotropy of properties is shown in glass-cloth-based laminates based on cord glass fabric. The properties of glass-cloth-based laminates based on hollow fabrics are lower; their strength is also lower (1500 kg m^{-3} versus 1750 kg m^{-3}) than that of laminates based on glass fabrics composed of continuous fibres exhibiting similar structure.

The highest glass plastic strength, stiffness and deformation anisotropy that meet the construction loading conditions can be attained using the method of winding the reinforcing filler, i.e. bundle, thread and strip. In this case, the strength and stiffness are determined mainly by the volumetric content and chemical composition of the glass-fibre filler. Table 5.16 presents the basic mechanical properties of unidirectional wound glass plastics. The properties of these materials can be improved using special processing techniques. The mechanical properties of unidirectional glass plastics based on glass fibres exhibiting different chemical compositions and epoxy binders are presented in Table 5.17.

Table 5.17 Mechanical properties of unidirectional glass plastics [1] (glass-fibre content is 70% by volume)

Type of fibre (diameter of 10 μm)	*Mean strength of fibres, from spool (MPa)*	*Fibre modulus of elasticity (MPa)*	*Glass plastic tensile strength (MPa)*	*Glass plastic modulus of elasticity (MPa)*	*Glass plastic compressive strength (MPa)*
Alkali-free aluminosilicate 'E'	2.80	74 000	1.85	58 600	1.50
Magnesium aluminosilicate	5.00	95 000	2.50	70 000	1.95
High-modulus	4.50	110 000	1.92	78 000	2.00

Figures 5.4, 5.5 and 5.6 present data characterizing the change of the basic properties as a function of temperature and time of exposure at 250 and 300°C (tests at the ageing temperature) of the STP-97K glass-cloth-based laminate widely used for manufacture of items operating at high temperatures. The compression and flexure characteristics are most sensitive to the heating temperature and time, i.e. the decrease of the glass-cloth-based laminate strength and stiffness is less in tension.

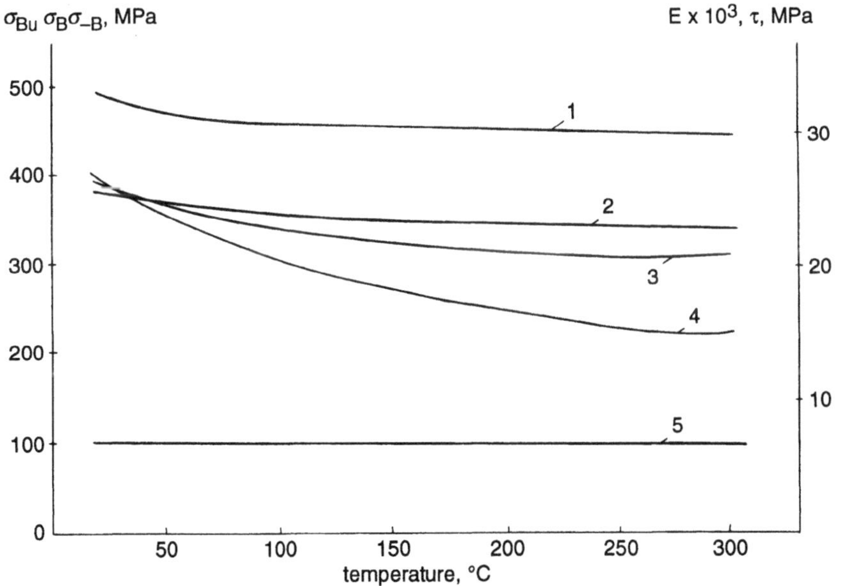

Fig. 5.4 Temperature effect on mechanical properties of STP-97K glass-cloth-based laminate: 1, tensile strength σ_b; 2, elasticity modulus in tension E; 3, static bending (flexural) strength σ_{bu}; 4, compressive strength σ_{-b}; 5, interlaminar shear strength τ.

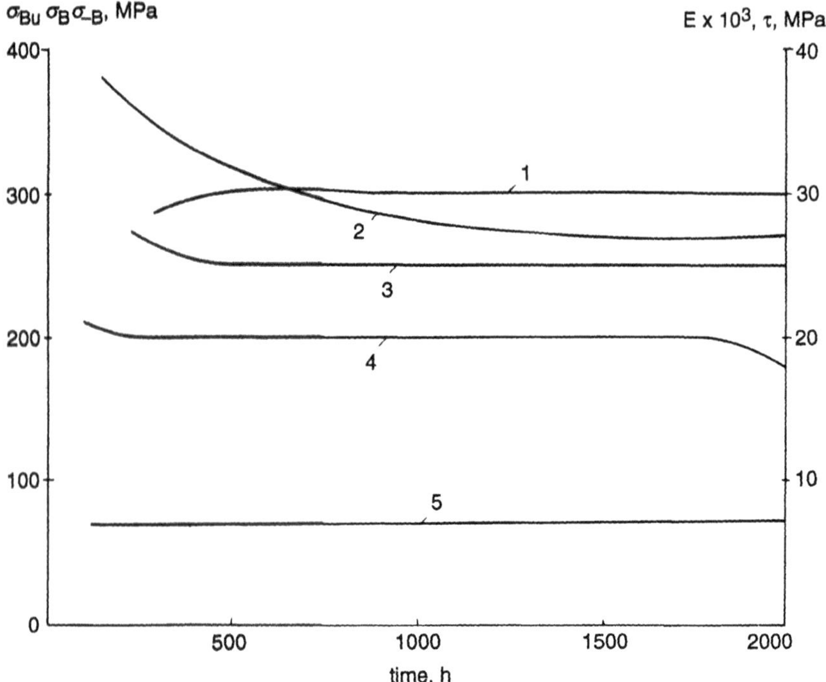

Fig. 5.5 Effect of thermal ageing at 250°C on mechanical properties of STP-97K glass-cloth-based laminate: 1, tensile strength σ_b; 2, elasticity modulus in tension E; 3, static bending strength σ_{bu}; 4, compressive strength σ_{-b}; 5, interlaminar shear strength τ.

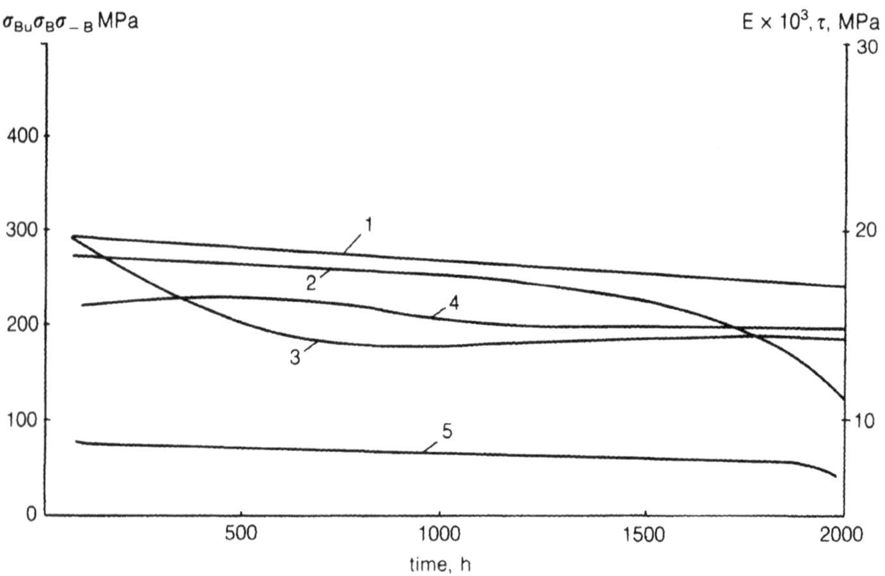

Fig. 5.6 Effect of thermal ageing at 300°C on mechanical properties of STP-97K glass-cloth-based laminate: 1, elasticity modulus in tension E; 2, compressive strength σ_{-b}; 3, static bending strength σ_{bu}; 4, tensile strength σ_b; 5, interlaminar shear strength τ.

Table 5.18 Properties of glass-cloth-based laminates in thermal cycling

Number of cycles	Flexural strength (MPa) at test temperature	
	20°C	*400°C*
Initial state	500	170
10	320	228
20	288	225
30	262	224
50	254	183
80	215	190
100	200	172

Of interest are data enabling one to estimate the behaviour of STP-97K glass-cloth-based laminate in thermal cycling. The glass-cloth-based laminate was exposed to 100 cycles at a temperature of 130°C for 3 h and, subsequently, at a temperature of 400°C for 20 min. The flexure test results are presented in Table 5.18.

The increased strength at the temperature of 400°C at the start of the thermal cycling requires additional explanation. Sufficiently high stability of the properties at elevated temperatures is exhibited by glass-cloth-based laminates based on organosilicon binders, but their level, however, is lower than in materials based on polyimides.

The properties of glass plastics based on three-dimensional (3D) strengthened fabrics and organosilicon matrices are presented in Table 5.19.

The application of three-dimensional strengthened sewing and multilayer fabrics ensures the increase in the strength of plastics in the case of interlaminar shear, which is important for matrices of all types but, in particular, for organosilicon matrices.

Table 5.19 Properties of glass plastics based on fillers exhibiting different structures and organosilicon binders

Type of filler	Material grade	Density $(kg\,m^{-3})$	Flexural strength (MPa)	Tensile strength (MPa)	Compressive strength (MPa)
TSP, sewing fabric	TSPK-101	1.650	90	75	30
	TSP-9FA	1.600	120	11	50
MKT, 3D fabric	MKTK-101	1.500	110	95	35
	MKT-9FA	1.600	100	12	38
All-knitted fabric	STANK	1.500–1.700	75	65	30

Using knitted fabric enables parts to be moulded with high length to width and height ratios.

The rational selection of a composite material for application in specific constructions is determined by the longevity index, i.e. long-term strength, creep and endurance. These characteristics determine the possible service life of the material.

Tables 5.20 and 5.21 present the long-term strength of glass-cloth-based laminates in tension and flexure.

Analysis of the indices of long-term strength of glass-cloth-based laminates in both tension and flexure demonstrated that the maximum strength level was achieved for glass-cloth-based laminates based on epoxy and polyimide binders, which determines their application in vital constructions. At elevated temperatures up to 200 and 350°C, respectively, the advantages of epoxy and polyimide glass-cloth-based laminates over other plastics are preserved. If, however, the quality of the glass plastics is evaluated in terms of the strength preservation factor determined as the long-term strength to ultimate strength ratio, the best value of this factor is observed for glass-cloth-based laminate based on the phenol–formaldehyde matrix. As regards this glass-cloth-based laminate, the tensile strength preservation reaches the value of 0.90 and 0.88 at temperatures of 20 and 200°C, respectively.

Endurance or fatigue breakdown of composite materials is a complex process conditioned by the difference of physical and mechanical properties of the material's

Table 5.20 Long-term tensile strength of glass-cloth-based laminates (MPa)

Glass-cloth-based laminate grade	Time (h)						
	0.1	1.0	10	10^2	5×10^2	10^3	2×10^8
VPS-21	400	385	370	355	345	340	335
VPS-24	400	370	335	305	225	275	265
ST-69N	465	450	425	400	380	375	–
STP-97S	400	370	330	300	280	270	–
STP-97K	480	450	430	400	380	360	–

Table 5.21 Long-term flexural strength of glass-cloth-based laminates (MPa)

Glass-cloth-based laminate grade	Time (h)					
	0.1	1.0	10	10^2	5×10^2	10^3
ST-69N	740	700	660	625	600	–
STM-F	480	450	420	390	320	190
STP-97S	450	410	370	330	300	295
STP-97K	350	330	300	280	260	245

components. Under the effect of fatigue loading, depending on the type of matrix and filler, a glass plastic exhibits fatigue failure and crack formation in the matrix and reinforcing filler, weakening of the adhesive bonds, and interlayer separation [17]. In doing so, glass-cloth-based laminates exposed to minor loading display first cracks, then peeling and cracking of the matrix with subsequent separation and breakage of the fibres at the points of intersection of the warp and weft threads. In unidirectional glass plastics, as in other composite materials, the extent of strength reduction at the moment of breakdown is lower than in orthogonal and cross-reinforced plastics.

The behaviour of composite materials, including glass plastics, depends on the type of deformation and loading rate. Increase of loading rate brings about a considerable heating up of the material, which may be an additional cause of material breakdown.

Table 5.22 presents the fatigue characteristics of glass-cloth-based laminates based on epoxy and polyimide binders. The indices of fatigue tensile strength of unidirectional glass plastics are specified below (Fig. 5.8).

Earlier, some authors [18, 19] demonstrated that a certain relation exists between the static and fatigue characteristics. This relation is quite clearly traced using the data of Tables 5.15 and 5.22. As humidity and temperature increase, a fatigue strength decrease was observed in glass plastics based on thermosetting binders. The temperature effect on the endurance limit for the STKM-F glass-cloth-based laminate based on the cord fabric and polyaminomaleimide matrix is represented in Fig. 5.7. The characteristics of the fatigue strength of the VPS-27 unidirectional glass plastic based on epoxy resin and glass thread with reinforcing filler content ranging from 68 to 72% are presented in Fig. 5.8. In this case, the minimum endurance limit takes place in flexural deformation.

Another very important characteristic of glass plastics is the cracking resistance in both fatigue loading and thermal cycling. For evaluation of crack resistance in fatigue loading, the characteristic crack resistance K_{Ic} adopted for evaluation of metals

Table 5.22 Fatigue strength of glass-cloth-based laminates (MPa)

Type of deformation	Glass-cloth-based laminate grade	Number of loading cycles			
		10^4	10^5	10^6	10^7
Tension	EDT-69N	380	300	230	180
	STP-97S	180	155	125	90
	STP-97K	265	220	175	130
	STP-97KP	80	60	50	40
Flexure	EDT-69N	410	360	300	240
	STM-F	310	270	220	175
	STP-97S	280	240	200	160
	STP-97KP	140	110	95	85

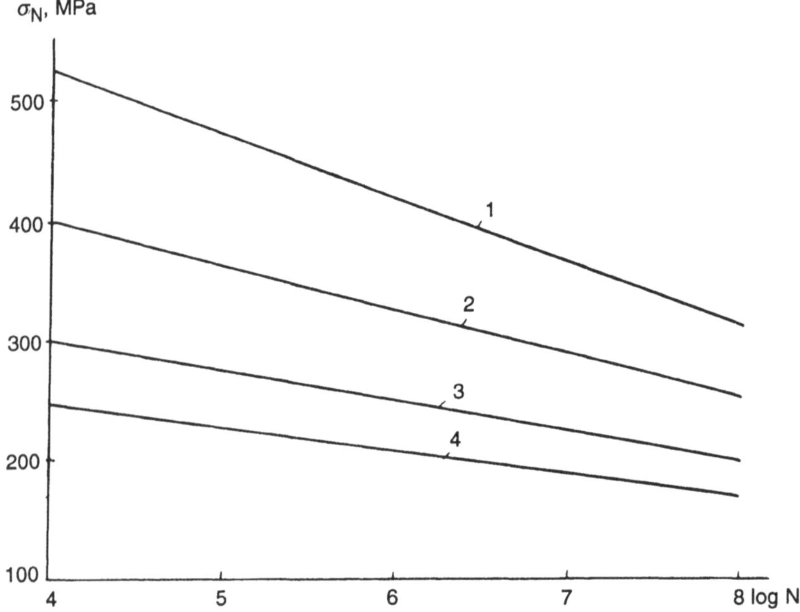

Fig. 5.7 Fatigue strength σ_N (N = number of cycles) of glass-cloth-based laminate in heating (loading rate of 25 Hz): 1, in tension, 20°C; 2, in flexure, 20°C; 3, in tension, 250°C; 4, in flexure, 250°C.

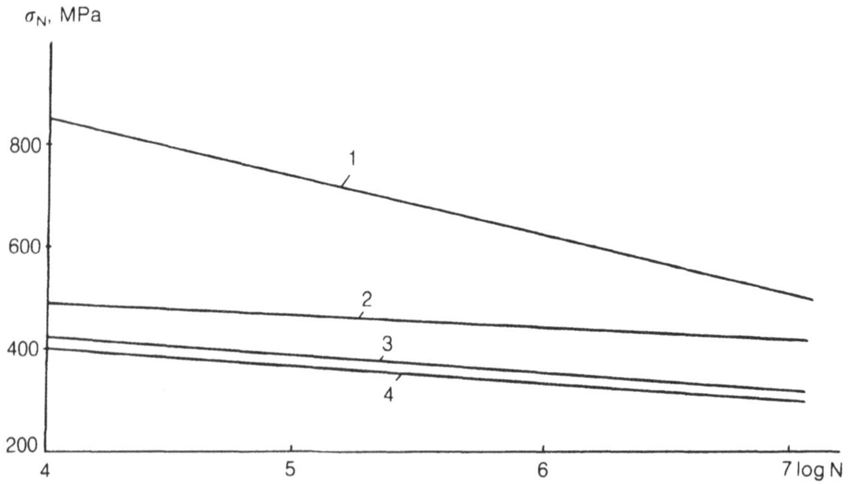

Fig. 5.8 Fatigue strength σ_N of VPS-27 unidirectional glass plastic for various types of loading (loading rate of 25 Hz): 1, in tension; 2, in compression; 3, in tension–compression; 4, in flexure.

Table 5.23 Effect of thermal cycling on crack resistance

Reinforcing fibre	Fibre surface treatment product	Crack resistance[a]	
		K_q (N mm$^{-3/2}$)	K_{qc} (N mm$^{-3/2}$)
Solid	Paraffin emulsion	60.2	48.0
Solid	-76 lubricant	51.0	40.3
Hollow	Paraffin emulsion	41.3	40.2

[a] K_q is crack resistance in initial state; K_{qc} is crack resistance after thermal cycling.

is gaining ever-growing acceptance. Depending on the type of filler and binder, index K_{lc} is within the limits of 80 to 200 kg mm$^{-3/2}$.

Table 5.23 presents the effect of thermal cycling on crack resistance of unidirectionally reinforced glass plastics. In this case, the crack resistance dependence on the type of lubricant and glass is shown [20].

The thermal cycling conditions are as follows: 5 min stay in liquid nitrogen at a temperature of $-196°C$, then a stay in air at a temperature of $20°C$.

On completion of the thermal cycling, the crack resistance of glass plastics is higher in the specimens reinforced with continuous fibres. The authors explain this fact by the higher stiffness as compared to that of specimens reinforced with hollow fibres. Noticeable is the effect of fibre surface treatment on crack resistance of glass plastics in the initial state and after thermal cycling.

5.4.2 Thermal properties

Knowledge of the thermal characteristics of glass plastics is of particular importance, as these materials are used in constructions wherein their heat-insulating properties are decisive. The operating temperatures of these constructions cover the broad range from cryogenic temperatures to those of matrix destruction.

The basic thermal characteristics of glass plastics over the temperature range wherein no practical destruction of the matrix takes place are presented in Table 5.24.

Data published earlier [21] and the indices presented in Table 5.24 enable one to draw the conclusion that the above characteristics depend on density of the material and, accordingly, on its porosity and the extent of binder curing. In the case of adequate impregnation of the glass-fibre filler, and the absence of voids and cracks, the glass plastic exhibits higher thermal conductivity and diffusivity.

The linear expansion coefficients for identical structures of the filler and the same binder content in the glass plastic are rather close.

Plastic materials with fibre filler exhibit different linear expansion coefficients in the thickness direction and in the plane of a sheet. The thermal expansion

Glass plastics

Table 5.24 Thermal properties of glass plastics

Properties	Type of binder	Glasscloth-based laminate grade	Test temperature (°C)					
			20	100	150	200	250	300
Thermal con-ductivity $(W\,m^{-1}\,°C^{-1})$	Polyester	VPS-24	0.29	0.29	0.30	0.30	–	–
	Phenol–for-maldehyde	FFA	0.22	0.23	–	0.24	–	0.25
	Epoxy	ST-69N	0.37	0.38	0.38	0.39	–	–
	Organosi-licon	SK-9X	0.34	0.34	–	0.35	–	0.36
	Polyimide	STP-97S	0.35	0.34	–	0.34	0.34	0.32
		STP-97K	0.31	0.31	0.31	0.31	0.31	0.31
		STP-CMK	0.38	0.39	0.39	0.39	0.40	0.40
		STM-F	0.32	0.32	0.32	0.32	0.30	0.32
Thermal diffusivity $(10^{7}\,m^{2}\,s^{-1})$	Polyester	VPS-24	1.7	1.6	1.55	1.55	–	–
	Phenol–for-maldehyde	FFA	1.05	1.25	–	1.10	–	1.20
	Epoxy	ST-69N	2.3	1.9	1.55	1.50	–	–
	Organosi-licon	SK-9X	2.2	–	–	1.47	–	1.58
	Polyimide	STP-97S	2.26	2.16	–	1.94	2.12	2.20
		STP-97K	1.90	1.80	1.70	1.70	1.70	1.70
		STP-CMK	2.35	1.95	1.60	1.50	1.70	1.70
		STM-F	1.55	1.35	1.35	1.35	1.40	1.60
Specific heat $(kJ\,kg^{-1}\,°C^{-1})$	Polyester	VPS-24	1.10	1.14	1.20	1.20	–	–
	Phenol–for-maldehyde	FFA	1.05	1.20	–	1.45	–	1.4
	Epoxy	ST-69N	0.85	1.05	1.30	1.35	–	–
	Organosi-licon	SK-9X	0.8	0.88	–	1.17	–	1.13
	Polyimide	STP-97S	0.85	0.85	–	0.83	0.75	0.72
		STP-97K	0.94	0.99	1.05	1.05	1.05	1.05
		STP-CMK	0.85	1.05	1.30	1.35	1.25	1.25
		STM-F	1.15	1.32	1.32	1.48	1.48	1.11

of laminated glass plastics in the thickness direction is usually higher than in the sheet plane owing to the higher linear expansion coefficient of the matrix and lower effect of glass-fibre filler on the above index.

In a unidirectional glass plastic, the linear expansion coefficients exhibit less difference along and across the sheet than in carbon and organic plastics.

As glass plastics are widely used as materials in the interiors of ships, railway cars and aircraft, of great importance are indices that determine their behaviour in buring, i.e. combustibility, smoke formation and toxicity in burning. These characteristics depend on the type of matrix.

Combustibility of glass plastics can be reduced by adding antipyrenes (fireproofing or fire-retarding agents) to the matrix or using oligomers containing halides as a binder. These modification methods are widely used in polyester and epoxy binders, thus enabling one to decrease their combustibility and transfer them into self-extinguishing binders. The smoke release in this case, however, drastically increases.

The minimum combustibility and smoke release are displayed by glass plastics based on polyimide and organosilicon matrices. These indices are somewhat worse for materials based on phenol–formaldehyde binders. It should be noted that, for various applications of glass-cloth-based laminate, there is no single procedure yet for evaluation of a material's combustibility.

5.4.3 Dielectric properties

The dielectric characteristics of glass plastics are determined by the chemical composition and volumetric content of the components, and depend on the electromagnetic field frequency and the material operating conditions, primarily temperature and humidity. In this case, the major contribution to glass plastic dielectric constant (ε) is made by the reinforcing filler, whereas the dielectric loss tangent (tan δ) is determined largely by the properties of the polymeric matrix [22]. However, varying the glass chemical composition or using special glass for production of fibres enables the glass plastic dielectric losses to be changed over a broad range.

Table 5.25 presents the dielectric and insulating characteristics of glass plastics based on binders of various type.

Figures 5.9 to 5.13 present data characterizing the change in dielectric properties with heating, higher-than-usual humidity and water.

As the temperature rises, tan δ and ε of glass plastics increase somewhat and then decrease. This may be caused by the removal of low-molecular-weight products and, in some instances, the initiation of matrix destruction. The higher the matrix thermal oxidation stability, the more stable are the properties of the plastics when heated.

The effect of long-term heating of glass plastics on the dielectric constant and the dielectric loss tangent is characterized by the data presented in Table 5.26.

The real operating conditions of items may involve the combined effect of high temperatures and higher-than-usual humidity. Glass plastics based on organosilicon and polyimide binders exhibit satisfactory properties under the combined effect of temperature and higher-than-usual humidity.

Under higher-than-usual humidity conditions and water, the dielectric and insulating properties of glass plastics degrade. Most sensitive to the moisture effect are the dielectric characteristics of the material at lower frequencies. The behaviour of glass plastics in these conditions largely depends on the rate of moisture sorption by the material. In doing so, the rate of material saturation with moisture in the direction

Table 5.25 Dielectric properties of glass plastics

Properties		Glass-cloth-based laminate grades								
	FFA	VPS-24	ST-69N	SK-9FA	SK-101	SK-9X	STP-97S	STP-97K	STP-CMK	STM-F
Dielectric loss tangent										
at 10^6 Hz	0.007	0.0146	0.0157	0.002	0.012	0.002	0.0084	0.0034	0.0110	0.0062
at 10^{10} Hz	0.015	–	0.0185	0.016	0.006	0.009	0.012	0.0081	0.0120	0.0144
Dielectric constant										
at 10^6 Hz	4.15	4.6	4.58	5.05	4.2	4.5	4.6	4.16	4.34	4.54
at 10^{10} Hz	4.43	–	4.47	4.83	2.96	5.04	4.71	4.57	4.25	4.70
Volume resistivity, ρ_v (ohm cm)	–	4.5×10^{13}	3.1×10^{15}	3×10^{13}	2×10^{13}	1×10^{14}	2×10^{13}	7×10^{13}	9.8×10^{13}	2.2×10^{12}
Surface resistivity (ohm)		1.5×10^{14}	5.3×10^{16}	1.2×10^{14}	4×10^{13}	5×10^{14}	2×10^{13}	8×10^{13}	3.4×10^{14}	8.1×10^{13}
Electric strength (kV mm^{-1})		–	–	–	–	9.0	8.0	14.2	–	42 to 50

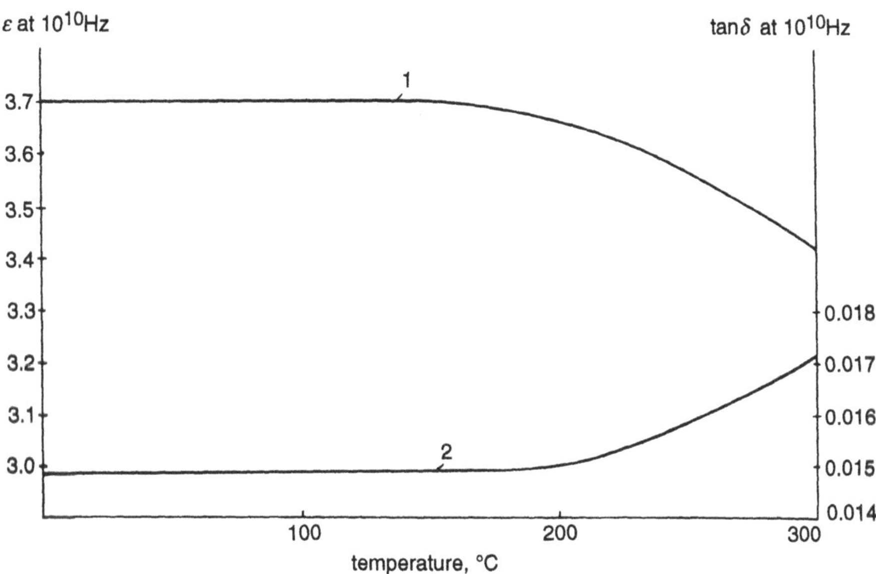

Fig. 5.9 Dielectric properties of SK-9XK glass-cloth-based laminate as a function of temperature at frequency of 10^{10} Hz: 1, dielectric constant (ε); 2, dielectric loss tangent (tan δ).

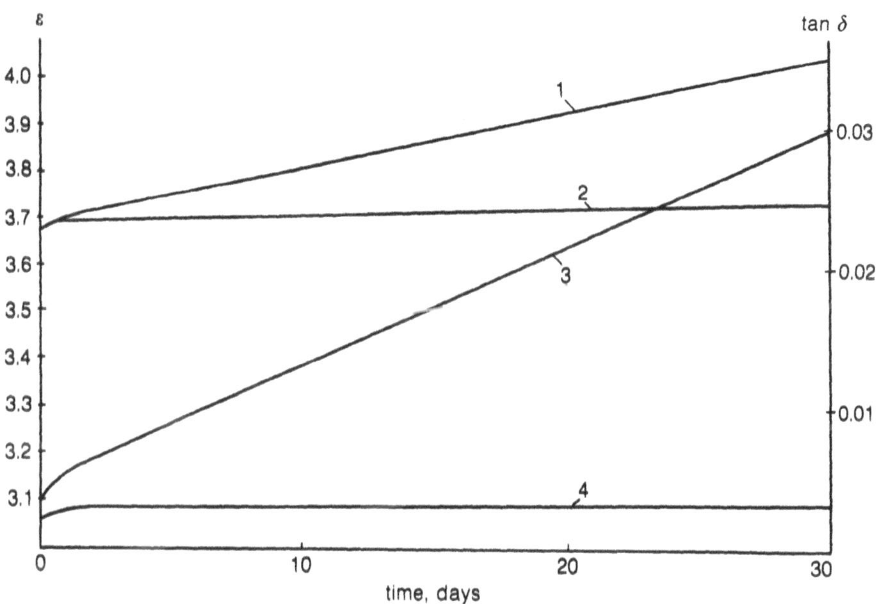

Fig. 5.10 Dielectric properties of SK-9XK glass-cloth-based laminate in humid medium at $\phi = 98\%$: 1, dielectric constant (ε) at 10^6 Hz; 2, dielectric constant (ε) at 10^{10} Hz; 3, dielectric loss tangent at 10^6 Hz; 4, dielectric loss tangent at 10^{10} Hz.

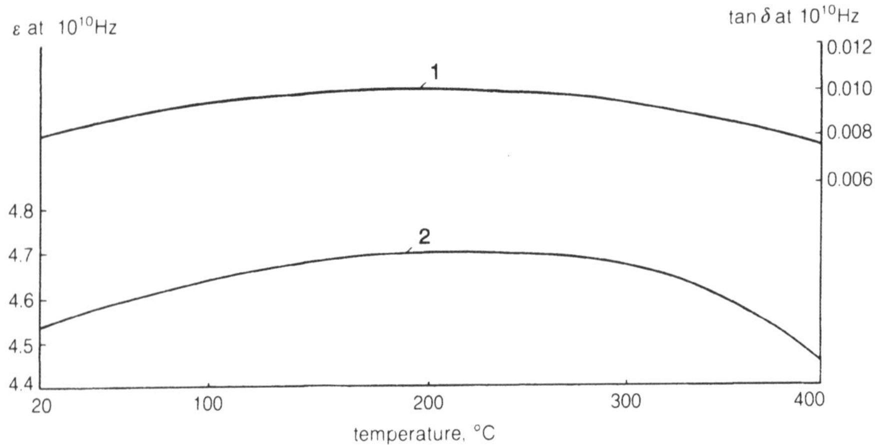

Fig. 5.11 Dielectric properties of STP-97K glass-cloth-based laminate at elevated temperatures: 1, dielectric loss tangent at 10^{10} Hz; 2, dielectric constant at 10^{10} Hz.

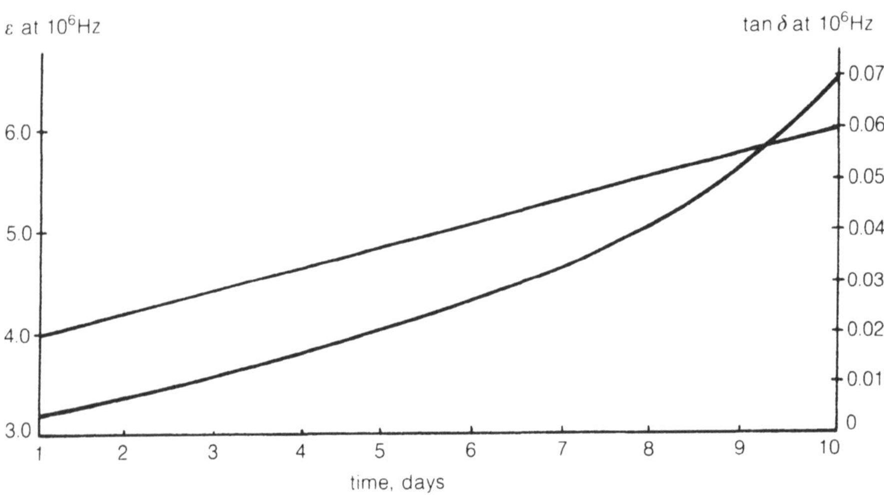

Fig. 5.12 Dielectric properties of STP-97K glass-cloth-based laminate after a stay in water at frequency of 10^6 Hz: 1, dielectric loss tangent at 10^6 Hz; 2, dielectric constant at 10^6 Hz.

parallel to the layers is higher than in the perpendicular direction. Hence, it is expedient in using glass plastics to protect the material end faces against moisture penetration.

5.5 GLASS PLASTICS TECHNOLOGY AND APPLICATIONS

Manufacture of glass plastics and items therefrom is effected using conventional methods that have also gained acceptance in the production of items from other composite materials. These methods include contact moulding, direct extrusion, vacuum and autoclave moulding, pressure impregnation and moulding in a closed

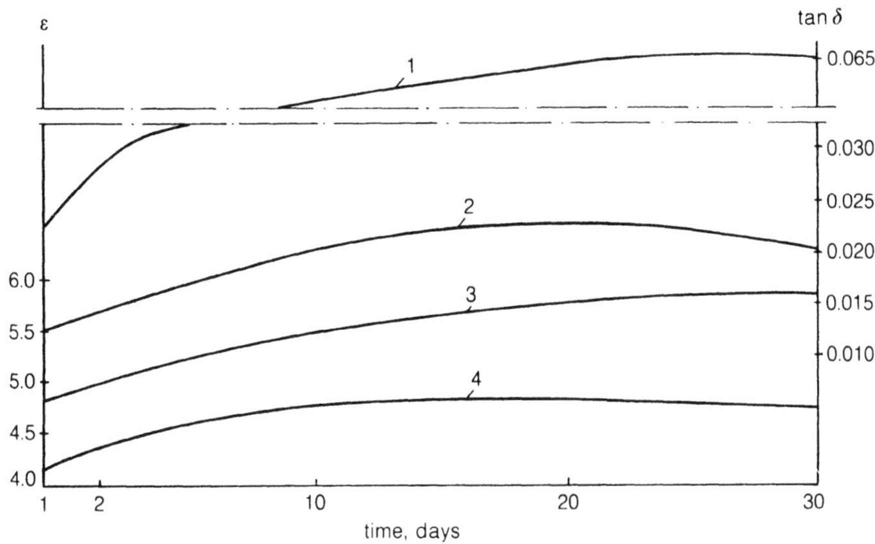

Fig. 5.13 Dielectric properties of STP-97K glass-cloth-based laminate after a stay in a humid medium at $\phi = 98\%$: 1, dielectric loss tangent at 10^{10} Hz; 2, dielectric loss tangent at 10^6 Hz; 3, dielectric constant at 10^{10} Hz; 4, dielectric constant at 10^6 Hz.

Table 5.26 Dielectric properties of STP-97K glass-cloth-based laminate after thermal ageing at 300°C

		Holding time (h)			
Properties	*Frequency (Hz)*	*100*	*1000*	*2000*	*2000 h, next 48 h at $\phi = 98\%$*
Dielectric	10^6	0.0028	0.0025	0.0030	0.0480
loss tangent	10^{10}	0.0053	0.0074	0.0078	0.0348
Dielectric	10^6	4.19	4.25	4.14	4.42
constant	10^{10}	4.22	4.23	4.17	4.58

mould, winding and pultrusion. In some instances, the above process techniques are combined.

Selection of one or other method in the production of glass plastics and items depends on the part's purpose, shape and overall dimensions, design requirements, scope of application, operating conditions and other factors.

Contact moulding is employed in the manufacture of a comparatively small number of large-sized and low-loaded items. Direct extrusion is used, as a rule, in the manufacture of large batches of relatively small-sized parts. Vacuum moulding enables the production of large-sized parts required in pilot-scale production. A large series of large-sized and complicated items exhibiting high strength is manufactured using the autoclave moulding method. Items wherein precise thickness of a wall should be maintained (for instance, fairings) are moulded in a rigid closed mould using the pressure impregnation method. The winding and pultrusion method is used in the mass production of high-strength materials.

Usually, an item's production process comprises the following major operations: preparation of the binder and the reinforcing filler (lay-out of the latter), filler impregnation, moulding and mechanical finishing of the items.

In recent years, the attention of process engineers has been concentrated not on the development of radically new process techniques but on the elaboration of existing methods enabling intensification of some process operations, improvement of ecological conditions and reduction of the power input of the manufacturing process due to reduction of moulding temperature and time.

If the production process operations of glass plastics and items are considered, it should be noted that the most severe conditions are observed in the course of glass-fibre filler impregnation and drying.

The production of laminated plastics makes wide use of the method based on filler impregnation with solutions of binders. The existence of organic solvents in the binder in the impregnation stage requires additional power input in their removal from the fillers and their further recovery in large-scale production. And this does not rule out the possibility of environmental pollution. The ecological imperfections of the existing impregnation processes are evident. Similar complexities arise in the production of items using the winding method.

To eliminate some of the drawbacks of the impregnation stage, ever-increasing use is made of high-viscosity binders or those with active solvents and powder-like and film binders.

In the production of glass-cloth-based laminates based on high-viscosity binders and those with active solvents, polymer melts require the application of methods for the forced impregnation of the glass fabric lamination with fast-curing binders with subsequent continuous extrusion. To improve the impregnation process, the forced impregnation is preceded by drying and vacuum treatment of the glass fabric so as to dispose of adsorbed moisture [23–25].

The application of powder-like binders to a glass fabric enables one to abandon completely organic solvents, which is particularly important when the use of high-boiling solvents is needed. Phenol–formaldehyde oligomers have been used as powder-like binders for the first time [26]. The use of powder-like polyaminoimide binders turned out to be much more effective. Usually, N-methylpyrrolidone is employed at the boiling temperature above $200°C$ as a solvent of this type. This powdered binder is applied to the glass fabric in an electrostatic field. The comparative properties of glass-cloth-based laminates based on polyaminoimide binder applied to the glass fabric either from solution or as a powder in an electrostatic field are presented in Table 5.27.

The application of powder-like binders enables noticeable improvement of the properties of glass plastics. The prepreg based on these binders is, however, quite

Table 5.27 Comparative properties of STM-F glass-cloth-based laminates on polyaminoimide binder

	Glass-cloth-based laminate strength (MPa)								
	Flexural			Compressive			Tensile		
Binder application method	20°C	250°C	250°C, 1000 h	20°C	250°C	250°C, 1000 h	20°C	250°C	250°C, 1000 h
From solution	690	425	260	650	390	360	475	270	210
In electro-static field	800	470	300	700	590	400	650	350	270

stiff and exhibits limited drapability. It is used mainly in the production of sheet materials.

Film binders feature high drapability and the possibility of use in the production of complex items.

The properties of glass plastics based on the PK-2a film binder are specified above. In the course of manufacture of items based on film binders, the operation of impregnation is combined with item moulding. Film binders based on polyimides exhibit lower drapability and higher fluidity temperature. To impart the required adaptability to manufacture with them, it is necessary to change the oligomer chemical structure [27]. These binders are more suitable for production of sheet materials.

The winding method is employed mainly for manufacture of shells, pipes, bottles and similar items featuring different overall dimensions and designations. In moulding using this method, the binder-impregnated continuous threads, bundles or strips are wound on the mandrel until a certain thickness is attained, cured and removed from the mandrel.

The winding method enables control to be achieved over a broad range of the content of the glass filler and the reinforcement pattern, which ensures the production of materials exhibiting different physical and mechanical properties. Equipment with computer control systems ensures the practical implementation of any reinforcement pattern.

The reinforcing filler either can be impregnated in the course of winding (the wet method) or use can be made for winding purposes of prepregs fabricated earlier from threads, bundles and strips (the dry method). Production of prepregs enables removal of the solvent and its recovery to be ensured. The process, however, gets somewhat complicated because of the need for prepreg heating so as to produce a solid material.

In this case, it is difficult to control the content of the filler in the item and to achieve its high volumetric content [28]. Depending on the type of binder, the prepreg working life is limited.

In the case of 'wet' winding, the reinforcing filler coming off the bobbin holder is formed as a strip, passes the guide and enters the impregnation bath tank.

If required, the filler is dried to remove the solvent and partially cure the binder and then wound on the mandrel.

To improve the impregnation of the reinforcing filler and to reduce its damageability, the impregnation process involves the use of the ultrasonic effect. The ultrasonic effect ensures the increase of structural homogeneity, improvement of binder processing characteristics and decrease of air inclusions therein [29].

For the purpose of reducing the abradability of fibres, the above authors [28] recommend that a strong pre-tension of the fibres is not created in winding-out, a tray with a toroidal surface is used as the strip-forming device instead of a pinned hackle, and that the guide rollers revolve and to prevent the binder proportioning using squeezers.

To improve the solidity of glass plastics in pultrusion and winding, it is expedient to use binders with a lower viscosity due to active thinners or heating, binder vacuum treatment, forced or repeated impregnation, and wind on a heated mandrel.

Compaction of wound items structure is contributed by tension adjustment, in particular, in manufacture of thick-walled items, and the use of the method for moulding in autoclaves or using thermoshrinking material.

In the transport industry, in particular, in ship, railway car and aircraft manufacture, widely used are three-layer panels exhibiting the coverings with a honeycomb, foam plastic and network filler placed in between. The honeycomb fillers are usually fabricated from glass fabrics, synthetic paper and other materials.

The three-layer construction panels enable one to decrease the mass of the items, increase stiffness, improve the heat insulating properties and achieve the required radio engineering characteristics.

The properties and application of the above three-layer constructions are described in special publications in a detailed manner. The last 10−15 years have witnessed the ever-growing application of three-layer glass plastics wherein the knitted fabric from the glass fibre has been effectively used as a filler [30].

The method for knitted fabric production is 7−10 times more efficient than that for fabric production. The knitted fabric is elastic and easily laid out to shape. The ring shape of the knitted fabric loops, arranged so as to have the butt face in the plane of the knitted fabric, determines its thickness. A knitted fabric soaks up binder extensively, and the excess may not be removed subsequently from the fabric by squeezing out without damage to the knitted fabric texture. Hence, to ensure the required content of binder, use is made of low-viscosity binder with passive solvent. Following impregnation and, in particular, drying of the knitted fabric, the compressibility of the fabric under pressure and its volumetric mass decrease.

Three-layer materials with the knitted fabric filler and constructions therefore can be manufactured using two methods, i.e. simultaneous moulding of the bearing coverings and filler, or precuring of the knitted fabric with subsequent moulding-on of the coverings. The application of the second moulding method enables the maximum possible thickness of the three-layer material to be obtained, but in this case, the strength of glueing together the coverings and filler is low, and the production process gets very complicated. The most reliable and least labour-intensive way of ensuring the material rational thickness at maximum filler-to-coverings joint strength is its simultaneous moulding under a pressure of 0.07−0.08 MPa. The best thickness of the three-layer material should be ensured by the filler rational density.

The properties of a glass knitted fabric filler based on an epoxy phenolic binder are as follows: compressive strength 10 MPa, and tensile strength 2.6 MPa.

The tensile and compressive strengths of the three-layer panels is determined mainly by the properties of the bearing coverings. As the thickness of the coverings and the density of the panels grow, the strength of the three-layer materials increases. The advantage of three-layer knitted fabric panels is sufficiently high strength at uniform breakaway of the coverings. The flexural stiffness of three-layer materials is 1.5 to 2 times higher than that of solid glass-cloth-based laminates, but at a much smaller mass.

The advantage of such three-layer glass plastics is the reduction of the labour input in their manufacture as compared to solid glass-cloth-based laminates and, in particular, honeycomb constructions owing to the simplicity of the knitted fabric lay-up and the adhesive-free jointing of the coverings and filler. In aircraft, these materials are used effectively for the interior parts (side and window panels, ducts); in small aircraft, these materials are employed for manufacture of engine nacelle panels, landing gear fairings and other parts.

As matters stand now, the application of glass plastics in various branches of engineering has already been made. However, broad-scale scientific and research efforts will contribute to the expansion of the range and redistribution in consumption as regards various types of items and branches of engineering. In spite of the growing use of composite materials exhibiting lower density, higher strength and stiffness, the scope of glass plastics applications in aircraft, aerospace engineering, electronics and other fields remains at a sufficiently high level. New developments are gaining primary acceptance in these branches of engineering.

Glass plastics are widely used in the transport industry, the building industry, production of chemical equipment and pipelines. This work now deals mainly here with a review of glass plastics use in aviation.

Glass plastics are the basic material for manufacture of the radar radomes and the surface antennas of airplanes, helicopters and rockets. They can be arranged in various sections of the aircraft, being load-bearing structural members, though relatively low-loaded. The glass plastics designated for these purposes should feature stability of dielectric characteristics in higher-than-usual humidity and heating conditions. Most suitable as materials for radio devices are glass plastics based on epoxy, polyimide and organosilicon matrices and various textures of fabrics composed of solid and hollow fibres of aluminoborosilicate glass and, in exceptional cases, quartz glass. The construction of fairings and, primarily, the walls is determined by the purpose and location of the fairing.

Owing to their high fatigue characteristics, glass-cloth-based laminates have become one of the major materials of the blades of helicopters and propeller fans. The blades of helicopters and propeller fans are exposed to a complex package of mechanical loads and deformations. They are used in the production of spars and blade coverings. In the first case, glass plastics are frequently used in hybrid materials in combination with carbon plastics and organites. To manufacture spars taking up the main load, use is made of unidirectional glass plastics. The blades are covered with glass-cloth-based laminate based on cord fabrics whose texture is selected depending on the loads applied to the covering. As a rule, epoxy polymers serve as a matrix in this case.

Depending on the location of the propeller fan in the aircraft and its possible position in the jet engine, additional requirements are imposed on the material of the blades in terms of preservation of mechanical properties at elevated temperatures. The composite material's erosion resistance is of no small importance.

Glass plastic is used for the manufacture of a broad range of interior parts, floor, side and window panels, of the aircraft. All parts of the interior must satisfy the requirements for refractoriness, low smoke release and low toxicity of gases liberated in burning. Though less stringent, these requirements are mandatory also for boats, ships and railway carriages. Glass plastics based on phenol–formaldehyde and polyimide matrices behave in the best way possible as interior materials. The past years have witnessed the particularly broad use of glass plastics and other materials based on poly(ester imides) for these purposes.

The refractoriness and low smoke release requirements are imposed on the ducts and air conduits of an air conditioning system. Depending on the requirements imposed thereon, the air conduits are manufactured using the vacuum or autoclave method. Some instances admit the use of strip or bundle winding.

In aircraft and rocketry, the winding method is applied for the manufacture of the large range of bottles with different designations, shapes and dimensions. High-pressure bottles are provided with a metal shell stiffened by glass bundle winding.

In rocketry, glass plastics based on silica fibres serve as external and internal heat-shield elements owing to their high ablation characteristics. As the matrix in heat-shield glass plastics, phenol–formaldehyde polymers are used. Very often, however, the structural members perform a number of combined functions in rocketry and aerospace engineering, being multifunctional load-bearing, radiotransparent and heat-shield material. Most suitable for these constructions are glass plastics wherein the matrix can be represented depending on the operating conditions by phenol–formaldehyde, organosilicon, epoxy and other polymers.

As noted above, glass plastics can be used in combination with carbon and organic plastics, thus ensuring impact and compression resistance.

Glass plastic in combination with thin duralumin sheet increases the fatigue characteristics of the material, thus ensuring a considerable reduction of an aircraft's construction weight.

REFERENCES

1. Trofimov N.N., Kanovitch M.Z., New achievements in the field of polymeric materials reinforced with glass and oxide fibres, *Journal of All-Union Chemical Society Named After D.I. Mendeleev*, 1989, **34**(5), 447–53.
2. Aslanova M.S., Gordon S.S., Dreitser V.I., Roginsky S.L., On reinforcement of glass plastics with hollow-structure fibres, *VNIISP Collected Papers 'Glass Fibre and Glass Plastics'*, 1970, pp. 159–67.
3. Lipatov Yu.S., Hybrid matrices for polymeric composites: alloys and mixtures of linear and network polymers, *Materials of the 14th Mendeleev Congress on General and Applied Chemistry 'New Structural and Functional Materials'*, Nauka, Moscow, 1989, p. 99.
4. Rosenberg B.A., Epoxy polymers and problems of creating polymeric matrices for high-strength composites, *Journal of All-Union Chemical Society Named After D.I. Mendeleev*, 1989, **34**(5), 453–9.
5. Kiselev B.A., Thermosetting binders and reinforced plastics based thereon, *Chemistry and Technology of High-Molecular Compounds*, vol. 11, VINITI, Moscow, 1977, pp. 163–204.
6. Kolomoyets G.A., Golubenkova L.I., Valetsky N.M., Vinogradova S.V., Stanko V.I., Korshak V.V., Phenol–formaldehyde oligomers with O-carborane groups, *Plastmassy*, 1974, **2**, 19–21.
7. Kolomoyets G.A., Golubenkova L.I., Valetsky N.M., Vinogradova S.V., Stanko V.I., Korshak V.V., Resites based on 1,2-bis(4-oxyphenyl)-carborane and phenol, *Plastmassy*, 1976, **10**, 30–2.
8. Shitikov V.K., Kiselev B.A., Stepanova V.N., Trofimenko V.V., Korshak V.V., Sergeev V.A., Phenolphthaleinhaltig Kopolymer, *Plast. Kautsh.*, 1974, **21**(10), 734–6.

9. Kiselev B.A., Sobolev I.V., Shpet V.S., Mayakov V.V., Ivanov G.A., Linear expansion coefficients of glass plastics based on polymethylsiloxane resin in conditions of heating up to 1000°C, *Polym. Mech.*, 1974, **3**, 531–5.

10. Gorbatkina Yu.A., *Adhesion Strength in Polymer–Fibre Systems*, Khimia, Moscow, 1987, p. 140.

11. Kiselev B.A., Kurochkina N.I., Grebneva T.V., Film organosilicon binder and glass-cloth-base laminate based thereon, *Plasticheskiye Massy*, 1990, **1**, 87–8.

12. Kireev V.V., D'yachenko B.I., Rybalko V.P., Ryzhov V.I., Polenov A.B., Bychkovskaya O.V., Tsvetaeva N.M., Structural ceramoplastics based on organosilicon oligomers of different composition, *Materials of the 14th Mendeleev Congress on General and Applied Chemistry*, 'New Structural and Functional Materials', Nauka, Moscow, p. 74.

13. Yanovsky Yu.G., Demidov Yu.M., Kerber M.L., Akutin M.S., Vinogradov G.I., Davydova I.F., Kiselev B.A., Korolev A.A., Effect of curing temperature conditions on structure and properties of polyaminoimide, *Plasticheskiye Massy*, 1983, **10**, 31–4.

14. Suslov A.P., Prokopkina V.A., Dolmatov S.A., Levshanov V.S., Kotukhova A.M., Glass plastics based on polybismaleimidamine, *Plasticheskiye Massy*, 1981, **7**, 27–8.

15. Pankratov V.A., Promising binders for high-strength heat-resistant polymeric composites, *Journal of All-Union Chemical Society Named After D.I. Mendeleev*, 1989, **34**(5), 459–68.

16. Mikhalsky A.I., Organofunctional finishing agents in filled polymeric systems, *Results of Science and Engineering: Chemistry and Technology of High-Molecular Compounds*, vol. 19, VINITI, Moscow, 1984, pp. 151–222.

17. Oldyrev P.P., Tomuzh V.P., Multicycle fatigue of composite materials, *Journal of All-Union Chemical Society Named After D.I. Mendeleev*, 1989, **34**(5), 545–52.

18. Tomuzh V.P., Protasov V.D. (ed.), *Breakdown of Composite Materials Structures*, Zinatne, Riga, 1986.

19. Oldyrev P.P., Multicycle fatigue of glass plastics in soft and rigid loading conditions, *Mechanics of Composite Materials*, 1981, **2**, 218–26.

20. Rudman I.R., Veselyansky Yu.S., Kanovitch M.Z., Method for determination of crack resistance of glass plastics after thermal cycling, *Plasticheskiye Massy*, 1990, **1**, 74–6.

21. Kiselev B.A., *Glass Plastics*, Khimia, Moscow, 1961.

22. Gurtovnik I.G., Sportsmen V.I., *Glass Plastics of Radio Designation*, Khimia, Moscow, 1987, p. 145.

23. Grutsenko V.F., Kerber M.L., Konev V.D., *State-of-the-Art and Prospects of Development of Methods for Production of Electrotechnical Laminated Plastics*, Informelectro, Moscow, 1988, p. 52.

24. Arsen'eva E.D., Grutsenko V.F., Kerber M.L., Binders for continuous process of production of foiled dielectrics, *Electrotechnical Production*, 1988, **3**, 23–6.

25. Shevchuk A.A., Leont'ev V.I., Kryzhanovsky V.K., Calculation of output of process for production of prepregs by forced impregnation of fabrics with polymer melts, *Plasticheskiye Massy*, 1990, **9**, 45–8.

26. Aleksandrovitch I.R., Kerber M.L., Akutin M.S., Kiselev B.A., Structural glass-cloth-base laminate with lower combustibility, based on powder-like phenol–formaldehyde binders, *New Polymeric Composite Materials in Machine Building, Proceedings of All-Union Scientific and Engineering Symposium*, Moscow, 1978, p. 24.

27. Koton M.M., Bolotnikova L.S., Svetlichny V.M., Davydova I.F., Kiselev B.A., Kudryavtsev V.V., *et al.*, Viscous and elastoviscous properties of meltable polyimides, *Plasticheskiye Massy*, 1986, **4**, 11–13.

28. Zelensky E.S., Kul'kov A.A., Kuperman A.M., Puchkov L.V., Wound plastics technology, *Journal of All-Union Chemical Society Named After D.I. Mendeleev*, 1989, **34**(5), 515–20.

29. Kolosov A.E., Karimov A.A., Replis I.A., Khazin V.G., Klyavlin V.V., Impregnation of fibre fillers with polymeric binder, *Mechanics of Composite Materials*, 1989, **4**, 724–31.

30. Karpova T.Ya., Parikovsky V.I., Kovalsky A.G., Landsman S.V., Glass knitted fabric filler for three-layer glass plastic constructions, *VNIISP Collected Papers 'Glass Fibre and Glass Plastics'*, 1975, **5**, 38–44.

6

Hybrid composite materials

B.V. Perov and I.P. Khoroshilova

6.1 INTRODUCTION

The most important criteria for the effective use of polymer composite materials are their life durability, such as fatigue and long-time strength, crack propagation resistance, plasticity reserve at long-term loading, resistance to impact loads, maintainability of articles therefrom, etc. The multifunctionality of the material is of prime importance: the combination of high elastic and strength properties with optimal thermophysical ones, radio-engineering indices, resistance to combustion, low smoke emission, non-toxicity and other specific characteristics. The diverse and contradictory requirements for modern composite materials has brought to the forefront the very complex problem of developing new types of composite materials, in which a combination of reinforcing layers from two or more types of fibres – the so-called hybrid composite materials – is used. Polymer hybrid composite materials (PHCM) have found the widest application to the present time.

Such an approach enables simultaneous realization of the advantages of heterogeneous composites in constructions and elimination of their undesirable features. For example, the introduction of layers of carbon filler into the composition of the material of rotor blades made from glass-reinforced plastic ensures a substantial increase of their torsional rigidity and fatigue strength and a multiple increase of their service life. And, vice versa, the use of carbon plastic articles often becomes impossible because of their low impact strength, without including coatings of organic or glass filler.

The use of PHCM in constructions also enables the solution of economic problems, which is of no small importance – an appreciable reduction of price of an article is possible without minimal or complete loss of quality. Five main types of hybrid composite materials are assumed to be distinguished [1]:

1. *Averaged.* Fibres of different types in such a PHCM are mixed in the whole mass of the material. No sections with macroconcentrations of any type of fibre are available.

2. *Intralaminar*. Fibres of different types in each layer of PHCM are regularly alternated. Referred to this type of PHCM are materials based on reinforcing hybrid fabrics. The layers can be 'interwashed off'.
3. *Interlaminar*. Each layer of PHCM consists of one type of fibre. As a whole, the PHCM is made up of layers alternating in a definite order from fibres of different types (e.g. a part from carbon plastic or organic fibre-reinforced plastic).
4. *Separate reinforcing elements*. These include straps and stiffening ribs.
5. *Superhybrids*. Layers of composites based on organic polymer and metal matrix and sheets of metal foil are laid in a definite pattern.

When manufacturing medium-loaded and heavily-loaded aircraft constructions, most often used are intralaminar and interlaminar PHCM based on carbon, glass and organic fillers. The basic problems to be solved are the following: enhancing the impact strength of carbon plastic constructions; increasing the rigidity, compression strength and moisture resistance of parts made from organic plastic; increasing the fatigue strength of articles made from glass-reinforced plastic; and reducing the price of the structural material as a whole.

Using the method of hybridization, it is possible to optimize, to a large extent, the anisotropic structure in the volume of PHCM in accordance with the stress–strain state of the part.

At the same time, an active search for uncommon solutions related to the science of materials is conducted with the aim of rational combination of fundamentally different hybrid compositions in a 'material construction' system. For example, such solutions include the combination of aluminium sheets with organic- and glass-reinforced plastic or titanium plates with carbon-filled plastic (the so-called laminated hybrid metal–polymer composites) in unified composite structures.

6.2 LEVELS OF HYBRIDIZATION AND STRUCTURE OF HYBRID COMPOSITE MATERIALS

6.2.1 Components of hybrid composite materials

The selection of the PHCM components relative to each specific case is determined by the purpose of hybridization, requirements imposed on the material or the construction being designed. Since the reinforcing elements bear the main load in composite materials strengthened by continuous fibres, the problem of selecting the types of compatible fillers and the level of their properties is of prime importance when designing and producing PHCM. From this point of view, the production of a reinforcement of high quality is one of the main trends in the field of perfection of PHCM properties.

High-strength high-modulus carbon fillers are the most promising and important components of the reinforcing systems for both structural and special-purpose PHCM. The range of carbon fibres differing in composition of the initial raw stock, texture, condition of the surface and level of physicomechanical characteristics is relatively

Table 6.1 Comparative properties of reinforcing fibres for PHCM

Property	E glass fibre	S glass fibre	Organic high-modulus fibre	Carbon fibre			
				High-strength	With enhanced elongation	Medium-modulus	High-modulus
Ultimate tensile strength (MPa)	2500	4550	3800	3400–3600	4500–7200	4700	2300–2800
Tensile modulus of elasticity (GPa)	75	87	130	240	240–300	300	400
Density (kg m^{-3})	2540	2490	1440	1780	1750	1770	1830
Elongation at break (%)	3.8	4.5	2.5	1.5	2.0–2.4	1.6	0.5–1.3
Coefficient of linear thermal expansion (10^{-6}K^{-1})	5	–	–2	–0.5	–0.1	–	–0.2
Moisture absorption at 20°C and relative humidity of 65%	0.1	–	3	0.1	0.1	0.1	0.1

great. Nevertheless, the development of new types and modifications is carried out rather actively. The new types of carbon fibres differ not only in high level of mean strength and modulus of elasticity (Table 6.1), but also in enhanced stability of properties. The expansion of the scale of using these fillers in PHCM has meant that special attention has had to be given to the condition of the surface of the carbon fibres. The high brittleness, abrasion and tendency to 'fluffing' of carbon fillers, and difficulties in their compatibility with fillers of other types, necessitated the preparation of the surface of the fibres (finishing, applying depletion layers) relative not only to the type of matrix (as in the case of monoplastics) but also to the type of the other fibres used in the same PHCM. The successful solution of this problem enabled production of PHCM with a high level of properties and improved technological effectiveness, and laid the foundations for the organization of mechanized and automated production of semifinished textile products intended for interlaminar PHCM (in the USSR, unidirectional tapes, woven tapes and cord fabrics of serge and linen weave) and combination fabrics for intralaminar PHCM. The distinguishing feature of carbon reinforcing fillers that specifies the actuality of hybridization as a method is their high cost. Carbon monoplastics remain one of the most expensive structural materials in the world, considering that, apart from the high price of the initial fibre, the cost of the composite includes the expenses for textile processing, preparation of the surface and purchase of expensive high-technology equipment.

Reinforcements based on organic polymeric fibres are of great interest for PHCM, as they most fully meet the requirements for the weight perfection of constructions

(Table 6.1), the good compatibility of organic (aramid) fibres with carbon ones (due to the close values of the coefficients of linear expansion) and glass fibres being of great importance. Their combination in a construction solves complex problems pertaining to the science of materials such as the reduction of the brittleness of carbon-filled plastic and the increase of flexibility and plasticity of glass-reinforced plastic.

Polyethylene fibres, the first from the family of organic fibres based on thermoplastic polymers formed as a gel, can be used successfully as coatings for brittle plastics. Polyethylene fibres have a density of $970 \, kg \, m^{-3}$, strength of about 2800–3000 MPa and high deformability; they are distinguished for their high fatigue strength, vibration and shock resistance, abrasion resistance, radiotransparency, and low values of dielectric constant [2, 3] and dielectric loss tangent (4×10^{-4}). All these advantages, however, because of the small compression strength of the monoplastic (about 60–70 MPa), can be most fully realized only in a hybrid material by combining polyethylene fibres with carbon or glass fibres.

To produce heat-resistant PHCM with an operating temperature up to 400°C, simultaneously possessing reduced combustibility and increased chemical resistance, with impact strength being close to that of organic plastics, it appeared necessary to develop polymeric fibres of hetero-organic type (polybenzimidazole, polybenzothiazole, polyoxathiazole, polyimide).

Despite the unique and diverse properties of carbon and organic fibres, which determined advances in the field of the modern science of materials, the traditional glass fillers, which have a history of many years of development, did not lose their significance. The accessibility of the initial raw material, the simple technological process of production, the low cost, the widest range and the fairly good mechanical characteristics afford them a strong position in practically all fields of modern technology. Glass filaments possessing high strength (up to 4.6 GPa), acceptable modulus of elasticity (up to 110 GPa) and good technological characteristics, and fabrics, tapes and other types of textile materials made therefrom, are one of the PHCM components used most often to reduce the cost of high-strength carbon or organoplastic constructions. All three main types of glass fibres – E (borosilicate fibre with low alkali content), S and R (with increased content of aluminium oxide) – and glass fillers with active finishing compositions applied on the surface of the fibres and fibre bundles are used in PHCM. Some typical properties of glass fibres of type E and S are given in Table 6.1. All the data in the table are averaged and may change depending on the supplier (country), the quality (grade) of the fillers and the method of testing.

The successful use of PHCM is determined by the chemical, mechanical and physical stability of the 'fibre–matrix' system. The role of the matrix in affording such stability is very important and diverse. The matrix provides the solidity of PHCM, its resistance to the action of environmental factors, the required level of operating temperatures (within the limits permissible by the thermal stability of the fibres incorporated into the PHCM composition) and chemical resistance. The mechanical properties of the matrix mainly determine the level of shear and compression strength of PHCM, at loadings of the material in directions that differ from the direction of orientation of the fibres, and contribute to the fatigue strength, torsional strength, creep, damping ability, and the work of fracture. Finally, the processability of PHCM depends only on the matrix composition, i.e. its ability to be processed into articles of a given shape and dimensions, the lifetime of semifinished products (prepregs, stacked blanks) and the temperature, time and pressure of moulding.

The matrices that are most widely used in modern commercial PHCM are compounds based on thermosetting epoxy compositions. It is these matrices that appear to be the prevailing ones in the field of aircraft materials science in practically all the countries that produce and employ PHCM, despite their high cost, toxicity in the course of processing and the inability of most of them to retain their properties after long-time service at temperatures above 150–180°C. But epoxy matrices ensure rather stringent requirements for the level of mechanical characteristics, enable production of materials with minimum porosity, and possess good physical and service properties (water, chemical and radiation resistance, non-toxicity in the hardened state, etc.) (Table 6.2). Owing to the great diversity of possible modifications, epoxy binders are of great interest for PHCM. There exist real ways for selecting epoxy matrices compatible with all types of fibres incorporated into the PHCM composition, capable of getting hardened under conditions necessary for reaching the optimal structure and solidity of the hybrid composite (low temperature and short hardening time) and possessing the technological possibilities that make the production substantially simple and cheap (long lifetime of the binders and prepregs and improved rheological properties of the matrices). But it should be borne in mind that not all epoxy matrices can be used in PHCM: the specificity of the method of hybridization imposes certain limitations on the range of possible selections. As an

Table 6.2 Typical properties of unfilled epoxy matrices

Properties	*Indices*
Density ($kg\,m^{-3}$)	1100–1300
Rockwell hardness	M100–M113
Volume shrinkage (%) at hardening	
with amines	7.5–8.5
with phenolic resins	3–3.5
Coefficient of linear thermal expansion (K^{-1})	$(5–8) \times 10^{-5}$
Ultimate tensile strength (MPa)	55–130
Ultimate compression strength (MPa)	105–210
Ultimate bending strength (MPa)	70–160
Tensile modulus of elasticity (MPa)	2800–5000
Shear modulus (MPa)	1500–2000
Poisson's ratio	0.20–0.39
Izod impact strength (specimen with a notch) ($J\,m^{-1}$)	0.1–10
Water absorption (%)	2.8–5.3

Table 6.3 Properties of unfilled epoxy binders ENFB, 5-211-BN and EDT-69

Properties	ENFB	5-211-BN	EDT-69
Ultimate tensile strength (MPa)	75	65	65–70
Ultimate compression strength (MPa)	165	150	155
Tensile modulus of elasticity (MPa)	3600	3300	3500–4000
Elongation at break (%)	2	3.5	2.5–2.8
Ultimate shear strength (MPa)	70	45	54
Glass transition temperature ($^\circ$C)	156	80–100	120
Crack resistance, K_{Ic} (kg mm$^{-3/2}$)	3.21	0.312	–
Maximum hardening temperature ($^\circ$C)	160	120(150)	130
Lifetime in prepreg at 20°C (months)	12	1	3

example we can consider the peculiar features of using in PHCM the three epoxy binders that are widely employed in production of composite parts and units: ENFB (based on epoxy novolac resin, monomeric active diluent and catalyst – a complex compound of boron trifluoride); 5-211-BN (based on a combination of standard and brominated bisphenol-type epoxy resins and hardener – aniline phenol–formaldehyde resin with increased content of methylol groups); and EDT-69 (based on a combination of aliphatic and aromatic epoxy resins and hardener – bis(N,N'-dimethylcarbamide)-diphenylmethane). The properties of these binders are given in Table 6.3.

The ENFB binder possesses good technological effectiveness, has an unusually long lifetime in prepreg at room temperature, does not need special conditions for transportation and storage, is hardened at 160°C within 4 h in accordance with the stepless operating conditions, and affords a high level of strength characteristics in monoplastics and PHCM. Owing to the features of the chemical structure of the matrix, its solidity and the high degree of phase contact in the material, PHCM based on the ENFB binder have an extremely high resistance to the action of environmental factors (water, moisture, fuel, organic solvents), are stable to thermal ageing at temperatures up to 150°C, and are tropic- and fungus-resistant. However, the presence of the monomeric diluent in the binder composition restricts the range of PHCM produced based on it by carbon glass-reinforced-, carbon boron- and boron-glass-reinforced plastics. The organic fillers introduced into PHCM, owing to the high sorption ability with respect to this component of the ENFB matrix, cause the disturbance of the optimal ratio of the components in the binder layers between the fibres. This results in the change of the course of the hardening reaction, the structure of the polymer network is disturbed, and the properties of the material, especially its heat and water stability, become worse.

In a number of cases one has to select for a hybrid structure the 5-211-BN binder, which is inferior to ENFB in its technological effectiveness and heat resistance, but meets the requirements for compatibility with all types of fibres. But even for this, practically universal matrix, it is necessary to correct the hardening conditions depending on the nature of the surface of the combined fibres and on what requirements are imposed on the structure. For example, for practically equal levels of mechanical characteristics, PHCM based on the 5-211-BN binder with service temperatures of 80, 120 or 150°C should be hardened at temperatures of 120, 150 or 180°C, respectively. PHCM containing carbon filler as a tape with the fibre surface activated by oxidation are produced according to lower-temperature conditions than PHCM containing finished carbon-fibre bundles.

The EDT-69 binder, developed specially as a universal one for PHCM (this explains its being multicomponent), is fit for practically any combination of heterogeneous fillers, though it needs correction of the hardening conditions depending on the condition of the surface of the reinforcing fibres. The binder has a fairly long lifetime in the prepreg and affords a high level of mechanical properties in the materials. The hardened matrix of EDT-69 has loose-packed crosslinked structure, which limits its service temperature down to 60–80°C.

The level of the mechanical and operating characteristics of PHCM based on commercial epoxy binders is not sufficient at the modern stage of technological development. To ensure full realization of the properties of the new reinforcing fibres, especially carbon and organic ones, new matrices are required that possess high cohesive strength in combination with high modulus of elasticity, crack resistance and universal compatibility with all types of reinforcing means. The ecological purity of the binders and the necessity for technological processes in their processing are of prime importance. The search in the field of new-generation matrices for PHCM has developed in the following directions: modifications of developed binders at a new level; synthesis of fundamentally new oligomeric systems (compositions that form interpenetrating networks in hardening, liquid-crystal polymers, etc.); production and employment of thermoplastic matrices. The new thermoplastic matrices (polysulphones, poly(phenylene sulphides), poly(ether ether ketone), poly(ether imides), etc.)

Table 6.4 Comparative properties of unfilled thermoplastic and epoxy binders

Properties	Thermoplastic				Epoxy
	Polysul-phone	Poly(phe-nylene oxide)	Poly(ether ketone)	Polybenzi-midazole	ENFB
Ultimate tensile strength (MPa)	72	77	100	77	75
Tensile modulus of elasticity (MPa)	2730	2590	3400	2380	3600
Elongation at break (%)	50–100	50–60	–	5	2
Glass transition temperature (°C)	174	190	143	215	156
Fracture toughness ($kJ\,m^{-2}$)	–	0.65	1	–	0.08

(Table 6.4) are of special interest for PHCM, providing simultaneous solution of several problems – improvement of the processability of matrices (practically unlimited lifetime in prepregs, fairly good compatibility with heterogeneous fillers, reduction of the moulding cycle); enhancement of the service properties of hybrid structures (much higher fracture viscosity than that of the thermosetting systems, good water and moisture resistance); improvement of the economic efficiency (maintainability of the constructions, good possibilities for utilization of the waste); reaching ecological purity of the materials. At the same time it should be pointed out that good compatibility with the fibrous fillers does not remove the problem of preparation of the surface of the fibres. The selection of finishing agents, barrier layers and other methods of affecting the surface (corona discharge, etc.) remain problems of high priority.

6.2.2 Role of interface in production of hybrid composite materials

The interface is indefinite, to a certain extent, but it is an extremely important part of any laminated material. Its role is rather significant in hybrid composites, where the formation of boundary structures occurs under the action of many factors that sometimes exclude each other.

The interface is the combination of regions where the matrix is bonded with the filler by chemical or mechanical means and with the polymer layer adjoining the fibre. This polymer may have a more ordered structure and its properties may differ from the properties of the unfilled binder owing to, for example, selective adsorption of the binder components by the fibre or the catalytic effect of active centres of different nature on the fibre surface on the chemical reactions in the process of binder hardening. If the interaction in the interface is very strong, the laminated material may become extremely brittle; if it is too weak, the composite will become useless as a structural material because of the difficulties in redistribution of load between the fibres. It is evident that a combination in PHCM of two or more types of reinforcing fibres interacting with the same polymeric matrix by various mechanisms, will inevitably result in the organization of the near-boundary layers in different structures. The behaviour of the PHCM produced, especially under the action of different media, becomes difficult to predict.

The mechanical properties of the reinforcing fillers, the hybridization effect obtained, the shear, compression and transverse tearing strength of PHCM, the fatigue characteristics of the material and its resistance to environmental action are realized in many respects, and sometimes even completely, by the nature and completion of the interfacial interaction of the components.

When estimating this problem from the standpoint of its development, it should be pointed out that specifications for PHCM quality were always free of theoretical representations about the mechanisms of formation of cohesive contact. Thus, the practical solution of the problem of improving moisture resistance and, consequently, stability of mechanical and radio-engineering characteristics of glass-filled materials caused the development of the technology of finishing hydrophilic glass fibres. The problem of increasing the shear and compression resistance of materials containing carbon filler brought about the development of methods for activating the surface of carbon fibres, intensification of the work devoted to the synthesis of polymeric matrices with increased cohesive strength and development of the molecular-kinetics approach to the problem of polymers [2].

The peculiar features in the structure and condition of the surface of different types of fibres determine the practical approaches to increasing the adhesive interfacial interaction. For example, for carbon fibres, despite the large area of the interface with the matrix, the hydrophobic property of the fibres, which prevents their wetting by binders with polar functional groups (the most widespread type), is the decisive factor. It is precisely because of this that oxidizing treatment – oxidation with air, nitric acid, anodic oxidation, etc. [3–6] – appeared to be very effective for them. There is a direct connection between the concentration of the oxygen-containing functional groups formed, as well as free radicals (paramagnetic centres), and cohesive strength in the 'fibre–matrix' interface.

The picture of the interfacial interaction in materials containing fibres is complicated also by the fact that the adhesion to them can be determined not only by the primary contribution of molecular or mechanical adhesion, but also by the diffusion of segments of macromolecules into fibre pores [7, 8].

The formation of phase boundaries in PHCM is inseparable from the processes of structurization in the binder. A condensed thermosetting matrix is heterophase and is a molecular medium in which particles of the colloidal dispersed phase are distributed, the phase occupying from 30 to 70% of the matrix volume [9–11]. They are formed at the stage of producing the prepreg and during hardening of the article blank. This process results in the local distribution of binder components. The described phenomenon, for example, is characteristic of binders in widespread production of PHCM, such as epoxy phenolic (type 5-211-B or 5-211-BN, containing bisphenol-type epoxy resin and aniline phenol–formaldehyde hardener). In alcohol–acetone solutions of such systems, phenol–formaldehyde resin dissolved in ethanol on a limited scale, is concentrated in the dispersed particles. The dispersive medium enriched in epoxy oligomeric component and in direct contact with the filler surface becomes incompletely hardened. Depending on the nature and surface energy of the filler, it absorbs oligomers and colloidal particles in a different way, changes the rates of microphase lamination and, in the long run, alters the density of the space network of the near-boundary layer. Some deviations of PHCM properties from the supposed levels, for example, the discrepancy between experimental data of static longevity in interlaminar shear and concepts of thermofluctuation theory, can be explained from the standpoint of the two-phase construction of thermosetting matrices. In such cases put in the forefront is not the 'fibre–matrix' interface, but the interface between the microphases, greatly exceeding it in area [6–13].

The thermodynamically caused microphase lamination in an oligomer–polymer system is a defect of existing PHCM because it runs under uncontrolled conditions. Therefore, along with the synthesis of new oligomers and binders, the optimization of the colloidal structure of compounds of known composition is a serious problem in the field of development of PHCM based on heterogeneous fillers and thermosetting matrices possessing enhanced adhesive ability.

The different sensitivity of the colloidal structures of a thermosetting binder to fillers that are different in chemical nature and morphology of the surface, and the selective wetting of these fillers with solutions or melts of oligomer compositions, make it difficult to obtain a defect-free interface in PHCM to a greater extent than in monocompositions. The differences in heat conduction and interfacial surface tension in the 'fibre–matrix' interface cause the manifestation of a thermocapillarity effect and may be responsible for the local de-resination and formation of voids in the articles. The development of a binder universally compatible with PHCM components is not the only way out of such a situation. Evidently, the development and application of one finishing agent or depletion layer on fibres of different nature is capable of converting almost any binder into a universal one.

The interface acquires primary importance in the case of PHCM based on thermoplastic matrices, which is caused by the high viscosity of the melts of the thermoplastic binders and difficult impregnation of the fibrous fillers with them. Each successful solution of the problem of wetting this or that combination of fibers with a thermoplastic means production of a new PHCM having a chance for successful introduction. The list of ways of realization of thermoplastic PHCM in technology is far from being complete: activation treatment followed by filler finishing; introduction of surface-active and plasticizing additives to the thermoplastic that reduce the surface tension and viscosity of the melt; perfection of processing methods with the emphasis placed on the procedure of applying dynamic action (ultrasound, vibration impregnation, etc.).

6.3 CALCULATION AND DESIGN OF HYBRID COMPOSITE MATERIALS

6.3.1 General information

As already mentioned, the combination of different reinforcing fibres in one material is a promising trend in the production of construction with high operation reliability and reduced mass. Several technological versions of the combination of continuous fillers (Fig. 6.1) are possible.

Three-component materials can be manufactured by the principle of homogeneous mixtures, fibres of different type being evenly distributed in the primary filament. According to the second version, layers of reinforcing fillers of different nature are alternated in this or that order. The third version makes provision for production of a multi component combination filler manufacture of fabric, mat or sheets of veneer by using different filaments and fibre bundles. Each of the said methods has its own advantages and disadvantages. The principle of making homogeneous mixtures provides a more uniform distribution of stresses in the fibres and matrix as the plastic is loaded, but the fabrication of the primary filament or fibre bundle from different fibres can be extremely difficult because of substantial differences in their processability.

The principle of alternation of layers can be effected most easily from the

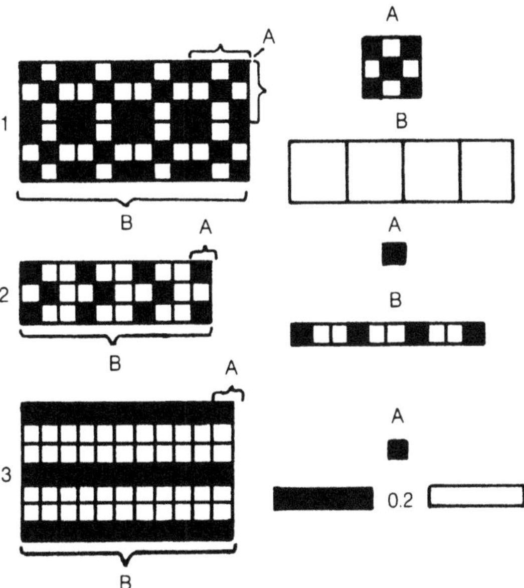

Fig. 6.1 Examples of structures of multifibrous PHCM with combination at different levels, i.e. at the level of primary reinforcing elements (1), monolayer (2) and stack (3): (□) fibres of one type; (■) fibres of another type; (A) primary reinforcing element; (B) monolayer.

technological point of view and it enables varying PHCM composition to be achieved within a wide range and ensures the high degree of realization of mechanical properties.

The fabrication of PHCM based on woven or unwoven combination filler is also a high-technology method, but in this case the fibres have a definite mutual orientation specified by the textile structure of the fabric. To a certain extent, this limits the freedom to correct PHCM properties, and the degree of realization of the mechanical properties of the fibres is reduced (by up to 15–20%) because of local distortions of the fillers. The process of production of combination fabrics is rather power-intensive and makes them expensive. PHCM monolayers can be classified by composition (Fig. 6.2(a)) and structure of reinforcing (Fig 6.2(b)).

Irrespective of the selected version of PHCM, differences in the thermophysical properties of the fillers cause internal thermal stresses in the processes of both moulding the material and its operation in the article. For example, when heated, glass fibres expand in the longitudinal and transverse directions, whereas carbon fibres have a negative coefficient of linear thermal expansion along the fibre axis and positive in the transverse direction. Therefore, on cooling, internal stresses inevitably arise in glass carbon-filled plastic. The tensile stresses may arise in the matrix under the action of the radial compression of the fibres.

The design of a PHCM is one of the most important aspects in production thereof. The morphological analysis of the approaches used can be carried out by means of a diagram (Fig. 6.3) [14].

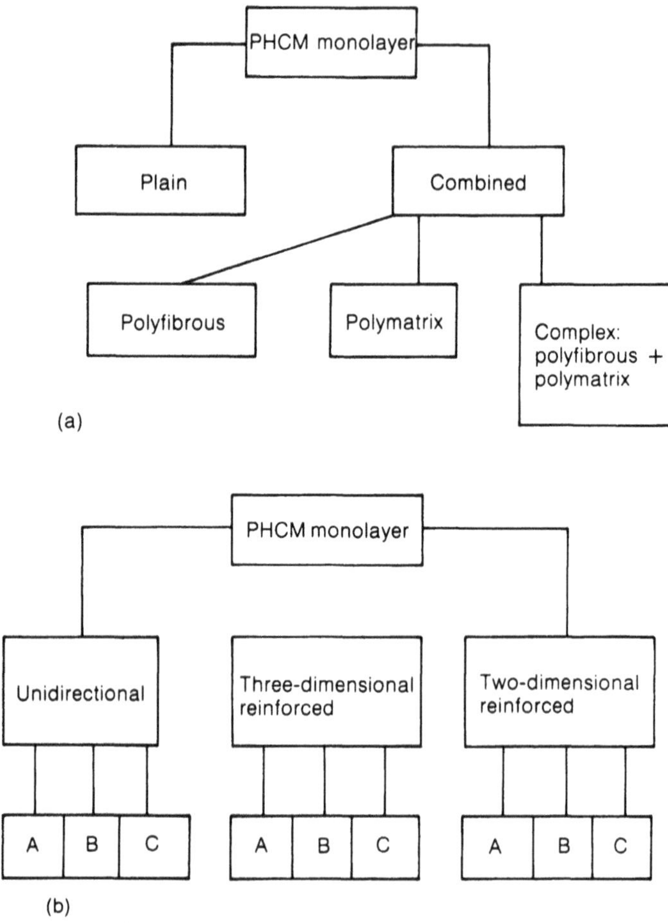

Fig. 6.2 Classification of PHCM layers by (a) composition and (b) reinforcing structure: combination (A) at the level of filament and fibre bundle; (B) at the level of tape and fabric; (C) at the levels of (A) and (B) simultaneously.

The left-hand front face of the cube characterizes the possible requirements imposed on PHCM: X, entrance. Versions: $X_?$, no requirements; $X_\approx$, requirements formulated approximately, $C_i^* \leqslant C_i \leqslant C_i^{**}$ $(i = 1, \ldots, k)$; $X_!$, requirements are accurate, $C_i = C_i^*$ $(i = 1, \ldots, m)$. L (upper face), design operator. Possible situations: L, no procedures; L_n, n procedures exist without evident preference, $n \in N$, $n < \infty$, $L_!$, there is a single procedure (e.g. instructions); *, **, numerical values of properties. The right-hand front face (Y) corresponds to classification for exit. $Y_!$, recommended the single existing PHCM; $\{Y_3\}$, several PHCM of the existing ones; Y_H, new PHCM.

The comprehension and perfection of the system of confirmations and system of actions connected with the method of combination (hybridization) at the modern stage necessitates the modelling of PHCM quality. In the simplest version the quality

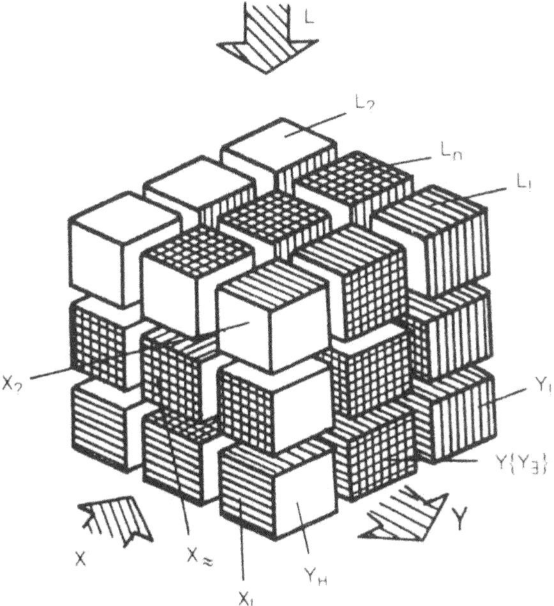

Fig. 6.3 PHCM design diagram: X, entrance; $X_?$, no requirements; $X_\approx$, requirements formulated approximately; $X_!$, requirements are accurate; L, design operator; $L_?$, no procedures; L_n, n procedures exist without evident preference; $L_!$, single procedure; Y, exit; $Y_!$, the recommended single one of the existing PHCM; $\{Y_3\}$, several of the existing PHCM; Y_H, new PHCM.

can be considered as the Boolean product of the functions of accuracy and rationality. The progressing concept requires the production of materials according to criterion $k = 1$. The combination (hybridization) expands the possibilities thereof. The existing materials can be subdivided into two interesting classes: accurate (A-I, class M_1) and rational (R-I, class M_2). The problem of production of PHCM can be defined as the problem of representing the rational materials (class M_2) into the accurate ones (class M_1) according to the criterion of quality: $M_2 \xrightarrow{k=1} M_1$. The production of PHCM according to this principle enables one to obtain an appreciable gain in time as compared with the materials of class M_2 (the question is about the total time spent on production of optimal versions of the material and article).

Encountered most often for statically defined systems in design are cases that are encoded by a generalized formula:

$$L_n(X_\approx V X_!,\ Y_! V \{Y_3\})$$

and for statically indefinite systems:

$$L_?(X_? V X_\approx,\ Y_! V \{Y_3\})$$

In mathematical modelling of PHCM properties, which is necessary for the modern methods of optimization, many mechanical and physical characteristics in the first approximation can be assessed using the linear average of the characteristics of the used components from the formula:

$$C^* = \sum_{i=1}^{N} C_i^* V_i, \qquad \sum_{i=1}^{N} V_i = 1$$

where C^* is property of PHCM (e.g. Young's modulus along fibres, Poisson's ratio, etc.) in the direction marked with an asterisk; C_i^* is the appropriate property of the *i*th phase (both reinforcing and bonding) in the same direction*; N is the number of reinforcing and bonding phases in the PHCM. With such modelling it is possible to use well developed methods of linear programming. Unfortunately, such a description is not always correct.

For many properties the contribution of the phases is not linear. In this case, one speaks about the presence of a hybrid effect of the described property. From this standpoint, it is possible to single out weak and strong positive and negative hybrid effects. The term 'synergism' is sometimes used for positive hybrid effects. The increase of crack resistance (K_{Ic}) of carbon-reinforced organic plastic in comparison with the limiting values of this characteristic of the initial carbon and organic plastic can serve as an example of synergism (Fig. 6.4). The same PHCM possesses a negative hybrid effect in the tensile strength index – the level of tensile characteristics peculiar to both organic and carbon plastic is not reached in the hybrid material. The strong positive and negative hybrid effects have been disclosed in many multiple-fibre

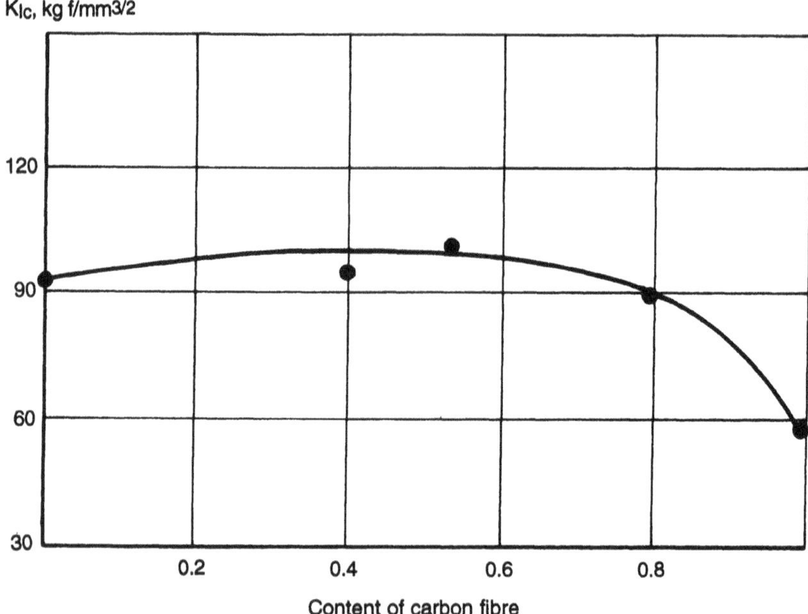

Fig. 6.4 Change of crack resistance (K_{Ic}) of carbon-reinforced organic plastic depending on the content of carbon fibre.

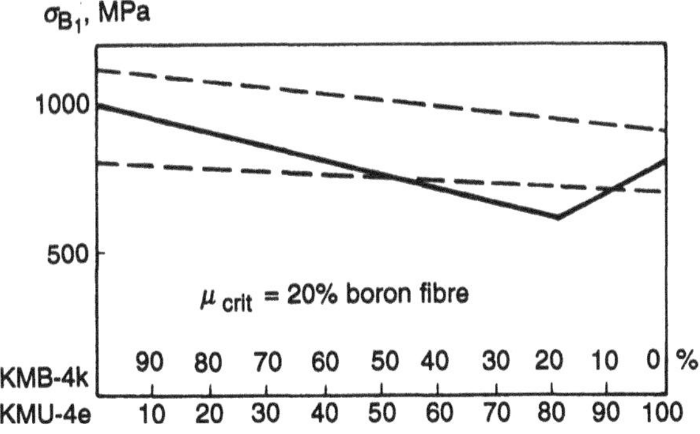

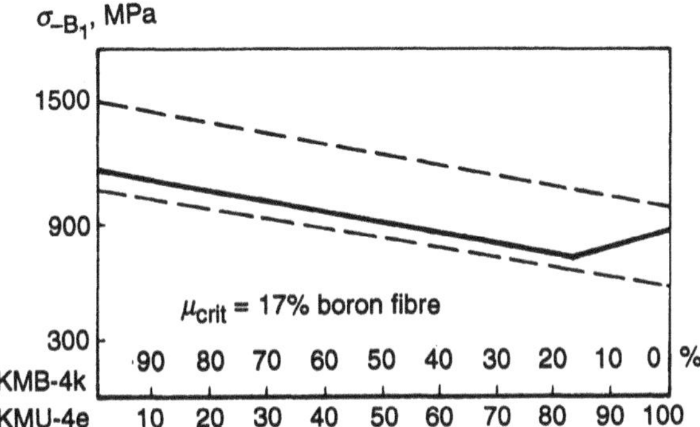

Fig. 6.5 Change of tensile and compression strength of carbon-reinforced boron plastic depending on the ratio of carbon and boron fillers.

composites. As a rule, they reflect the static and dynamic strength of materials. Most of the positive hybrid effects have not yet found a theoretical explanation enabling reliable quantitative estimates to be made.

Figure 6.5 gives design and experimental data on the tensile and compression strength along the fibres of a unidirectional carbon–boron plastic obtained by layer alternation of carbon plastic KMU-4e and boron plastic KMB-4k depending on their volume content. The behaviour of the material in tension obeys the 'mixing rule'. The design in line with the classical theory of laminated plates (solid line with a kink) predicts the presence of a strong negative hybrid effect, which does not conform to the experimental results. The straight broken lines correspond to linear averaging by the maximum and minimum values of strength of the combined monolayers of monoplastics. Amenable to linear averaging in this case is not only the mathematical expectation of the tensile strength of PHCM, but also the dispersion of this value. The experimental points display the presence of a positive hybrid effect for some versions of the ratio of components. The conversion of this characteristic to the specific one results in still more pronounced synergism. The use of the traditional scheme of the classical theory of laminated plates does not give any result for predicting this value. The great extent of the hybrid effects can be disclosed in the analysis of the strain diagram of PHCM. The necessity to account for hybrid effects, however, does not always complicate the design of PHCM. Unlike monoplastics, PHCM often possess non-linearity of properties. The design of PHCM containing monolayers of composites possessing different moduli (Young's tensile modulus of unidirectional carbon plastic may exceed almost by 30% the appropriate index of compression in the same direction) is possible. The combination of such materials with components that do not produce the effect of different moduli enables it to be neglected in design.

6.3.2 Interlaminar hybrid composite materials

To design the strength and elastic properties of laminated composites, the known dependences of the theory of elasticity are used, which enables the elastic and strength properties of a composite to be designed with any complex structure of reinforcing by the characteristics of monolayers. Similar dependences for designing not only the elastic and strength charcteristics, but also some special ones (e.g. thermophysical characteristics), have been developed and found wide application, by the properties of the components included in the monolayer. The problems of designing interlaminar PHCM of different composition have been investigated rather widely [4–22 and others]. Figure 6.6 shows a diagram of the basic stages of designing constructions from PHCM [19].

The adopted procedure of designing a construction and PHCM optimal for it envisages the work to be performed in three stages:

1. Formulation of the complex of specifications for PHCM.
2. Selection of the PHCM appearance.
3. Improvement and optimization of the PHCM structure.

The main operation of the first stage is ranking of the requirements for the material by their degree of significance, as a result of which a sequence is determined, where the parameters of the specifications will be considered when developing the PHCM design. The ranking is made with the use of the method of dual comparisons [23] with separation of the operations of coarse and fine ranking. The first of them consists of breaking down the requirements into equivalence classes by isolating the outlines of the graph of the 'predominance–indifference' structure and subsequent linear ordering of the classes according to their significance [24]. The second consists of ranking in each equivalence class the parameters by specific weight through comparing the components of the proper eigenvector.

The second stage of the procedure includes several operations: functional selection of components; selection of the grade of reinforcement and binder; selection of the technological process; and selection of the tentative structure of PHCM.

In the 'functional selection of components' operation one must consider the algorithm of the non-obvious enumeration of the structural parameters [25]. This operation is characterized by the active use of databanks on the qualitative and

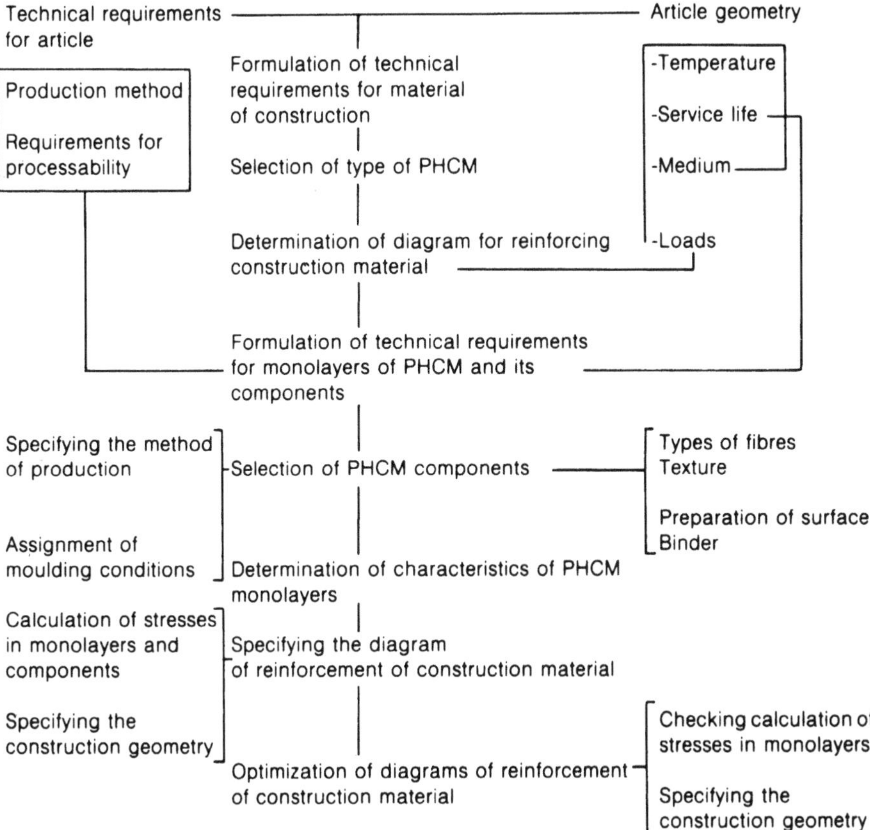

Fig. 6.6 Diagram of basic stages for designing constructions from PHCM.

quantitative comparison of the combination components and the databank of their functional logical connectives. The results of the operation consist of selecting components enabling the specifications for the combination PHCM to be satisfied in the adopted problem setting.

The selection of the grade of reinforcement and binder is united into one structural unit, since the reinforcement is selected according to its properties obtained by conversion of the experimentally determined characteristics of unidirectional plastics containing fibres of this type in the selected binder. The grade of the binder is determined by the requirements for the unified binder for definite types of fillers. The binder for PHCM should provide the joint work of all the reinforcing elements and possess good processability.

The selection of the technological process should provide the required volume contents of the binder and reinforcing fillers, the thickness of the monolayer, the low level of residual stresses and the solidity of the material.

The next operation consists of selecting the tentative structure of PHCM; the possibility of fulfilling the specifications is analysed due to creation of an interlaminar

PHCM with layer alternation of the disoriented monolayers of the previously developed materials (glass, carbon, organic plastics, hybrid monolayers, etc.). The elastic-strength characteristics of the material are predicted on the basis of the theory of laminated materials [26] with the polynomial criterion of layer strength [27]. The methods of non-linear programming are used for the optimization of the number of monolayer misorientation angles [28].

The third stage of the procedure is connected with the improvement and optimization of the design of an interlaminar PHCM that most accurately satisfies the specifications. The numerical methods of non-linear programming and mathematical modelling of the physicomechanical properties of PHCM comprise the basis of this stage.

On the basis of accumulated experience in designing structures from interlaminar PHCM, some general features typical of them have been disclosed. It was indicated that the tensile, compressive and bending moduli of elasticity of three-component materials (carbon glass-reinforced and carbon organic plastics) are close to each other and they increase linearly with the increase of the high-modulus filler content in PHCM: the more it rises, the higher is the modulus of elasticity of the latter. The Poisson's ratio decreases linearly. The tensile strength of a three-component PHCM with multi-modulus fillers does not obey the mixing rule, and changes linearly with increase of the content of fibres having reduced modulus of elasticity to some critical value. The elongation abruptly changes at this value, followed by the rise of the value peculiar to the low-modulus monoplastic. The tensile strength for PHCM has minimum value at the critical content of the high-modulus carbon fibres: for carbon plastics 26–27%, for carbon organic plastics 22–24% [29]. The nature of the dependence of PHCM compression on the content of high-modulus fibres differs from that in tension. As its content grows, the compression strength increases, since critical stresses in PHCM depend on modulus of elasticity of the matrix and fibre in the process of compression at constant degree of reinforcement [30]. On the basis of the developed design procedures, a great number of types of PHCM have been produced, studied and employed in practice; as a rule, these are three-component systems based on different combinations of carbon, glass and fibre. When using these procedures, however, it is necessary to take into account some of their features disclosed in practice. For example, the nature of the distribution of the high-modulus fibres across the PHCM section appreciably influences its elastic properties, especially when bending stresses arise in it. With the high-modulus fibres positioned closer to the external layers, the modulus of elasticity of PHCM can greatly exceed the design value. In this case an analogy can be drawn with the three-layer construction of low-modulus filler and skins from high-modulus monoplastic. The successive alternation of the layers of organic and carbon plastics also enables one to obtain higher values of the compression moduli of elasticity and strength of PHCM than the design values. It is achieved owing to the effect of the additional reinforcing of the matrix positioned between the layers of carbon filler by the organic fibres.

In carbon-filled glass the bending modulus of elasticity changes according to the additivity rule in the entire interval of content of carbon and glass fibres, but the appearance of the layer is of great significance here. With the 'carbon/glass' ratio equal to 3/2, the bending modulus of elasticity varies from 85 to 275 GPa depending on whether the carbon layer is positioned above or below. Thus, when designing interlaminar PHCM, not only the quantitative ratio of the heterogeneous fillers should be considered, but also the mutual position of the layers therefrom. Characteristically, the succession of layers in PHCM is not significant for the realization of fatigue characteristics.

The fatigue strength of PHCM increases linearly as high-modulus fibres are introduced (as a result of reducing the stresses in the matrix); the design of the structure is conducted according to the additivity rules. The higher the modulus of elasticity of the reinforcing fibres introduced into PHCM, the greater is the increment of its fatigue strength. Thus, the introduction of carbon fibres into plastics enables the fatigue strength of the material to be increased 1.5–2 times. The orientation of the fibres produces a substantial influence on the fatigue strength of PHCM. Unidirectional materials have fatigue properties. With the increase of the number of layers and orientation of the filler (especially the high-modulus one) of $\pm 45°$ and $90°$, the fatigue characteristics of PHCM go down. Unidirectional PHCM are usually not used in real constructions, but instead one uses materials with complex structure of reinforcing. In this case, when designing interlaminar PHCM, it is expedient to use the formulae of the theory of elasticity for complex reinforced systems. In absolute value, the fatigue strength of three-component PHCM amounts to 50–60% of the strength in static loading. Interlaminar PHCM possess better fatigue properties than intralaminar PHCM, the damping ability being reduced about two-fold. When introducing into glass-reinforced plastic carbon fibres misoriented relative to the glass ones up to $\pm 30°$ to $\pm 45°$, the damping ability of the material increases 1.5–2 times and reaches, for example, 4% with 16 vol% of carbon fibres with an angle of reinforcement equal to $\pm 45°$. By optimization of the composition, it is possible to design a PHCM with a high vibration resistance for different types of fatigue load. For example, the vibration resistance of carbon-reinforced glass plastic increases 1.5–2 times as the fraction of glass fibres misoriented relative to the carbon ones at angles of $\pm 30°$ to $\pm 45°$ increases by up to 25–30%. A small misorientation (up to $\pm 15°$) causes a reduction of the vibration resistance of PHCM irrespective of the ratio of the fibres therein [20].

The main incentive in designing parts of constructions from PHCM consists most often in combining strength and rigidity with a fairly good impact strength, determining, in the long run, the operating reliability of the article. The increase of resistance of a laminated composite to impact fracture is attained by the method of including into the high-modulus material some other high-strength fibres possessing greater work of fracture. In accordance with the additivity rule, the impact strength of PHCM based on carbon fibres increases as stronger and less rigid fibres are introduced into its composition (i) in direct proportion to their volume fraction, (ii) in proportion to the square of the ratio of the realized strengths of the low- and high-modulus fillers and (iii) in inverse proportion to the ratio of their moduli of elasticity. The method of laminating the stack is of great importance. Comparison of two types of laminating (layer alternation and lamination by bundles, when the layers of low modulus are positioned symmetrically relative to the middle plane of PHCM) indicated that higher impact strength is achieved in the layer lamination [31].

The correction of PHCM properties and optimization of their required parameters in the plane of reinforcing are achieved by orientation of the layers of different-modulus fibres in the stack height. With the increase of the angle of deviation of carbon in PHCM, properties such as impact viscosity, fatigue strength, modulus of elasticity, tensile, compressive and bending strength change abruptly in the interval of the angle of misorientation of $\pm 7°$ to $\pm 35°$, then the intensity of reduction of these characteristics is slowed down and after misorientation by $45°$ they remain practically constant. An exception concerns the tensile strength in carbon-reinforced glass plastic, which depends from the very beginning only on the amount and strength of the glass fibres. Characteristics similar to those indicated above in the transverse direction gradually increase at angles of reinforcing up to $\pm 30°$, after which the intensity of their growth rises and with misorientation by an angle of $90°$ the PHCM properties are determined by the properties and content of the fibres. The values of the shear modulus in the reinforcing plane, modulus of elasticity and strength in the diagonal direction, and logarithmic mechanical damping index pass through a maximum with misorientation of the layers at an angle of $\pm 45°$, increasing 1.5–4 times as compared with the initial values [16].

The purpose of producing PHCM may consist not only in perfection or optimization of the complex of mechanical characteristics, but also in enhancing its physical, physicochemical, electrical, thermal and frictional properties.

There are not yet examples of correcting the special properties of composites by polyreinforcing, but the advantages and effectiveness of PHCM obtained by such a method are already evident. By changing the structure and ratio of the carbon and glass fillers in the interlaminar PHCM, it is possible to increase its density 1.5 times, coefficient of linear thermal expansion 10 times, volume electrical resistance by two orders, heat capacity 1.5 times and heat conduction several times [20]. The possibility of varying the coefficient of linear thermal expansion within a wide range is essential for designing the elastic and strength properties of PHCM. This possibility is connected with the difference of the coefficient for the heterogeneous fillers and intrinsic anisotropy (e.g. for carbon tape this coefficient in the temperature range from 20 to 200°C is negative weftwise and positive warpwise). When producing interlaminar PHCM, the difference in the coefficient of linear thermal expansion causes the onset of additional stresses owing to the combination of layers with different thermo-elasticity. The calculation of the tangential stresses arising between the layers indicates that in the temperature range from 20 to 100°C the internal shear stresses between the layers in carbon-reinforced glass plastic, for example, can reach 2 MPa. To reduce these stresses, the rational lamination of adjacent layers that are deformed in a different way under the action of temperature is necessary. The optimal version is selected by means of previously constructed dependences of the mean values of the coefficients of linear thermal expansion of the combined materials on the angle of reinforcement. Thus, to prevent the onset of additional thermal stresses in carbon-reinforced plastic, it becomes sufficient to stack the carbon fibres at an angle of 35–40° with respect to the glass fibres (the coefficients of linear thermal expansion of the glass-reinforced laminate with an angle of reinforcement equal to 0° and carbon-reinforced plastic with an angle of reinforcement of 35–40° are approximately equal).

The example of solving the problem of designing PHCM intended for a load-carrying space-telescope truss combining high physicomechanical properties with minimum temperature is given elsewhere [32].

For calculation of the basic thermophysical characteristics of PHCM (the coefficients of linear thermal expansion, specific heat capacities, thermal conductivity) and designing articles therefrom on the basis of these calculations, the self-coordination method is offered with the use of a three-phase model of a composite reinforced by anisotropic fibres of different types. Finite analytical formulae expressing the dependence of the elastic thermophysical properties of PHCM on the elastic and thermophysical properties of the components and their volume content have been worked out and tested [33].

A very important problem arose when polymer composites started to be used in aviation, i.e. the necessity of protecting the parts therefrom against the action of a lightning discharge. One effective way for solving this problem is the design and production of PHCM based on the combination of reinforcing fillers of structural purpose with metallized carbon- or glass-reinforced fibres, and additional reinforcement of the monoplastic (e.g. epoxy carbon plastic) with wire from high-carbon steel or aluminium foil. The use of the metal or metallized addditional reinforcing fillers in PHCM certainly reduces the weight coefficient of the construction, but apart from the proper increasing of heat and electrical conduction, and lightning resistance, it most often also enables improving the mechanical properties (increasing the bending strength by up to 25%, impact strength two-fold) and erosion resistance of the material.

The electrical and heat conduction of PHCM is also increased by introducing carbon fibres into the glass-reinforced and organic plastics. The carbon plastics are characterized by high thermal conductivity along and across the fibres, up to 13 and within 0.54–$0.80 \, W \, m^{-1} K^{-1}$, respectively. Their coefficients of thermal conductivity in the transverse direction and heat capacity are higher than those of glass-reinforced laminates. The electrical properties of the carbon plastics are much higher than those of glass-fibre-reinforced laminates [34]. The electrical conductors of PHCM and associated effects depend on the distribution of the carbon fibres in the stack. Thus, thin layers of them positioned on the article surface are used for removing static electricity, and they impart radio-absorbing properties. Layers and bundles of carbon fibres inside the stack can be used as heating elements. Performing the function, for example, of a source of radiation heating, these additional reinforcing means facilitate and optimize the hardening process of PHCM, accelerating it or reducing the thermal stresses. The radiotransparency of articles is afforded by a PHCM of another structure: combination of glass and organic fabric fillers. By introduction of polycrystalline fibres (for example, from aluminium oxide) to the organic plastic, these fibres provide an additional increase of compression strength to PHCM.

Despite the fact that the carbon plastics possess a complex of unique properties and in many cases they do not need modifying by other reinforcing means, they have a serious shortcoming, i.e. corrosion activity with respect to metal surfaces in contact with carbon plastics. In this case the surface should be thoroughly protected. Corrosion can be prevented by introducing glass fibres into external composites. A similar technique of reverse nature (introduction of carbon fibres into the external layers of the glass-reinforced plastic) is the case when the construction should have increased chemical resistance. Articles and PHCM of such structure can be operated in chemically aggressive media (acids, alkalis, salt solutions). Layers of organic plastic can be introduced into the PHCM structure with the same purpose.

6.3.3 Intralaminar hybrid composite materials

The scope of using PHCM in constructions, in particular of large size and of complex configuration, depends on not only the level of the mechanical and service characteristics of the material, but also its processability. From this standpoint, one of the most promising trends in production of PHCM is the use of combination fabrics of different types of weave as fillers, despite the fact that their production process is rather labour-consuming and the fabrics are expensive. Up till now, the structure of reinforcing fabric has been selected most often on the basis of simplified methods of calculation, engineering experience and successive improvement of design. Despite the successes gained by using the traditional methods of designing monomaterials considering the heterogeneity of the reinforcement, the boundary layer and other factors [16, 17, 35, 36], design methods based on the principle of material and construction unity have found a wide application lately [2–6, 37, 38]. It should be noted that the structure of a composite based on reinforcing fabrics, as compared with other types of reinforcing materials, is one of the most complex and unamenable to mathematical prediction.

The different types of fibre-reinforced composite materials can be effectively used only on the basis of scientifically substantiated methods of designing the structure. The optimal structure of a laminated fabric should be designed for each specific use. To expand the range of mechanical properties of composites based on the fabrics, various types of hybrid structures are used at present – fabrics woven from hybrid filaments containing fabrics woven from plain filaments of several types. The process of determining experimentally the optimal structure of different types of materials based on fabrics is labour-consuming and very expensive. Practically, the optimal structure of a laminated fabric can be determined only theoretically on the basis of general statements of the structural mechanics of reinforced plastics, the statements being united by the integral design procedure.

The design procedure of intralaminar PHCM was developed in the USSR in 1986, and this procedure implemented both approaches. Problems dealing with the search, simulation modelling of hybrid effects and optimal discrete design of the PHCM structure were solved with due consideration for the specificity of aircraft materials and constructions (new criteria of PHCM design quality, systematic approach to the

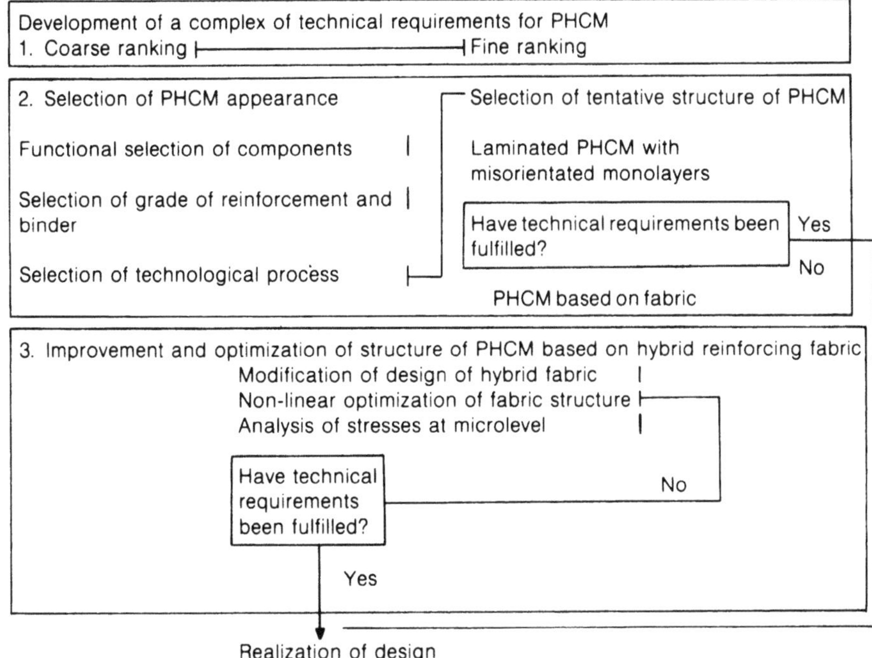

Fig. 6.7 Block diagram of procedure for optimal design of hybrid reinforcing fabrics for PHCM with a complex of preset properties.

formulation of specifications, etc.). For advancing the work, a procedure of optimal design of hybrid fabrics was elaborated enabling the optimal structure of reinforcing to be selected, proceeding from the requirements for anisotropy of the basic elastic and strength characteristics of the plastic. A simplified block diagram of the procedure is given in Fig. 6.7. As in the case of interlaminar PHCM, the design procedure makes provision for the work in three stages:

1. The formulation of the complex of specifications for PHCM.
2. The selection of the PHCM appearance.
3. The improvement and optimization of the structure of PHCM based on the hybrid reinforcing fabric.

The execution of work aimed at designing intralaminar PHCM coincides with the design of the interlaminar PHCM at the first stage and some part of the second stage.

If it becomes impossible to fulfil the specifications by production of a PHCM with layer alternation of suitable monolayers, the transition is made to selection of the appearance of the hybrid reinforcing fabric providing production of the composite that most fully meets the level of requirements. In this case the basic parameters of the fabric structure are determined – position of the filaments (fibres, bundles) of different types warpwise and weftwise, angle between the filaments in the case of oblique weaving, approximate volume content of each of the components, and number of filaments per unit length in the warp and weft. Then the tentative optimization of the design parameters is effected by the methods of linear programming [39] with the use of simplified mathematical models of the 'mixing rule' type for determining PHCM characteristics.

The third stage of the procedure deals with the improvement and optimization of the design of hybrid reinforcing fabric and obtaining the design of the composite that most fully meets the specifications. This stage comprises the following operations:

1. Improvement of the design of the hybrid fabric.
2. Non-linear optimization of the structure of the combination material based on the hybrid reinforcing fabric.
3. Mathematical modelling of the stress–strain state of PHCM with the use of universal numerical methods of finite [40] and boundary [41] elements, the method of elementary cells [42], etc.

The design of a hybrid fabric is improved by selecting the fibre weave pattern on the basis of estimation of the composite characteristics by simplified non-linear models of the structural theory of reinforced plastics [26, 43, 44] with the approximation of distorted fibres by broken line.

The optimization of the structure of fabrics is made by using the methods of non-linear programming with varying volume contents of the composites, angles of distortion of the warp and weft fibres, and type of fabric weave.

Universal numerical methods of finite and boundary elements and elementary cells are used for the analysis of the strain and stress fields in the matrix and fibres of a fabric. Analysis of the stresses at the microlevel enables zones of stress to be determined, regions in the material that are dangerous from the standpoint of strength to be disclosed and the influence of the changes in fabric structure on these regions to be seen. At this stage the improvement of the design of a hybrid fabric can be reached, first of all, by changing the type of fabric weave, varying the geometrical parameters of the fibres, and rationally changing some properties of the reinforcing elements, which should lead to the reduction of the dangerous concentrations of stress.

If necessary, all three operations of the third stage of the procedure are repeated in succession, until an acceptable design of the hybrid reinforcing fabric is achieved.

The last point of the design of a hybrid fabric is consideration of the technology of textile processing, which imposes a number of limitations caused by the technological possibilities of the weaving equipment available.

The final checking of the design is effected after producing a hybrid reinforcing fabric and plastic based thereon, and conducting a series of technological and experimental investigations required for establishing the conformity of the properties obtained to be specifications for the material.

The suggested design procedure of hybrid fabrics intended for PHCM possessing a complex of specified properties allows:

1. Enhancing the quality of PHCM designs, owing to expansion of the number of design alternatives under consideration, more detailed and all-round analysis of

each design, and solution of fundamentally new problems dealing with the systematic approach to the integrity of the material and constructions.

2. Reducing the time for developing materials that are optimal for specific articles, owing to the optimization and automation of the information flow between designers, to the computer-aided design and to diminishing the number of errors in designing.

3. Reducing the expenses for development, owing to the reduction of the number of specialists engaged in routine work, decreasing the scope of expensive and labour-consuming technological and experimental investigations, considerable reduction of the costs connected with the recommendation of unsatisfied or erroneous designs, etc.

4. Enhancing the level of scientific organization of labour and specific weight of creative work in the design process, owing to the use of the same procedure, delegating a number of uncreative operations to the computer, etc.

The realization of such schemes allows a computer-aided design (CAD) system to be developed for designing composite materials. It is closely connected with the production of new highly efficient equipment providing the manufacture of parts under automatic operating conditions by the methods of programmed winding and laminating. As an example, within the limits of the CAD system based on the mathematical prediction of the elastic and strength properties of laminated fabrics, design calculations were conducted of structures of combination fabrics for a composite that meets the requirements given in Table 6.5. The results of the calculations of

Table 6.5 Technical requirements for PHCM based on hybrid fabric

Characteristics of material	Technical requirements	Characteristics of material	Technical requirements
Shear modulus, G_{12} (MPa)	5000–7000	Transverse tensile modulus of elasticity, E_2 (GPa)	17.00
G_{12}/ρ	350	Longitudinal ultimate compression strength, σ_{-B1} (MPa)	500
Density, ρ (kg m^{-3})	1300–1900	Transverse ultimate compression strength, σ_{-B2} (MPa)	200
Longitudinal tensile modulus of elasticity, E_1 (GPa)	40.00–60.00	Ultimate interlaminar shear strength, τ_{12} (MPa)	70
Fatigue strength, σ_{-1}, based on 10^7 cycles (MPa)	250	Transverse ultimate tensile strength, σ_{B2} (MPa)	150
Longitudinal ultimate tensile strength, σ_{B1} (MPa)	1000		

Table 6.6 Design structure of hybrid fabric

Name of fabric parameters	Version 1	Version 2
Type and linear density of filaments of		
warp	Carbon fibre bundle UKN from 2500 filaments (200 tex)	Carbon fibre bundle from 5000 filaments (410 tex)
weft	Complex glass filament VMPS-6-7.2 × 1.0 × 2.0 × 76	
Density (dm^{-1}) of		
warp	90	43
weft	1600	1600
Width (m)	0.300–0.900	0.300–0.900
Thickness of fabric at $P = 1\,MPa$ (mm)	0.316	0.316
Weaving	Four-filament satin, serge, hopsack	

structures of hybrid reinforcing fabrics are presented in Table 6.6. Samples of combination fabrics manufactured in compliance with the design structures were tried out in PHCM with the use of two types of binders having different elastic and strength characteristics: epoxy binder of type I with elongation of 34%; and epoxy binder of type II with higher strength characteristics, but more stringent elongation of 1.2–1.5%. The PHCM samples were manufactured by direct moulding according to the operating conditions optimal for the above-mentioned binders. The test results of the PHCM samples are presented in Tables 6.7 and 6.8. The tension–compression fatigue strength ($R = -1$) on the basis of 10^7 cycles for the PHCM based on the binder of type I is equal to 230–250 MPa and for the PHCM based on the binder of type II it is 240–260 MPa.

Comparing the results obtained, it should be noted that the PHCM, in which the fabric of satin weave was used has properties that are 5–20% higher than in the other materials irrespective of the binder used. This fabric is more convenient from the technological standpoint too. Thus, for example, the fabric with filament weave of hopsack type is intensively stretched in impregnation, and it is laid out with difficulty, etc.

The use of a carbon-fibre bundle with half the number of filaments in the fabric at equal tensile and compressive strength enables the shear characteristics of PHCM, to be enhanced somewhat.

The elastic and strength characteristics of PHCM based on the binder of type II are higher in tension, compression and shearing (also the fatigue strength) than in materials based on the binder of type I. At the same time the strength in interlaminar shearing of materials based on the binder of type I is less by 30% and the shear strength in the plane of the sheet is twice as high as in materials based on the binder of type II. The actually realized strength of a variety of carbon-fibre bundle UKN

Table 6.7 Properties of PHCM based on epoxy binder of type I, using three different weaves of hybrid fabric

Characteristic[a]	Carbon fibre bundle of 200 tex			Carbon fibre bundle of 410 tex		
	Satin	Serge	Hopsack	Satin	Serge	Hopsack
σ_{B1} (MPa)	608	623	525	630	626	612
σ_{-B1} (Mpa)	539	543	501	417	385	425
σ_{B2} (MPa)	664	498	590	670	535	541
σ_{-B2} (MPa)	323	275	286	355	358	349
E_1 (GPa)	61	63	56	62.5	57	64
E_2 (GPa)	29	28	28	29	29	27.5
G_{12} (MPa)	5800	5700	5400	–	–	–
τ_{12} (MPa)	86	77	73	–	–	–
τ_{13} (MPa)	42	41	38	–	–	–
ρ (kg m^{-3})		1600–1685			1600–1700	
t (mm)		0.330–0.346			0.360–0.390	
G_{12}/ρ	362	356	338	–	–	–

[a] Longitudinal and transverse ultimate tensile strengths, σ_{B1} and σ_{B2}; longitudinal and transverse ultimate compressive strengths, σ_{-B1} and σ_{-B2}; longitudinal and transverse tensile moduli of elasticity, E_1 and E_2; shear modulus, G_{12}; interlaminar and in-plane ultimate shear strengths, τ_{12} and τ_{13}; density, ρ; monolayer thickness, t.

Table 6.8 Properties of PHCM based on epoxy binder of type II, using three different weaves of hybrid fabric

Characteristic[a]	Carbon fibre bundle of 200 tex			Carbon fibre bundle of 410 tex		
	Satin	Serge	Hopsack	Satin	Serge	Hopsack
σ_{B1} (MPa)	661	562	537	647	573	558
σ_{-B1} (Mpa)	539	484	488	469	–	541
σ_{B2} (MPa)	427·	419	563	338	296	421
σ_{-B2} (MPa)	249	239	219	284	232	173
E_1 (GPa)	68	65	66	71	70.5	63
E_2 (GPa)	25	23	28	27	22	23
G_{12} (MPa)	7300	7170	6200	–	–	–
τ_{12} (MPa)	62	59	56	–	–	–
τ_{13} (MPa)	28	27	26	–	–	–
ρ (kg m^{-3})		1600–1700			1600–1700	
t (mm)		0.320–0.330			0.330–0.360	
G_{12}/ρ	456	448	387.5	–	–	–

[a] Longitudinal and transverse ultimate tensile strengths, σ_{B1} and σ_{B2}; longitudinal and transverse ultimate compressive strengths, σ_{-B1} and σ_{-B2}; longitudinal and transverse tensile moduli of elasticity, E_1 and E_2; shear modulus, G_{12}; interlaminar and in-plane ultimate shear strengths, τ_{12} and τ_{13}; density, ρ; monolayer thickness, t.

has been assessed in the process of work on the basis of the obtained experimental data. Calculated on the basis of these data was the desired strength of carbon-fibre bundle at which the specification will be fully satisfied. The carbon-fibre bundle strength realized in PHCM should amount to 4000 MPa. The effectiveness of the procedure of optimal design of combination fabrics has been confirmed many times by experimental developments of fabrics of various structure, containing in the warp combinations of carbon-fibre bundles with organic (Armos) fibre bundles or glass filaments, and in the weft glass filaments and organic filaments SVM.

The principle of producing intralaminar PHCM can also be used for correcting the special properties of PHCM, e.g. electrophysical ones. A hybrid fabric designed from aramid fibre with low density and high elastic and strength properties and from metallized carbon or glass fibres possessing high electrical conduction is used to protect flying vehicles constructed from composite materials against lightning strikes. The PHCM made from these fabrics are very effective for protecting construction skins both in the case of a direct lightning strike and from the side effects (influence on aircraft control systems by low-frequency electromagnetic fields arising during the discharge).

6.4 PRODUCTION OF HYBRID COMPOSITE MATERIALS AND ARTICLES THEREFROM

As a method, hybridization allows not only production of a composite material with optimal properties, but also optimization of a technological process. For example, use of low-modulus carbon, organic or glass filament as a flimsy weft in tapes from high-modulus carbon or boron fibre enhances the processability of the unidirectional material, as it promotes the retention of fibres orientation in the layer, simplifies the processes of impregnation and lamination of the stack and makes the production process of the article more economic. The combination of different fibres in hybrid fabrics produces woven fillers from brittle fibres that are unamenable to textile processing.

The type and shape of the tentative PHCM and the technology of production are selected to attain the desired quality in the article with minimum cost. For the solution of this problem, constructions from PHCM should not copy similar constructions made from traditional materials; they should be developed giving consideration for the complete realization of the nature of each component and its fracture characteristics. The principal feature of all the technological processes of production of articles from monoplastics and PHCM is the combination, simultaneously, of the process of production of the material and construction. Such an approach has its own history, and the technology of processing of high-strength high-modulus PHCM is based at present on methods and technology developed in the course of introducing glass plastics into production, the PHCM technology being of a higher level than the technology of the appropriate traditional two-phase composites.

The basic technological operations that form a unified cycle for producing this or that specific part or construction from PHCM comprises the following operations:

1. Manufacture of semifinished products (prepregs).
2. Lamination of a stack from the prepregs in accordance with the designed block diagram of the article.
3. Forming the material with the use of direct moulding, vacuum moulding with flexible diaphragm, autoclave moulding, etc.
4. Winding with fibres, bundles previously impregnated with the binder or tapes – prepregs with subsequent hardening on the mandrel.

The method of pultrusion is used for manufacturing long-profile article cross-sections (flat, box-shaped and other complicated configurations). The method of manufacturing parts by pressure impregnation (injection method) is cheap, rather simple and comparatively clean from the ecological point of view.

The modern production of high-quality articles is characterized by the efforts of enterprises using PHCM to receive from their specialized suppliers not fibres and binders but semifinished products (prepregs, stacked blanks, etc.) with a strictly guaranteed ratio and quality of the components.

Manual stacking followed by direct moulding is the simplest and oldest method of moulding reinforced plastic materials. The method indicated is now used especially in the manufacture of pilot samples of articles, its advantages being low cost and absence of limitations on the dimensions of the produced parts. The disadvantages include labour consumption and the impossibility of ensuring really high quality of the material, since gaps, overlaps and misorientation of the filler, which have a decisive bearing on the level of a construction's characteristics, depend in this case on the workers.

The high sensitivity of high-modulus components of PHCM to factors causing a deviation from the specified direction of reinforcement results, for example, in an orthogonally reinforced material with layers stacked at an angle $8-15°$ to the axis of loading, in the change of the modulus of elasticity in tension by 1.5 times at the expense of misorientation of the fibres only by $\pm 5°$ from the given direction.

With stacking of layers at an angle of $30-40°$ and the same error, a similar change may happen with shear modulus in the plane of reinforcing. Therefore, the basic trend of the modern science of composite materials is the transition to semi-automated and automated production. Machine stacking promotes the removal of subjective factors, improvement of the quality of the laminated material and reduction of its cost due to the rise in productivity. By the level of automation the variety of the devices used is very great: from tape-laying heads with manual control to completely automated multiple-axial units, including numerically controlled ones that enable production of flat stacks from any reinforcing fillers with any structure of reinforcing as well as shaped three-dimensional stacks of medium complexity (by means of the flexible membranes or self-adjusting mechanical compression devices).

For moulding articles from PHCM, including large-size ones of complicated configuration, the method of autoclave moulding is used primarily (this method is used mainly in production of parts intended for aircraft equipment). In the process of autoclave moulding a prepared stack provided with an elastic diaphragm is subjected to vacuum processing (at the initial stages of the hardening process), so as to remove the volatiles and air inclusions, and then it is hardened with simultaneous application of pressure and heating. The selection of vacuum depth and the values of the temperature and pressure with consideration for the basic parameters of the processes of hardening of the binders and their fluidity is very important. As a result, it becomes possible to produce solid articles of the required composition and structure, of even thickness and with high quality of the surface. The main disadvantages of the method of autoclave moulding comprise the high cost of the equipment, the periodicity of the process and the necessity to use manual labour. The prospect of reducing the cost of the technological process consists in production of universal quickly hardening binders for PHCM, development of materials for strong heat-resistant vacuum bags, as well as the mechanization and automation of as many operations as possible, including the fitting out of autoclave complexes with systems for automatic control and program correction of the technological parameters using microprocessor equipment.

One of the basic prerequisites for expanding the use of parts and constructions from PHCM is the development of cheap technologies – winding with filaments or tapes – prepregs, manufacture of two- and three-dimensional fabric and woven structures, pultrusion and injection moulding.

The method of winding articles from PHCM is similar to the methods for articles from carbon-reinforced and glass plastics. Its main features are as follows: maximum realization of the properties of the fillers, possibility of a high degree of automation of the process, vast possibilities for designing articles due to the diversity of fibre orientation patterns, stability of the properties of the material obtained and comparatively low cost (about half that of a PHCM produced by manual lamination).

The effectiveness of the winding method in production of parts from PHCM depends greatly on the level and condition of the winding equipment. In connection with the production of new reinforcing materials, expansion of the range of articles that differ by purpose, shape and dimensions, and increase of the scope of PHCM use, are necessary for its permanent modernization. The main trends in modernization of winding equipment and winding technology are as follows: expansion of the technological possibilities of the machines (equipping them with arrangements for winding with tape–prepreg in five coordinates, with laminating devices); increase of the maximum tension of the tape up to 400–500 N per centimetre run; toughening the tolerance by the values of the technological parameters to be checked and increase of their number; raising the output (due to the increase of the travel rate of the actuating members of the machine to 30–40 m min^{-1}, at the expense of combining the operations and reduction of the time of readjustments of the machines from one type of reinforcing filler to another); enhancing the reliability of the equipment (e.g. protection of electrical and electronic equipment from current-conducting dust formed as a result of 'fluffing' of high-modulus reinforcing components of PHCM); and using new numerically controlled systems based on microprocessors.

Pultrusion (a continuous process of production of structural solid and hollow profile articles from single-axially oriented PHCM) is a process similar to the process of extrusion of thermoplastic materials. The main stages of this method are as follows: impregnation of the fibre bundle with a binder; pressing out the excess binder and imparting the specified section to the material by drawing the bundles through a spinneret of the ceramic heated extrusion head (resin hardening occurs at the same time); and slicing the profile article into elements of given length. The purposeful selection of quick-hardening high-strength binders and the use of high-frequency or microwave heating have enabled the sharp growth of pultrusion units. At present the method has been tried out well relative to the PHCM of various structures and enables wide variation of the properties of structural articles of the profile type consisting of various combinations of carbon, organic and glass fibres. A stringent check of the tension and orientation of their permanent content in the material gives rise to enhanced elastic and strength properties of articles produced by the method of pultrusion, as compared with those manufactured by other methods of moulding. To produce the profile articles having complex reinforcing schemes, the following are used: drawing of previously impregnated single-oriented tapes or crosslinked bands with diagonal position of the filaments, fibres or fabrics; and methods of manufacturing tubular articles combining helical winding and drawing. At present many types of complex-reinforced parts from PHCM, like stiffening ribs, panels, pipes, blades, girders of different section, etc., have been successfully produced by the method of pultrusion.

Articles from hybrid plastics based most often on glass and carbon fibres are produced by the method of pressure impregnation, which schematically comprises the following operations: the filler (dry fibres, bands, fabrics) is laminated according to the chosen scheme of reinforcing into the mould holes intended for introducing the binder and removing air; the mould is closed; injected into it is the binder, which, as it penetrates into the reinforcing material, pushes the air from it and impregnates the fibres; and finally the material is hardened. Of great significance are the rheological characteristics and the binder hardening rate, enabling good impregnation of the filler without disturbing the orientation of its layers at relatively low pressure (about 10^{-1} MPa) and hardening the material within 15–45 min. The advantages of the injection method of moulding consist in the low cost of the equipment, ecological purity (at all stages of the technological process the binder is out of contact with both the atmosphere and attending personnel), small power consumption, the possibility of producing parts with a good surface and precise dimensions, and the possibility of automation of the process. The difficulties of the method comprise the necessity for making special formulations of binders, the former combining compatibility with heterogeneous fibres, high strength and heat resistance with enhanced processability. Apart from this, it is rather difficult to attain the degree of material reinforcement.

The selection of the process of production of constructions from PHCM is determined by a number of factors in each specific case: the economical considerations, technical and operational requirements, presence of skilled labour and available equipment. If the optimal versions of the technological processes are selected according to the criterion of the technological realization of the properties of the initial components in the construction, not only the empirical approach is possible, but also the estimation of these versions by means of mathematical models of each stage, blocks of stages or the entire technological process [45].

6.5 CHARACTERISTIC FEATURES AND FIELDS OF APPLICATION

The experience of production and application of monofibrous composites based on glass, carbon and organic reinforcing fillers enabled an all-round and critical estimation of the utilization qualities of articles from these materials to be made.

In comparison with other composite materials, the main advantage of glass-reinforced plastics is their low cost. They are used more in chemical, transport, agricultural machine building, ship building and construction industries (especially in cases when large size mass-production articles are to be produced). Carbon plastics possess the most advantageous combination of low density, high strength, modulus of elasticity and a whole complex of unique special properties, but they are relatively brittle and expensive. They are mainly used in the aerospace industry, where the weight characteristics are very important, as well as the sports and leisure goods industries, where considerations of a prestige nature are of great significance. The organic plastics possessing very high specific characteristics in tension and impact strength are also successfully used in aircraft equipment for production of high-pressure vessels and in production of consumer goods, but it is necessary to take into account the low shear and compression characteristics of organic plastics and the need to protection them from the action of moisture.

The problems of manufacturing constructions for which the requirements exceed the possibilities of mono-reinforced plastics can be solved by making PHCM of various compositions and structures. The diversity of the possible forms of PHCM is very great. Some possibilities of the method of hybridization are shown in Tables 6.6–6.8 on the examples of interlaminar and intralaminar PHCM. The data indicate that the moduli of elasticity of the first type of PHCM increase linearly with the rise of the high-modulus component therein. In this case Poisson's ratio linearly decreases from the value that is characteristic of the more high-modulus filler. The tensile and compression strength goes down with the increase of the content of the more low-modulus fibres to some critical value, after which, as a result of an abrupt change of elongation, it starts going up again.

The impact strength of PHCM increases with the introduction of reinforcing fillers with smaller modulus of elasticity and higher strength into its composition. The nature of fracture of PHCM in impact bending is more complicated than that of the mono-reinforced materials. If carbon plastic undergoes brittle fracture, the splitting and pulling of the fibres out of the matrix and lamination along the interface of the material layers take place on the surface of PHCM, e.g. carbon-reinforced glass plastic. As the content of the more low-modulus fibres increases, the time of deformation and work of fracture grow appreciably. With the increase of the content of the low-modulus fibres several times, the time of fracture, elongation in impact tension and deflection in impact bending increase. The area below the strain curve for PHCM containing glass or organic fibres in addition to carbon ones considerably exceeds the area beneath the strain curve of the carbon plastic.

The estimation of the mechanical properties of PHCM having complex schemes of reinforcing is more complicated. For complete understanding of the dependences that connect them, it is necessary to accumulate experimental and design data, the calculations being made with the use of the theory of laminated media by the criteria of ultimate strains or stresses.

As a criterion for estimating the reliability of brittle materials the crack resistance characteristic determined by the value of the stress intensity factor K_{Ic} or effective crack resistance EK_{Ic} can be used, which implies the ratio of crack resistance to tensile strength value:

$$EK_{Ic} = K_{Ic}/\sigma$$

where

$$K_{Ic} = \sigma_m^*(\pi l)^{1/2}$$
$$\sigma = K_k \sigma_m^*$$

l is half-length of crack, σ_m^* is strength of plastic with crack, and K_k is stress concentration factor.

Crack resistance characterizes the degree of rise of the stresses in the material, when approaching some defective region (crack). As a result of exceeding the critical value of K_{Ic} the crack starts opening. Hybridization makes it possible to increase K_{Ic} and effective crack resistance of a structural material [46]. The introduction to a carbon plastic, including that with a complex scheme of reinforcing, of longitudinal layers of glass fibres causes a considerable increase (synergism) of the fracture viscosity, changing its mechanism (from brittle fracture in the case of monoplastic to viscous lamination in the case of PHCM); this material does not lose its load-carrying capacity even with a great number of cracks. The organic fibre has the same kind of influence on the effective crack resistance of PHCM. Thus, substitution of about one-third of the layers in carbon plastic for organic plastic enables increasing its K_{Ic} by 30–35%.

Important for practical use is the possibility (by means of PHCM) of solving problems connected not only with economy, mechanical or physical properties of constructions, but also with such problems as non-toxicity, ecology, etc. An example is to substitute a fairly good polyester or phenolic organic plastic (but with high toxicity of the smoke) used in aircraft interiors by a carbon-reinforced glass plastic based on the same binders without this shortcoming, but possessing even higher compression, shear and tear-out characteristics.

At present, there is practically no field of technology where PHCM are not used or could not be used. However, they are most efficiently used in the aerospace industry. At the first stage (in the 1970s and 1980s) they were used mainly in auxiliary constructions (radomes, hatch covers, landing gear doors, rudders, ailerons, etc.); with the accumulation of experience of design and operation of parts from PHCM, as well as advances in the field of the level of properties and quality of their components, the application of PHCM in medium- and heavy-loaded constructions was started and has been successfully developed (parts and elements of the wings and fuselage of aircraft; helicopter rotor blades; blades of cooling fan of jet engines; elements of the solar batteries; compartment doors; and other constructions of satellites and space vehicles). Examples of the effective application of PHCM in this direction are the carbon-reinforced glass and carbon organic plastic units and constructions that are available in practically all types of flying vehicles for passenger, transport and military purposes. PHCM became practically indispensable in the helicopter industry. The service life of rotor blades, the most vital unit of a helicopter, made from carbon-reinforced glass and carbon organic plastic is practically unlimited (in comparison with the service life of the vehicle as a whole).

The application of monofibrous composites and PHCM in the aircraft industry provides a reduction of construction weight by 20–40%, increase of service life 1.5–3 times, reduction of labour input and power consumption in production of parts and units by up to 50% and savings of expensive metallic materials. For example, the use of one ton of PHCM in the construction of the 'Ruslan' aircraft provides a saving of up to five tons of aluminium alloys and up 12 tons of steel, as well as a saving of fuel of up to 80 tons per year (design parameters) at the expense of the load ratio and reduction of losses caused by corrosion.

The reduction of the number of parts in the units by the proper design of a construction from the minimum number of large-size undetachable parts jointed in the process of moulding is an essential advantage of PHCM. Owing to this, the number of joints is reduced, and the solidity and reliability of constructions increase. It becomes possible to obtain a smooth aerodynamic surface in the process of their production and, consequently, to enhance the flight performance of the aircraft due to the air resistance factor.

Despite their still high cost, PHCM based on carbon, organic and glass fibres are of great interest not only for the aerospace industry. Their excellent strength characteristics in combination with low density gave rise to their wide introduction in the production of sports goods (frames of tennis rackets and bows, hockey sticks, golf clubs, skis and ski sticks, oars), the building of sports boats and yachts (mats and spars, boards for windsurfing), in the aircraft industry (frames for hang gliders) and in the production of the bodies of racing cars. In Japan, for example, about 70% of all produced carbon plastics and PHCM based thereon are used for sports goods. The expendiency of using PHCM in machine building has been practically confirmed for production of parts of high-speed centrifuges, looms (rapiers) and other parts of machines with high rotational speed and reciprocating motion (rotors, turbine blades, flywheels). The high cost restricts the use of PHCM in the automotive industry for the time being, though from the technical viewpoint their prospect in this field is beyond any doubt. In addition to the basic parts of vehicle bodies, the most likely uses for PHCM may be in Cardan shafts, leaf springs, engine parts and frames for trucks.

Good biocompatibility of carbon fibres has not yet been realized for medical purposes. For the independent use as implants, carbon plastics are too rigid and brittle, and the PHCM based thereon, though they possess the required service properties, have much worse biocompatibility. Nevertheless, PHCM are used in medicine rather successfully for producing light-weight prostheses, wheelchairs for invalids and in other medical equipment. Carbon-reinforced organic plastics are used for protecting the patients and attending personnel from excessive doses of X-ray radiation.

Despite the fact that PHCM have obtained recognition, and found and consolidated their place in the general flow of technical progress, it should be noted that these unique materials have not been mastered properly; their production and applications are, if not at the initial stage of development, very close to it. However, it is the hybrid materials that have the broadest and most reliable future, since only they make it possible to solve successfully the most complex technical problems – to develop easily producible and economically effective constructions with an optimal complex of mechanical, physical and service properties for each specific case.

REFERENCES

1. Lubin J. (ed.), *Handbook of Composites*, vol. 1, Van Nostrand Reinhold, transl. from English, Mashinostroyeniye, Moscow, 1988.
2. Vakula V.I., Pributkin L.N., *Physical Chemistry of Adhesion of Polymers*, Khimiya, Moscow, 1984.
3. Kobets L.P., Poliakova N.V., Kuznetsova M.A., *et al.*, Influence of surface treatment on the properties of carbon fibres and plastics, *Mechanics of Polymers*, 1978, **4**, 579–82.
4. Kobets L.P., Golikova L.A., Strebkova T.S., *et al.*, Investigations of the lifetime of activated carbon fibres, *Physics and Chemistry of Material Treatment*, 1977, 129–35.
5. Khokhonov Kh.B., Kazdarov A.A., Karamurzov B.S., Kobets L.P., Influence of chemical treatment on the state of surface oriented carbon materials, in *Physics of Interfacial Phenomena*, Kabardin-Balkar State University, Nalchik, 1984, pp. 30–9.
6. Bednarick J., Les fibres de carbone et les nouveaux materiaux pour l'industrie aeronautique, *Materials et Techniques*, Fevric, N Horse = Serie, 1988, pp. 112–19.
7. Lubin J. (ed.), *Handbook of Composites*, vol. 2, Van Nostrand Reinhold, transl. from English, Mashinostroyeniye, Moscow, 1988.
8. Kobets L.P., Nikitenko J.T., About defects on the phase boundary in polymer composites, *Mechanics of Composite Materials*, 1982, **3**, 546–7.
9. Novikov V., Kobets L., Structurization in epoxy matrix reinforcing carbon fibres, *Mechanics of Composite Materials*, 1987, **4**, 706–12.
10. Deyev J.S., Kobets L.P., Microstructure of the epoxy matrix, *Mechanics of Composite Materials*, 1986, **1**, 3–8.
11. Perov B.V., Kobets L.P., Deyev J.S., Structuring in epoxyphenolic matrix, *Mechanics of Composite Materials*, 1988, **1**, 81–5.
12. Kobets L.P., Mikhailov V.V., Nadiozhina O.N., About the destruction mechanism of carbon and boron plastics under influence of interlaminar shear stresses, *Mechanics of Composite Materials*, 1983, **2**, 251–6.
13. Lipativ Yu. S., Peculiarity of structure of polymer hybrid matrix stipulated by the mechanism of microphase division, *Mechanics of Composite Materials*, 1983, **5**, 771–80.
14. Perov B.V., Stroganov G.B., Belkin S.V., Methods of designing hybrid composites, Paper presented at *5th All-Union Conf. on Composite Materials*, MGU, Moscow, 1981.
15. Tsyplakov O.G., *Designing Articles from Composite–Fibrous Materials*, Leningrad, 1984.
16. Gunyaev G.M., *Structure and Properties of Polymeric Fibrous Composites*, Khimiya, Moscow, 1981.
17. Rabinovich B.L., *Introduction to Mechanics of Reinforced Polymers*, Nauka, Moscow, 1970.
18. Collection, *Modern Problems of Designing and Producing Constructions for Flying Vehicles from Composite Materials*, Moscow, 1985.

19. Gunyaev G.M., Designing of high-modulus polymer composites with tailor-made properties, in *Composite Materials*, Nauka, 1981, Moscow, pp. 24−8.
20. Gunyaev G.M., Rumyantsev A.F., Fedkova N.N., Mitrofanova E.A., Tchekina Z.F., Stepanychev E.J., Mechmutov J.M., Optimization of composition and structure of bi- and three-component composites, in *Composite Materials*, Nauka, Moscow, 1981, pp. 223−7.
21. Maksimov R.D., Plume E.Z., Ponomarev V.M., Stress−strain properties of the unidirectional reinforced hybrid composites, *Mechanics of Composite Materials*, 1984, **1**, 35−41.
22. Rikards R.B., Snisarenko S.J., Impact deformation of the beam from hybrid composites, *Mechanics of Composite Materials*, 1985, **1**, 97−103.
23. Yerokhina L.S., *Methods of Forecasting the Development of Structural Materials*, Mashinostroyeniye, Leningrad, 1980.
24. Zykov A.A., *Theory of Finite Graphs*, Nauka, SO AS USSR, Novosibirsk, 1969.
25. Shepfild J., *Mathematical Logic*, Nauka, Moscow, 1975.
26. Skudra A.M., Bulavs F. Ya., *Strength of Reinforced Plastics*, Khimiya, Mocow, 1982.
27. Malmeister A.K., *et al.*, *Resistance of Rigid Polymeric Materials*, Zinatne, Riga, 1972.
28. Himmelblau D., *Applied Non-Linear Programming*, Mir, Moscow, 1975.
29. Skudra A.M., Plume E.Z., Gunyaev G.M., Yartsev V.A., Belyaeva N.A., Properties of the glass plastics reinforcing high-modulus fibres, *Mechanics of Polymers*, 1972, **1**, 68−74.
30. Brautman L. and Krok R.M. (ed.), *Modern Composite Materials*, transl. from English, Mir, Moscow, 1970.
31. Gunyaev G.M., Rumyantsev A.F., Fedkova N.N., Mashinskaya G.P., Bardina N.P., Stepanychev E.J., Mechmutov J.M., Polyfibrous composites, in *Aviation Materials*, ONTJ, Moscow, 1977, pp. 46−55.
32. Afanasenko V.J., Afanasiev Yu. A., Gluschenko A.G., Kotlov V.J., Sokolova I.I., Tabaldiev S.R., Design of the bearing truss for space telescope made of hybrid composite, *Mechanics of Composite Materials*, 1988, **4**, 705−14.
33. Kochetkov V.A., Effective elastic and thermal properties of the unidirectional reinforced hybrid composite, *Mechanics of Composite Materials*, 1987, **2**, 250−5.
34. Kobets L.P., Gunyaev G.M., *Carboplastis in Plastics of Structural Purpose*, ed. E.B. Trostyanskaya, Khimiya, Moscow, 1974, pp. 204−45.
35. Obraztsov N.F., Vasillyev V.V., Bunakov V.A., *Optimal Reinforcement of Shells of Revolution from Composite Materials*, Mashinostroyeniye, Moscow, 1977.
36. Vasillyev V.V., *Mechanics of Constructions from Composite Materials*, Mashinostroyeniye, Moscow, 1988.
37. Teters G.A., *et al.*, *Optimization of Shells from Laminated Materials*, Zinatne, Riga, 1978.
38. Kalnin I.L., Surface of carbon fibres, modification and influence of it on the destruction of high-modulus carboplastics, in *Proc. 1st Soviet−American Sym.*, Zinatne, Riga, 1979, pp. 221−30.
39. Murtaf B., *Modern Linear Programming*, Mir, Moscow, 1975.
40. Zinkevich O., *Method of Finite Elements in Technology*, Mir, Moscow, 1975.
41. Benerdji P., Batterfield R., *Method of Boundary Elements in Applied Sciences*, Mir, Moscow, 1984.
42. Kuzovkov E.G., *Graph Model of Elastic Body. Calculation of Stressed−Strained State*, Preprint, ONTI AS USSR, Kiev, 1985.
43. Skudra A.M., Bulavs F. Ya., *Structural Theory of Reinforced Plastics*, Zinatne, Riga, 1978.
44. Zhigun I.G., Polyakov V.A., *Properties of 3D-Reinforced Plastics*, Zinatne, Riga, 1978.
45. Kalinnikov V.A., Use of a linear statistic model for solving the problem of optimization of technological processes for constructions from composites, in *Composite Material*, Nauka, Moscow, 1981, pp. 236−40.
46. Shalin R.E., Perov B.V., Composites in aviation technique and national economy, Paper presented at *1st int. Conf. on Composite Materials*, Moscow, 1990.

7

Principles of developing organic-fibre-reinforced plastics for aircraft engineering

G.P. Mashinskaya and *B.V. Perov*

7.1 INTRODUCTION

The lightest polymeric composites – organoplastics based on reactive or thermoplastic matrices, reinforced with polymeric fibres – occupy a special place among polymeric composite materials. Judging from the position of formal classification, organoplastics should be referred to as a class of reinforced plastics. But the similar polymeric nature of both components, predetermining diffusive chemical interaction between the matrix and the fibre and, consequently, the absence of a physical boundary between the components, permits organoplastics to be referred to as a group of the same name (the so-called 'homonymic' composites), including such systems as polymers reinforced with polymers, metals reinforced with metals, etc. These systems differ in principle from compositions of different names (the so-called 'heteronymic' composites), combining components of different kinds, such as glass–polymer or carbon–polymer, characterized by a clear physical phase separation boundary between phases [1].

Organoplastics such as polymer–polymer composites of the same name are multi-component multiphase compositions, and like any complex polymeric system may be subjected to various modifications with the purpose of further perfection of their properties and expansion of their functionality. Variation of the composition and structure of fibres, as well as that of polymeric matrices and compositions based on them, may be made over a broad range, as a result of chemical, photochemical, radiation, physicochemical and thermochemical reactions, the processes of selective sorption and diffusion of low-molecular-weight and oligomeric products, the actions of mechanical, electrical, ultrasonic and other fields, the introduction of modifying extraneous powdery or fibrous fillers, and the variation of reinforcement schemes and structure of reinforcing fillers, etc. These represent an inexhaustible source for perfection and optimization of composition and structure of organoplastics in accordance with the imposed requirements. The polymeric nature of the basic components (fibre and matrix) provides wide potential possibilities for modification of organoplastic properties. This is, first of all, bound up with the fact that the number of chemical compounds, as well as the topological and supermolecular structure of a wide class of polymers, which are able to form various polymer–polymer compositions, is practically unlimited.

This is confirmed by the whole history of development of polymer–polymer composites, giving a possibility for objective evaluation of the present achievements. It is also true that cloth laminates based on phenol–formaldehyde resins and organic natural fibres, produced in the 1930s and widely used up till now, are the universal 'long-livers' among polymeric materials and the 'ancestors' for the class of polymer–polymer reinforced composites – the organoplastics. The first-generation organocloth laminates based on natural fibres (cotton, flax, etc.) have been widely used up till now for the electric power industry and for various branches of machine-building (Table 7.1) [2]. Comparing the level of the material properties achieved during the 1930s with the up-to-date level, we may affirm with certainty that neither material is able to compete with organoplastics in terms of growth of basic properties: the strength of organocloth laminates was increased by 5–7 times and unidirectional composite materials are at present matchless in respect of specific tensile strength among modern structural materials. Composites with a specific tensile strength in excess of 200 km were produced on the basis of high-strength aramid and polyethylene fibres.

The period between the first cloth laminates and modern organoplastics based first

Table 7.1 Comparison of the properties of organocloth laminates based on natural and synthetic fibres

		Reinforcing fibres	
Properties	Cotton	Poly(ethylene terephthalate) (Lavsan)	Aramid (SVM)
Density ($kg\,m^{-3}$)	1250–1350	1250–1300	1270–1310
Working temperature (K)			
(for a long time)	353–362	353–363	423–453
Ultimate tensile strength, σ_b (MPa)	43–58	145	600–750
Ultimate bending strength, σ_{bb} (MPa)	160	140	380–500
Ultimate compressive strength,			
σ_{-b} (MPa)	150	75	200–300
Young's modulus, E (MPa)	5400	5000	35000–45000
Volumetric electrical resistance,			
ρ_v ($\Omega\,cm$)	10^{12}–10^{13}	5×10^{13}	10^{14}–10^{15}
Surface electrical resistance, ρ_s (Ω)	10^{12}–10^{13}	1.2×10^{14}	10^{14}–10^{15}
Mechanical loss tangent, $\tan\delta$, at			
frequency of 10^6 Hz	0.03	0.019	0.018–0.032
Water absorption, 24 h (%)	0.1–0.8	0.13	0.25–0.4

of all on high-strength high-modulus aramid fibres was a period of accumulation of experience in creating and mastering composite materials, based on carbon and hetero-chain fibre-forming polymers – aliphatic polyamides, poly(ethylene terephthalate), polyacrylonitrile, poly(vinyl alcohol). Being combined with phenol–formaldehyde, epoxy or polyester resins, these fibres gave a possibility to produce composite materials with high thermophysical and dielectric properties, good chemical resistance and sufficient strength [3, 4]. So, for example, the epoxy organoplastic based on poly-(ethylene terephthalate) fabrics and characterized by high and stable dielectric pro-perties in the case of long-term moistening has been successfully used for many years for the manufacture of critical parts in the electrical and radio engineering industry (see Table 7.1). The organoplastic based on phenol–formaldehyde binder and poly-amide fibres is up till now considered a classic example of an effective heat-shielding material.

Creation of modified poly(vinyl alcohol) fibres with higher water resistance gave a possibility for the first notable step for improvement of the elastic and strength properties of organoplastics [5]. However, a principally new solution in the sphere of creating high-strength high-modulus polymeric fibres for reinforcing of plastics was the realization of the idea to produce fibres of a rigid-chain fibre-forming polymer, being at the stage of forming fibres in the mesomorphic state [6–8].

At present, it is possible to determine at least two tendencies of work in this area:

1. Creation of high-strength organoplastics based on aramid fibres similar to poly(*p*-phenylene terephthalamide) (PPHTA) and polyheteroarylene (PHA) with a strength of 4000–5000 MPa.
2. Creation of high-modulus organoplastics based on aromatic heterocyclic fibres of poly(*p*-phenylenebenzobisthiazole) or poly(*p*-phenylenebenzobisoxazole) with modulus of elasticity of 300–350 GPa.

The work in mastering the technology for production of high-strength, high-modulus fibres of flexible-chain polymers and, first, polyethylene fibres are of special interest based on convincing successes in improvement of the mechanical properties of fibres from rigid-chain polymers [9]. Thus at present there are favourable conditions for realizing the next jump in the development and perfection of polymer–polymer materials.

The present chapter is dedicated to generalization of experience on creating, studying and practical use of organoplastics based on aramid polyheteroarylene fibres and thermoreactive binders, since materials of such type are widely used in various branches of industry.

7.2 PECULIARITIES OF THE PROPERTIES OF ARAMID FIBRES USED AS FILLERS FOR POLYMERIC COMPOSITES

At present, high-strength high-modulus fibres are produced from aramids related to rigid-chain polyheteroarylenes (PHA) (e.g. Terlon*) [7, 10]. SVM fibres are most widely

---

*Terlon is an analogue of Kevlar-49 produced by Du Pont, USA.

intrafibrillar zones decreases the transverse strength of aramid fibres, which is low even without it. Intrafibrillar pores and microcracks adjacent to the fibre surface participate in selective transportation of monomeric and oligomeric components of liquid binder during production of composites.

According to the X-ray diffraction data of PHA fibres, Vnivlon and SVM are mesomorphic and Terlon is crystalline. Mesomorphic fibres of PHA are characterized by highly oriented structure.

For example, the angle of misorientation of polymeric chains for SVM is 2° [11] and for PPHTA it is up to 10°, the strength of the latter being considerably less as a result. Amorphous sections practically do not differ from crystalline ones in the degree of orientation of macromolecules, but they contain a large share of loaded chains. The extremely low strength of intrafibrillar bonds causes the premature splitting of organic fibres and an extended zone of fracture in all kinds of mechanical action [12]; also their theoretical strength is extremely high (25 GPa) and the activation energy of the initial stage of the fracture process is close to the dissociation energy of chemical bonds $(210-250\,\text{kJ}\,\text{mol}^{-1})$ [13, 14].

In contrast to glass and carbon fibres, the limit elongation of which does not practically depend upon the rate of loading and is mainly a consequence of defect accumulation, the elastic and stress–strain properties of aramid fibres do not depend upon

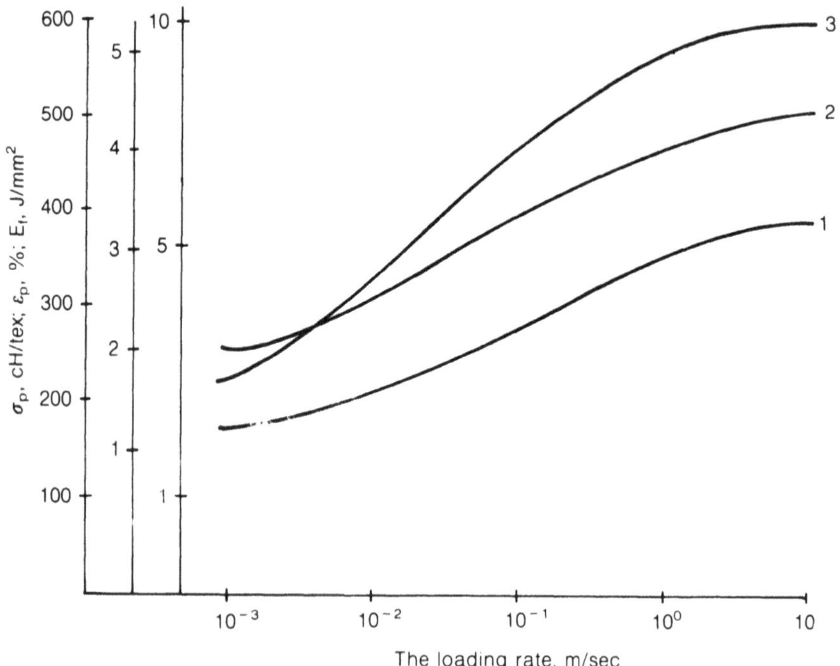

Fig. 7.2 Strength σ_p (1), elongation at break ε_p (2) and specific fracture energy E_f (3) of SVM aramid threads as the function of fracture rate.

the loading time (Fig. 7.2). This testifies to the essential role of viscous flow in formation of bearing ability of reinforcing fillers based on such fibres.

The compressive strength of high-strength organic fibres is six times and the transverse strength (calculated from the results of testing unidirectional organoplastics) is two times less than the axial tensile strength. Since the ratio of normal and shear moduli for these fibres reaches 100, they are at present the most anisotropic ones [15].

Under normal conditions the limit deformation ε_p of aramid fibres ($\varepsilon_p = 2-4\%$) is determined by the maximum straightening of polymeric chains. Up to 15% of macrochains of SVM fibre disintegrate when deformation reaches $(0.8-0.9)\varepsilon_p$. In cases of higher deformations, accumulation of rupture becomes avalanche-like [16].

The stress–strain curve of SVM fibre (Fig. 7.3) is linear up to $\sigma = 0.5\sigma_b$ (σ_b is ultimate strength) when usually there is a reverse bend, separating the elastic and the forced elastic states. Subsequent loading is accompanied by oriented strengthening of the fibre. The area of elastic deformation increases with decrease of temperature and Young's modulus reaches 140–150 GPa. On heating, the elastic and strength properties of SVM fibre decrease comparatively slowly, and at a temperature of 473 K, corresponding to the temperature of 'defreezing' of the flexural vibrations of

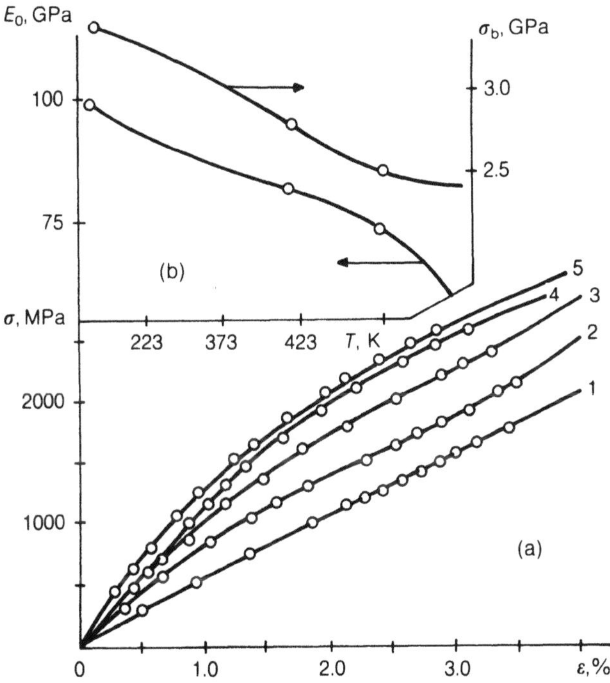

Fig. 7.3 (a) Stress–strain (σ–ε) diagram of SVM threads at temperatures of 293 K (1), 423 K (2), 473 K (3), 523 K (4) and humidity of 45%, and at a temperature of 293 K and humidity of 75% (5). (b) Temperature dependence of strength σ_b (1) and original elasticity modulus E_0 (2) for SVM thread.

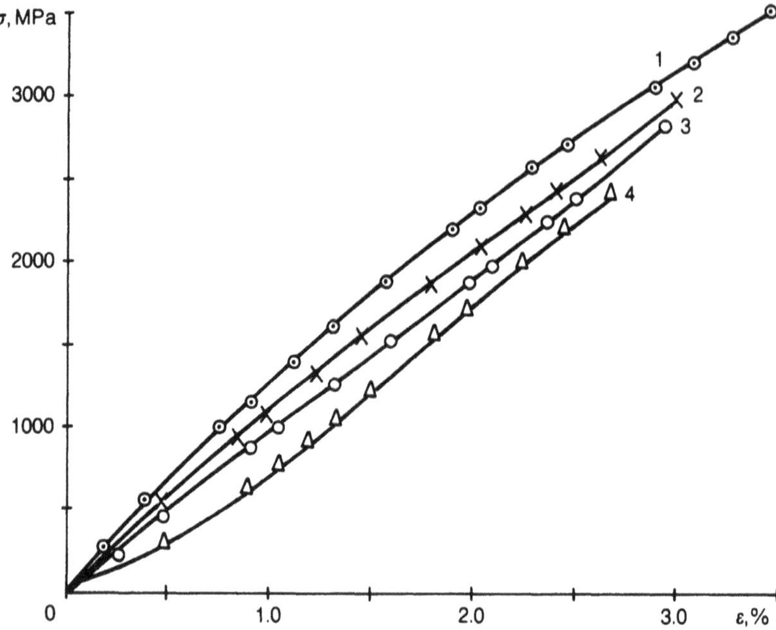

Fig. 7.4 Stress–strain diagrams of Armos thread at different temperatures and relative humidity $\phi = 45\%$: (1) 293 K; (2) 423 K; (3) 473 K; (4) 523 K.

the polymeric chain [17], the decrease of strength is equal to 20–40% when modulus of elasticity is equal to 15–20%. The limit working temperature of fibres (Fig. 7.3(b)) may be easily determined by the temperature curve of initial modulus of elasticity. Modified PHA fibres (Armos) [7] differing from SVM in the composition and microstructure of the fibre-forming polymer are characterized by higher values of initial modulus and ultimate strength in the whole range of temperatures (Fig. 7.4). At high temperature in the moistened state the total deformation of the fibre is increased and the share of its elastic component is decreased.

The heat stability of aramid fibres and, especially, of polymeric composites based on them is to a considerable degree determined by their thermal deformation properties, which depend essentially upon the type of fibre-forming polymer and its moisture content. So, for example, crystalline PPHTA fibres in the range of 293–423 K are characterized by shrinkage – the linear thermal coefficient of expansion α in the axial direction is equal to $(-6$ to $-13) \times 10^{-6} K^{-1}$. In contrast, mesomorphic PHA fibres are characterized by prevailling expansion of $(6$ to $10) \times 10^{-6} K^{-1}$, and shrinkage occurs in the initial stage of heating and at temperatures exceeding 573 K [18, 19].

Analysing thermal deformation properties of initial SVM fibres, one may see that, in contrast to glass and carbon fibres, characterized by low and practically constant values within the limits of $(0–5) \times 10^{-6} K^{-1}$ within the range of temperatures from

293 to 673°C, the SVM threads are characterized by a more complicated dependence of α deformation with temperature: elongation and positive α during heating to T_{max}; and shrinkage and consequently negative α at temperatures above T_{max} (Fig. 7.5).

Heating organic fibres to temperatures close to T_{max} (Fig. 7.5(a)) causes a variation of sign of α; elongation of fibres is continued in this case at the cooling stage. After heating above T_{max} deformability of fibres disappears up to 473 K (Fig. 7.5(b)).

Drying of threads over silica gel (Fig. 7.5(c)) as well as estimation of the dependence of their size in the case of holding in a medium with various moisture (Table 7.3) showed that positive deformation (elongation) of SVM threads is bound up with their moisture content in the initial state and subsequent desorption of moisture from the fibre.

After initial heating, SVM fibre contains mainly bound water and, evidently, just this water effects the flexibility of polymeric chains of the amorphous phase. As bound water molecules are removed, chains of macromolecules straighten (elongation of thread) at temperatures exceeding T_{max}; when we observe a deeper desorption process and beginning of water molecules removal from monolayers, their new conformational rearrangement takes place. Fibre shrinkage according to such a mechanism is observed in the case of heating to 523 K, i.e. just until the beginning of a second desorption process (Fig. 7.6).

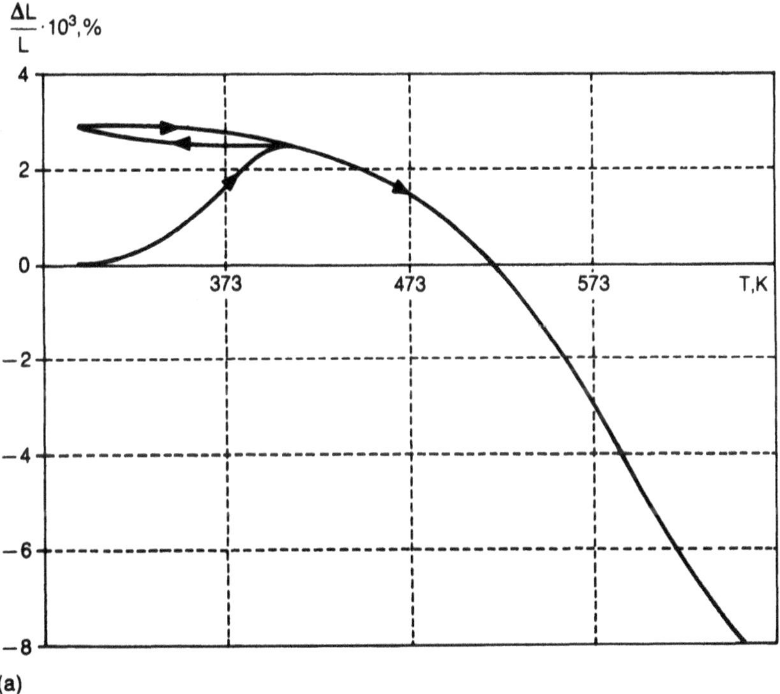

Fig. 7.5 Heat deformation curves of SVM fibres during cyclic heating to (a) 423 K and (overleaf) (b) 523 K with following cooling and (c) after drying fibres over silica gel.

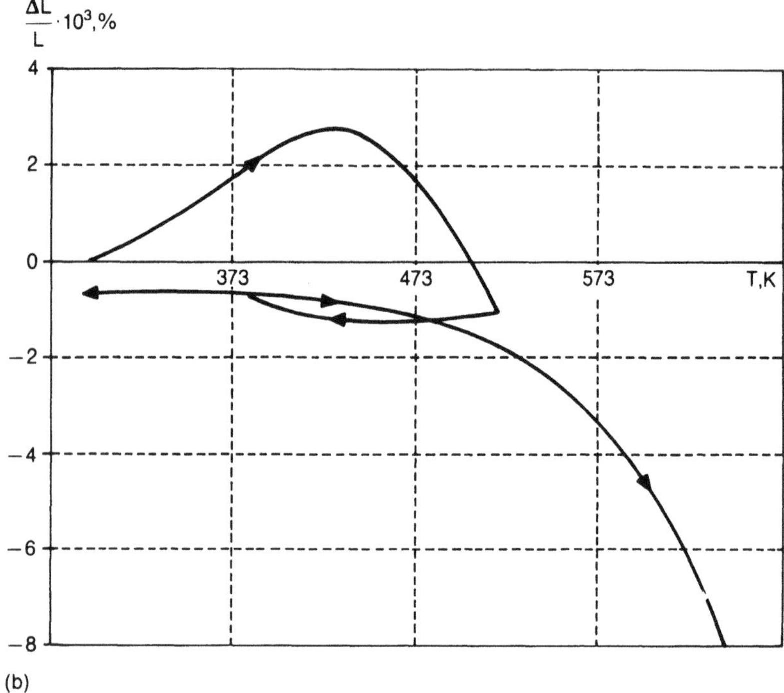

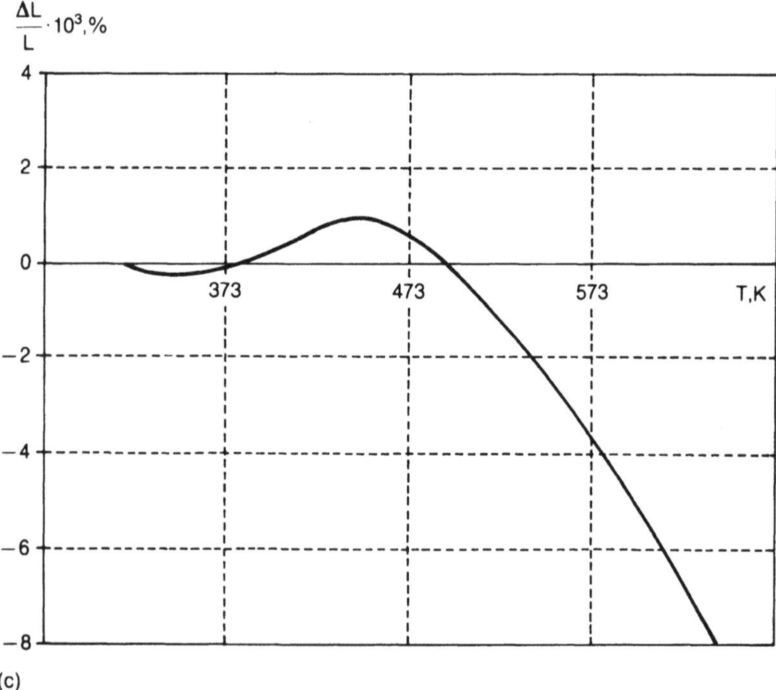

Fig. 7.5 (*continued*)

Table 7.3   Effect of medium humidity on variation of mass and length of SVM fibres[a]

Medium humidity, ϕ (%)	$(\Delta m/m) \times 100$ (%)	$(\Delta l/l) \times 100$ (%)
50 (room)	0	0
30	− 0.8	+ 0.6
65	+ 0.6	− 0.3
98	+ 16.0	− 3.0

[a]Specimens were held in a desiccator for no less than 7 days.

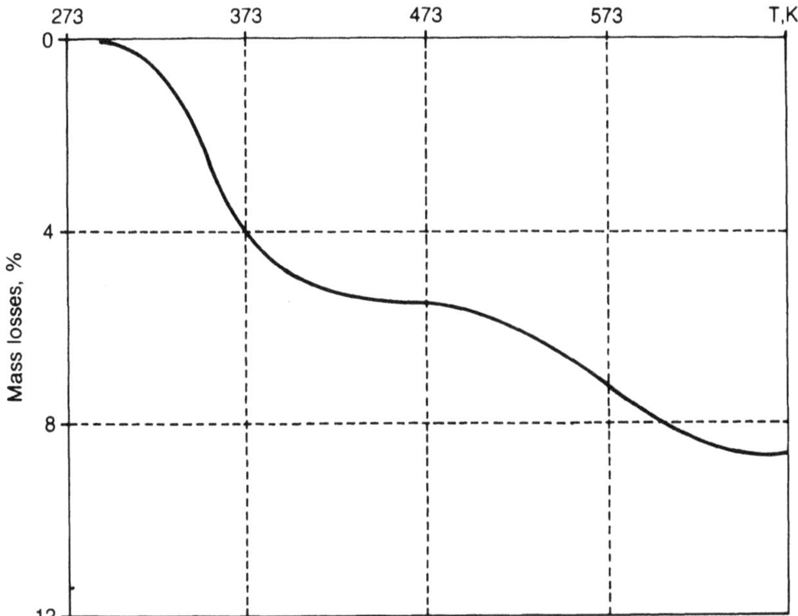

Fig. 7.6 Thermogravimetric analysis (TGA) curve for SVM fibres.

Such behaviour of PHA fibres is determined by their more active interaction with water compared with that of PPHTA fibres: partial amorphization of mesomorphous structure of PHA fibres takes place as a result of moisture absorption, which causes a decrease of strength and stiffness, distortion of configuration of macromolecules and reduction of fibre length. Increased moisture content in PHA fibre is bound up with sorption of water molecules by not only amide but also hydrophilic groups of heterocycles. Hydrogen bonds are hydrolysed on both amorphous and crystalline sections of structure when equilibrium moisture content is reached, resulting in 5−7% increase of elastic and strength properties. Also the fibre properties are restored after moisture elimination; in the case of durable loading we observe the progress of creep, increasing when the moisture content exceeds 1% [20, 21]. It should also be noted that holding in water for 100 h reversibly decreases the SVM strength by 20–30%.

Passage from expansion of the polymer fibre to shrinkage is observed when the water content is about 1% (correlates with the maximum strength of elementary fibres under these conditions). In the case of complete absence of moisture, the thermal coefficient of expansion of PHA fibres becomes negative, $(-7$ to $-10) \times 10^{-6} \mathrm{K}^{-1}$.

Heat deformation variations of aramid fibres in transverse direction $(\alpha_\perp)$ exceed the longitudinal value $(\alpha_\parallel)$ by 25 times. The thickening of dry fibre takes place while heating it to 473 K, when $\alpha_\perp$ reaches $58 \times 10^{-6} \mathrm{K}^{-1}$. The moistening of PHA fibres contributes to decreasing of their cross-dimensions under heating; in this case $\alpha_\perp$ reaches $(-300$ to $-500) \times 10^{-6} \mathrm{K}^{-1}$.

Service properties of an organoplastic are detemined by longevity and creep of aramid fibres, the level of which depends on the mechanism of accumulation of defects and fracture of fibres. The latter is in general coordinated with the thermal fluctuation of strength but it is of two-stage nature (Fig. 7.7). Owing to difference of lengths of filaments in a fibre under stress, which are less than a certain critical value (for SVM fibre, 2000 MPa), shorter fibres are strengthened under the action of forced elastic deformation; as a result the service life of the fibre is increased considerably.

Comparison of the kinetic characteristics of aramid fibre deformation with the results of studying the molecular processes that take place at the respective stages of

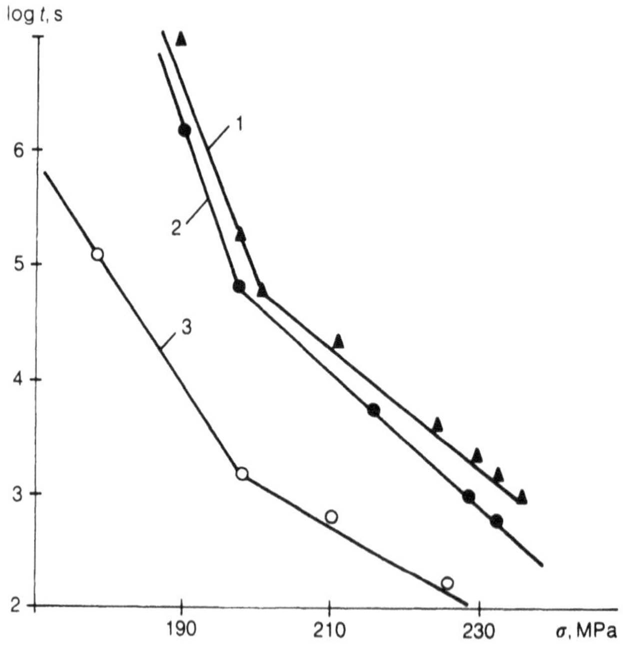

Fig. 7.7 Durability of SVM thread at different temperatures: (1) 293 K, (2) 373 K, (3) 473 K.

elongation has made it possible to determine that the value of creep depends on the concentration of rupture of fibre-forming polymer valence bonds [16]. In this case there are also two separate mechanisms determining the kinetics of aramid fibre deformation: in the range of deformation up to 2.3% (about 75% of the limit value), accumulation of rupture of valence bonds develops very slowly and the factor determining creep is mainly molecular regrouping; further growth of deformation is accompanied by an increase of intensity of bond rupture up to the critical concentration corresponding to fracture. The leading role of the conformational mechanism of deformation gives the possibility to define SVM aramid fibres as hereditary materials characterized by non-linearity of viscoelastic properties [16].

Selecting the load of preliminary static drawing, the temperature and duration of loading in such a way that the rate of stress relaxation would certainly exceed the rate of accumulation of damage in filaments, one may considerably increase the initial strength of fibres. This method has been successfully used for winding articles of shell type on a split mandrel, as well as for production of high-pressure cylinders, consisting of a thin shell and an external layer of high-strength aramid fibres. Preliminary static deformation of unidirectional organoplastics and articles made of them has a positive effect on the strength, deformation and fatigue characteristics of the material. So, for example, the destructive pressure of a shell increases by 10–12% and the fracture stress of the material of a shell manufactured by winding on a split mandrel increases by 16–20% with increase of axial force of unclasping [22].

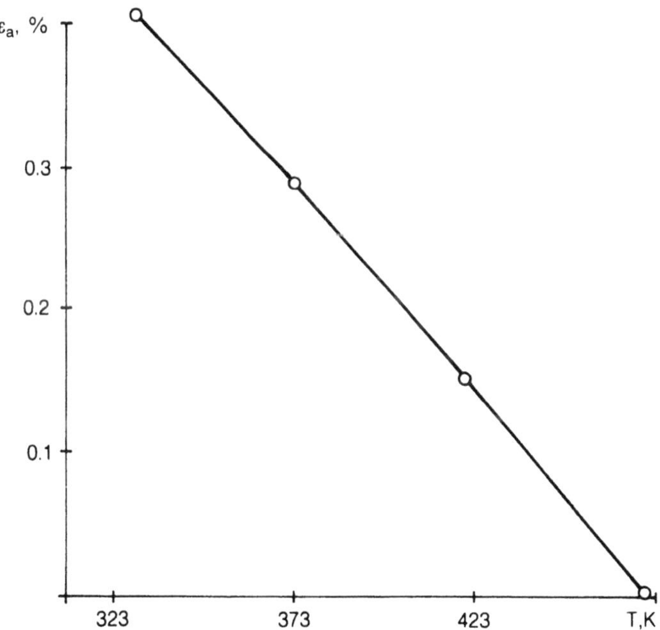

Fig. 7.8 Temperature dependence of secure values of vibrostrain amplitudes for SVM threads.

Under real service conditions, static loads are combined with vibrations [23]. Safety values of the amplitude of cyclic deformation component (ε_a), when the formation of irreversible deformations is excluded and moduli of elasticity and toughness–elasticity are constant, decrease monotonically with increase of temperature (Fig. 7.8). In the range of amplitudes covered by this curve, vibrations initiate irreversible deformations and decrease of equilibrium stress. Consequently, the vibration effect on fibres is analogous to the thermal effect in respect of degree of influence: internal self-heating of a specimen takes place. It was determined that destruction of polymer macromolecules does not occur under such action, and irreversible deformation at 293 K, increasing deformational stiffness, is determined by orientational rearrangement of structure.

To stimulate relaxation processes, contributing to the improvement of elastic strength properties of aramid fibres and composites based on them, one can use the joint action of static and vibrational loads on the reinforcing filler. As can be seen from the data given in Table 7.4, an increase of secant modulus of toughness–elasticity of aramid fibre accompanies the increase of vibration action amplitude ε_a; in this case preliminary vibration action practically does not affect the strength properties of the thread. In the case of combined action of vibration (ε_a from 0.15 to 0.88%) and static deformations (ε_{st} from 0.5 to 2.5%) on fibre SVM 29.5 tex, we observe a vibration effect, the essence of which consists of isothermal acceleration of the relaxation process. The vibration effect increases with increase of amplitude and is revealed during shift of the relaxation process towards short times (Figs. 7.9(a), (b)).

The polymeric limit-oriented structure of aramid fibres predetermines their principal difference from glass and carbon fibres, which is first of all in the relaxational nature of their strength and high sorption ability. In this connection, the most important task in the field of perfecting aramid fibres is a decrease of dimensional variability and provision of property stability during the process of service. This problem may be solved both by means of improvement of technology of production and processing and by perfection of aramid fibre structure.

Table 7.4 Effect of combined application of static and vibrational (ε_a) components of deformation[a] on value of secant modulus of toughness–elasticity E_e of aramid fibre (29.4 tex; $\sigma_b = 2.5\,\text{GPa}$, $\varepsilon = 3.6\%$)

	ε_a (%)				
E_e	0	0.27	0.38	0.57	0.88
$E_e \cdot$ (GPa)	40	42	46	50	55
ΔE_e (%)	–	5.0	15.0	25.0	37.5

[a] Action period – 16 min.

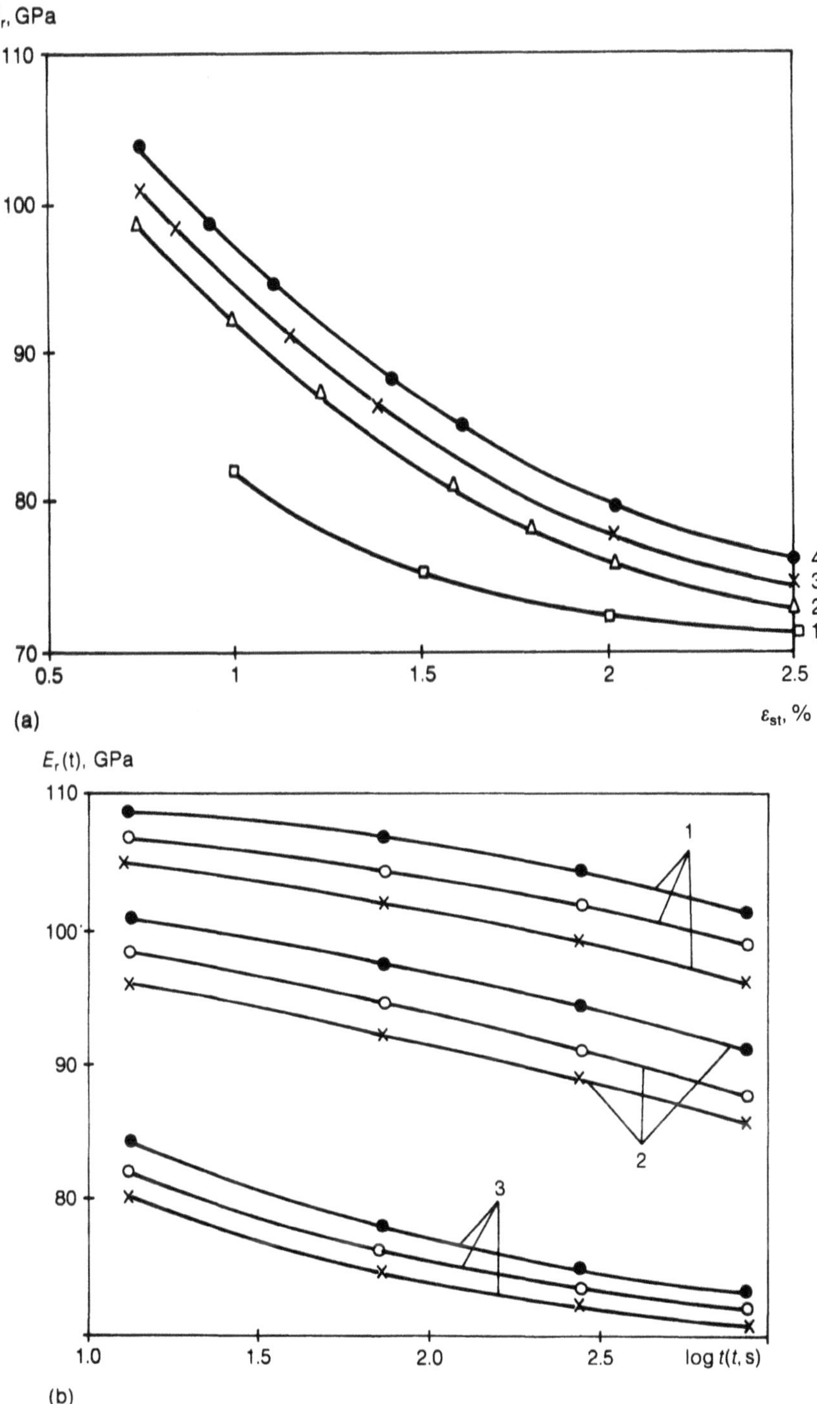

Fig. 7.9 (a) Isochronal value of relaxation modulus E_r as a function of residual strain at 293 K and different amplitudes of vibration action (%): (1) 0, (2) 0.15, (3) 0.25, (4) 0.88. (b) Relaxation modulus E_r at 293 K, for various amplitudes of vibration action (%): ($\bullet$) 0, ($\circ$) 0.15, ($\times$) 0.27; and different static strain ε_{st} (%): (1) 1.0, (2) 1.5, (3) 2.5.

7.3 THERMOREACTIVE BINDERS FOR ORGANOPLASTICS

There are two groups of requirements for binders for organoplastics: traditional requirements for binders for fibrous polymeric composites; and specific requirements dictated by the peculiarities of polymeric fibres as reinforcing fillers. The first group of requirements includes: good wetting power and adhesion to polymeric fibre; minimal shrinkage in the process of curing, which does not cause microcracks; high adhesive strength and elongation, which exceeds the elongation at break of the reinforcing fibres; comparatively high value of modulus of elasticity; and high elasticity and dynamic modulus of shear. An important technological requirement is also maintaining a long lifetime of binder in the prepreg composition; there is the possibility for quick curing at comparatively low temperatures (373–393 K) without releasing volatile products.

The second group includes requirements for binders determined by the polymeric nature of the reinforcing fibres. Aramid fibres are characterized by high chemical stability, but they may swell in some binder components, accompanied by considerable misorientation and decrease of elastic and strength properties, which is especially revealed under the action of water and steam on the material.

For production of structural organoplastics based on aramid fibres, most often used are epoxy matrices based on polyfunctional aromatic compounds forming close-joined heat-resistant networks under curing. Epoxy resins (epoxydiane, epoxynovolac, chlorine- or bromine-containing resins, etc.) are used as an epoxy component of such matrices. Amine-type compounds (diaminodiphenylsulphone, triethanolaminotitanate, anhydrides or latent hardener like complexes of boron trifluoride, modified urea, etc.) are usually used as hardeners (Table 7.5).

Taking into account the service and technological requirements, epoxy binders are modified by means of various additions of rubbers, poly(vinyl acetals), polysulphones, etc.

Epoxydiane binders with amine hardener are used for manufacturing articles by means of the method of wet winding of bundles, as well as for manual impregnation of cloths in layer-by-layer assembly of a stack. The preliminarily forced material made of this binder prepreg is characterized by excessive stickiness and low viability (2–3 days), and its application is limited by this fact.

Epoxyaniline phenol–formaldehyde binder is used in the form of a solution and gives the possibility to produce dry prepregs that may be conveniently used for manufacture of flat article stacks, first of all stacks of sheet materials. In the case of some decrease of drying temperature and increase of content of volatile products, it turns out to be possible to produce, based on this binder, elastic prepregs with satisfactory technological stickiness for assembling articles of intricate configuration. Viability of their prepregs is limited.

Epoxy binder with modified urea as a hardener in the form of solution or melt is used for impregnation of tapes and fabrics. The produced prepregs are characterized by a long period of viability and good technological properties in the case of assembly of large-scale part stacks, including those with layers of reinforcing fillers of various fibres (aramid, carbon, glass). The basic advantage of this binder is the lower curing temperature.

Table 7.5 Binders used for production of organoplastics

| Type of binder | Final temperature of curing (K) | Working temperature of material (service) (K) | Period of storage (days) | | Combustibility of organoplastic | Purpose |
			293 K	278 K		
Epoxydiane	433	−333 to 373	−	−	Combustible	Wet winding and manual lay-out
Epoxyaniline phenol−formaldehyde	438	−333 to 453	15	45	Self-extinguishing	Prepregs for sheet materials and cellular glued panels
Epoxy with modified urea	398	−333 to 353	90	45	Self-extinguishing	Prepregs for sheet materials, cellular glued panels and hybrid structural materials
Epoxyisocyanate	423−448	−333 to 373	30	90	Self-extinguishing	Prepregs for laminated materials and glue-free cellular panels
Epoxynovolac with aromatic diamine	453	−333 to 423	90	180	Combustible	Prepregs for laminated materials and cellular panels, including glue-free ones
Tetrafunctional epoxy with aromatic diamine	453−473	−333 to 453	30	60	Self-extinguishing	Prepregs for laminated materials, cellular panels and hybrid structural materials
Epoxypolysulphone (film)	443	−333 to 393	180	360	Combustible	Prepregs for laminated materials and cellular panels with high strength of joining
Modified phenol−formaldehyde	433	−333 to 373	30	−	Hardly combustible	Prepregs for laminated materials and glue-free volumetric constructions (cellular and knitted ones)

Epoxyisocyanate binder and modified phenol–formaldehyde binder are used for obtaining technological sticky prepregs based on fabric and knitted linen, and are used for manufacture of laminated and honeycomb constructions of aircraft interiors. Sufficiently high cohesive and adhesive strengths of these compositions give the possibility to manufacture cellular panels on their basis without additional application of adhesive to honeycomb faces.

Epoxynovolac binder (with aromatic diamine as hardener) is characterized by a smooth process of curing, which may be stopped practically at each stage up to 50–60% degree of conversion of binder polymer and continued after performing additional technological operations until the material is cured completely. This peculiarity of epoxynovolac binder makes it possible to use incompletely cured semifinished products and elements of structures in the case of assembly of complicated modular aggregates, including those in the combination of prepregs with adhesive films. Prepregs may be used for manufacture of cellular structures without the use of adhesive.

Tetrafunctional epoxy binder is used in the form of melt or solution. Various prepregs were produced based on it, including unidirectional structures of 'veneer' type of fibres and bundles. Organoplastics of sufficient heat resistance that self-extinguish under the action of a flame that are manufactured of such prepregs.

Prepregs of epoxysulphone film and epoxy rubber adhesive are used for manufacture of skins of cellular panels that are highly leakproof and have good strength of adhesion with fillers. The properties of organoplastics based on these binders and fabric of satin interweaving from SVM fibre are given in Table 7.6.

Table 7.6 Properties of organoplastics based on various binders[a] (fabric of satin interweaving 8/3 of SVM fibre, 14.3 tex, test temperature 293 K)

Binders	Density $(kg\,m^{-3})$	σ_b (MPa)	E (GPa)	σ_{-b} (MPa)	σ_{bb} (MPa)	τ^b (MPa)
Epoxy with modified urea	1300	700	30	220	450	30
Epoxyaniline phenol–formaldehyde	1300	730	32	220	470	83
Tetrafunctional epoxy with aromatic diamine	1340	730	38	300	530	40
Epoxyisocyanate	1250	650	34.3	180	370	30
Modified phenol–formaldehyde	1180	440	22.0	150	350	24
Epoxynovolac with aromatic diamine	1300	740	34	240	470	40

[a] Volume content of matrix 50–55%.
[b] Ultimate interlaminar shear strength.

Formation of organocloth laminates from prepregs based on the above-mentioned binders is performed under various specific pressures. Vacuum formation is usually used for manufacture of large-scale low-loaded constructions. This method is used, for example, in manufacturing of aircraft salon parts: ceiling, board, illuminators and similar panels, one-layer ones or with light fillers (honeycombs, knitted fabric).

Parts used for construction of external frames on aircraft are usually manufactured by means of the autoclave method under a specific pressure of 0.2–0.5 MPa. Organocloth laminates produced by pressing under high specific pressures (5–8 MPa) are usually used for flat load-bearing panels of floors and partitions. The effect of forming pressures on the properties of organocloth laminates is shown in Table 7.7.

Table 7.7 Effect of compression moulding pressure on organocloth laminate properties based on epoxyaniline phenol–formaldehyde binder and SVM cloth of satin structure 8/3, fibre 14.3 tex (fibre content 50%)

Compression moulding pressure (MPa)	Density (kg m^{-3})	σ_b (MPa)	E (MPa)	σ_{-b} (MPa)	σ_{bb} (MPa)	τ (MPa)
0.1	1200	550–650	27000	190	400	20
0.2	1250	650–700	28000	200	430	25
0.5	1300	700–720	32000	220	460	30
2.0	1350	730–750	35000	230	470	30
8.0	1370	800–820	40000	250	500	30

Table 7.8 Effect of binder on properties of unidirectional organoplastics (SVM fibre)

Binder	Method for manufacture of plastics	Density (kg m^{-3})	σ_b (GPa)	E (GPa)	σ_{-b} (GPa)	τ (GPa)
Epoxydiane with amine hardener	Wet winding: fibre 29.4 tex, bundle 600 tex	1280–1300 1260–1300	2.5–2.8 2.0–2.2	95 87	0.35 0.30	0.055 0.04
Epoxyaniline phenol–formaldehyde	Dry winding: prepreg tape, width 10 mm of fibre 29.4 tex	1280–1300	1.8–2.0	85	0.40	0.03
Epoxynovolac with aromatic diamine	Dry winding: prepreg tape, width 10 mm of fibre 29.4 tex	1280–1300	1.8–2.0	85	0.38	0.05

When manufacturing articles from unidirectional organoplastics by means of the method of winding, combination of reinforcing fibres with binder is performed directly before winding (wet winding), or preliminarily impregnated-type prepreg (dry winding) is used for winding. The properties of materials are given in Table 7.8.

7.4 PECULIARITIES OF THE PHYSICOCHEMICAL INTERACTION OF ARAMID FIBRES WITH THE COMPONENTS OF THERMOREACTIVE OLIGOMERIC BINDERS

7.4.1 Influence of the components of epoxy oligomeric binders upon the structure of aramid fibres

Realization of the properties and load-carrying ability of fibrous reinforcing fillers in composite material, level of service characteristics and their stability within the tailor-made period of storage and use of materials are determined by the character of the interphase interaction of components, specific properties and the structure of the 'fibre–matrix' boundary in composites. By analogy with polymeric composite materials based on impermeable glass and carbon fibres, the main attention in the case of studying the physicochemical processes of interaction of the components in organoplastics should be paid to wetting and capillary impregnation of fibrous filler with liquid binder.

As is obvious from Table 7.9, contact angles and thermodynamic work of adhesion of various polymeric fibres with epoxydiane binder are approximately similar, which is evidently bound up with the similarity of values of surface tension of adhesive and substrate, and also with the levelling role of water, which is sorbed by the fibre surface.

In the case of contact of an oligomeric binder with polymeric fibres, there is a possibility for swelling of the latter, followed by change of linear dimensions of the sample; the better the structural regularity of a fibre-forming polymer, the less it interacts with an active liquid medium (Fig. 7.10). On the other hand, the chemical

Table 7.9 Contact angles and thermodynamic work of adhesion W_a in 'polymeric fibre–epoxydiane binder' system

Fibre	Fibre diameter (μm)	Contact angle (deg)	$W_a \times 10^3$ ($N\,m^{-1}$)
Polycaproamide (Capron)	21	10	88.7
Poly(*m*-phenylene isophthalamide) (Fenilon)	15	17	87.5
Aramid			
Terlon	12	11	88.6
SVM	13	15	87.9

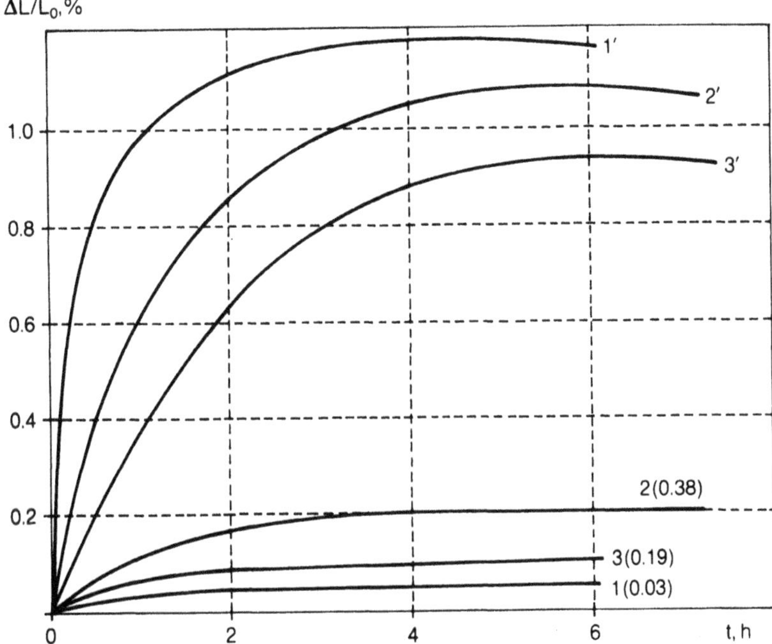

Fig. 7.10 Kinetic curves of axial swelling of polymeric fibres in components of epoxydiane binder EDT-10: (1) TEAT; (2) DEG-1; (3) ED-20 ($T = 363$ K); (1′), (2′), (3′), polycaproamide fibre; (1), (2), (3), aramid fibre SVM. Increase of fibre mass is indicated in brackets (%).

composition, structure and molecular mass of oligomeric product affect the intensity of fibre swelling in it. It is obvious from the example of epoxydiane binder with amine hardener (triethanolamine titanate (TEAT)) that the hardener causes the lower swelling of fibre, and diepoxyethylene glycol (DEG-1) is characterized by a noticeable effect on fibres.

Wide-angle X-ray diffractograms of SVM fibre are characterized by the presence of rather narrow meridional and blurred equatorial reflections and absence of 'inclined' reflections, which testifies to the high orientation of crystalline macromolecules, i.e. the liquid-crystalline (LC) structure of SVM fibre. Processing of fibre with components of epoxy binder causes an increase of intensity of meridional reflections, the angular positions and halfwidths of which are constant (Fig. 7.11). X-ray diffractograms taken in the equatorial direction were not subject to essential changes after processing with binder components. The increase of intensity of the meridional reflections may be determined by additional orientation of macromolecules in LC areas or a decrease of structural defectiveness, i.e. by increase of fraction of crystalline molecules. The average angle of misorientation of macromolecules in processed molecules is increased somewhat (Table 7.10) and consequently changes in X-ray diffractograms of SVM fibres may be caused by increase of LC content.

Noting the general features of the variation of structural parameters of SVM fibre

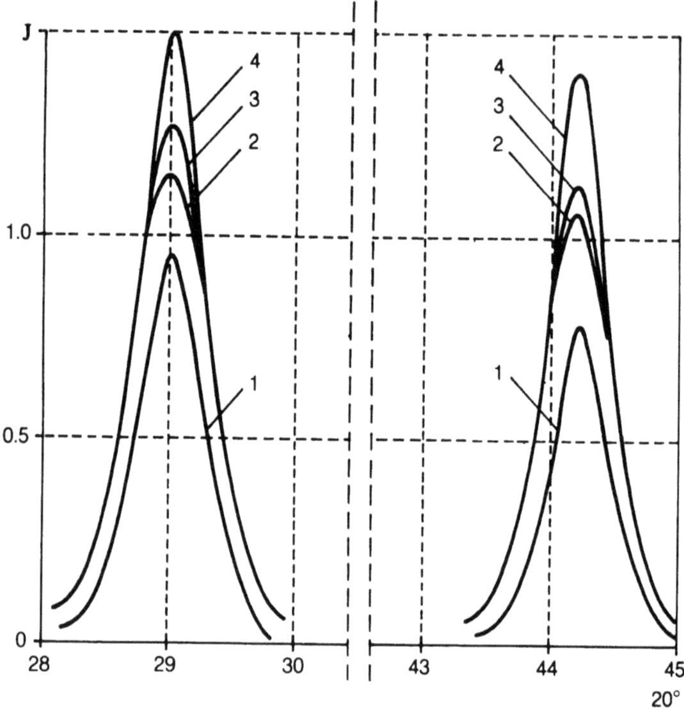

Fig. 7.11 X-ray diffractograms of initial SVM aramid fibre (1) and fibre treated with EDT-10 binder components: epoxydiane resin, ED-20 (2); diepoxyethylene glycol, DEG-1 (3); and triethanolamine titanate, TEAT (4). Meridional reflections are recorded and shown.

Table 7.10 Characteristics of X-ray diffraction of SVM fibre before and after treating with EDT-10 epoxydiane binder and its components

Specimen	Relative intensity of meridional reflection	Misorientation angle of fibre structure (deg)
Initial fibre	1.0	2.7
Fibre, treated with EDT-10 binder components:		
epoxydiane resin (ED-20)	1.2–1.3	3.4
diepoxyethylene glycol (DEG-1)	1.3–1.4	3.6
triethanolamine titanate (TEAT)	1.5–1.7	3.0
Fibre, treated with liquid		
binder EDT-10	1.0–1.1	2.9

during its processing with binder components, it can be seen that TEAT causes the maximum increase of regularity and minimum misorientation of macromolecules; the effects of ED-20 and DEG-1 on fibre structure are identical.

Thus the X-ray diffraction data testify to the ability of all components of epoxy binder to penetrate into SVM fibre volume and to the different degree of each component's effect on the fibre-forming polymer structure.

This conclusion is also confirmed by the estimate of variation of fibre mass after the action of binder components on it at temperatures below the gelation temperature of the binder.

So, in particular, after processing SVM fibre with EDT-10 binder components for 2 h at 373 K and subsequent extraction, it was found that the binder components are removed from the fibre incompletely – particles of incorrect form with dimensions 0.02 to 0.5 mm were found on its surface. DEG-1 and resin ED-20 are held by SVM fibre in large quantities, TEAT less so. The fibre surface is not fractured in this case, and its oriented structure remains.

The character of binder component interaction with SVM fibre at temperatures up to the beginning of devitrification of fibre-forming polymer was refined by the method of differential scanning calorimetry (DSC) under conditions of dynamic heating. On thermograms of initial SVM fibre and epoxydiane resin ED-20 there were no transformations found.

In the case of DEG-1 and TEAT during heating there were marked clearly expressed thermal effects: an exothermic process, actively passing in DEG-1 within the range of temperatures of 473 K, is bound up with reaction between epoxy and hydroxy groups; and an endothermic effect while heating TEAT in the range of temperatures of 373–453 K is determined by decomposition of associates, formed by TEAT molecules by means of hydrogen and coordinate bonds.

The thermograms of threads impregnated with binder components are of quite another form. In the case of SVM fibre combination with DEG-1, a small endothermic effect washed away within 373–453 K appears instead of an intense exothermic peak. The exothermic maximum of intensity, less than that for pure DEG-1, is shifted towards high temperatures by 60 K. Still more considerable changes took place in SVM–TEAT system. The endothermic process is completely absent; the exothermic process begins above 373 K and proceeds intensively after 473 K, with the maximum being at 483 K. Thermal effects were not noticed on thermograms of thread processed with ED-20 resin. This testifies to the fact that ED-20 diffusion is not accompanied by any essential energetic or chemical interaction. Essential changes in thermograms of TEAT and DEG-1 in combination with SVM fibre are evidently caused by interaction of active groups and atoms of these compounds with fibre-forming polymer, the macromolecules of which contain benzimidazole rings. TEAT is related to the class of tertiary amines, characterized by strong basic properties. That is why OH groups of TEAT can solvate NH groups of the imidazole rings and the tertiary nitrogen can form strong H bonds with them. To a certain extent, these comments also refer to NH groups of the amide group. From the DSC data it can be concluded that hydroxyl groups of DEG-1, as a result of their interaction with the active fibre

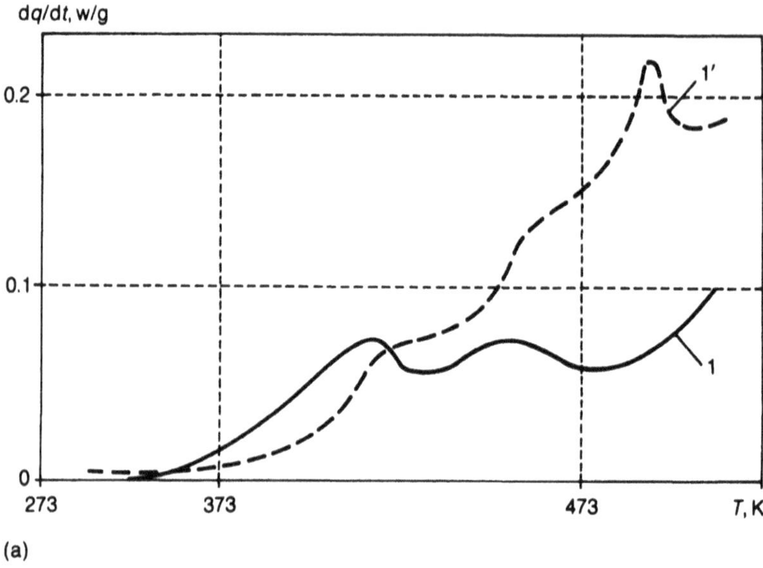

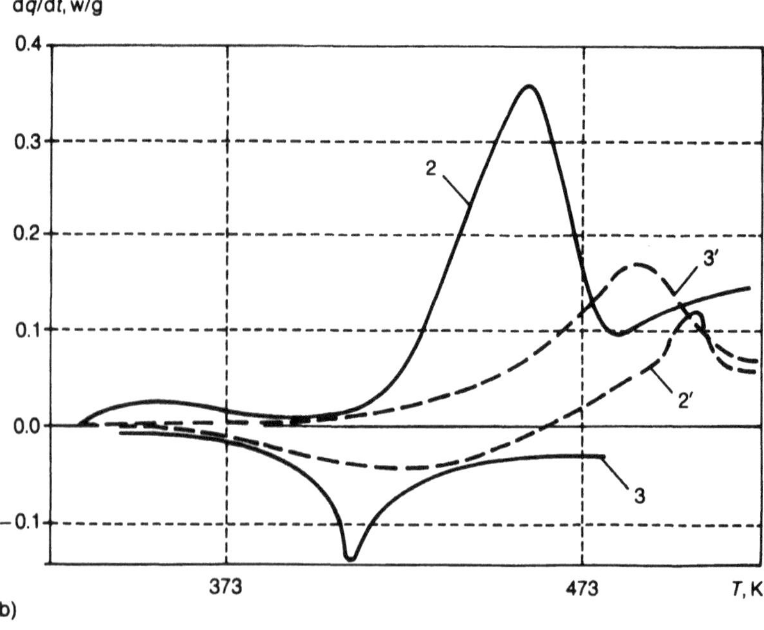

Fig. 7.12 DSC thermograms of (a) epoxydiane binder EDT-10 (1, 1′) and (b) its components DEG-1 (2, 2′) and TEAT (3, 3′): (1, 2, 3) binder EDT-10 and its components; (1′, 2′, 3′) composition based on combination of SVM fibre with EDT-10 (1′), DEG-1 (2′) and TEAT (3′).

groups, practically do not react with their own epoxy groups. This corresponds to a slight (3%) conversion of DEG-1 epoxy groups.

Thus DEG-1 is diffusing into fibres most actively, ED-20 resin somewhat less, and TEAT is characterized by minimal diffusion.

TEAT reveals the maximum activity in the process of interaction by forming strong hydrogen bonds, DEG-1 reveals less activity and ED-20 resin is completely inert to fibre. Binder components diffusing into the fibre and plasticizing it contribute to a decrease of internal stresses, conformational rearrangement, regulation of macromolecules and increase of LC content. Formation of various physicochemical bonds between fibre-forming polymer and binder components at high temperature may compensate the sequences of fibre plasticization and lead to intensification of intermolecular and interfibrillar interaction.

The results of SVM fibre processing with some components give the possibility to understand the character of the processes occurring during the manufacture of real material. The X-ray diffraction data obtained on microplastic testify to the insignificant effect of EDT-10 binder on the state of liquid-crystalline areas (see Table 7.10). When comparing thermograms of EDT-10 binder curing and its compositions with SVM fibre (Fig. 7.12), we revealed the additional exothermic process in the range of temperatures of 473–493 K, which may be explained by interaction of TEAT with polymeric fibre.

The complex diffusional–chemical interaction of SVM fibre with EDT-10 binder components makes it possible to explain the high realization of aramid fibre strength in combination with this binder, exceeding 100% limit in a number of cases.

Evidently the diffusional penetration of comparatively low-molecular-weight DEG-1 of linear structure into fibre favours relaxation of internal stresses on the core–shell interface and some regulation of fibre structure. The binder, on curing and interacting with the fibre, cements the thin surface layer and provides reliable adhesive connection of matrix and reinforcement. Different intensity of diffusion of components gives the possibility to assume the presence in the surface layer of fibres of a plasticized layer, including segments of molecules of binder components. This supposition is confirmed by results of investigation of thermodynamic compatibility of EDT-10 binder components with SVM fibre.

7.4.2 Thermodynamic interaction parameters of solutions of oligomeric multicomponent binders with aramid fibres

The depth and trend of physicochemical processes of interphase interaction in organoplastics depend to a high extent upon the primary contact of components at the stage of impregnation of fibrous filler and obtaining prepreg.

The study of diffusive–capillary phenomena in the case of organoplastic formation gives the possibility to determine kinetic parameters and thermodynamic features of the interaction of solutions of oligomeric binders with fibrous fillers, as well as to use them for optimizing the composition, structure and technological parameters for obtaining prepreg and organoplastic based on it.

The 'fibrous filler–solution of binder' system may be considered as thermo-dynamically multiphase, the compatibility of the individual components of which may be determined on the basis of general thermodynamic criteria of compatibility.

Admissibility of this statement is determined by the fact that the process of impregnation for reinforcing aramid fibre filler with alcohol–acetone solutions of epoxy binders is determined by mass transfer of binder components in fibre volume, the process being practically reversible at the stage of impregnation, and under concrete conditions the distribution of binder components among the fibre and solvent phases reaches equilibrium.

To estimate the interaction of polymeric fibrous filler and binder solution at the stage of impregnation, the following effective parameters have been chosen:

1. Thermodynamic affinity, $-\Delta\mu^0$ (kJ mol^{-1}), between binder component and fibre, characterizing the motive power of component diffusion into fibre volume

$$-\Delta\mu^0 = RT\ln(C_b/C_s) \tag{7.1}$$

2. Heat of interaction process, $-\Delta H^0$ (kJ mol^{-1}), characterizing the strength of the bond of a sorbed molecule with fibre

$$\Delta H^0 = \mathrm{d}(\Delta\mu^0)T/\mathrm{d}(1/T) \tag{7.2}$$

3. Entropy of interaction process, $-\Delta S^0$ (J mol^{-1}K^{-1}), characterizing the degree of orientation and compression of sorbed molecules in the case of localization of them in fibre in comparison with solution

$$\Delta S^0 = (\Delta H^0 + \Delta\mu^0)/T \tag{7.3}$$

Here R is the universal gas constant, T is absolute temperature, and C_b and C_s are the concentrations of binder in fibre and solution at the moment of obtaining equilibrium.

The above-mentioned thermodynamic parameters are calculated on the basis of analysis of kinetic curves of sorption of this substance by fibre. Coefficient of diffusion (D) and activation energy of process (E) are calculated by these curves, as well as concentrations C_b and C_s in equation (7.1). The obtained thermodynamic ($\Delta\mu^0$, H^0, S^0) and kinetic (D, E) characteristics give the possibility to describe the process of interaction of binder component with polymeric fibres sufficiently fully and to make the required conclusions about the usefulness of one or other binder formula for organoplastics.

This method was used for determining thermodynamic and kinetic interaction parameters of components in organoplastics, covering a wide range of substances used at present for obtaining thermoreactive binders of epoxy type.

Studies have showed [24–26] that the epoxy resins and hardeners most often used at present in composition of binders of epoxy type are placed in a certain order in terms of the value of their thermodynamic affinity to the reinforcing fibre. The position of one or other product in this order depends upon the presence of functional groups in its structure that are able to interact with active centres of fibre-forming polymer. For example, the molecule of diepoxyethylene glycol resin DEG-1, being a flexible aliphatic chain, has low thermodynamic affinity with fibre, and high rate of diffusion and depth of penetration into aramid substrate (Table 7.11). The presence of phenyl rings in epoxydiane resin molecules determines their high stiffness, and decreases the rate of diffusion and depth of penetration, but thermodynamic affinity with fibre is greater.

Triethanolamine titanate, a hardener of amine type, the molecule of which has a complicated structure and contains a large quantity of active hydroxyl groups, is characterized by high indices of thermodynamic affinity between heat and entropy of interaction process; in this case the rate of diffusion and depth of penetration of this product are minimum owing to steric hindrance. Even higher affinity to aramid fibre is typical for phenol–formaldehyde resins, which are often used as hardeners of epoxy composition (see Table 7.11).

In multicomponent oligomeric binders, the kinetics of mass transfer in the system 'fibre–mixture of oligomers' is affected by the interaction of binder components not

Table 7.11 Thermodynamic and kinetic characteristics of process of interaction of epoxy binder component solutions with aramid fibre SVM

Components of binder	$-\Delta\mu°$ $(kJ\,mol^{-1})$ at $T = 293\,K$	$-\Delta H°$ $(kJ\,mol^{-1})$	$-\Delta S°$ $(J\,mol^{-1}\,K^{-1})$	Diffusion coefficient, $D \times 10^{-14}$ $(m^2\,s^{-1})$	Depth of penetration[a] (μm)
Epoxydiane binder EDT-10					
Epoxydiane resin (ED-20)	0.82	1.78	3.20	13.6	50 ± 5
Diepoxyethylene glycol resin (DEG-1)	0.18	0.73	1.87	17.0	70 ± 5
Triethanolamine titanate (TEAT)	1.17	2.75	5.39	3.5	20 ± 5
Epoxyaniline phenol–formaldehyde binder 5-211B					
Epoxydiane resin (ED-20)	0.82	1.78	3.20	13.6	50 ± 5
Epoxydiane tetrabromide-containing resin	1.19	2.75	5.32	1.2	20 ± 5
Aniline phenol–formaldehyde resin	1.34	3.53	7.47	3.5	30 ± 5

[a] Comparative depth of component penetration was determined on films cast from fibre-forming polymer.

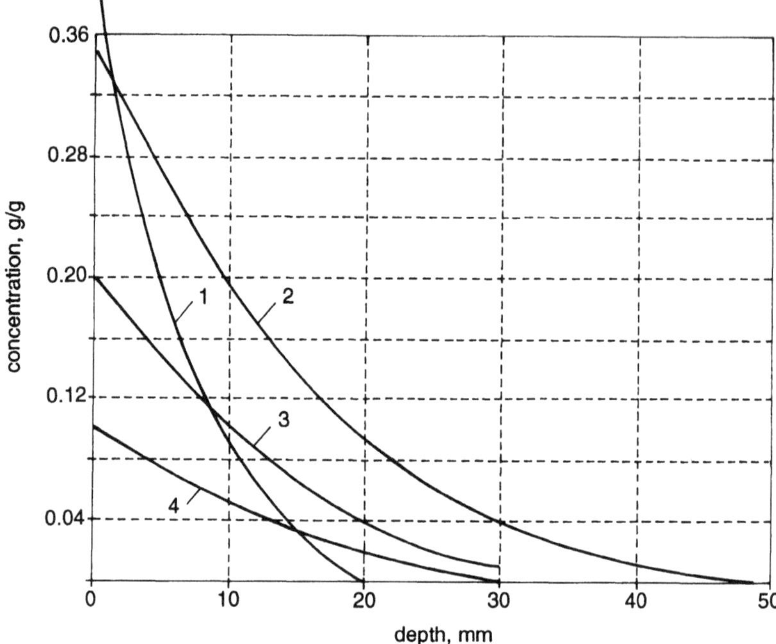

Fig. 7.13 Curves of distribution of triethanolamine titanate (TEAT) concentration with depth of stack of aramid films; adsorption from 5% solution ($T = 293$ K): (1) TEAT; (2) mixture of TEAT with diepoxyethylene glycol, DEG-1 (1:1); (3) mixture of TEAT with DEG-1 and epoxydiane resin ED-20 (1:1:9); (4) mixture of TEAT with ED-20 (1:9).

only with fibre but also between themselves. Figure 7.13 contains the curves of distribution of epoxydiane binder components in a multilayer stack of films, produced of fibre-forming polymer. As is obvious, the rate of diffusion and depth of TEAT penetration in the presence of aliphatic epoxy resin DEG-1 are the maximum ones. The low affinity of DEG-1 to fibre-forming polymer ($-\Delta\mu^0 = 0.18$ kJ mol^{-1} at 293 K) shows a trend to subsequent quick decreasing on negligible elevation of temperature ($-\Delta\mu^0 = 0.13$ kJ mol^{-1} at 313 K), favours lowering of adsorption interaction of TEAT and aids its transition inside fibre-forming polymer in the process of mutual diffusion. In this case molecules of DEG-1 perform the role of 'carrier' of hardener molecules, decreasing the level of residual stress and helping to increase realization of the properties of the fibre in composite material.

When choosing a formula of multicomponent oligomeric binder for organoplastics, it is important that the indices of thermodynamic affinity of some components of binder are similar and these components are adsorbed by fibre at an equal rate and in the same proportion in which they were in the original solution. Otherwise the component characterized by high affinity and low rate of diffusion inside the fibre will be easily adsorbed by the surface layer of the fibre. In contrast, the component characterized by low compatibility is insufficiently held by physicochemical structures of fibre and penetrates to sufficient depth or even completely fills the free volume of fibre. In those cases in which the binder composition includes components whose interaction with fibre is described by these end schemes, there is the possibility for development of complex processes, the kinetics of which will be determined by thermodynamic affinity of binder components. So, for example, a component adsorbed by the surface layers of fibre may prevent diffusion of a more mobile incompatible component inside the fibre.

The most dangerous is the situation when the process of mass transfer and distribution of concentrations causes variation of binder composition, as a result of which technological properties deteriorate, as well as kinetics and fullness of curing of binder in the free state. It is obvious that the principle of additivity, developed in examples of heteronymic polymeric composites, cannot be applied to organoplastics. Actually, studies showed that there is no unambiguous correlation between the matrix vitrification temperature in bulk and the α transition temperature into organoplastic; compared with matrix vitrification temperature the latter may be shifted towards either lower or higher temperature. The value of this shifting amounts to $60-100°C$ for hardeners of catalytic type, $40-50°C$ for anhydride hardeners and $0-20°C$ for hardeners of aromatic diamines.

The effect of partial evacuation of binder components from reaction volume and variation of kinetics and completeness of composition curing accompanying this process are reflected in the service properties of the material. So, for example, modifier and hardeners of type epoxyfurfurolic ether (EFU) and aminoethoxyphosphazene (AETPh) are characterized by low affinity with SVM fibre ($-\Delta\mu^0 = 0.1$ and $0.23\,kJ\,mol^{-1}$ respectively), owing to which they penetrate easily into fibre, decreasing the quantity of hardener taking part in curing of composition.

Deterioration of binder formula and plasticization of fibre by adsorbed low-molecular-weight substances is followed by increase of water adsorption and decrease of stability of properties in the case of moistening organoplasics based on them (Table 7.12, Figs. 7.14 and 7.15).

Table 7.12 Basic properties of organocloth laminate based on epoxy-type binders of varying thermodynamic compatibility with SVM fibre

| Binder | $-\Delta\mu$ (kJ mol^{-1}) of hardeners | Strength (MPa) | | | | Water sorption, for 24 h (%) |
		σ_b	σ_{-b}	σ_{bb}	τ	
Epoxynovolac with modifier EFU	0.1	650	170	350	27	1.1
Epoxyaniline phenol–formaldehyde	1.34	700	220	460	35	0.4

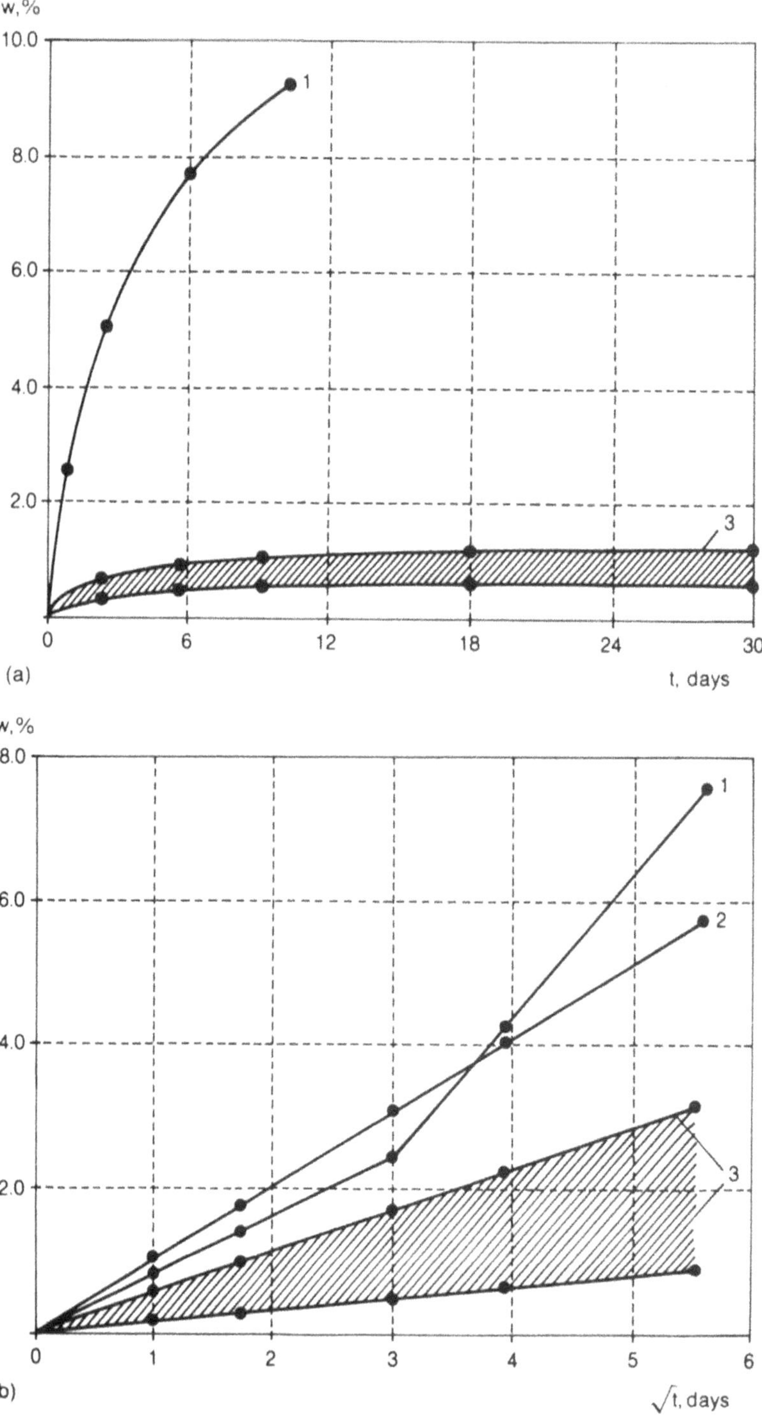

Fig. 7.14 Kinetics of moisture (a) and water (b) sorption of organocloth laminates, based on SVM fibre and epoxy binders with hardeners and modifiers: (1) aminoethoxyphosphazene (AETPh); (2) epoxyfurfurolic ether; (3) range of properties of organoplastics used in serial production.

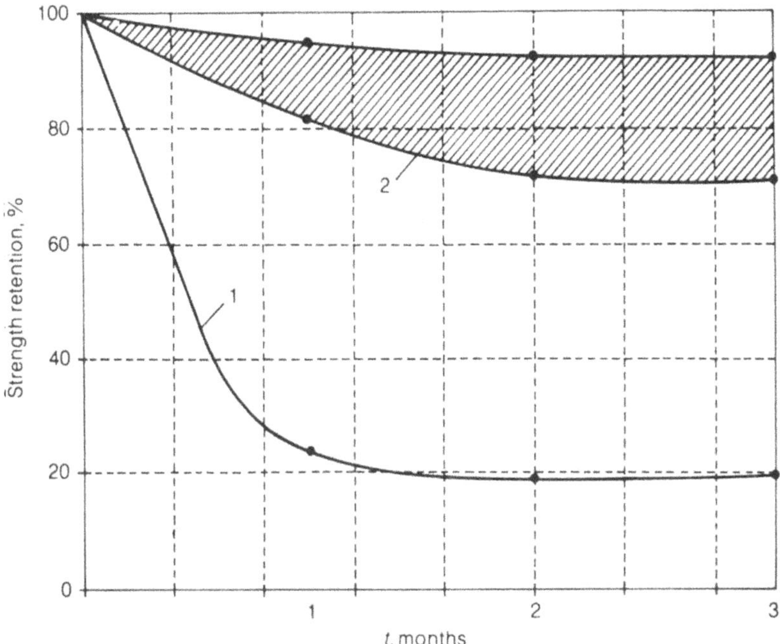

Fig. 7.15 Preservation of tensile strength of organocloth laminates during holding in a tropical climate chamber: (1) organocloth laminate based on epoxy binder with hardener AETPh; (2) property range of organocloth laminates used in serial production (specimens without varnish and paint protection).

On the other hand, binder components that penetrate into aramid fibre in the liquid-crystalline state may change its viscoelastic properties and structure (Fig. 7.16). The temperatures of beginning (T_1) and end (T_2) of the devitrification process, which are equal for the original SVM fibre to 511 and 551 K, respectively, are decreased to low temperaure by 25–45 K for fibre placed in a composition of organoplastics based on epoxy binders of various types (Table 7.13). The increase of moulding pressure of organoplastic increases the degree of binder penetration into fibre, which in turn causes additional decrease of characteristic temperatures of SVM fibre in organoplastic composition. Application of thermoplastic polymeric matrix does not affect variation of the range of temperature of SVM fibre devitrification in the composite material, which emphasizes the specific role of diffusion processes that take place in the case of using low-molecular-weight oligomeric binders.

Diffusion of epoxy binder components into SVM fibre is followed by the decrease of dynamic shear modulus (G_d) of fibre in organoplastic composition. For example, at 450 K the G_d value of monofibre SVM is equal to 0.98 GPa, and for fibres with different penetration degree of epoxy binder components G_d is decreased to 0.33–0.52 GPa.

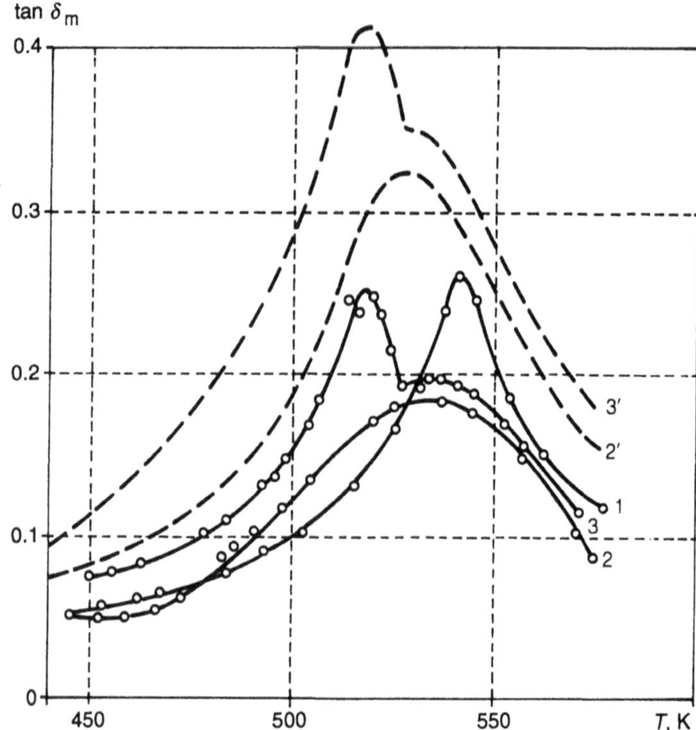

Fig. 7.16 Temperature curve tan δ_m (mechanical for SVM filament (1) and organoplastic based on epoxyaniline phenol–formaldehyde binder (2,3). Broken curves are calculated dependence tan $\delta_m = f(T)$ for fibre placed in an organoplastic composition (by additivity). Moulding pressure for organoplastic is 0.5 MPa (2,2') and 10 MPa (3,3').

Decrease of devitrification temperature range and the temperature of the α peak on the temperature curve of the mechanical loss tangent, and decrease of dynamic shear modulus for SVM fibre under the effect of components of oligomeric binders – all these changes may be explained by molecular and structural plasticization of SVM fibre in organoplastic composition.

7.5 PECULIARITIES OF THERMOREACTIVE BINDER CURING IN THE PRESENCE OF ARAMID FIBRES

The presence of active groups on the surface of aramid fibres makes it possible to relate this type of reinforcing filler to the category of catalytically active components of composite materials. Figure 7.17 shows DSC curves for the process of catalytic curing of binder containing epoxydiane resin in the presence of aramid fibre SVM (curve 1) and non-filled binder (curve 2). Curing is a very complicated process. On the DSC curve the process of curing of non-filled binder is accompanied by two partially overlapping exothermic peaks with maxima at temperatures of 391 and 427 K (when

Table 7.13 Characteristic temperatures of aramid fibre SVM in initial state and in organoplastic composition

Polymeric matrix	Regime of organoplastic formation	Temperatures of devitrification processes (K)		Temperature (K) of peak on curve tan $\delta_m = f(T)$
		Beginning, T_1	End, T_2	
Filament				
without matrix	–	511	551	538
Epoxy novolac	433 K, 0.5 MPa, 4 h	504	541	531
Epoxydiane modified with diepoxyethylene glycol	433 K, 0.5 MPa, 4 h	492	535	521
Epoxyaniline phenol–formalde-hyde	433 K, 0.5 MPa, 4 h	496	537	525
	433 K, 10 MPa, 4 h	466	527	512
Polycarbonate	503 K, 1.5 MPa, 4 h	511	551	538

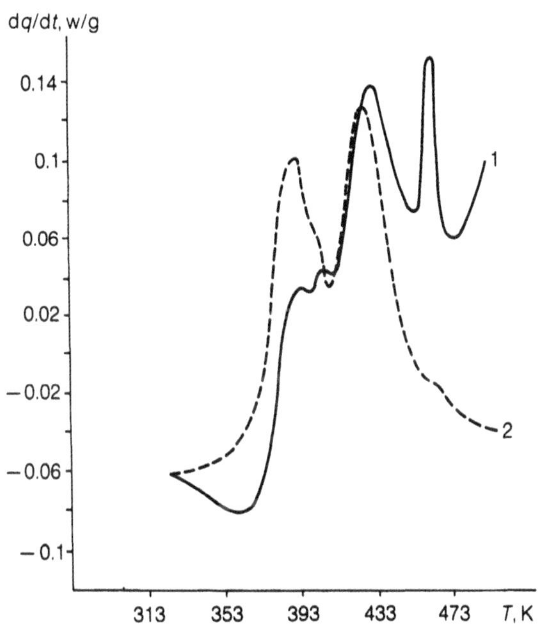

Fig. 7.17 DSC curves for the process of cure of epoxy resin in the presence of aramid fibres SVM (curve 1) and non-filled (curve 2). Heating rate, 2 K min^{-1}.

the heating rate is $2\,\mathrm{K\,min^{-1}}$). In the presence of aramid fibre SVM as a filler, the first peak intensity is considerably decreased, the second maximum remains and an additional exothermic peak with maximum at a temperature of 463 K appears. These changes testify to the aramid fibre surface effect on epoxy binder curing process. Depending upon the composition of the cured system, the character and the degree of aramid fibre surface effect on curing process and cured matrix properties may be different. Especially, this depends essentially upon the content of diluents used, as well as upon the quantity and the effectiveness of curing catalysts. In a number of cases the curing process is affected by moisture contained in the filler, and therefore it is necessary to pay serious attention to drying aramid fillers before impregnation with binder. The presented DSC thermograms illustrate the difference in character and degree of SVM aramid fibre surface effect on processes of curing of various thermoreactive epoxy binders.

Figure 7.18 contains DSC thermograms of the process of curing a binder, containing a mixture of various epoxy resins, with an aromatic diamine without any accelerators in the presence of filler (curve 1) and in pure form (curve 2). In this case the effect of aramid fibres on the curing process is not observed. The form and temperature characteristics of the exothermic peak on the DSC curve are not changed on adding aramid fibres to the cured system.

A more complicated situation is observed when using highly effective accelerators

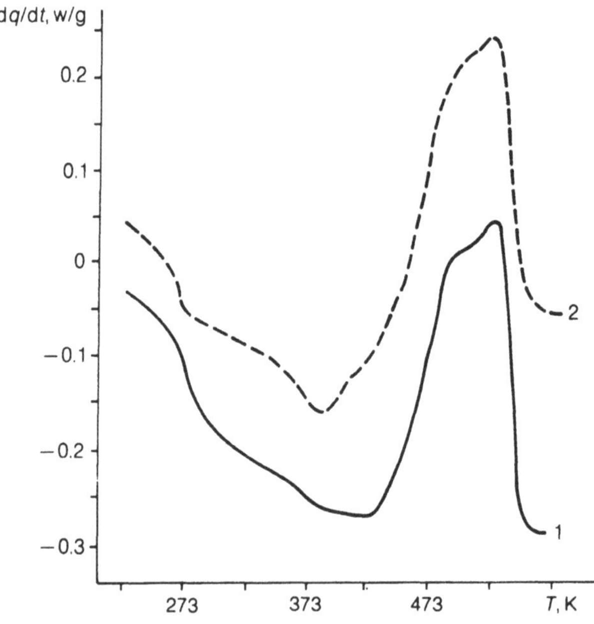

Fig. 7.18 DSC curves for the process of cure of a binder, containing a mixture of epoxy resins, with aromatic diamine: (1) in contact with SVM fibres; (2) non-filled compositions. Heating rate, $10\,\mathrm{K\,min^{-1}}$.

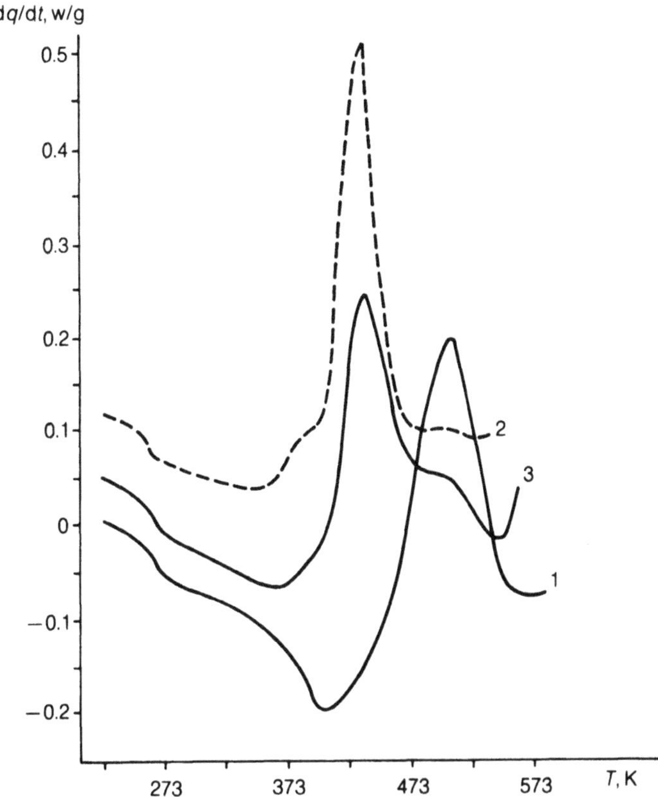

Fig. 7.19 DSC curves for the process of cure of a binder, containing epoxy resin, with aromatic diamine with accelerator: (1) in contact with SVM fibres; (2) non-filled compositions; (3) in contact with SVM fibres whose surface is preliminarily treated with a protective composition. Heating rate, $10 \, \text{K} \, \text{min}^{-1}$.

of cure. Figure 7.19 shows DSC curves of the process of epoxy resin curing with aromatic diamine containing a small quantity of amine complex of Lewis acid as an accelerator. In the presence of filler (aramid fibres SVM) the temperature of the exothermic peak maximum of the curing process (curve 1) is shifted to high temperatures by more than 70 K comparing with non-filled composition (curve 2), i.e. the action of curing catalyst (especially at the stage of gelatinization) is completely blocked by active groups of filler and is noted only at the final stage of cure. This effect may be partially or fully eliminated by means of preliminary treatment of aramid fibre surface. One may see on the DSC thermogram of binder curing process in the presence of preliminarily treated aramid fibres (Fig. 7.19, curve 3) that the activity of curing process at the gelatinization stage remains at the same level as for non-filled binder.

The opposite effect is observed in the situation when aramid fibre, used as reinforcing filler, increases the activity of the curing composition. Figure 7.20 contains

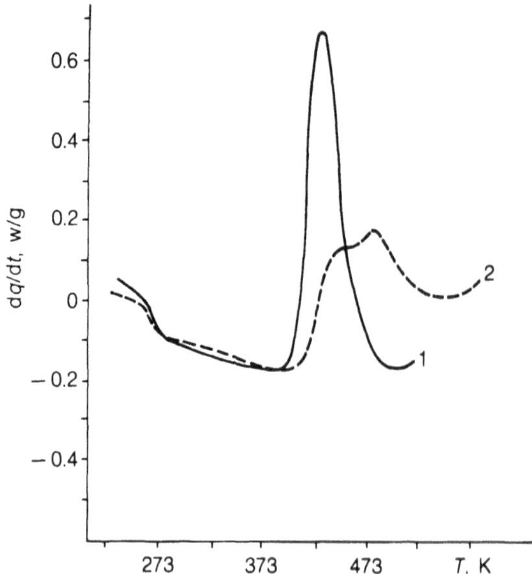

Fig. 7.20 DSC curves for the process of catalytic curing of a binder, containing a mixture of diane and epoxytriphenol resins, in the presence of aramid SVM fibres (curve 1) and non-filled (curve 2). Heating rate, $10\,\mathrm{K\,min^{-1}}$.

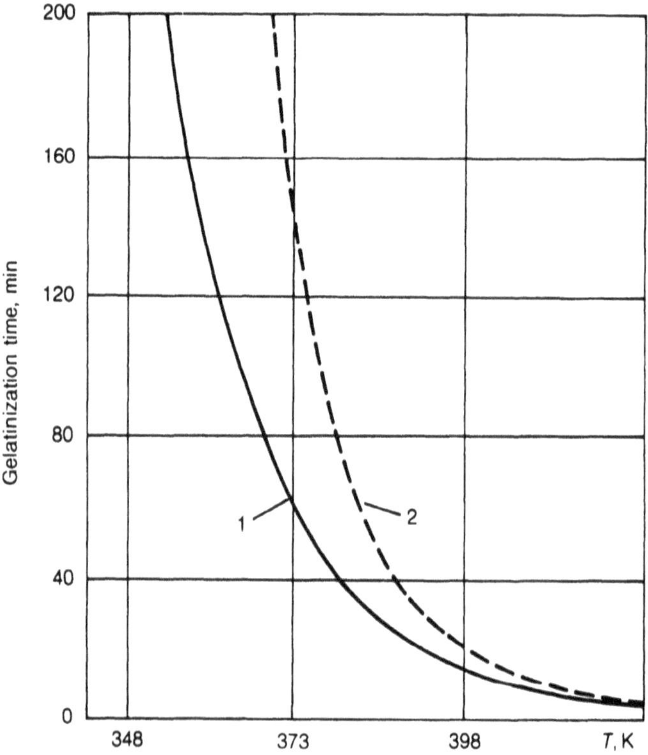

Fig. 7.21 Temperature dependence of gelatinization time of binder based on diane and epoxytriphenol resins in the presence of SVM fibre (curve 1) and non-filled binder (curve 2).

DSC curves of the curing process for a binder, containing a mixture of epoxydiane and epoxytriphenol resins, cured with a product based on urea as initiator, in the presence of SVM fibre (curve 1) and when pure (curve 2). When adding aramid fibre to the curing system, the cure process intensity increases. It is well seen also when comparing the temperature dependence of gelatinization periods of prepreg and non-filled binder (Fig. 7.21).

Thus the kinetics of the curing process of polymeric matrices on aramid fibre surface depends considerably upon the character and the depth of physicochemical interaction of binder components with fibre-forming polymer [27, 28].

7.6 ADHESIVE INTERACTION IN POLYMERIC FIBRE–POLYMERIC CURED MATRIX SYSTEMS

It is evident from previous sections that physicochemical interaction of components at all stages of organoplastic formation is complex and differs principally from that for other polymeric composites (fibreglass and carbon plastics) by the presence of processes of mass transfer between components and by active mutual effect of fibre and matrix, propagated for considerable part of their volume.

To understand this problem, one should consider fundamental works in the field of studying adhesive interaction of polymer–polymer compositions. Such works include the scientific papers of a number of Russian scientists: S.S. Voyutsky, B.V. Deryagin, Yu. S. Lipatov, V.N. Kulezniev, A.A. Berlin, V.E. Basin, E.B. Trostylanskaya, Yu.A. Gorbatkina, *et al.* [29–38]; and those of other scientists: D. Paul, S. Newman, E. Plueddemann, *et al.* [39, 40].

As long ago as the 1960s S.S. Voyutsky showed the role of chain molecules or their sections in the formation of an interphase layer, causing the disappearance of a clear boundary between an adhesive and a substrate and the formation of a strong bond due to maximum molecular contact. Diffusion of adhesive into substrate changes in principle the character of adhesive interaction of polymers: in the case of mutual diffusion to a depth of 5–8 Å the molar contact area may be increased by 3–5 times.

The role of adhesive phenomena is most essential during formation of adhesive interaction of compatible polymers. However, noticeable diffusion may also be observed for incompatible polymers that are sharply different in their polarity. A great contribution to the development of these ideas was made by V.N. Kulezniev, who showed the role of segmental solubility of contacting polymers in formation of a diffusion layer [34, 35]. Conclusions about the compatibility of the majority of polymers at the level of segments, even in the case of their incompatibility at the macromolecular level, have a great value for consideration of adhesive interaction of the components in organoplastics.

However, in this case it should be noted that the complication of the physico-chemical interaction of a multicomponent oligomeric curing composition with a polymeric fibre will evidently not permit at present the use of any of the numerous theories of adhesion for a detailed description of adhesive contact in organoplastics. A fuller picture for studying this problem may be obtained based on the main statements of a number of theories, which do not contradict, but supplement, each other.

Investigation of adhesive interaction in organoplastics showed that, in contrast to polymeric composites based on impenetrable fibres (glass, carbon, etc.), the interphase interaction of components in organoplastics differs in the fact that formation of a traditional boundary layer on the surface of an impenetrable fibre in this case gives way to a volumetric interaction of components in interphase diffusion area. As a result, as shown in sections 7.4 and 7.5, there is the possibility for essential changes of chemical composition of the matrix, depth of cure and structural orientation of the surface of maximum oriented penetrable polymeric sublayer (which is aramid fibre) due to selective sorption and diffusion of binder component into fibre.

On the other hand, absorptive–diffusive and chemical processes, accompanying production of semifinished products (prepregs) and formation of polymer–polymer system during curing, may cause misorientation of structure and decrease of elastic strength properties of reinforcing fibres under the action of the components of a thermoreactive oligomeric binder. The temperature and humidity of the surrounding medium, as well as the duration of preliminary contact of components in the non-cured state, essentially predetermine the intensity of their interaction and the degree of difference of component properties in the composite material and in the free state. All these processes directly affect the strength of adhesion and realization of properties of components in composites, and consequently the elastic strength and service characteristics.

The study of adhesive strength of aramid fibres with polymeric matrices of various chemical composition was carried out by use of methods suggested by Yu.A. Gorbatkina and G.S. Shul. The methods are based on measurement of elementary fibre shear force relative to a layer of cured matrix, connecting the fibre under test with two boron fibres – binder carriers [37, 41–43].

Structural investigations of the fracture zone showed that, besides the purely adhesive character of fracture, we often meet specimens that have fibrillar formations near the edge of the hole, formed after drawing out the tested fibre from the matrix. Those formations are residues of aramid fibre shell. Evidently in the case of action of shear stresses, the adhesive strength of structural fragments of fibre shell with polymeric matrix is in some cases sufficiently high to cause pull-out of a part of the surface layer of fibre. The studies showed that the character of the interphase zone of fracture depends upon the composition and the structure of the fibre, as well as upon the intensity of physicochemical interaction of fibre with matrix (Fig. 7.22).

Removal of aramid filaments from the cured matrix is accompanied by their splitting in the axial direction, which brings to mind the character of fracture of these fibres during the strength test.

The strength of adhesion of crystalline aramid fibre Terlon with epoxy polymers of various compositions is intermediate between the higher-strength adhesion of SVM aramid fibre and lower-strength adhesion of glass fibre (Table 7.14). The structure of SVM fibre does not include tightly packed crystalline formations, which makes it more accessible for development of adsorptive–diffusive processes, contributing to

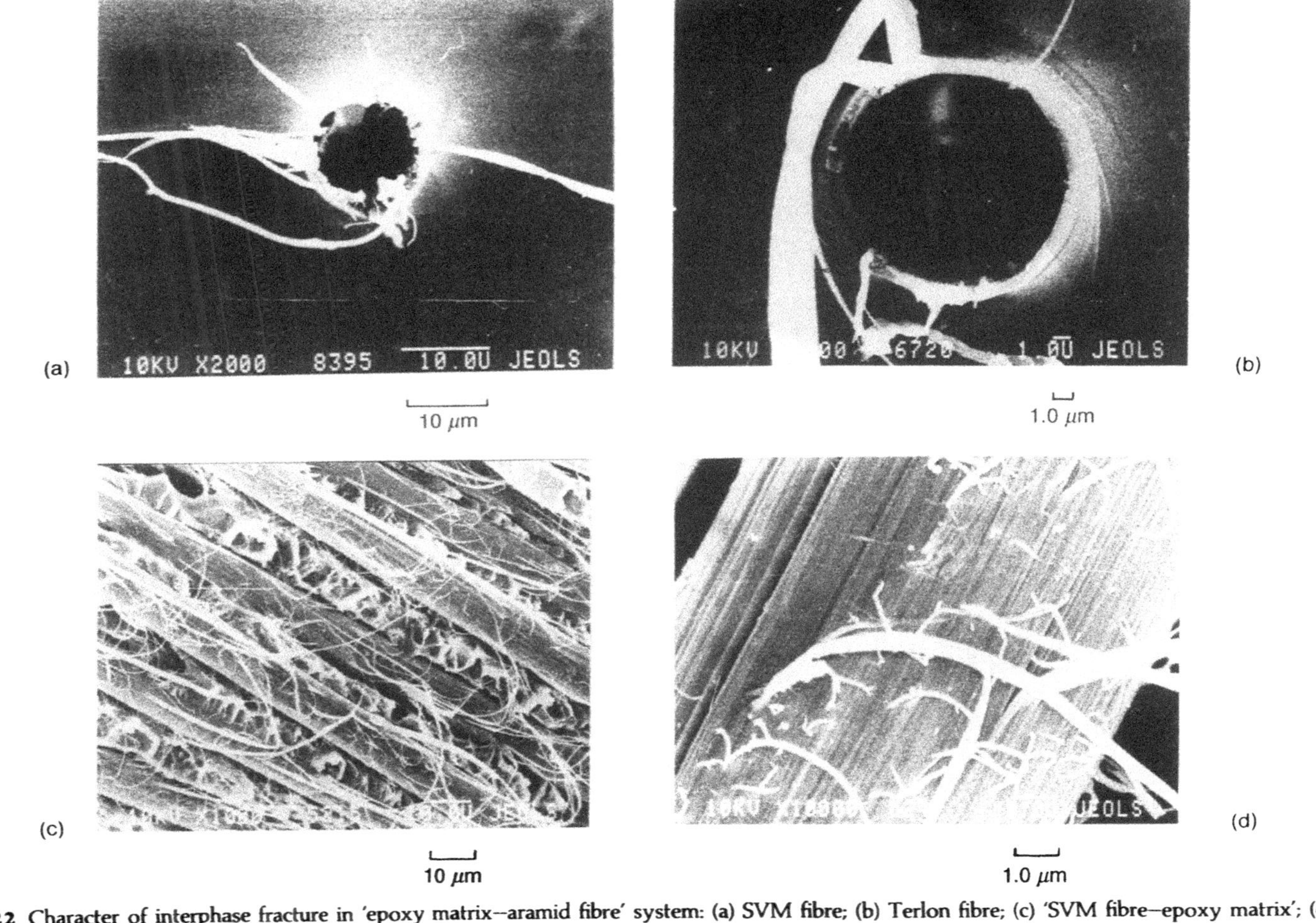

Fig. 7.22 Character of interphase fracture in 'epoxy matrix–aramid fibre' system: (a) SVM fibre; (b) Terlon fibre; (c) 'SVM fibre–epoxy matrix'; (d) SVM fibre on rupture zone of composite.

Table 7.14 Strength of adhesion of epoxy matrix with aramid fibres of various compositions

		Number of fractures		Strength of adhesion, for $S = 6 \times 10^{-3}\,mm^2$
Fibre	*Matrix*	*Adhesive*	*Cohesive*	*(MPa)*
Aramid SVM	Epoxydiane with amine hardener	40	50	57
	Epoxyaniline phenol–formaldehyde	41	59	59
	Epoxynovolac with aromatic diamine	37	54	52
Aramid Terlon	Epoxydiane with amine hardener	130	46	48
Polycaproamide	Epoxydiane with amine hardener	39	23	24[a]
Glass	Epoxydiane with amine hardener	109	122	43

[a] Gluing area $S = 11 \times 10^{-3}\,mm^3$.

propagation of the zone of interphase interaction and increase of adhesion with polymeric matrices. The value of adhesive strength of polymeric fibres with epoxy polymers not only depends upon physicochemical interaction between them: more active in this respect, but less strong, polycaproamide fibre is characterized by lower strength of adhesion with epoxy matrices, compared with strong aramid fibres.

The adhesive strength of various aramid fibres with polymeric matrices correlates with the strength of organoplastics based on them. Under similar conditions the compressive and shear strengths of organoplastics, reinforced with SVM fibres, amount to 215 and 34 MPa respectively. When reinforcing the compositions with less active crystalline Terlon, the strength of components adhesion is decreased, which causes a respective decrease of compressive strength of composite to 170 MPa and shear strength to 21 MPa (Table 7.15).

As already noted above, the kinetics of physicochemical interaction in 'aramid fibre–oligomeric binder' system depends upon the temperature and duration of contact between the components until accomplishment of process of curing.

Application of various epoxy compositions showed that, in the case of increase of duration of prepreg storage, one may observe a trend to increase of strength of aramid fibre adhesion with the matrix: in 10–14 days the process of structuring of binder in the prepreg composition reaches saturation.

An increase of prepreg storage period above a certain critical value, owing to a decrease of its viability, may be accompanied by deterioration of the mechanical properties of a composite, and first of all its interlaminar strength while keeping high strength of adhesion of components in interphase 'fibre–matrix' area (Table 7.16).

Table 7.15 Effect of nature and strength of adhesion of aramid fibres with epoxy matrix on mechanical properties of organoplastics

	Fibres		
		SVM	
Properties	*Terlon*	*Initial*	*Sized*
Strength of adhesion (MPa)	48	56	63
Compressive strength (MPa)	170	215	260
Shear strength (MPa)	21	34	46
Realization of tensile strength (%)	67–75	85–95	95–110

Table 7.16 Effect of prepreg storage duration on strength of adhesion of components and mechanical strength of composition based on SVM fibre and epoxyaniline phenol–formaldehyde matrix

Prepreg storage duration (days)	*Strength of adhesion[a] (MPa)*	*Number of fractured specimens*		*Relative strength of organocloth (%)*	
		Adhesive	*Cohesive*	*In case of compression*	*In case of static bending*
1	59	28	38	100	100
5	–	–	–	122	104
10	–	–	–	135	112
14	64	23	28	122	110
30	63	34	35	124	94

[a] Area of gluing of elementary fibre is $S = 4.5 \times 10^{-3}\,\text{mm}^3$.

Under the usual service conditions practically all types of reinforcing fibres contain a certain quantity of moisture, sorbed from the surrounding medium, which in a number of cases negatively affects the strength of adhesion of these fibres with polymers. Moisture is uniformly distributed in aramid fibres both on the surface and in the volume of fibre-forming polymer. The quantity of sorbed water may reach 10%, which, as is known from experience of working with fibreglass plastics, is one reason for the decrease of adhesive strength of components and deterioration of material quality. It follows from the data given in Table 7.17 that, in the case of change of moisture content of aramid fibre from 10% to practically absolutely dry

Table 7.17 Effect of water content in aramid SVM fibre on its wetting with epoxydiane binder

Conditions for fibre wetting: relative humidity of air	Regime of fibre heat treatment	Equilibrium water content (mass%)	Wetting angle (deg)	Thermodynamic work of adhesion $W_a \times 10^3$ $(N\,m^{-1})$
0.98	–	10.0	17	88.0
0.63	–	5.0	15	88.5
0.30	–	3.0	9	89.5
0.50	373 K, 1 h	1.0	8	89.6
0.50	423 K, 5 min	0.6	20	87.3
0.50	453 K, 1 h	0	18	87.8

state, the wetting angle and thermodynamic work of adhesion of binder to fibre go through extrema. Optimal conditions for interaction of components are ensured when the moisture content in SVM fibre is within the limits of 1–3%, which provides the best conditions for contact between them and allows a composite with higher properties to be obtained. Since the quantity of water held by active centres of SVM fibre is equal to 2.3%, it is possible to come to the conclusion that the best conditions for forming good contact of organoplastic components are created in the case of removal of free (capillary condensed) water from fibre and retention of bound water.

In contrast to this, when manufacturing organoplastics it is necessary to remove water sorbed on the fibre surface completely. Not only at the stage of obtaining material but in service the interphase area in organoplastics is less vulnerable to the action of water than that in fibreglass plastics, which ensures, in turn, considerably higher stability of polymer–polymer composite properties during service under conditions of higher humidity. From the data shown on Fig. 7.23 it is clear that the strength of adhesion of epoxy matrix with aramid fibre is retained in the process of long-term moistening at higher level compared with that for fibreglass plastics. Subsequent drying of moistened specimens at 373 K for 1 h causes practically complete restoration of the adhesive strength of components.

The temperature dependence of the strength of adhesion of epoxy matrices with aramid fibre depends upon the character of physicochemical and physicomechanical interaction between them (Fig. 7.24).

Decrease of the strength of component adhesion within the range of temperatures below 273 K is, evidently, bound up with the action of thermal stresses [41]. When the temperature is higher than the glass transition temperature of a polymeric matrix, one can observe a regular decrease of strength of its adhesion with fibre; the higher the heat resistance of the matrix, the higher is the level and higher are the temperatures at which the values of strength of adhesion of components are retained.

For cyclic alternation of temperatures within the limits 293 ± 77 K, the strength of adhesion of aramid fibres with epoxy matrices is kept at a high level, which testifies

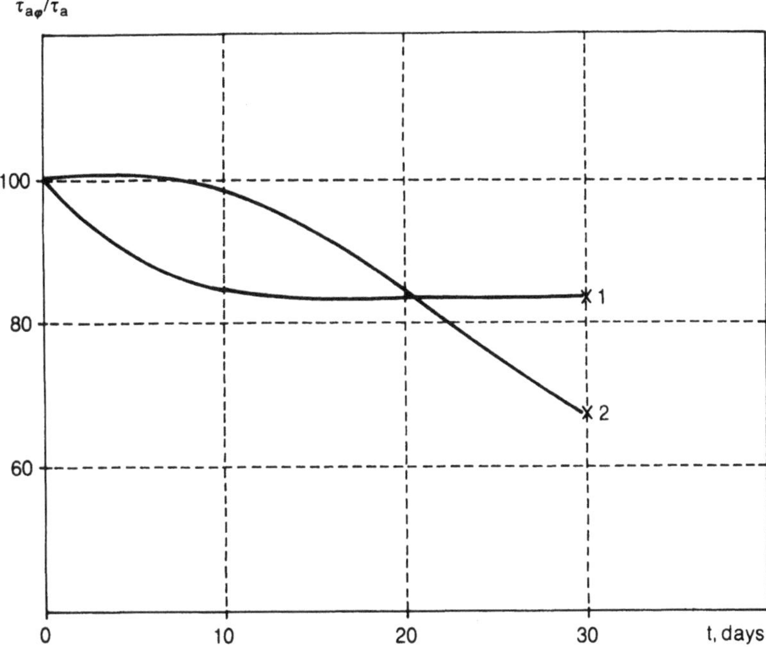

Fig. 7.23 Effect of moistening duration when relative humidity $P/P_s \sim 0.98$ on relative strength of adhesion of epoxy polymer with surface of fibres: (1) SVM aramid fibre; (2) glass fibre.

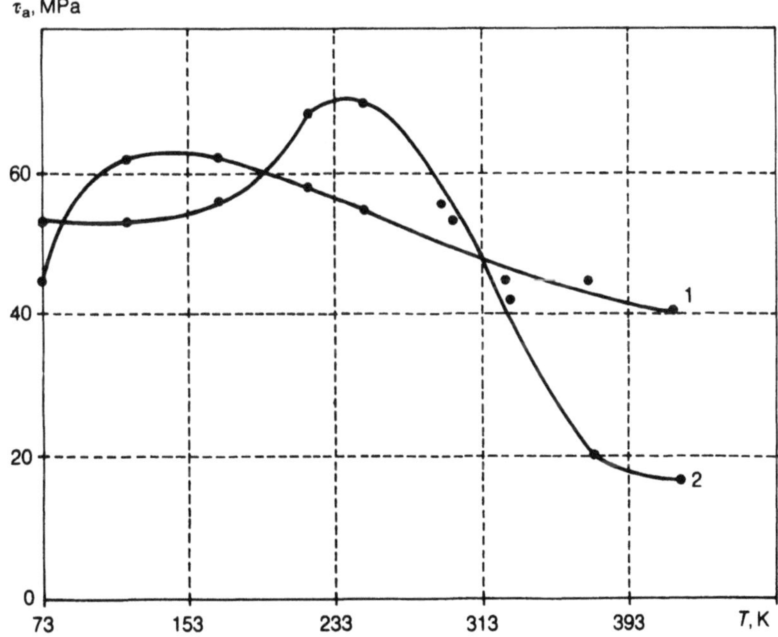

Fig. 7.24 Temperature dependence of the strength of adhesion in 'SVM aramid fibre–epoxy matrix' system: (1) epoxyaniline phenol–formaldehyde; (2) epoxydiane with amine hardener.

Table 7.18 Effect of cyclic alternation of temperature within the limits $293 \pm 77\,K$ on the strength of adhesion in 'fibre–epoxy matrix' system

Matrix	Fibre	Number of cycles	Relative preservation of adhesive strength
Epoxydiane with amine hardener	SVM	0	1.00
		100	1.04
Epoxyaniline phenol–formaldehyde	SVM	0	1.00
		100	0.95
	Glass	0	1.00
		100	0.86
	Carbon	0	1.00
		100	0.91

to the slower accumulation of defects and active relaxation of stresses in 'aramid fibre–epoxy matrix' systems, compared with stiffer compositions reinforced with glass and carbon fibres (Table 7.18).

7.7 RELAXATION PROCESSES IN ORGANOPLASTICS

Two intensive relaxation processes were revealed in thermoreactive-binder-based organoplastics [44–47]. Comparison of the temperature dependence of mechanical loss tangent for a block of cured binder and an organoplastic based on it permits one to relate the first maximum to the onset of segmental mobility in polymeric matrix of composite material composition.

Depending upon the chemical composition of the matrix and temperature and time conditions of its curing, the α_1 relaxation process intensity and the temperature of its maximum may change. So, for example, physicochemical processes of interaction between fibre and binder components in 'cold' cure polyester organoplastic have no chance for development due to the rapid growth of viscosity and quick curing of binder. As a result, the transition temperature from glassy state to highly elastic one in polyester polymer is equal in bulk and in composite material compositions, but the intensity of the α_1 relaxation process in bulk is considerably higher. Decrease of relaxation process intensity in matrix polymer in the presence of reinforcing fibres testifies to the essential change of effective dimensions of kinetic segments, responsible for the course of the relaxation process (Fig. 7.25(a)).

Application of modified epoxydiane polymer (the components of which interact actively with aramid fibres) as the matrix essentially changes the character of the relaxation loss spectrum: the position of the α_1 process is shifted towards higher temperatures and the vitrification temperature of matrix in organoplastic composition increases (Fig. 7.25(b)).

As shown in the section covering the analysis of thermodynamic compatibility of organoplastic components, triethanolamine titanate (TEAT) hardener of epoxydiane

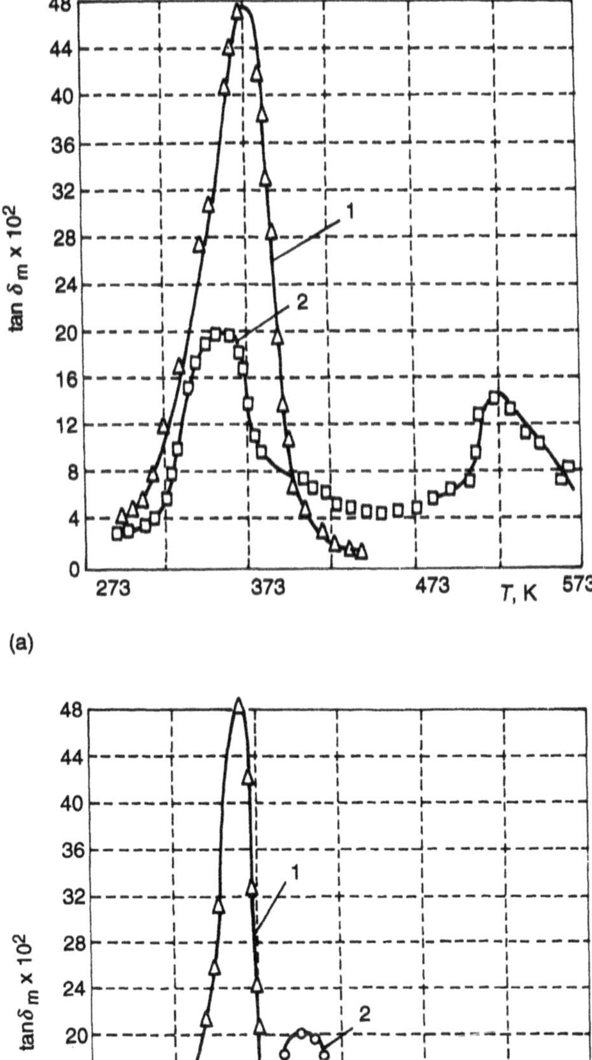

Fig. 7.25 Temperature dependences for coefficients of mechanical loss tangent (tan δ_m): (a) polyester binder (1) and organoplastic based on it (2); (b) epoxydiane (1) and unidirectional organoplastic based on it (2).

binder has a high affinity to fibre. An increase of hardener concentration on fibre surface contributes to an increase of epoxy polymer crosslinking thickness in the boundary layer. The increase of T_g of the matrix in a composite material is also caused by exhaustion of binder neighbouring the fibre layer with low-molecular-weight aliphatic epoxy resin DEG-1, which is characterized by high rate of diffusion in fibre volume and consequently its concentration in the matrix layer neighbouring the wall will be somewhat decreased.

On the contrary, binders in which the hardener or other chemically active reagent is able to diffuse into the fibre from the reactive volume of matrix are characterized by a decrease of the vitrification temperature of polymer in composite material. Such a situation is observed, for example, during cure of epoxy composite material containing aminoethoxyphosphazene as the hardener, which, as shown above, is characterized by high rate of diffusion into the fibre.

Dynamic characteristics of organoplastics permit estimation not only of the properties of the polymeric matrix in a composite material, but also of the strength of its adhesion with fibre. For example, the temperature of transition from glassy state into highly elastic one and the dynamic Young's modulus in highly elastic state of epoxynovolac polymer are considerably higher than those of epoxydiane (Fig. 7.26).

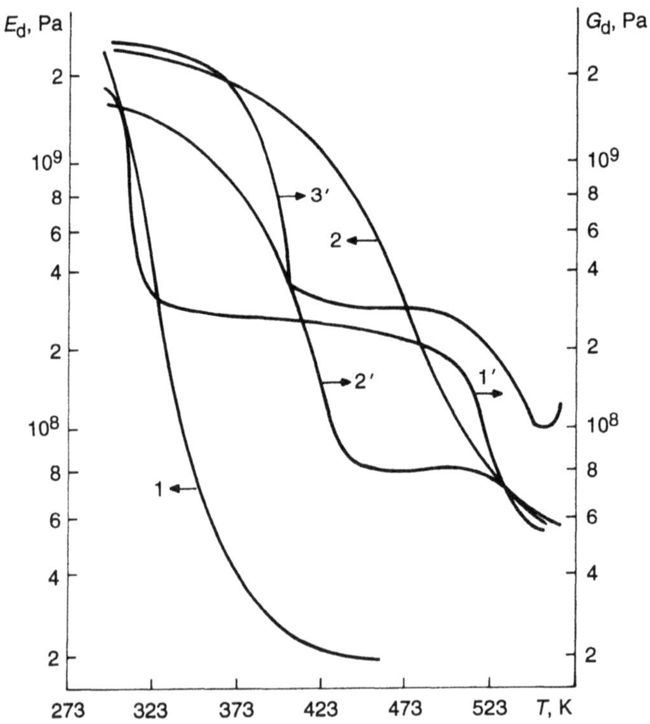

Fig. 7.26 Temperature dependences of dynamic Young's modulus (E_d) for block polymer of epoxydiane (1) and epoxynovolac (2), and shear modulus (G_d) of organoplastic based on epoxydiane (1'), epoxynovolac (2') and epoxyaniline phenol–formaldehyde (3') binders.

However, this regularity is not observed for these polymers in organoplastic compositions: the high temperature of epoxynovolac polymer transition corresponds to the minimum value of dynamic shear modulus in highly elastic state. At the same time (as shown above) epoxydiane and epoxynovolac binder are characterized by the maximum strength of adhesion with aramid fibre, which decreases with increase of temperature, particularly for epoxydiane polymer, considerably more intensively than when using epoxyaniline phenol–formaldehyde binder.

Owing to this it is possible to make a conclusion that variation of dynamic modulus of elasticity of an organoplastic in case of polymeric matrix transition from glassy into highly elastic state is predetermined by variation not only of intermolecular interaction during matrix cure, but also of strength of matrix adhesion with fibre.

The above materials may be ordered in terms of decreasing value of dynamic shear modulus in the highly elastic state of an organoplastic polymeric matrix, determined mainly by strength of component adhesion: epoxyaniline phenol–formaldehyde > epoxydiane > epoxynovolac (Table 7.19).

The second intensive α_2 relaxation process in organoplastics occurs in the temperature range of 503–555 K and is bound up with onset of large-scale molecular mobility and variation of intermolecular interaction in aramid fibres. Decrease of shear modulus G_d in the α_2 relaxation process is the result of intermolecular interaction variation in fibre on increase of temperature. This change may be characterized by

Table 7.19 Dynamic characteristics of organoplastics based on epoxy matrices and SVM aramid fibre cloth

	Epoxy matrices		
Characteristics	*Epoxydiane*	*Epoxyaniline phenol–formaldehyde*	*Epoxynovolac*
Dynamic shear modulus, $G_{20} \times 10^{-9}$ (Pa)	1.76	2.0	1.55
Dynamic shear modulus at temperature corresponding to respective high-elasticity state of matrix in organoplastic, $G' \times 10^{-8}$ (Pa)	1.7	2.7	0.9
Dynamic shear modulus at 573 K (corresponding to output of G_d on plateau), $G_{573} \times 10^{-7}$ (Pa)	5.4	10.0	6.0
Value of change of intermolecular interaction in fibre: $\Delta G_{\alpha_2} = G'/G_{573}$	3.2	2.7	1.5
Temperatures of relaxation process maxima: α_1 process, T_{α_1} (K)	322	433	425
α_2 process, T_{α_2} (K)	519	523	533

the equation:

$$\Delta G_{\alpha_2} = G'/G_{573}$$

where G' is the value of dynamic shear modulus at a temperature corresponding to the high-elasticity state of an organoplastic's polymeric matrix, and G_{573} is the value of dynamic shear modulus at 573 K (corresponding to G_d output on plate). Decrease of α_2 relaxation process temperature and increase of ΔG_{α_2} testifies to the plasticizing effect of binder on organic fibre as a result of chemical interaction and diffusional penetration of binder components into fibre volume (Table 7.19).

The presence of a specific interaction between reinforcing fibre and binder in organoplastic causes formation of a washing-out transient layer, which is characterized by broadening of maxima of mechanical losses, predetermined by molecular mobility in both plastic components; absence of a clear plateau of high elasticity in the temperature dependence of moduli G_d and E_d testifies to the smooth transition from matrix to reinforcing filler.

Combination of chemical interaction of components and their mutual diffusional penetration is to a higher extent typical for organoplastics based on epoxyaniline phenol–formaldehyde binder. Diffusional penetration of binder components into fibre volume causes some plasticizing of the latter. At the same time, active chemical interaction determines a rather high shear modulus at a temperature corresponding to high-elasticity state of polymeric matrix. In this case, new levels of structural organization arise in organoplastic volume. The molecular mobility of these structures is revealed in the form of α'_1 and α'_2 relaxation processes in the temperature ranges of 433–453 and 483–503 K respectively (Fig. 7.27).

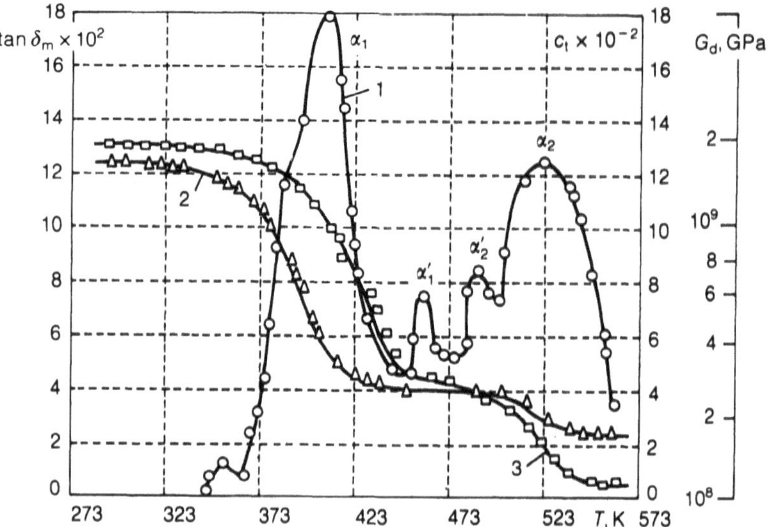

Fig. 7.27 Temperature dependence for coefficient of mechanical loss tangent (tan δ_m) (1), low-frequency sound velocity c_t (2) of organoplastic based on epoxyaniline phenol–formaldehyde and dynamic shear modulus G_d (3).

Manifestation of multiplicity of relaxation processes in organoplastics may be determined by pressure of four macroscopic levels of structural organization. The first level is bound up with the polymer matrix in plastic volume far from the fibre surface, and the α_1 relaxation process corresponds to it. The second level corresponds to the matrix layer structure near the surface of reinforcing fibre formed as a result of physical and chemical interaction of components; it appears as the α' relaxation process. The third level is bound up with the surface layer of organic fibre, modified with diffusing component of binder; this level shows itself as the α'_2 relaxation process. The fourth level corresponds to the structure of organic fibre core, and the α_2 relaxation process is determined by it.

Thus the structure and properties of an organoplastic, based on organic fibres and thermoreactive binders, is predetermined by the molecular levels of composite organization, depending upon the components and the specific forms of their interaction.

The revealed structural inhomogeneity of organoplastics and high level of mechanical losses in these materials permit one to forecast longevity and high rupture energy of these materials, as well as inclination to accumulation of local defects without formation of a critical crack. It is possible to believe that the developed relaxation processes predetermine a considerable reserve of reliability and high service life characteristics of organoplastics as structural materials.

7.8 MECHANICAL PROPERTIES OF ORGANIC-FIBRE-REINFORCED PLASTICS

7.8.1 Peculiarities of OFRP behaviour on static loading

The polymeric nature of OFRP components predetermines the viscoelastic behaviour of these materials and the significant role of relaxation phenomena in the process of structural changes on OFRP loading. The peculiarities of structural changes of organoplastics on loading depend on the anisotropy and maximum orientation of the the reinforcing fibres, the fibrillar character of their fracture, and the extent to which fibrils should be considered as the structural unit of aramid reinforcing filler.

The peculiarities of OFRP relaxation properties and mechanical behaviour depend, to a great extent, on the fact that stresses and strains in these materials are distributed between the reinforcing filler and the polymeric matrix through the developed heterogeneous interphase layer.

The mechanical properties of OFRP can be varied over a wide range, depending on the reinforcing filler type and structure, reinforcing scheme, material production technology, etc. Table 7.20 shows the properties of standard structural OFRP, in which fabrics of SVM aramid fibre of different structure and weaving are used as the reinforcing filler, as well as the properties of unidirectional OFRP based on filaments and tows of aramid fibres produced by winding and pultrusion methods.

The polymeric nature of OFRP affects the character of their strain and relaxation properties. Owing to the aramid fibres, high tensile strength is combined in OFRP with comparatively high straining ability: relative elongation at break of unidirectional OFRP is 2.5–3%, and for organic cloth laminates it is 3–3.5%.

Table 7.20 Properties of OFRP based on SVM fibres and epoxy matrices

		Properties (MPa)				
Reinforcing filler	*OFRP production technology*	σ_b	$E \times 10^{-3}$	σ_{-b}	σ_{bb}	τ
Filament, 29.4 tex	Wet winding	2500–2800	95	320	760	40
Tow, 600 tex	Wet winding					
Tape prepreg of 29.4 tex filament	Dry winding	2000	95	400	730	40
Filament, 58.8 tex	Pultrusion					
Sheet prepreg of filament	$P_{sp} = 0.5$ MPa[a] (lay-up)					
Unidirectional fabric (plain) of 29.4 tex filament (prepreg)	$P_{sp} = 0.5$ MPa (lay-up)	1200–1500	65–70	260–300	600	35
Fabric weave 8/3, filament, 14.3 tex (prepreg)	Lay-up:[a] $P_{sp} = 0.5$ MPa	740	34	240	420	40
	$P_{sp} = 8$ MPa	800	42	260	480	42

[a] Specific moulding pressure.

The high tensile strength and stiffness of such fibres are associated with the high chemical bond dissociation energy in the initial polymer chain and high degree of orientation of macromolecules along the fibre axis. The fracture mode of maximum oriented fibre under uniaxial tension is due to polymer microinhomogeneity, the result of which is non-simultaneous work of the separate fibre structural elements. The difference in the stressed condition of neighbouring structural elements results in shear stress generation in the boundary area, the generation of which is accompanied by fibre microfibrillization and splitting. As a result, on uniaxial fibre tension, orientational intergrowth of the interfibrillar cracks occurs along the direction of the tensile force, the fibre splits to a certain extent and becomes able to take the axial tensile load. Fibre splitting is accompanied by the breaking of individual, most stressed, fibre fragments, after which the load is taken by other fibrils, tow, etc., up to complete fracture, finishing with intensive splitting and crushing of the fibre break zone.

Thus, the process of damage initiation and accumulation in OFRP passes through the stage of fibre interfibrillar splitting with the successive breaking of individual longitudinal structural elements. Interfibrillar splitting of aramid fibres accompanies the fracture of OFRP based on them under exposure to practically all loading types [48].

For this reason, much energy is consumed in the fracture of OFRP, which is spent in the formation of the developed fracture surface and the straining of the split fibres before and after fracture. The process of OFRP fracture is more energy-consuming compared to the fracture of composites based on brittle glass and carbon fibres. This predetermined many important mechanical properties of OFRP.

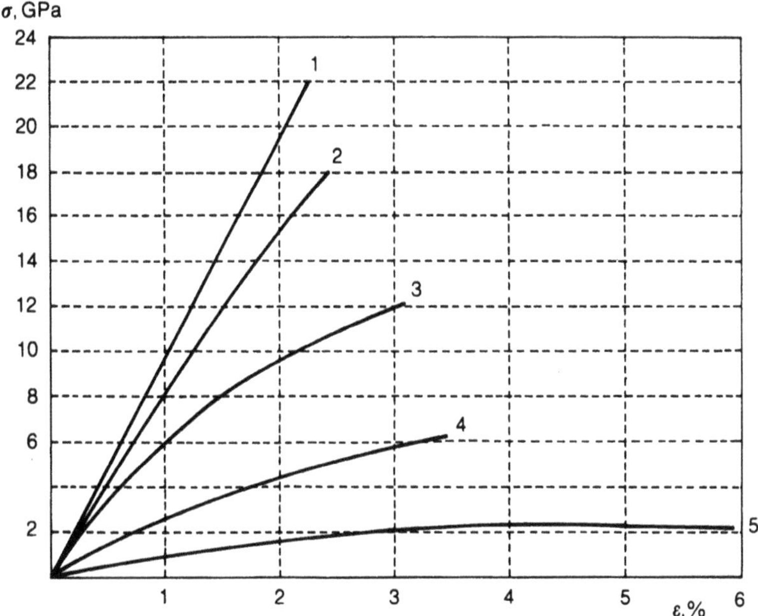

Fig. 7.28 Stress–strain diagrams for: unidirectional OFRP, produced by winding with filament (1) and prepreg (2); unidirectional organic cloth laminate (3); and organic cloth laminate of equal strength in the warp and weft fabric in the direction of reinforcement (4) and at an angle of 45° to the fibre direction (5).

The stress–strain diagram of high-strength woven OFRP, based on untwisted high-modulus aramid filaments, is practically straight up to fracture (Fig. 7.28). However, in the general case the σ–ε diagram for these materials consists of straight-line sections, the transition zone between them characterizing the change of elastic strain to resilient-elastic one. In this case one is able to speak about the bimodality of such materials; the resilient-elastic modulus in this case can be twice as low compared to the elasticity modulus.

The use of more complex structures (tows or unidirectional tapes) instead of filaments as the reinforcing filler for unidirectional OFRP results in a clearer manifestation of elasto-elastic strains: their straining diagram decreases its slope and is curved on achieving stress exceeding 70–80% of the breaking value. This is due to the increase of structural heterogeneity and the length differences of the fibres in the complex bundle, which results in strength reduction by 15–20% compared to OFRP based on fillers.

Unidirectional OFRP produced by prepreg (tape or fabric) lay-up because of the increase in fibre curvature and the growth of their length difference, have lower strength and increased fraction of elasto-plastic strain.

The strain in organic cloth laminates based on woven structures is characterized by a developed high-elasticity straining zone, transforming to a zone of conditionally plastic straining, which indicates the acceptability of considerable overloadings. In their texture and behaviour under mechanical loading, organic cloth laminates differ principally from unidirection OFRP produced by the winding method. The characteristic feature of fabrics as reinforcing fillers is defined by the fact that in the process of their production two perpendicular filament systems (warp and weft) are bent and acquire a wavy configuration while crossing each other. In the process of prepreg moulding, based on these fabrics, considerable decrease, sometimes by 30–50%, of the fabric thickness occurs in the plastic. As a result, load-bearing filaments in the plastic monolayer become curved at least in two planes; these curvatures are of regular character. Besides, there are additional reasons causing the deviation of the reinforcing fibres from the straight line: filament twisting, filament misorientation because of the fabric moving apart, etc. Presently, in the field of plotting the micromechanics models of these materials, dependences are established that allow one to take into account the effect of the fabric structure on reinforced plastic properties [49].

The other characteristic feature of textile reinforcing fillers is defined by the fabric structure: its ability to deform in the direction of the external loading, not only to take but (similar to the matrix) to redistribute the stresses generated under mechanical exposure, and to inhibit crack propagation.

The straining diagrams for organic cloth laminate based on the equal-strength fabric weave 8/3 of SVM fibre and epoxyaniline phenol–formaldehyde binder are characterized by non-linearity (see Fig. 7.28). On the straining diagram for organic cloth laminate, the initial linear section up to a load of 40–50% of the breaking value can be separated, i.e. the proportionality limit of the material is $(0.4–0.5)\,\sigma_b$. The elasticity modulus calculated by tangent slope angle at the stress $\sigma = 0.95\sigma_b$ comprises 36% of the initial value. For cross-reinforcement at an angle of $45°$, the strain curve can be approximated by two linear sections with different slopes. The elasticity modulus for the second section is 0.05% of the initial value. The behaviour of organic cloth laminate with reinforcement scheme $\pm 45°$ in tension is close to the strain of an ideal elastoplastic material. This reinforcing scheme is characterized by the presence of significant residual strain and also tensile reduction of an area in the specimens associated with the high value of Poisson's ratio ($\mu_{\pm 45} = 0.718$).

Figure 7.29 shows $\sigma–\varepsilon$ diagrams for organic cloth laminate, obtained under various loading rates. As can be seen, the material is rather sensitive to the effect of the rate; strain diagrams diverge practically from the very beginning of loading, which indicates the manifestation of viscous effects and damage accumulation.

The structural peculiarities of high-strength polymeric fibres affect the fracture mode of OFRP under tensile loading in the direction perpendicular to the reinforcing fibres.

The transverse strength of aramid fibres is lower compared to that of the polymeric matrix. That is why on transverse tension of unidirectional OFRP the fibres will be fractured first. On tension of orthogonally reinforced plastic in one of the reinforcement directions, two-stage material fracture is usually observed. The layer reinforced in the direction perpendicular to straining is the first to be fractured. In the case of GFRP and CFRP, fracture of polymeric matrix or bonds between the fibres and the

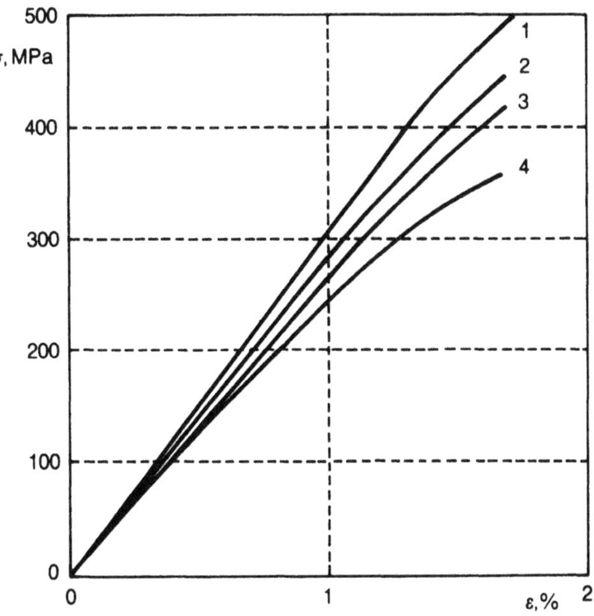

Fig. 7.29 Experimental σ–ε diagrams obtained at various loading rates (times) under conditions of $\dot\sigma = $ const: (1) $\dot\sigma = 127\,600\,\text{MPa s}^{-1}$, $t = 4.7 \times 10^{-3}$ s; (2) $\dot\sigma = 4000\,\text{MPa s}^{-1}$, $t = 1.3 \times 10^{-1}$ s; (3) $\dot\sigma = 2.5\,\text{MPa s}^{-1}$, $t = 1.8 \times 10$ s; (4) $\dot\sigma = 0.1\,\text{MPa s}^{-1}$, $t = 3.5 \times 10^{9}$ s.

matrix is observed, and in the case of OFRP the fibres are fractured due to transverse tension. At this moment the material loses its integrity and the characteristic break is formed in the straining diagram.

The character of OFRP behaviour under loading depends, to a great extent, on the technology of the material. The relaxation nature of organic fibres contributes to this, along with the structural sensitivity of their mechanical properties. Figure 7.30 shows that variations in filament technological tension in winding of unidirectional OFRP allow one to increase their strength and stiffness considerably, the straining diagram being changed in this case; the proportionality limit is increased from $0.6\sigma_b$ and on tension, comprising 35% of the filament breaking load, it coincides with the material ultimate strength.

The polymeric nature and high anisotropy of reinforcing fibres predetermines the comparatively low resistance of OFRP to compressive stresses. In the reinforcement direction the compressive strength of these materials is much lower compared to their tensile strength. The ratio of these characteristics for high-strength unidirectinal OFRP is 0.1–0.2; for organic cloth laminates, depending on the degree of anisotropy of the woven filament, it is 0.2–0.4.

On compression of OFRP specimens in the direction of the warp and weft filaments of reinforcing fabric, a non-linear dependence between the stress and the strain is observed under stresses exceeding 100 MPa. The average compression strength of

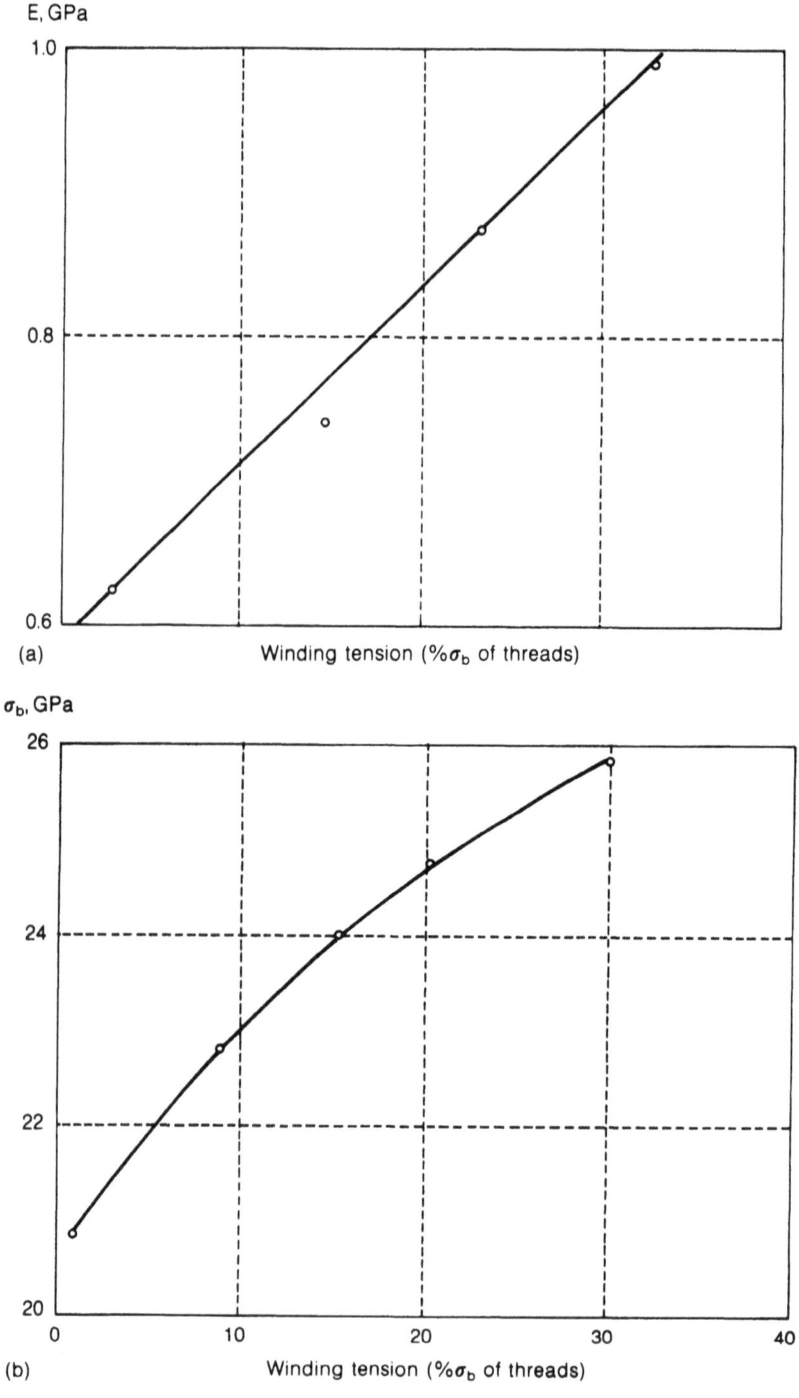

Fig. 7.30 Elasticity modulus (a) and tensile strength (b) of unidirectional OFRP as functions of winding tension value of SVM threads.

the organic cloth laminate of the weft and warp is practically the same, being 155 MPa. The breaking strains of the specimens on compression in the direction of the fabric warp and the weft are practically the same, being equal to 4.0–4.1%. The elasticity modulus under loading along the warp direction is 15 500 MPa, and along the fabric weft it is slightly lower at 13 000 MPa, which is due to high bending of the weft filaments in the fabric, compared to the warp filaments.

Figure 7.31 shows the σ–ε diagrams under compression of organic cloth laminate specimens in the direction of 30°, 45° and 60° to the reinforcing fabric warp. It can be seen from the given data that variation of the angle between the loading direction and the direction of the reinforcing fabric warp only has a small effect on the material strength and the ultimate strain value.

The analysis of fractured OFRP specimens shows that fracture under compression is due to shear; however, the reinforcing fibres are not cut but only curved. In woven fillers the fibre curvature is inherent to the structure itself, which is why the material load-bearing capacity becomes still lower. In the process of compression loading of OFRP, the non-linear elastic strain is observed first; quickly increasing plastic strains of the matrix result in stability loss of initially plastically strained fibres. The single fibre curvature, accompanied by the break of weak interfibrillar bonds, leads to the formation of release strips in the compressed zone and to axial cracking in the tensile zone. The fibre curvature under axial compression of OFRP beyond the ultimate

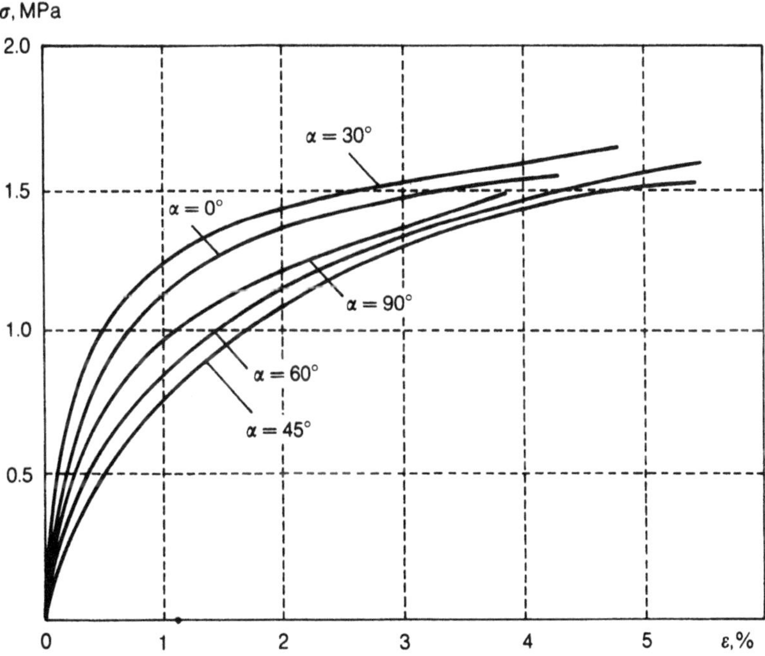

Fig. 7.31 Compressive stress–strain diagrams of organic cloth laminate specimens of equal-strength SVM fabric, cut at different angles to the fabric warp.

yield strength is accompanied by considerably transverse strains, promoting the weakening of interfibrillar bonds in the fibre and matrix delamination from the fibre; this delaminatin can have complex adhesive–cohesive character with the separation of the fibrillar formations from the fibre shell.

OFRP are characterized by the common mechanism of formation of release strips under compression both on the microlevel, when losing fibril stability in the filament, and on the macrolevel, when losing stability in the fibre tow. The peculiarities of release strip formation, and also the coincidence and possibly the mutual strengthening of macro- and micromechanisms of OFRP fracture under exposure to compression loads, are most probably the main reason for their very low compression strength, compared to other polymeric composites.

If this statement is true, and the weak point in OFRP, responsible for their low compression strength, is the realization of interfibrillar shear and stability loss of the filament structural elements, in order to block this mechanism it is necessary to create conditions that eliminate the possibility for its realization. These conditions include, first of all, the increase of aramid fibre transverse strength and the use of a polymeric matrix with elevated mechanical properties.

7.8.2 Long-term and fatigue loading

When calculating the load-bearing capacity of OFRP structural elements, these polymer–polymer materials are usually approached from the positions of conventional structural materials, for which the working zone is mainly defined by the elasticity limit. OFRP are principally different from all other composites in their non-linear behaviour under long-term and repeated loading, high work of rupture, low sensitivity to stress concentrators, and formation of an extended prefractured zone under fatigue loading. The high service life characteristics and considerable reliability of OFRP are conditioned by the serviceability of these materials in the extended range of the high-elasticity condition, where relaxation processes play the active role and damage accumulation is not accompanied by the formation and propagation of dangerous cracks.

Experimental study of the mechanical behaviour shows that these materials are capable of long-term resistance to static loading. Tensile stress rupture of these materials on the basis of 1000 h is 70–90% of the breaking stress (Fig. 7.32). The linear correlation between the values of long-term and short-term strength of OFRP based on various polymeric fibres becomes broken with increase of the material strength (Fig. 7.33).

OFRP are linear elastic materials only at low and short-term loads. Under long-term exposure to high enough loads, the creep developed in OFRP is the consequence of preferably viscous flow of the material. Damage accumulation in the process of OFRP creep plays a less significant role compared, for example, with GFRP or CFRP.

OFRP, like many other polymeric materials, are characterized by two creep stages. At first the high rate of strain growth is stabilized with time (Fig. 7.34). On load relaxation after long-term loading the specimen strain is sharply reduced. On repeated

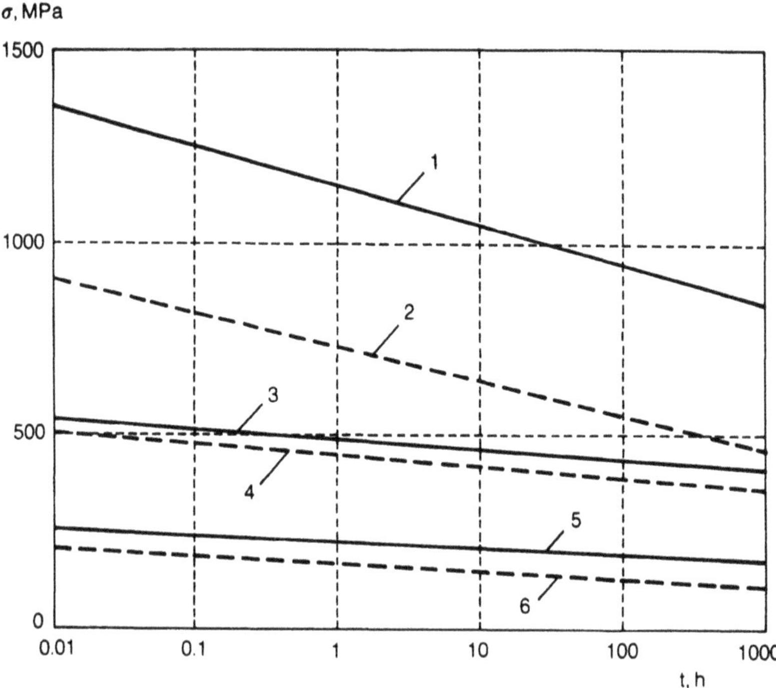

Fig. 7.32 Long-time tensile strength curves of aramid OFRP. (1, 2) unidirectional epoxydiane OFRP at 293 K (1) and 373 K (2); (3, 4) organic cloth laminate based on epoxyaniline phenol–formaldehyde matrix at 293 K (3) and 373 K (4); (5, 6) organic cloth laminate based on polyamide matrix at 293 K (5) and 373 K (6).

long-term loading of non-fractured specimens, both the rate and the total creep level are decreased. Such material behaviour is explained by highly elastic molecular processes, which are developed in the reinforcing fibres and are due to the regrouping of the polymeric macromolecules without breaking chemical bonds [50]. These processes are accompanied in time by stress redistribution between the OFRP components to the side of stress concentration. With the increase of stress or temperature the creep strain and creep rate of OFRP are noticeably increased. The lower the elasticity modulus and the higher the relative elongation at break, the more intensively does the creep process in OFRP proceed; the major role in this case is played by the chemical composition of the material and the reinforcing filler structure.

The value of crack resistance in the area of precritical crack growth is important for estimating the reliability of structural materials, since in the case of organoplastics the crack resistance at the storage of precritical damage actively affects the construction safety.

The study of organoplastic specimens with initiating cut (crack) showed that in some cases the crack is not increased practically in the process of their durable loading with constant load. However, it is possible to observe crack opening, which is intensive

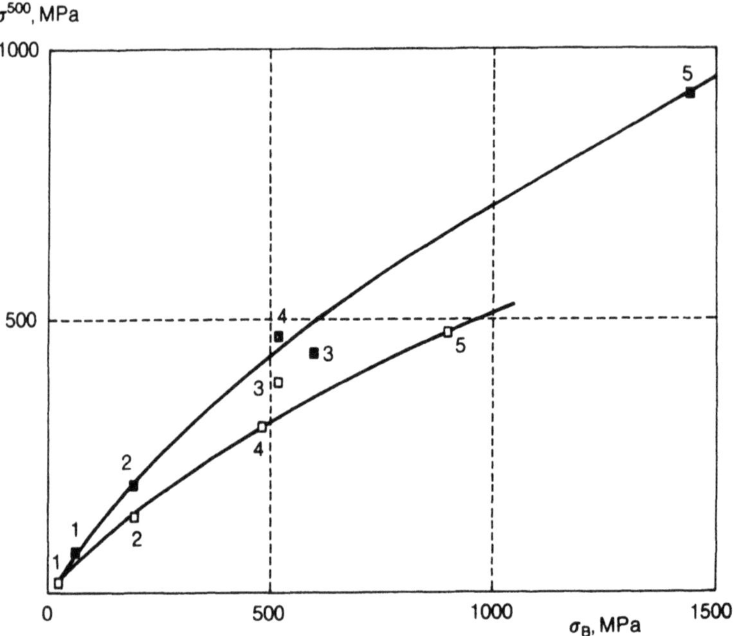

Fig. 7.33 Correlation between the long-time strength (σ^{500}) and tensile stress rupture of OFRP: (1) epoxydiane organic cloth laminate based on poly(ethylene terephthalate) (PET) fabric; (2) polyamide organic cloth laminate based on SVM fabric; (3, 4) organic cloth laminates based on SVM and epoxyaniline phenol–formaldehyde (3) and phenolphthalein (4) matrices; (5) unidirectional epoxydiane OFRP (SVM threads); (■) 293 K, (□) 353–373 K.

during the initial period (for the first 1–2 h) and stabilized (with constant rate) during subsequent loading test of specimens. The value of crack opening increases with growth of stress level in the specimen.

Dependence of the crack opening rate upon the stress is shown on Fig. 7.35 for organocloth laminates. The curve of crack opening rate is of complex character, probably determined by the multistage process of material fracture, beginning at the end zone of the crack. The rate drop at the upper branch of the curve is bound up with the increase of the damaged zone volume, as a result of which the process of crack opening is slowed.

Crack opening is bound up with both damage zone value and the extent of damage, since formation of a composite material network of microcracks is developed at places of low resistance to fracture. The extent and the volume of damage depend not only upon the value and duration of load action but to a great extent upon its composition and structure.

The direct relation of crack opening to material damage in its end zone was determined on the basis of obtained experimental results. A criterion of damage extent or crack resistance in this case may be the rate of opening of the crack-initiating cut in the specimen [51].

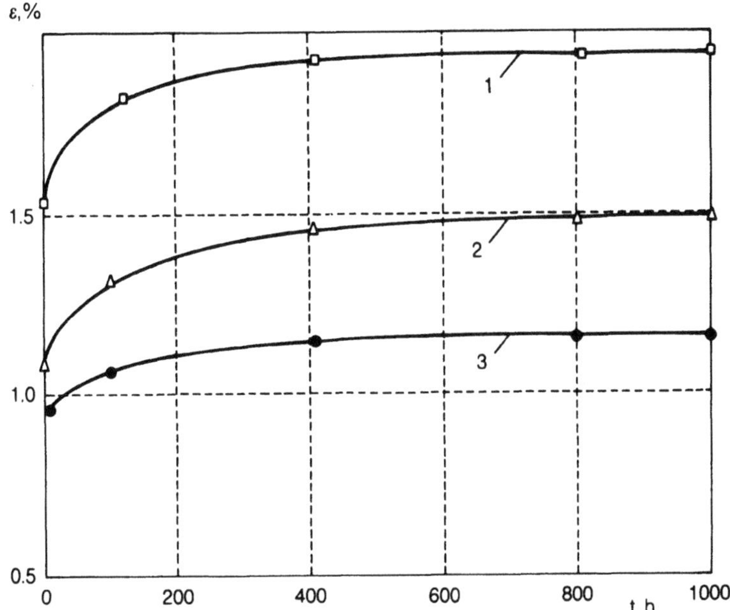

Fig. 7.34 Long-time creep curves of epoxy OFRP at various σ values: (1) 300; (2) 250; (3) 200 MPa.

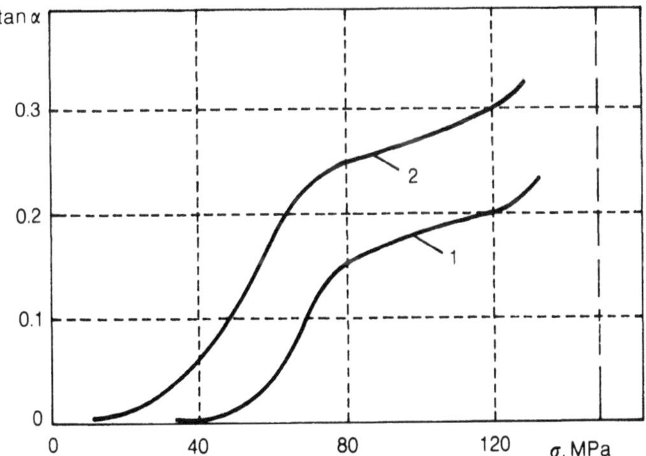

Fig. 7.35 Dependence of crack opening rate upon stress for organocloth laminate based on SVM satin cloth and epoxyaniline phenol–formaldehyde matrix.

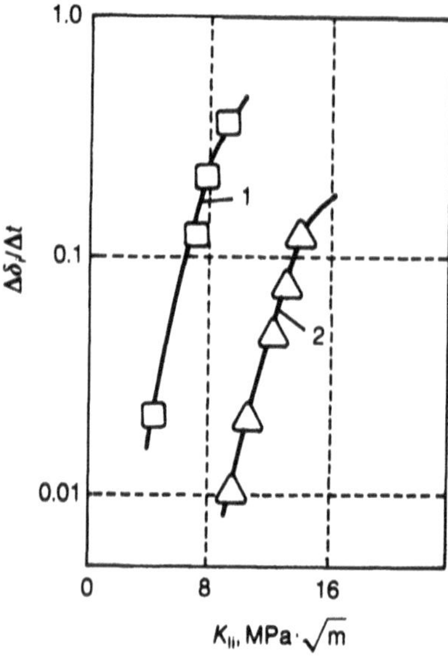

Fig. 7.36 Diagrams of slow fracture at long-term static tension (see Table 7.21) of OFRP, reinforced with aramid fabrics: (1) satin 8/3; (2) unidirectional fabric.

Figure 7.36 shows the diagram of slowed rupture under long-term static tension, demonstrating the dependence of crack opening rate upon the stress intensity coefficient. The linear sections of the diagram are described by the equation given in Table 7.21. Insulation of these sections determines the crack opening rate: $\tan\alpha = \Delta\delta/\Delta t$.

Thus organoplastic fracture is characterized by the multiple character of damage and fracture in the supercritical state, which hampers interpreting the fracture criteria for these materials. At the same time, the slow development of defects in OFRP and

Table 7.21 Indices of mechanical properties of long-term tensile crack resistance[a] of organo-cloth laminates

Aramid cloth structure	σ_b (MPa)	$E \times 10^{-3}$ (MPa)	$\varepsilon(\%)$	σ^{500} (MPa)	K_{1c} ($MPa\,m^{1/2}$)	K_{1i}	n	Coefficients of equation[b] $C \times 10^3$ ($mm^{5/2}\,h^{-1}\,kgf^{-1}$)
Sati 8/3	520	30	2.4	405	27.7	11.8	9.2	0.5
Unidirectional	1300	70	1.1	1000	39.9	6.8	9.2	1.9

[a] Specimens 30 mm wide with a side cut; ratio of cut length to specimen width, $\lambda = 0.20–0.25$.
[b] Equation: $\Delta\delta/\Delta t = CK_{1i}^n$ where $\Delta\delta/\Delta t$ is crack opening rate, K_{1i} is precritical stress intensity, and C, n are coefficients, depending upon material properties.

rupture in the supercritical condition ensure the high service reliability of these materials.

In order to describe the non-elastic behaviour of OFRP on long-term and short-term loading, Ju.V. Suvorova has suggested the non-linear model of inherent type, based on the notion of the instantaneous strain curve and the defining integral equation with Abel core. It is shown in some papers [52–55] that on OFRP loading two processes are simultaneously developed in them: reversible viscous flow and irreversible damage accumulation. On repeated static loading in the tension mode, both processes are added. Unloading can be accompanied by closing of partial defects. If these defects are small enough, they are closed completely after load relaxation, which results in reduction of breaking strain. The formation of large defects is a completely irreversible process. If one assumes that the quantitative characteristic of these defects is defined by the maximum strain value ε^*, reached at the moment of unloading, the recovery function $F(\varepsilon^*)$ can be introduced (Fig. 7.37) [53]. As can be seen, there exists a certain maximum strain value, up to which only small defects are formed, completely closing after load relaxation. It is shown on the example of the organic cloth laminate that at cyclic strain in the range of 0–0.8% each loading cycle is reversible, as the defects under these conditions do not accumulate. At $\varepsilon^* \sim 1.5\%$, residual strain accumulation results in process stabilization. In this case the number of newly formed defects is equal to the number of disappearing ones. On further increase of ε^*, there begins the avalanche merging of small defects into larger ones, which are incapable of closing

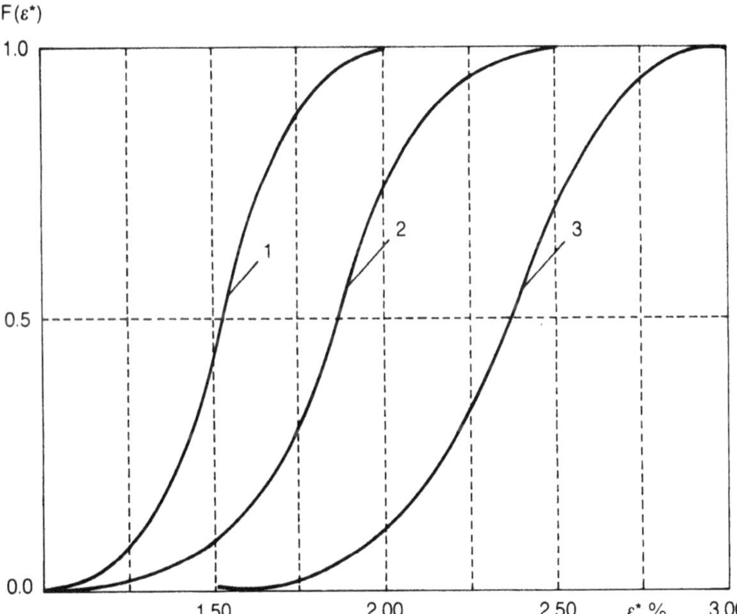

Fig. 7.37 Dependence of the recovery function $F(\varepsilon^*)$ on the limit strain ε^* of epoxy OFRP, reinforced with SVM fabric; at test temperatures: (1) 293; (2) 333; (3) 373 K.

and the merging of these larger defects fracture sources after a number of cycles necessarily leads to material fracture.

The increase of temperature or moisture content in OFRP within definite limits promotes the growth of the limited strain, which is associated with the reduction of the danger of formed defects acting as stress concentrators. Depending on the test and repeated static loading conditions, the maximum strain value of epoxy organic cloth laminate can reach 2.5%.

OFRP possess high tensile fatigue strength (Fig. 7.38). Unidirectional wound epoxy OFRP based on SVM filaments ($\sigma_b = 2500$ MPa) withstands without fracture 60 000 cycles at a stress of 1800 MPa, the residual strength being 2150 MPa. The endurance limit of organic cloth laminates under low cycle tension is 75–79% of the ultimate strength value.

The peculiarities of OFRP composition and structure predetermine their specific behaviour under exposure to bending loads. Contrary to CFRP and GFRP, OFRP are capable of demonstrating ductility on bending, like metals, without being subjected to critical fracture. The ability of OFRP to retain their integrity on bending, particularly on testing of specimens of small thickness, characteristic of aircraft engineering structural elements, ensures high endurance under sign-variable bending loads on low- and high-frequency loading (Fig. 7.39).

The fracture mechanism under exposure to long-term constant and cyclic loads in OFRP has its own specific features: high plasticity of the fibres in the transverse direction, low defectiveness in the interphase area, and the absence of a clearly defined

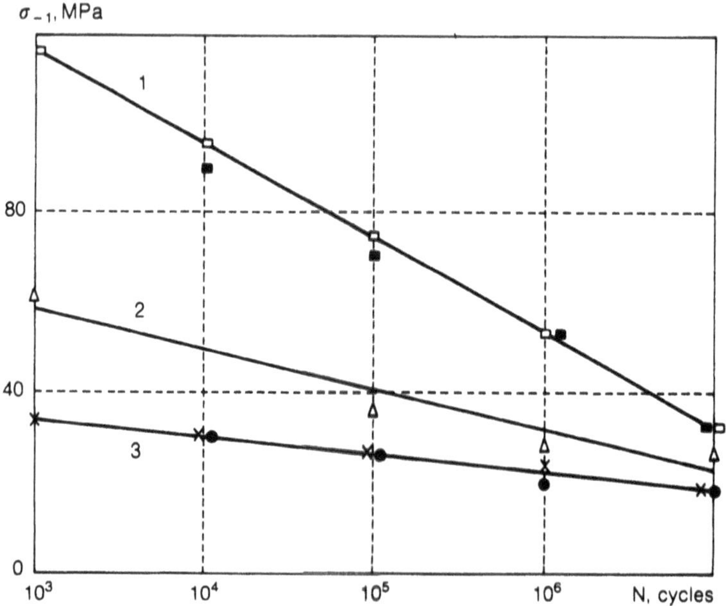

Fig. 7.38 Tensile fatigue strength of epoxy OFRP: (1) unidirectional OFRP; (2, 3) organic cloth laminates, produced at a specific moulding pressure of 8 MPa (2) and 0.5 MPa (3).

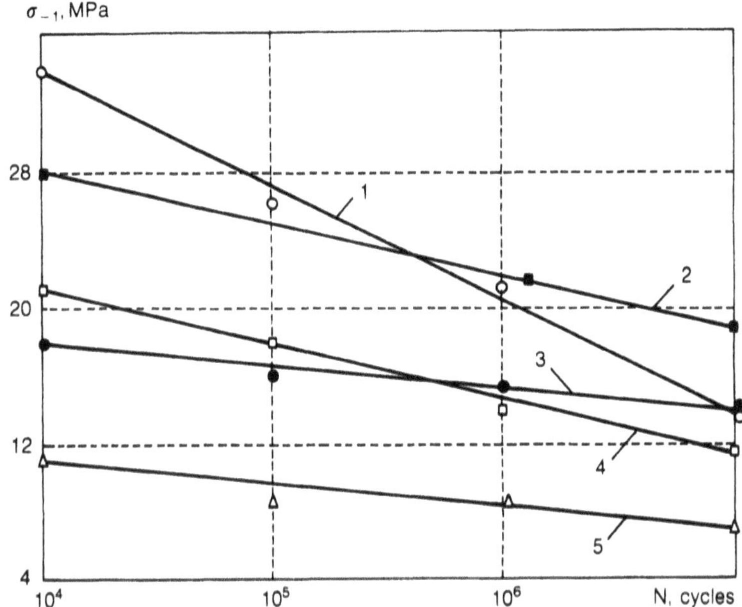

Fig. 7.39 Fatigue strength under symmetric bending of OFRP: (1, 2) Composites based on a unidirectional fabric of SVM filaments (1) and a combination of aramid (70%) and glass (30%) filaments (2) with epoxyaniline phenol–formaldehyde binder. (3, 4, 5) Composites based on equal-strength SVM fabric satin 8/3 and different binders: (3) epoxypolysulphone, (4) epoxy-novolac, (5) epoxyisocyanate.

interface between the components. All these complicate the process of crack initiation and growth, and the formation of the main crack, resulting in fracture.

In principle, the following ways of preventing crack propagation are realized in OFRP:

1. *Relaxation of elastic potential in maximum loaded system.* For all polymeric materials, the relaxation of stresses and strains is the key factor in increasing crack resistance. The proper use of relaxation process kinetics and their intensity control contribute to the improvement of OFRP performance characteristics, where the relaxation processes take place in both polymeric components and interact between each other.
2. *Creation of extended elastic compression fields.* The production of OFRP based on a reinforcing filler, stressed by tensile loading, results in the generation of its own internal compression stresses in the polymeric matrix, which hinders the onset of crack propagation under exposure to external tensile loading.
3. *Creation of barriers on the crack propagation path.* The realization of this possibility is inherent in the nature of fibrous composites, where the reinforcing fibre is the main barrier to the growing crack. However, for OFRP, crack inhibition on interaction with the fibre is complicated by the latter's longitudinal splitting, as a result of which the growing crack, instead of single filament fracture, has to

Table 7.22 Properties of prepreg based sheet cloth laminates

	Specific moulding pressure (MPa)	
Properties[a]	8.0	0.5
Density ($kg\,m^{-3}$)	1350	1270
Ultimate stress (MPa)		
tension	800	700
compression	250	220
bending	510	470
shear	41	33
CF on basis of 10^4 cycles	50	35
FCPR when $\sigma_{max} = 180\,MPa$ ($\mu m/cycle$)	1–3	27

[a] Low-cycle fatigue on pulsed tension, CF; fatigue crack propagation rate, FCPR.

A sheet semifinished product behaves practically until destruction as a linearly elastic material, the limit of proportionality in the case of tension being equal to 80–90% of the fracture stress. A favourable contribution to fracture is made by macrocracks developed across the specimen, formation of this crack taking place as a result of the combination of microdefects in a limited zone. An organocloth laminate that is similar in composition and formed under comparatively low pressures (0.5 MPa) behaves in the case of tension as a viscoelastic material, the limit of proportionality of which is equal to 50–60% of σ_b. The zone of fracture for this material is large; cracks develop both across and along a sample, with lamination of material and formation of shear cracks inside the layer.

A specific feature of sheet semifinished products is formation of a stability loss zone on the specimen in the case of 'instant' unloading of material after tension rupture. This zone is characterized by formation of a crumpling fold or local stress-whitening of samples in a narrow zone (formation of internal microcracks). The sheet organocloth laminate modulus of elasticity by warp and weft is equal to 35–40 GPa, and at an angle of 45° it is equal to 9.0–11.0 GPa. The value of modulus of elasticity and ultimate limit is increased monotonically when the material thickness is increased to 0.5–0.6 mm.

A typical feature of sheet organocloth laminates is high fracture viscosity and resistance to crack development. The organocloth laminate sensitivity to holes was tested on specimen strips $300 \times 25\,mm^2$ with ratio of centre hole diameter to specimen width equal to 0.2. The ratio of the strength of the material with hole to the strength of material without hole does not depend upon the angle between the tension direction and the axis of symmetry of the material, but remains constant and equal to 0.8.

The behaviour of organocloth laminate in the case of compression is determined mainly by the loss of reinforcing fibre stability. Organocloth laminate formed under low specific pressure is characterized by practically equal compressive strength of specimens cut out at different angles to the material warp ($\sigma_{-b} = 190\,\text{MPa}$). The ultimate compressive strength value of more rigid and monolithic sheet semifinished product increases to 250 MPa; the difference of cloth warp strength (250 MPa) and the strength at a angle of $45°$ to warp is 10–15%. Values of deformations, corresponding to the limit of proportionality in cases of compression (120–140 MPa) and tension (550 MPa), are equal and amount to 1.5%.

Serviceability of organocloth laminate in the case of loading with shear forces acting in the reinforcement plane is to a large extent determined by the structure of reinforcing textile filler, resistant to considerable deformation and the structure of which may be reconstructed in the direction of external force action. In this case shear deformation may reach 5–10%. Shear deformation of specimens cut out at an angle of $45°$ to the warp is increased by almost two times. When manufacturing organocloth laminates, the increase of specific moulding pressure affects the shear stiffness in the reinforcement place insignificantly. At the same time the shear strength of sheet semifinished product exceeds the strength of ordinary organocloth laminate by approximately 30% and reaches a value of 80–140 MPa (depending upon the size of cross-shaped specimen).

Low sensitivity of organocloth laminates to stress concentrators is also confirmed by the results of studying the shear strength in the reinforcement plane. So when the ratio of the central hole diameter to the specimen working part side length was equal to 0.1 during the test in a hinge frame with working area $100 \times 100\,\text{mm}^2$, organocloth laminate samples were destroyed along a line not going through this hole.

The test for interlayer shear by the short-beam method shows that the interlayer shear strength of sheet organocloth laminate is equal to 41 MPa and does not depend upon the direction of specimen cutting out.

The bending strength of sheet organocloth laminate is equal to 520 MPa. Fibre fracture is not observed in the stretched zone in the case of specimen fracture, but the compression zone is characterized by loss of stability of surface layers of material.

Long-time strength of sheet organocloth laminate in static tension based on 800 h is equal to 80% of the fracture stress. The creep deformation of the material in the case of tension stress of 400 MPa reaches 1.55% for 500 h.

Organocloth laminates are characterized by high fatigue strength and low fatigue growth rate. The fatigue tensile strength on the basis of 500 cycles is equal to 240 MPa. The fatigue strength growth rate for sheen organocloth laminate is equal to 0.05 mm/1000 cycles, which is 40 times lower than AFSGR (average fatigue strength growth rate) of D16AT aluminium alloy in the case of test under similar conditions (maximum cycle stress 120 MPa).

7.9 EFFECT OF WATER AND ITS VAPOUR ON ORGANOPLASTIC PROPERTIES

7.9.1 Sorption of water by organoplastics

In contrast to fibreglass and carbon plastics, the water sorption of the major components of which differ essentially and the reinforcing fibre practically does not absorb

Table 7.23 Values of equilibrium water sorption (M_∞) and diffusivity (D) for various fibres, polymeric matrices and composite materials based on them

Materials	$M_\infty (\%)$	$D \times 10^{-9} (cm^2 s^{-1})$
Fibres		
Terlon	2.0–3.5	–
Armos	3.5–5.0	–
SVM		
initial	14.0	–
modified	4–7	–
Aluminoborosilicate glass	0.3	–
Epoxydiane polymeric matrix	5.0	3.6
Epoxy organoplastics based on SVM fibre		
(initial)	9.4	3.6
Epoxy fibreglass plastics	1.1	0.4

moisture, organoplastics are subjected to the action of water over the whole volume of material. The equilibrium values of sorption M_∞ in glass fibres and carbon fibres are more than 10 times lower than those of respective polymeric matrices (Table 7.23). The difference of water sorption by organoplastic components is even less; sorption of water by polymeric fibres is in many cases higher than that for cured binders.

Sorption of water by synthetic fibres depends upon the chemical nature of the polymer, its structure, presence of admixtures and technology. So for example, during development of work in this direction the equilibrium water sorption of aramid fibres decreased from 14% to 2.0–3.5% at $P/P_s \sim 0.95$.

Figure 7.40 contains isotherms for sorption of water vapour by synthetic fibres of various chemical nature. The fibre water sorption varies within a wide range – from 0.5 to 14%. The largest quantity of water is sorbed by SVM fibre: when relative humidity is 0.65, sorption of water vapour amounts to 3–7%. The sorption properties of SVM fibre are in many cases determined by its production history (Fig. 7.41).

According to modern ideas, water sorbed by polymers is present in them in the forms of 'monolayer' water and free water, which is not bound to the polymer. Molecules of the latter are characterized by a trend to aggregation, with cluster formation inside the polymer. Thus the total content of sorbate is expressed by the total quantity of bound water molecules and their aggregates. In the case of the presence of pores and cracks in the polymer blending of aggregates and formation of capillary–condensed 'phase' water takes place. Monolayer water appears in polymers at P/P_s values that do not exceed 0.4, but aggregation of molecules begins at different values of this index and depends upon their polarity, structure, degree of crystallinity, etc. (Table 7.24).

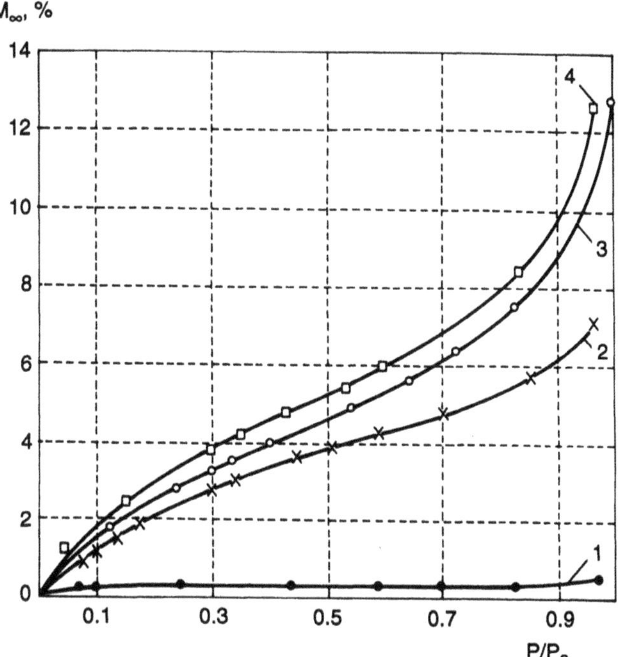

Fig. 7.40 Isotherms of water vapour sorption by various fibres: (1) polypropylene; (2) Fenilon; (3) SVM; (4) Terlon.

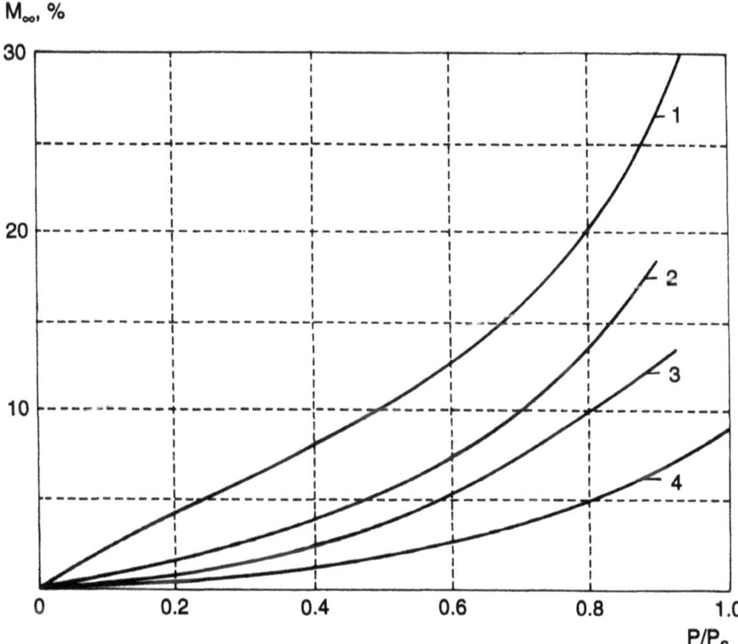

Fig. 7.41 Isotherms of water vapour sorption by SVM fibres produced by different technologies with various acidities: (1) pH 3, (2) pH 3.85, (3) pH 6.2, (4) pH 2.5.

Table 7.24 Quantity of bound water in polymeric fibres

Fibre	Quantity of bound water (%)	P/P_s value, under which aggregation of water molecules begins
Poly(ethylene terephthalate)	0.25	0.73
Aramid SVM	2.3	0.54
Aramid Terlon	3.0	0.62
Fenilon	4.3	0.85

Sorbed water is movable, and in the case of change of temperature and humidity conditions it is easily desorbed from the fibre. The kinetics of the desorption process was estimated by means of a thermogravimetric method under conditions of dynamic heating (Fig. 7.42). SVM fibres were studied, which were first conditioned at $P/P_s = 0.65$ until complete equilibrium (see Fig. 7.41, curve 4). The desorption process was most active on heating up to 473 K. Mass losses at 473 K correspond to quantity of water sorbed at $P/P_s = 0.65$ and amounted to 5.7%, which might be exclusively bound up with water molecule desorption. Analysis of the TGA curve shows the possibility to separate two temperature ranges: from 353 to 393 K and from 433 to 473 K. Capillary-condensed water is eliminated from SVM fibre at temperatures up to 393 K, and then at 433–473 K monolayer, energetically more strongly bound, water desorbs. The process of sorption–desorption is a complex one, which is testified by peak asymmetry on DTA curve [56]. It should be noted that the temperature range of water desorption corresponds to temperatures at which epoxy and phenol–formaldehyde binders widely used for reinforced plastics are cured. Desorption of moisture in the process of binder curing may cause formation of pores in the material. To clarify this, microplastics based on SVM fibre were studied, moistened before impregnation with epoxy binder under $P/P_s = 0.3$, 0.65 and 0.98, which determined the different contents of water in it.

Isotherms of sorption and desorption of water vapour were taken on microplastics; sorption of liquid water was studied (Fig. 7.43). The presence of a wide area of hysteresis between curves of sorption and desorption, as well as considerable differences in quantity of sorbed vapour and liquid water (Shreder's effect), indicate the presence of pores in plastics, formed as a result of water desorption from the fibre in the process of curing.

Binder components – resin and hardener – sorb water also, desorbing it during heating. Water may be released in considerable quantities and in particular during curing of polycondensation resins. Sorbed water affects the rate of binder curing and its crosslink density.

Thus curing composition is characterized by diffusion processes, causing an equalization of equilibrium water concentration in components, and an effect on formation of structure and properties of composite material.

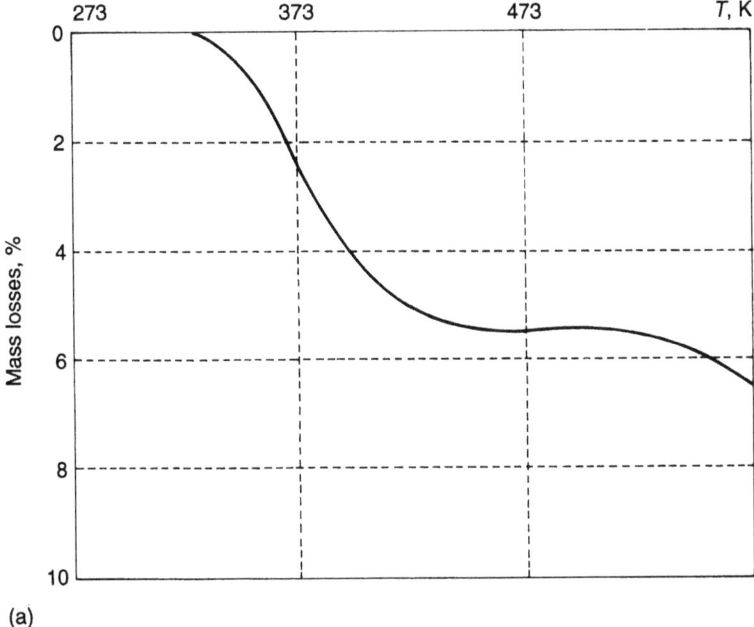

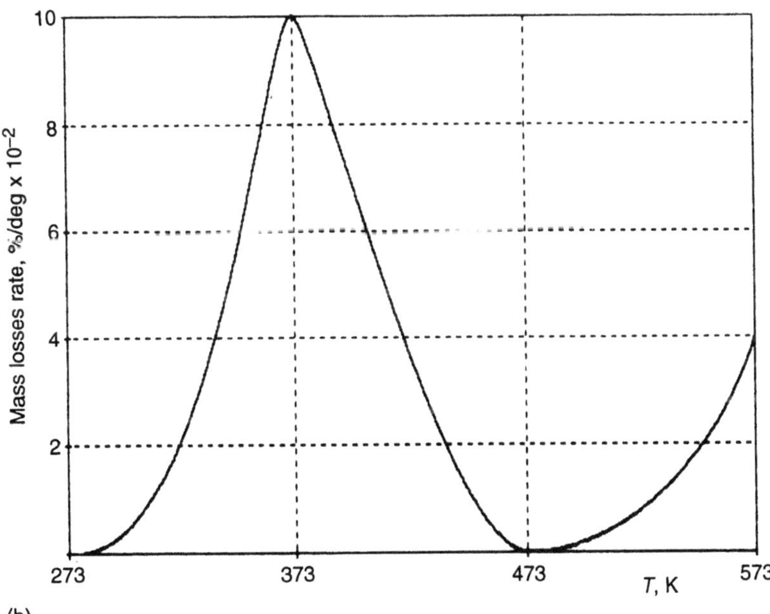

Fig. 7.42 (a) TGA and (b) DTA curves for SVM fibre.

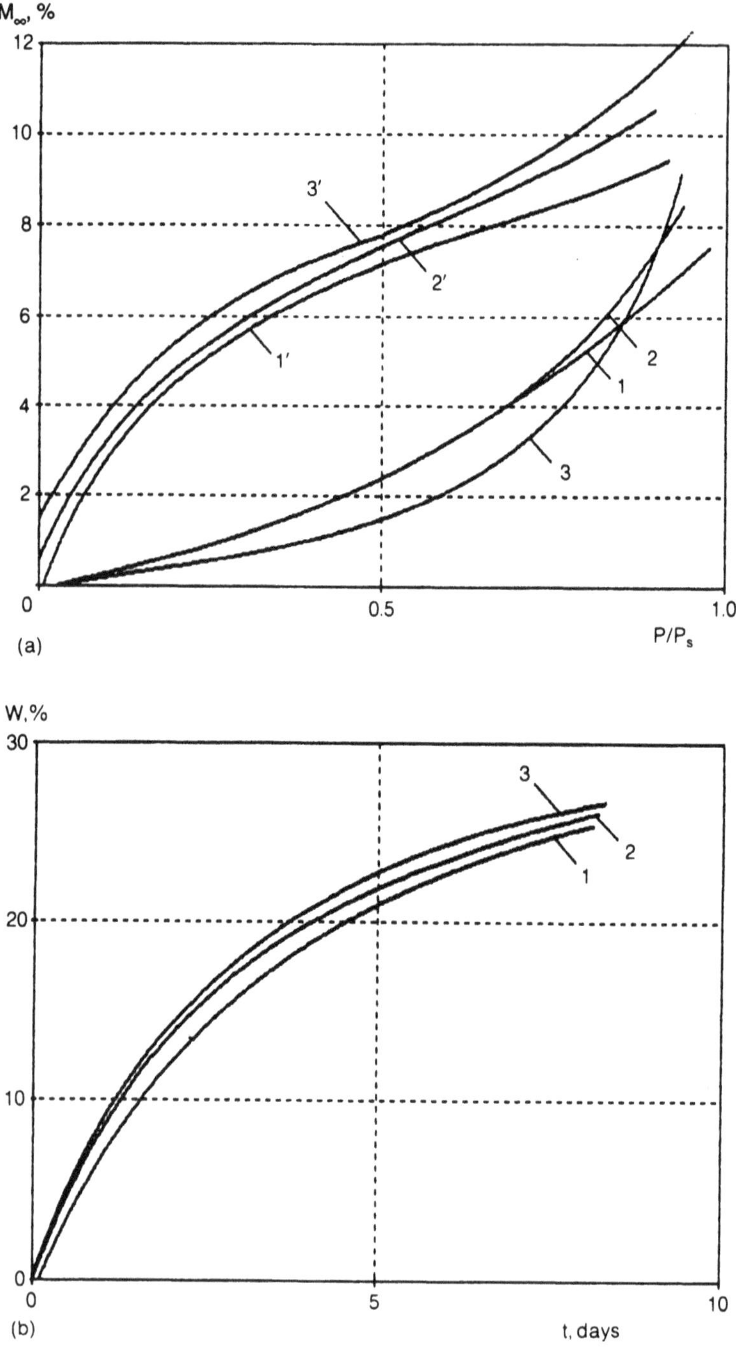

Fig. 7.43 (a) Isotherms of sorption (1–3) and desorption (1'–3') of water vapour and (b) kinetics of water sorption by microplastic based on epoxydiane binder and SVM fibre. The fibre was preliminarily moistened under $P/P_s = 0.3$ (1, 1'), 0.65 (2, 2') and 0.98 (3, 3').

7.9.2 Reversibility of transformations in organoplastics on exposure to water or moisture

The process taking place in the material in the case of moisture uptake from the surrounding medium may be conveniently subdivided into the groups. The first group covers reversible processes similar to swelling, during which the structure and properties of the material are changed due to the presence of sorbed water in it, but they return to the initial state during drying. The second group includes irreversible processes – chemical (hydrolysis, oxidation, etc.) and structural rearrangements of relaxation type, the after-effects of which do not disappear after removal of water. When estimating the resistance of organoplastics to temperature and moisture effects, and forecasting their service period, it is important to take into account the role of both processes, as well as their relative contribution to common effects of change of properties.

It was shown using the example of organocloth laminates based on epoxyaniline phenol–formaldehyde binder formed under various specific pressures (under 0.5 and 8 MPa) that these materials practically do not differ in the moisture absorption rate (Fig. 7.44). Moisture absorption is accompanied by change of linear dimensions of specimens (Table 7.25).

On the one hand we observe an increase of thickness h, i.e. dimension in the direction perpendicular to the reinforcement plane. These changes are practically completely reversible and the material is brought to its initial thickness after drying.

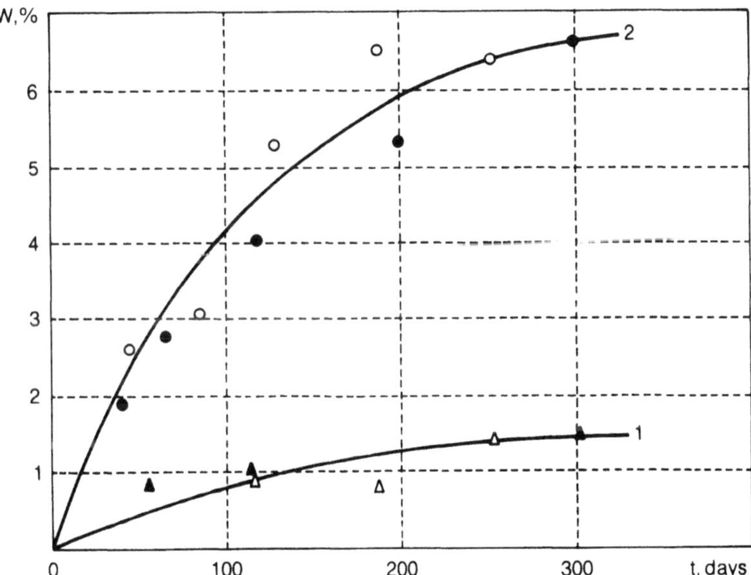

Fig. 7.44 Kinetic curves of water sorption by organoplastics at temperature of 333 K and relative humidity $\varphi = 80\%$ (1) and 100% (2). Organocloth laminates, formed under specific pressure: (○), (△), 0.5 MPa; (●), (▲), 8.0 MPa.

Table 7.25 Variation of linear dimensions of organocloth laminate specimens in the process of temperature and moisture attack[a]

Variation of dimensions (%)	State of specimens	Duration of exposure (days)				
		30	80	120	180	250
Thickness increase, $(h - h_0)/h_0$	Moist	2.8	4.7	5.8	7.4	6.5
	After drying	0.2	1.1	0.9	1.2	0.6
Decrease of width, $(b_0 - b)/b_0$	Moist	0.13	0.23	0.35	0.39	0.42
	After drying	0.16	0.31	–	0.56	0.70

[a]At 333 K and $\varphi = 100\%$; matrix, epoxyaniline phenol–formaldehyde.

An increase of thickness is linearly bound up with the quantity of sorbed moisture (Fig. 7.45). In this case the straight line crosses the abscissa axis at point $W \sim 1\%$. One may suppose that such a quantity of moisture is absorbed by pores inside the material. The rest of the moisture is evidently sorbed preferentially by the reinforcing

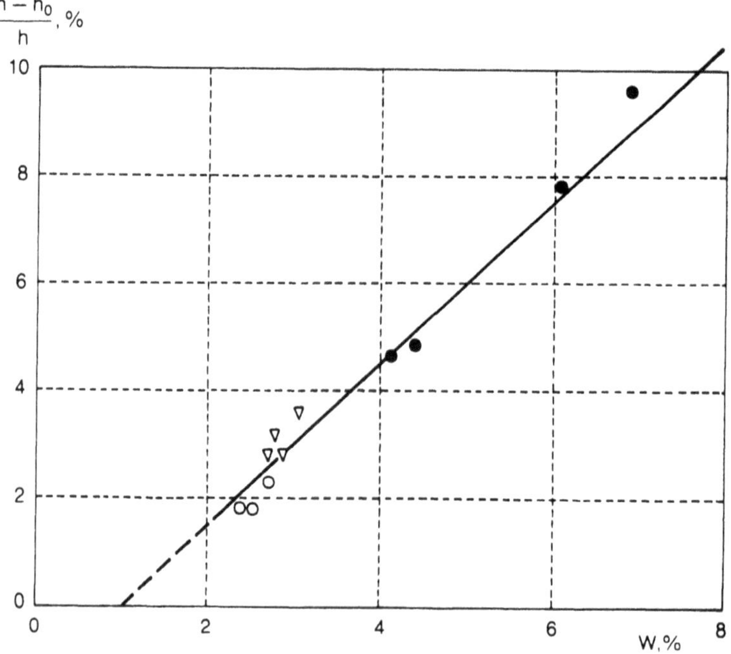

Fig. 7.45 Dependence of specimen thickness increase $(h - h_0)/h_0$ upon the quantity of sorbed moisture $W(\%)$ for organocloth laminates ($P_{sp} = 0.5$ MPa) at a temperature of 333 K and relative humidity $\varphi = 100\%$ during a day: (o) 30; ($\triangledown$) 80; (●) 120; (×) 180.

fibre, causing its swelling and respective increasing of specimen thickness. The dimensions of specimens in reinforcement interface are decreased as a result of moisture absorption, but not so essentially as in the case of thickness increase, but in contrast to the latter they are irreversible. As is seen from Table 7.25, drying of specimens leads not to a decrease, but to an increase of shrinkage in the reinforcement plane. Thus in the case of water sorption by organoplastic one can observe both reversible effects, disappearing in the case of moisture elimination, and irreversible ones, evidently bound up with structural rearrangements of relaxation type. These rearrangements occur also under the effect of moisture, but depend not only upon its content but also upon the test duration.

The tensile strength depends upon the content of moisture in organocloth laminates formed under different pressures, and changes in different ways, in spite of the fact that these materials have similar characteristics of water sorption (Fig. 7.46). Whereas an organocloth laminate with a density of $1270\,\text{kg m}^{-3}$ ($P_{sp} = 0.5$ MPa) is characterized by a rapid drop of strength in the beginning with subsequent slowing, the strength of material with density of $1400\,\text{kg m}^{-3}$ formed under $P_{sp} = 8$ MPa is increased, passing through a maximum, and then is decreased rather quickly. Analogous trends are observed for compressive and bending strength (Fig. 7.47).

Figure 7.46 (curves 2 and 4) shows that the difference in strength during temperature and humidity tests of two studied materials is predetermined considerably by the

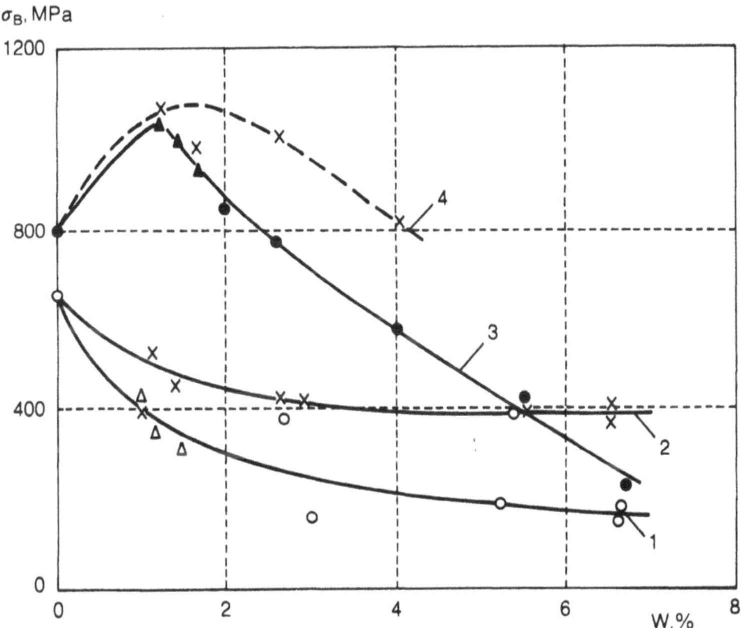

Fig. 7.46 Dependence of strength of organocloth laminates formed under $P_{sp} = 0.5$ MPa (1, 2) and $P_{sp} = 8.0$ MPa (3, 4) in the case of exposure to a quantity of sorbed moisture: in moist state (1, 3) and after drying (2, 4).

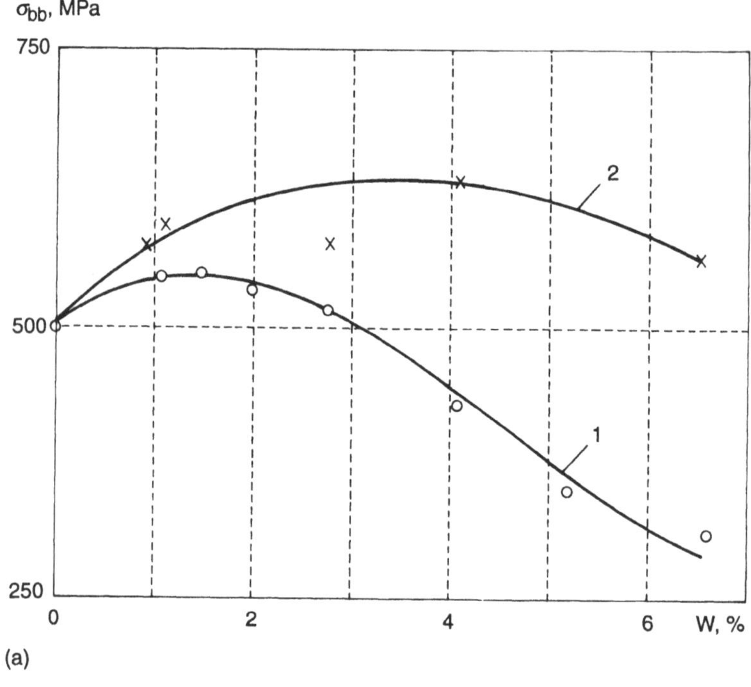

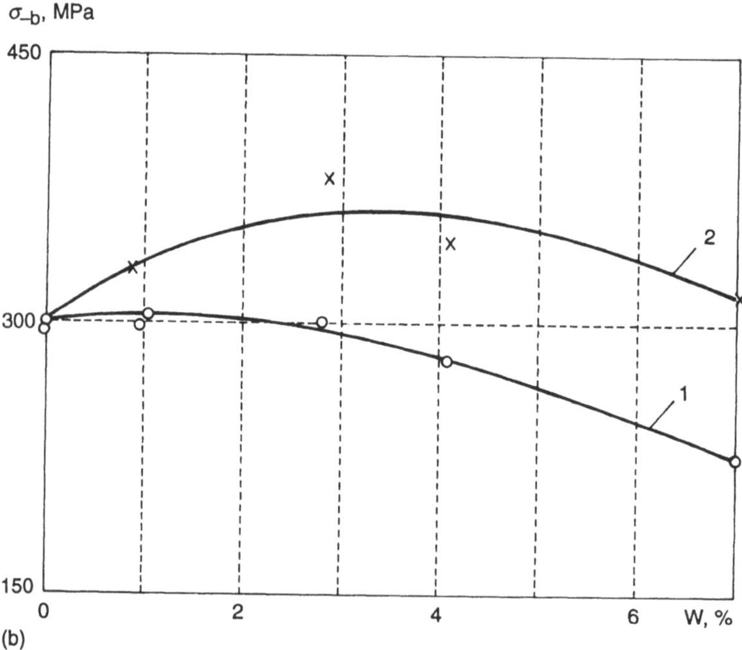

Fig. 7.47 Dependence of bending (a) and compressive (b) strength of organocloth laminate ($P_{sp} = 8.0\,\text{MPa}$) upon the quantity of sorbed moisture in moist state (1) and after drying (2).

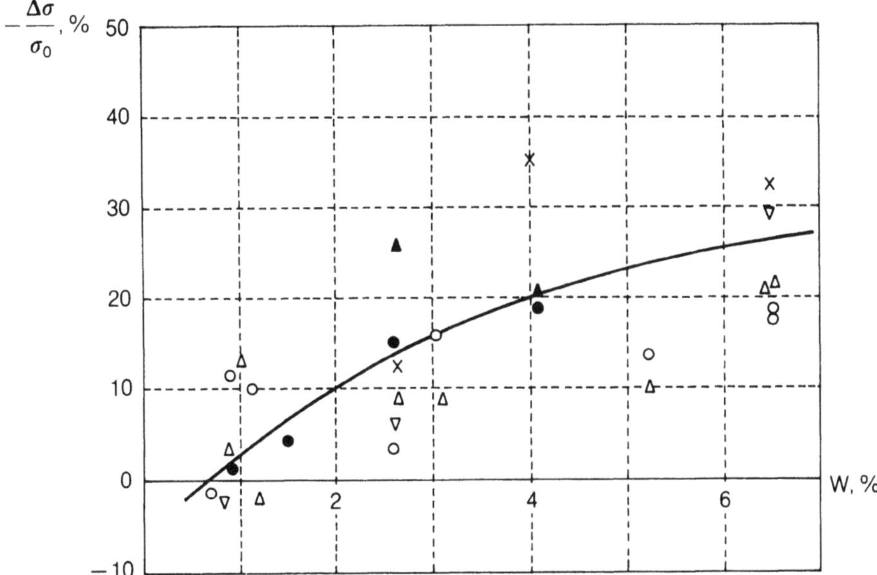

Fig. 7.48 Dependence of reversible effect of strength variation $-\Delta\sigma/\sigma_0$ (%) for organocloth laminates formed under $P_{sp} = 0.5\,\text{MPa}$ (o, ▲) and $P_{sp} = 8.0\,\text{MPa}$ (•, ▼, ▲, ×) in case of extension (•, ▼, o) and compression at 298 K (o, ▲, •, ▲) and 353 K (▼, ×) upon the quantity of sorbed moisture W (%).

different characters of the irreversible effects, bound up with the structure and technological history of the material. At the same time, the reversible effects for both materials depend upon a certain common trend (Fig. 7.48).

The contributions of reversible and irreversible processes to the common effect of strength variation under chosen conditions of temperature and moisture tests are approximately similar. Also in the case of change of conditions, the proportion of these contributions may be changed; it is evident that, when estimating the resistance of these materials to the action of temperature and moisture and, which is more important, while forecasting their service periods, it is impossible to ignore irreversible effects and to consider that the strength depends only upon the moisture content. The character of irreversible phenomena is different even in materials that are similar in composition of materials and differ only by the technology of production.

7.9.3 Organoplastic ageing mechanism

The specific features of the ageing mechanism of aramid organoplastics may be demonstrated on specimens of organocloth laminate based on SVM satin cloth and epoxyaniline phenol–formaldehyde matrix. Ageing of samples was carried out under natural conditions of subtropical climate in the city of Batumi. Organocloth laminate plates of 1.8 mm thickness were exposed on an open site for six years. The combination

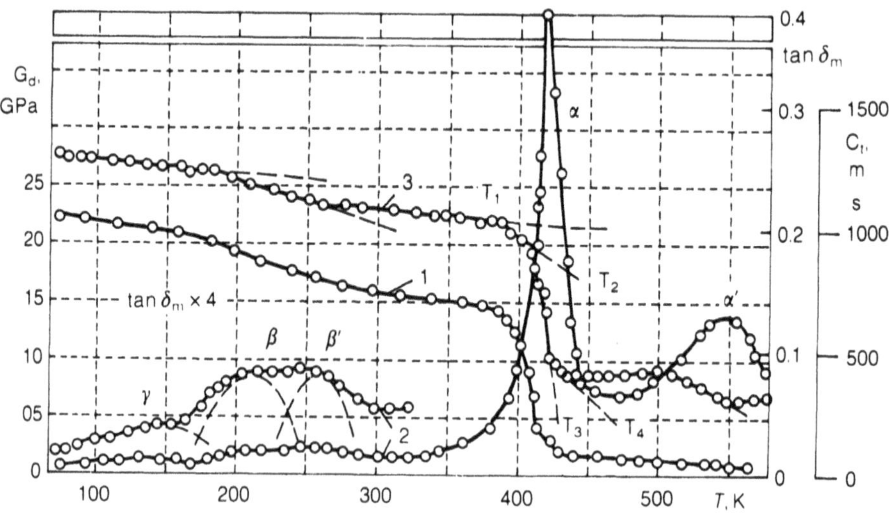

Fig. 7.49 Temperature curves for G_d (1), tan δ_m (2) and c_t (3) of organocloth laminate in initial state.

of such aggressive factors as high average annual relative humidity (80–85%), large quantity of precipitation (2500–2800 mm per year), very high level of solar radiation, the immediate neighbourhood of the sea and increasing content of chlorides and sulphates in the atmosphere determines the high aggressiveness of the damp subtropical climate of Batumi.

Organocloth laminate specimens exposed to various periods of atmospheric ageing were conditioned in a vacuum chamber at room temperature until the mass was stabilized. Then these specimens were subjected to detailed study of the dependence of dynamic shear modulus G' and mechanical loss tangent (tan δ_m) in a wide range of temperatures (77–570 K). These characteristics were determined by means of a reversed torsional pendulum with an error varying from 2–3% at low temperatures to 5–8% in a high-temperature area.

Figure 7.49 illustrates temperature curves for G_d, tan δ_m and low-frequency velocity of shear waves c_t calculated as per formula $c_t = (G_d d)^{1/2}$, where d is density of initial specimens of composite material.

Long-term exposure of organocloth laminate noticeably changed the character of temperature curves of G_d and tan δ_m.

For comparison, Fig. 7.50 contains high-temperature parts of curves of viscoelastic characteristics of composite material in the initial state (curve 1, 1′), as well as those after exposure for four (2, 2′) and six (3, 3′) years under natural conditions of subtropical climate. As is clear from the figure, the character of curves $G_d = f(T)$ and tan $\delta_m = f(T)$ is changed considerably with increase in duration of atmospheric ageing. More detailed kinetic dependences of some characteristics, determined on relaxation curves are shown on Fig. 7.51.

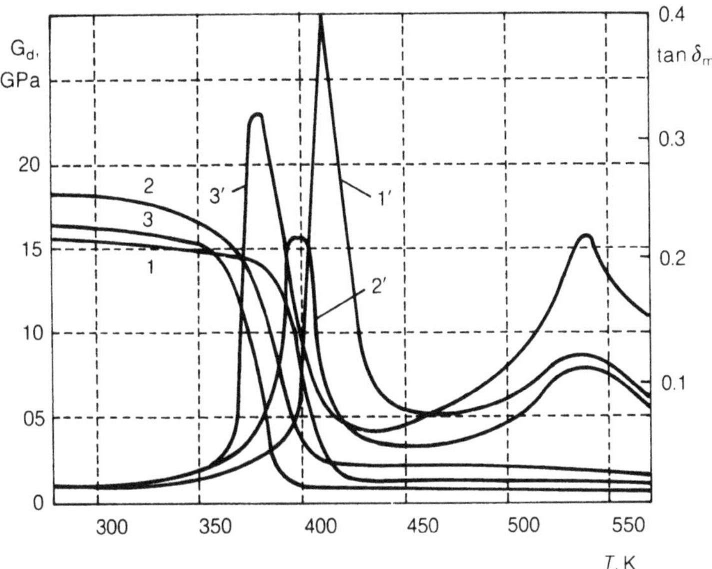

Fig. 7.50 Curves $G_d = f(T)$ (1–3) and $\tan \delta_m = f(T)$ (1'–3') for organocloth laminate in initial state (1, 1') and after atmospheric ageing for four (2, 2') and six (3, 3') years.

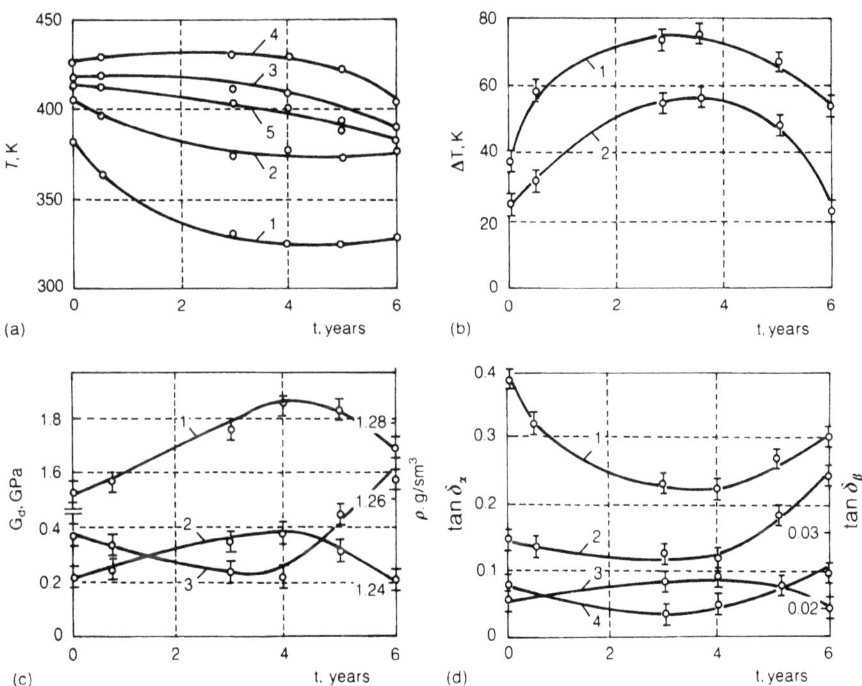

Fig. 7.51 Influence of atmospheric ageing time on characteristics of dynamic mechanical spectra of organocloth laminate. (a) Transition temperature T_1^* (1), T_2^* (2), T_3^* (3) and T_4^* (4) in α relaxation area and α maximum temperature (5). (b) Width of transition zone ΔT_{13} (1) and ΔT_{24} (2). (c) Dynamic shear modulus at $T < T_4 = 300\,\mathrm{K}$ (1), $T > T_4 = 450\,\mathrm{K}$ (2) and density (3). (d) Height of α (1), α' (2) β (3) and β' (4) maxima of $\tan \delta_m$.

Analysis of the given kinetic curves shows the possibility to make a number of suggestions for the mechanism of atmospheric ageing of organocloth laminate.

Attention should be paid to the fact that transition temperatures T_1 and T_2 are changed with increase of ageing time t (Fig. 7.51(a)). Exposure of composite material for four years at $30-50\,K$ decreases the temperatures T_1 and T_2, when at the same time T_3 and T_4 remain almost constant. The further increase of t causes the opposite effect: T_1 and T_2 remain constant and T_3 and T_4 decrease by $25-30\,K$: Thus if the area of α relaxation is to be characterized by two overlapping areas $\Delta T_{13} = T_3 - T_1$ and $\Delta T_{24} = T_4 - T_2$, the time dependence, as is clear from Fig. 7.51(b), of these parameters is extreme. When $t = 3.5-4$ years, binder zones of passage from glassy to rubber-like states reach a maximum and then decrease.

Dependences $G_d = f(t)$ (Fig. 7.51(b)) in glassy (curve 1) and rubber-like (curve 2) states of binder are also of extremal character, the respective extremum being also equal to four years.

It is known [57] that an increase of modulus of elasticity of crosslinked polymer in rubber-like state testifies to the increase of cross-bonding degree. Since in our case parameters characterizing the properties of organic fibre (temperature and height of α' peak) are not changed practically for the first four years of ageing (Figs. 7.50 and 7.51(d)), variation of dynamic shear modulus G_d at $T > T_4$ is evidently caused by the increase of density of the network of binder adhesion. The result, given of Fig. 7.51, bears additional arguments in favour of this proposition.

So the growth of G_d at $T > T_4$ is accompanied by an essential decrease of the height of α maximum of mechanical losses (Fig. 7.51, curve 1). The height of α peak after $3-4$ years of atmospheric ageing is decreased almost twice. Usually a decrease of α maximum height is considered as a criterion of cross-bonding degree increase [57]. For the first $3-4$ years of atmospheric ageing, simultaneous influence of moisture, ultraviolet radiation and temperature creates favourable conditions for achievement of higher extent of binder curing. This evidently takes place at the expense of both hardener residues and active groups, formed in the case of fracture of most stressed sections of macrochains.

Each of two levels of supermolecular organization of amorphous polymers may be considered as a set of microfields, differing by degree of molecular or intermolecular interaction effectiveness [57]. The wider and more diversified this set, the longer is the area of passage from glassy into rubber-like state.

At first it seems strange that an increase of density of the network of binder adhesion for the first four years of ageing, causing an increase of modulus of elasticity in glassy state, is accompanied by a decrease of vitrification temperatures T_1 and T_2. However, it becomes clear if we assume that, simultaneously with the cross-bonding process, prevailing for the first years of atmospheric action, we also observe the processes of binder destruction, playing the main role when $t > 4$ years of tests.

Therefore expansion of transition zones ΔT_{13} and ΔT_{24} observed on Fig. 7.51(d) goes initially at the expense of T_1 and T_2 decrease due to formation of a section of macrochains with lower effectiveness of intermolecular interaction. The sharp decrease of temperatures T_3 and T_4 at the second stage of atmospheric ageing proves that sections of macrochains placed in most tightly packed polymer sections at each of two levels of crosslinked binder supermolecular organization are involved in the fracture process. This determines the extremal character of the graph $\Delta T = f(t)$ (Fig. 7.51(b)).

Thus in the case of exceeding 3.5–4 years of atmospheric ageing, the crosslinking effects become inessential or do not show themselves at all, and all changes of viscoelastic characteristics of binder are determined by the fracture process. A convincing confirmation of this conclusion is a decrease of G_d in highly elastic and glassy states of binders (Fig. 7.51(c)), as well as an increase of α maximum at the second stage of ageing (Fig. 7.51(d), curve 1).

If our proposals about the prevailing effects of binder crosslinking process at the first stage of ageing and fracture processes at the second stage are true, it may be expected that a macroscopic characteristic of composite material such as density will first decrease, since the presence of a denser network of adhesion will prevent packing of macrochains at $T < T_g$, and free volume will be increased in the glassy state. The results of density measurement (Fig. 7.51(c), curve 3) confirmed the expected effects. An interesting indirect argument in favour of the conclusion about free-volume increase at the first stage of ageing is the time dependence of β peak height (Fig. 7.51(d), curve 3), characterizing molecular mobility of local type. Mobility of small kinetic elements of binder macrochains is reduced with free-volume increase with growth of β maximum of aged specimens. When $t > 4$ years, fracture of cross-bonds again permits more dense packing of macrochains to be achieved; therefore the β maximum intensity will be reduced again.

In conclusion, let us not forget that the studies carried out by us gave the possibility to reveal also the influence of a damp subtropical climate on the elastic properties of reinforcing filler–organic fibre cloth.

For the first years of atmospheric ageing polymeric binder served evidently as a reliable shield, protecting the organic fibre from the action of aggressive climatic factors.

As already noted above, the position of α' peak on the temperature scale and its height remain constant with increase of ageing time up to four years (Figs. 7.50 and 7.51(d), curve 2). In the case of further increase of test time, the height of α' maximum is increased sharply and after six years of ageing it is almost double the respective value of initial specimens. Such an essential growth of α' maximum at constant temperature may be caused only by an increase of the number of kinetic elements of fibre macrochains participating in mobility of micro-Brownian type, and it is possible only in the case of loss of high initial order along the organic fibre axis, i.e. in case of its misorientation. The observed growth of height of β' maximum of mechanical losses when $t > 3$–4 years (Fig. 7.51(d), curve 4) is also an additional confirmation of this conclusion. It may be expected that misorientation of organic fibre macrochains will be the reason for the sharp decrease of organocloth laminate strength.

The above-presented data of atmospheric ageing mechanism of organocloth laminate are used for estimating property variation of organoplastic of various compositions in the process of long service under high-humidity conditions [58, 59].

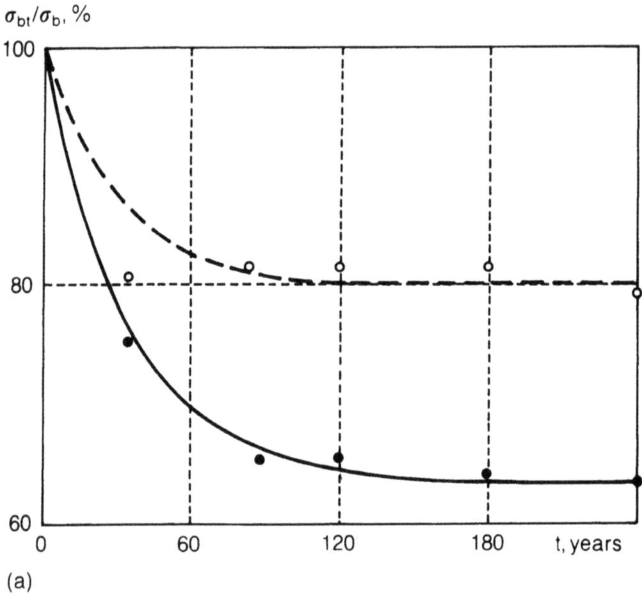

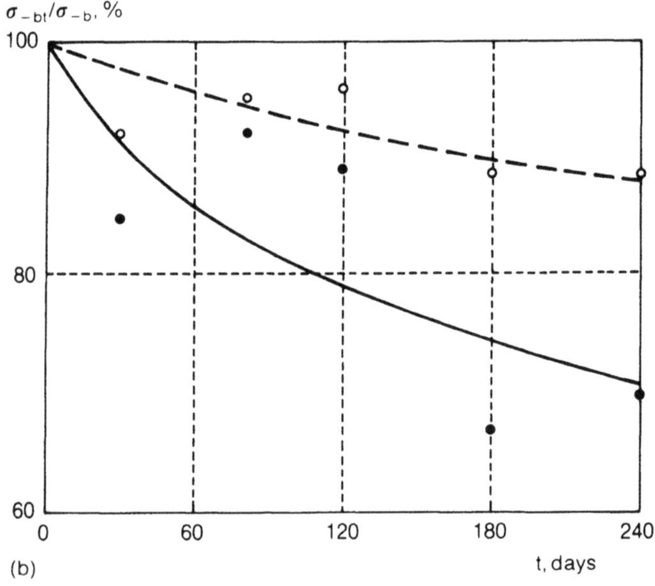

Fig. 7.52 Effect of temperature and humidity ageing on relative (a) tensile σ_{bt}/σ_b (%) and (b) compressive σ_{-bt}/σ_{-b} (%) strengths of organocloth laminate: (————) in moist state ($T = 333$ K, $\varphi = 100\%$); (— — —) after drying at 333 K.

7.9.4 Ageing of organoplastics due to heat and humidity

Unprotected aramid fibres are very sensitive to the action of ultraviolet light and moisture. The study of SVM fibre cloth strength variation during ageing under conditions of hot dry (Erevan) and damp (Batumi) climates showed that in a damp climatic zone after 30 days of ageing only 50%, and in a dry zone 80%, of strength remain. Decrease of strength of unprotected SVM cloth under these conditions is caused by the action of temperature and moisture factors and ultraviolet light. The character of variation of SVM cloth properties in composite material is determined by the essential screening effect of polymeric matrix. Let us consider, as an example, the ageing of aramid organocloth laminate based on epoxyaniline phenol–formaldehyde matrix. The mechanism of humidity ageing for this material was considered in the previous section. Material ageing was carried out on specimens free of varnish and paint coatings and with unprotected faces.

Heat and humidity ageing of organocloth laminate at $\varphi = 100\%$ within 240 days causes a decrease of compressive strength by 20% and tensile strength by 30%; subsequent drying of specimens causes partial restitution of compressive strength (up to 80–90%) and of tensile strength up to 80% (Fig. 7.52). One may judge the influence of cyclic action of media of various humidity on the properties of organocloth laminates by data obtained as a result of testing this material in a tropical climate chamber (Fig. 7.53) under the following modes: $T = 323\,K$, $\varphi = 100\%$, 8 h; $T = 293\,K$, $\varphi = 100\%$, 12 h; drying at 293 K. Exposure of specimens under these conditions for three months is accompanied by a decrease by 30%; with an increase of holding

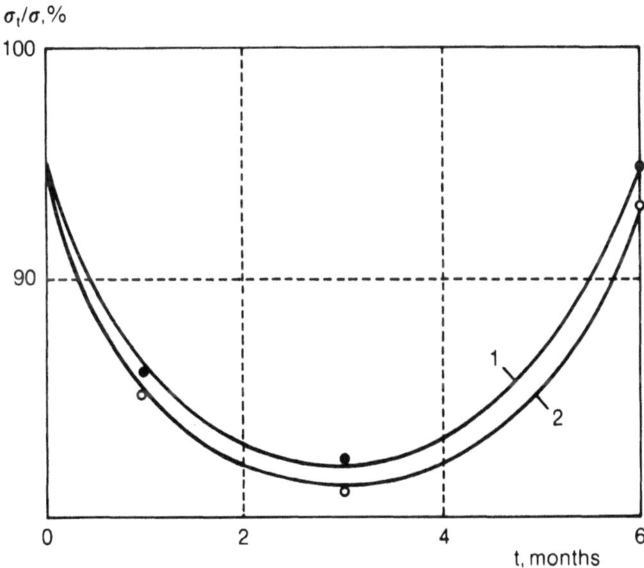

Fig. 7.53 Effect of duration *t* of holding in tropical climate chamber on relative tensile σ_t/σ (%) (1) and compressive (2) strengths of organocloth laminate.

duration to six months, strength is partially restituted: its level amounts to 95% of the original one.

The absolute values of medium temperature and humidity are determined by the climatic zone and other service factors, bound up with the purpose of the article. In this connection, estimation of variation of material service properties should be carried out in the process of imitation tests. One year of tests imitates the action of a medium with the following parameters: $T = 333\,K$ and $\varphi = 80\%$, 13 days; $T = 333\,K$, 8h (one cycle), subsequent holding at 293 K, 1h and at 293 K, 1h (45 cycles).

It was determined that after imitation tests the tensile and compressive strengths of specimens of organocloth laminate free of varnish and paint coatings and with unprotected faces remain at the level of 80–90% at 293 K; the bending strength remains constant (Fig. 7.54).

After the action of temperature of 353 K, the tensile and bending strengths decrease to 60–70%, and the compressive strength remains at the level of the original one.

Figure 7.55 illustrates the results of organocloth laminate tests for bending action of thermal and heat–humidity ageing factors in free and stressed states under a load equal to 30% of the destructive one. As is seen in the case of heat–humidity ageing, the bending strength is decreased to 70–75% in both free and stressed states; in the case of thermal ageing, it remains at the level of the original one in the free state and at the level of 90% in the stressed state.

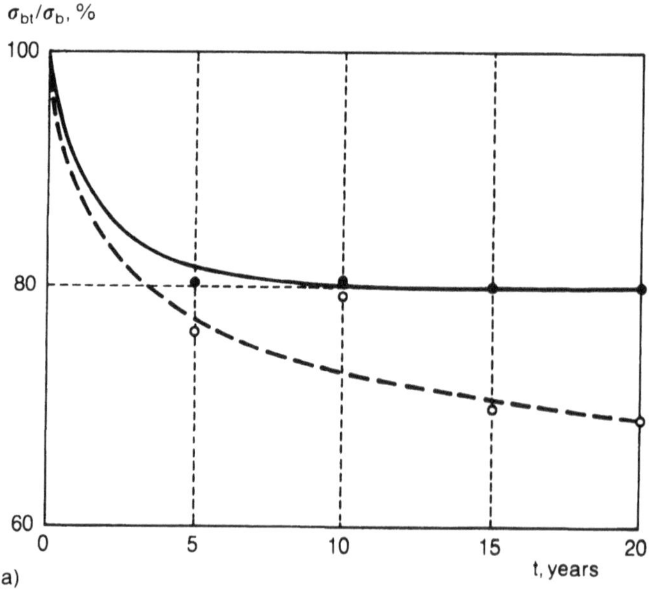

Fig 7.54 Effect of imitation tests on relative (a) tensile σ_{bt}/σ_b (%), (b) compressive σ_{-bt}/σ_{-b} (%) and (c) bending σ_{bbt}/σ_{bb} (%) strengths of organocloth laminate. Test temperatures: 293 K (———) and 353 K (– – –).

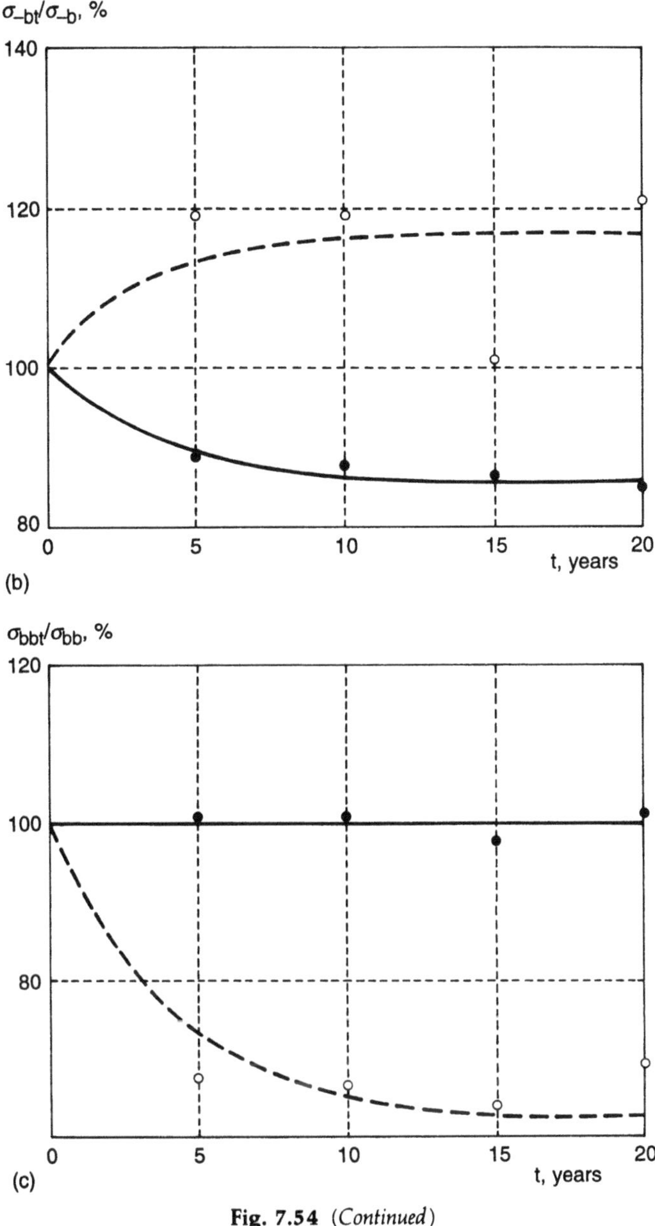

Fig. 7.54 (*Continued*)

Exposure of organocloth laminate on the deck and in the hold of a research ship during two voyages with total duration of 12 months does not affect the tensile strength, and only in the case of holding the material in a tank with sea water is it decreased to 75% of the original one (Fig. 7.56).

Figure 7.57 illustrates the variation of organocloth laminate strength under the action of a hot damp climate (Batumi) in the open and in a shed. On an open

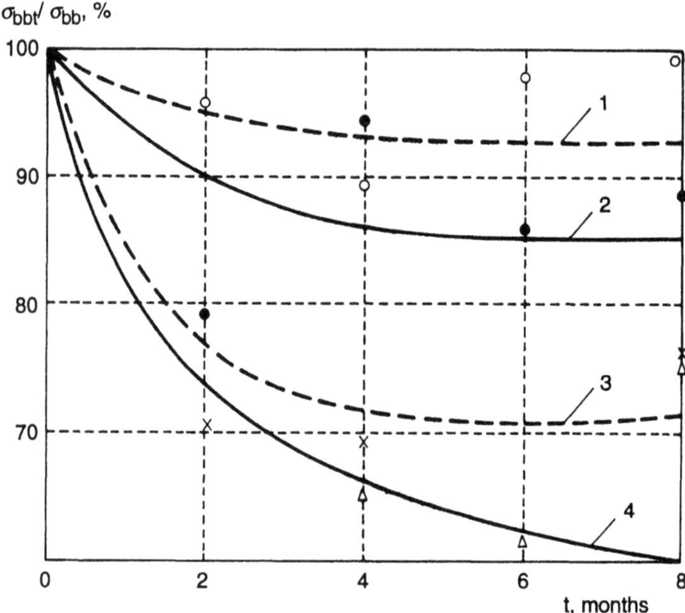

Fig. 7.55 Variation of relative bending strength σ_{bbt}/σ_{bb} (%) of organocloth laminate in stressed (———) and free (– – –) state after thermal (1, 2) and heat–humidity ($T = 313\,\text{K}$, $\varphi = 100\%$) ageing (3, 4).

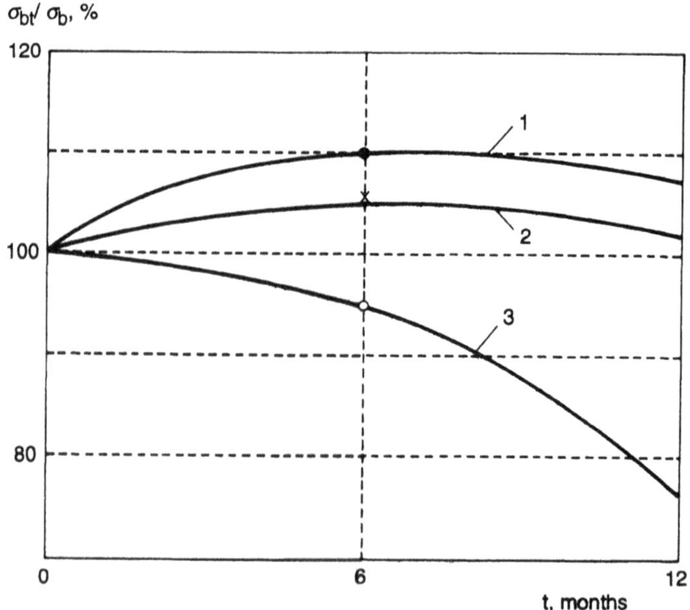

Fig. 7.56 Variation of relative tensile strength σ_{bt}/σ_b (%) of organocloth laminate during exposure on a ship: on deck (1), in the hold (2) and in a tank with sea water (3).

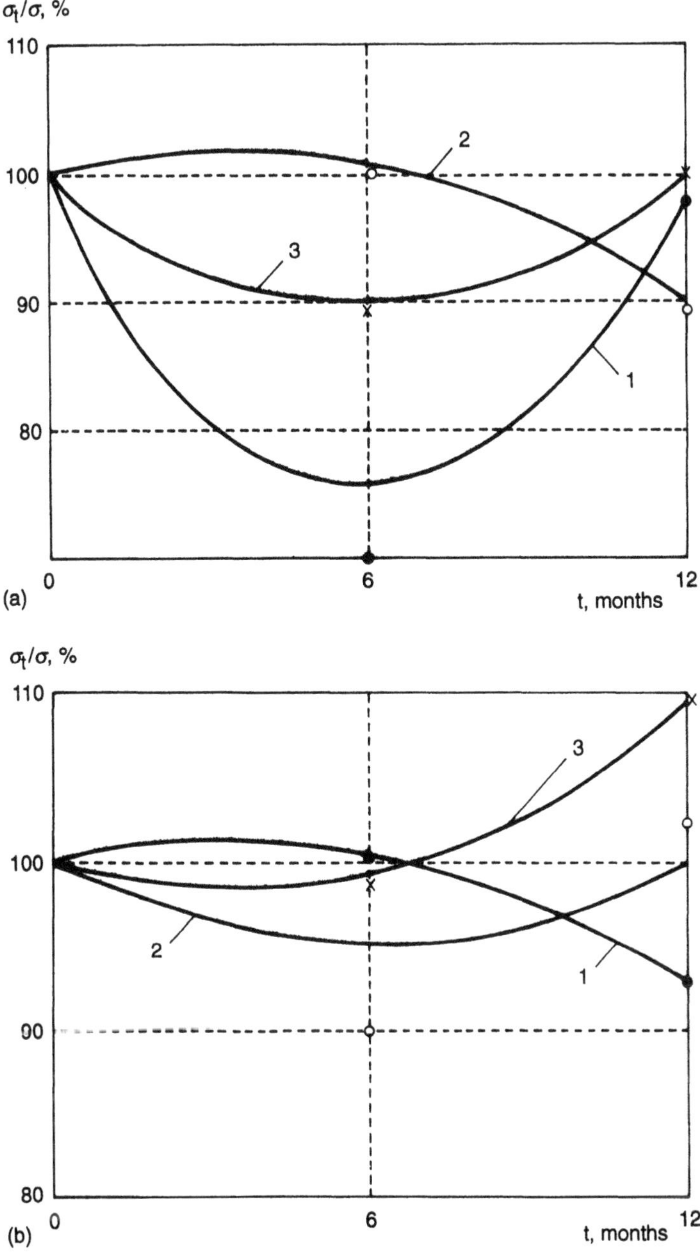

Fig. 7.57 Variation of relative tensile (1), compressive (2) and bending (3) strength σ_t/σ (%) of organocloth laminate after exposure under conditions of a hot damp climate on an open stand (a) and in a shed (b).

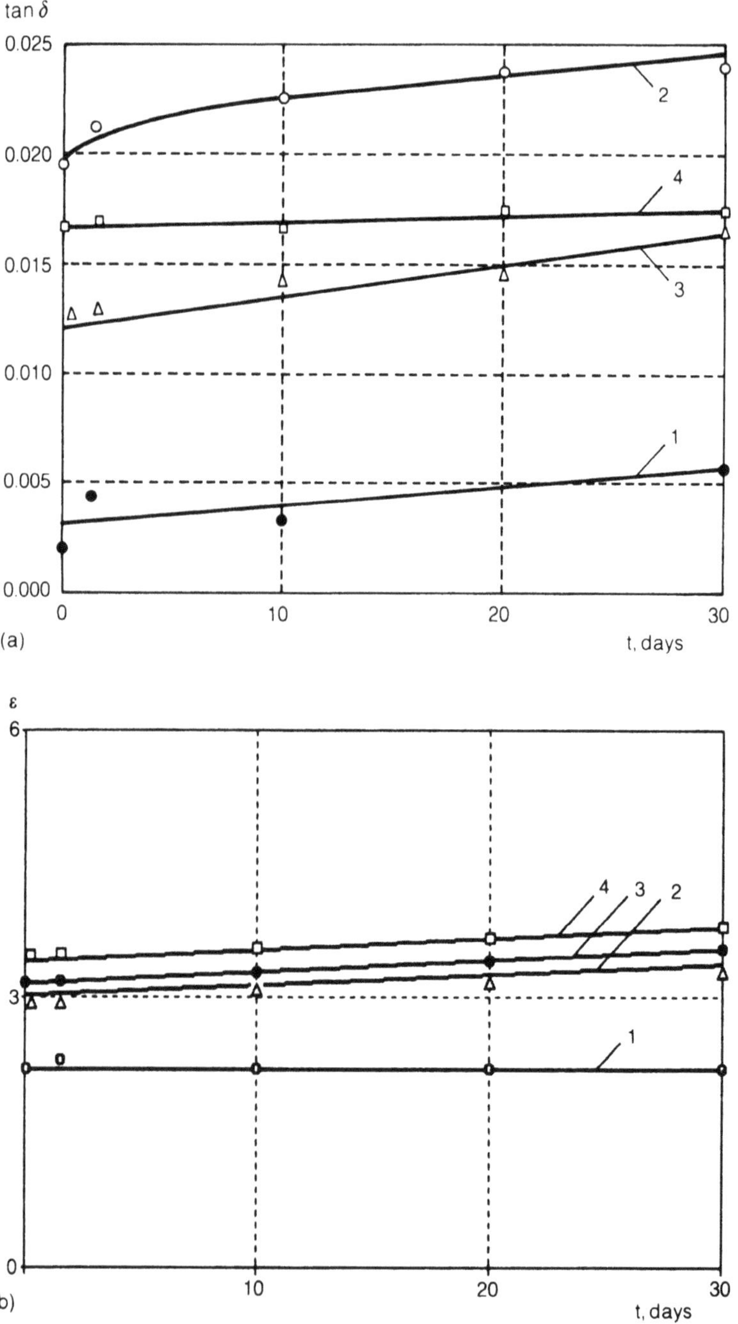

Fig. 7.59 Variation of (a) dielectric loss tangent tan δ and (b) dielectric permittivity ε of organocloth laminate in process of lengthy moistening, when $\varphi = 98\%$: (1) allyl binder, reinforced with polyolefin fibres; (2) epoxydiane binder, reinforced with poly(ethylene terephthalate) fibres; (3, 4) phenolfurfurolic (3) and epoxyaniline phenol–formaldehyde (4) binder, reinforced with aramid fibres.

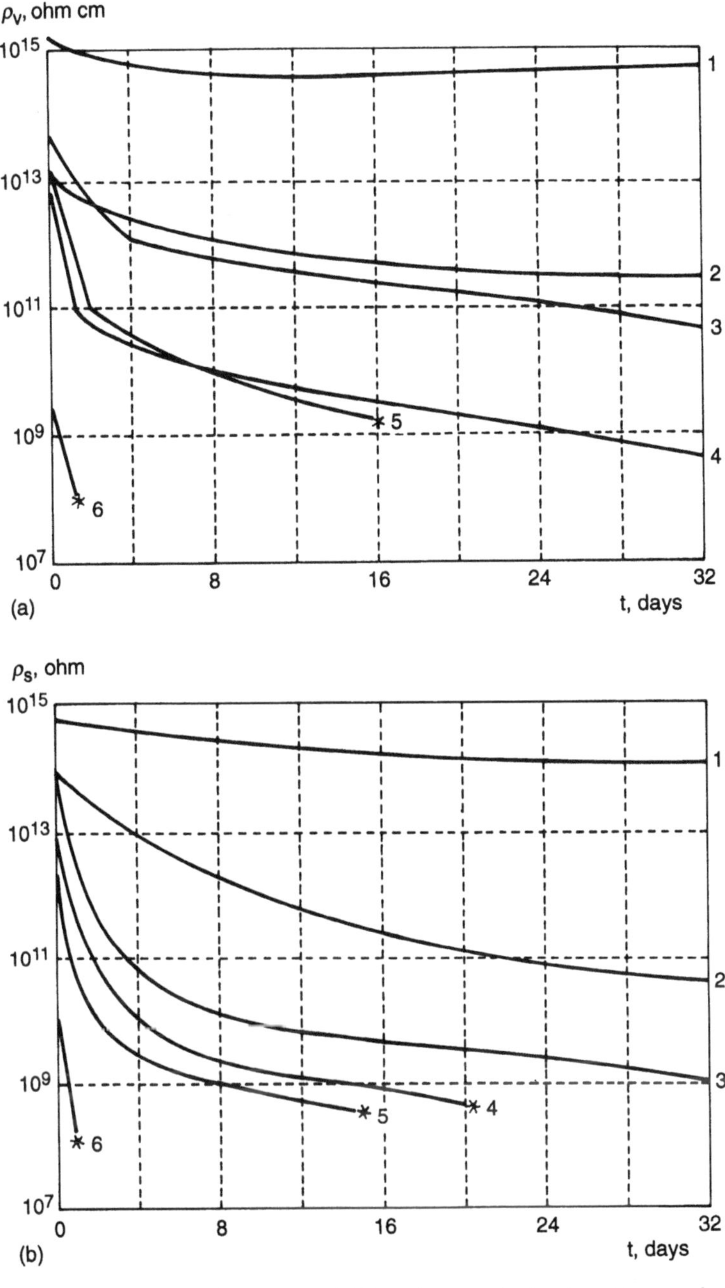

Fig. 7.60 Variation of (a) volume ρ_v and (b) surface ρ_s electrical resistance of composite materials in process of holding in an atmosphere with $\varphi = 98\%$, $T = 293$ K: (1) organocloth laminate based on polypropylene cloth and allyl binder; (2, 3) epoxy organocloth laminate based on aramid cloth SVM (2) and poly(ethylene terephthalate) cloth (3); (4, 5) epoxy fibreglass cloth laminates based on cloth with chemical (4) and paraffin emulsion lubricant (5); (6) electrotechnical cloth laminate based on cotton cloth and phenol–formaldehyde binder. The asterisks indicate that ρ_v and ρ_s are not measured for longer times.

action of humid medium ($\varphi = 100\%$) and temperature up to 323 K, which simulates the conditions of a tropical climate.

The best one in this case is an organocloth laminate reinforced with polyolefin fibres, the dielectric properties of which after three months in the tropical climate chamber are practically not changed.

The dielectric properties of aramid organocloth laminates are increased more in the case of moistening, which correlates with the higher water sorption of aramid fillers.

Organoplastics as dielectric materials are used for the manufacture of radio-transparent antenna shields of aircraft and helicopters. Such shields are of monolithic or cellular structure. Long service of shields manufactured of organoplastics showed the high stability of their radio engineering parameters under the action of a medium with high (up to 100%) humidity and at a temprature of 303–313 K, as well as under the action of low temperatures (213 K), salt (sea) mist and condensed precipitation.

7.11 THERMAL PROPERTIES OF ORGANOPLASTICS

Under the action of high temperatures, aramid fibres have insignificant shrinkage, reaching $\sim 2\%$ at a temperature of about 673 K. In the case of the mutual action of water vapour and temperature of about 373 K, shrinkage of SVM and Armos fibres reaches 0.5%, and the dimensions of crystalline Terlon fibre practically do not change.

On elevation of temperature over the vitrification temperature (Table 7.27) shrinkage phenomena are accompanied by thermal destruction of fibre-forming

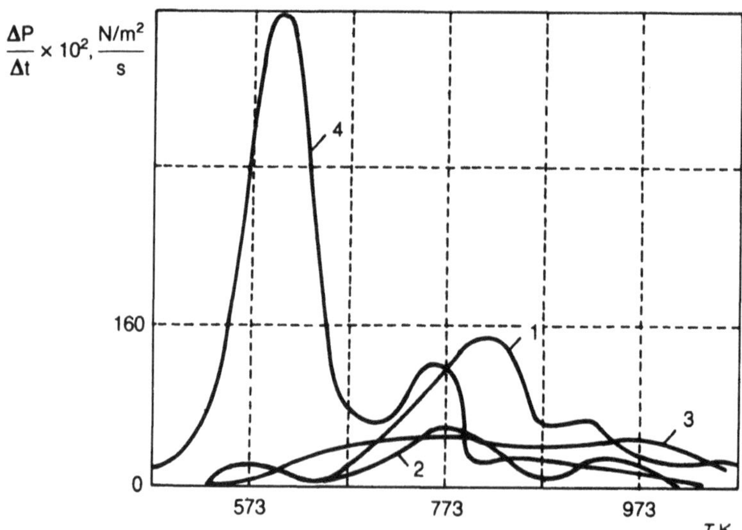

Fig. 7.61 Differential curves of gassing of aramid fibre and polymeric matrices when heating at a rate of 4°C min^{-1}: (1) Terlon; (2) SVM; (3) phenolphthaleinfurfurol polymer; (4) epoxydiane polymer (with amine hardener).

Table 7.27 Thermal properties of aramid fibres [7]

	Fibres		
Properties	SVM	Armos	Terlon
Density ($kg\,m^{-3}$)	1420–1450	1420–1450	1450
Tensile strength of thread ($kN\,tex^{-1}$)	160–235	150–290	160–220
Relative strength (%) at:			
77 K	80–90	–	–
373 K	85–90	85–90	–
473 K	70–75	70–75	65–75
523 K	65–70	65–70	55–60
Temperature (K) of:			
vitrification	504–523	433–483	618–673
decomposition	723–793	773–793	773
ignition	853–878	768	718
glowing	658–673	673	678
Relative strength and stiffness after 100 h			
of holding at 473 K (%):			
strength	80–90	80–90	60–65
modulus of elasticity	80–90	80–90	95–97
Coefficient of linear thermal expansion, $\alpha \times 10^{-6}(K^{-1})$ in the range of temperatures of 293–573 K:			
along fibre	± 1	–	$-$(1 to 2)
across fibre	60–65	–	30–60
Specific heat at 293 K ($J\,kg^{-1}K^{-1}$)	1420	–	1420
Coefficient of heat conduction ($W\,m^{-1}K^{-1}$)			
along fibre	0.045	–	0.040
across fibre	–	–	0.050

polymer [60]. Differential curves of gas discharge of aramid fibres were determined on a special stand for thermogravimetric analysis and are given in Fig. 7.61 [61].

Destruction of aramid fibres has much in common with destruction of polymeric matrices, and is a multistage process, accomplished with formation of highly carbonized coke residue as a result of pyrolysis. Thermal destruction of SVM fibre occurs in three stages with a low rate of gas emission in each of them and activation energy of 260–270 kJ mol^{-1}. Insignificant gassing within the temperature range of 523–598 K is bound up with elimination of solvent residues and other low-molecular-weight products. SVM fibre destruction is accompanied by, a high degree of carbonization (output of solid residue is 53%), hydrogen being eliminated practically completely and preferentially carbon and nitrogen remaining in the skin. From mass-spectral analysis data the basic products of SVM fibre destruction are CO, CO_2, H_2O, H_2 and benzene.

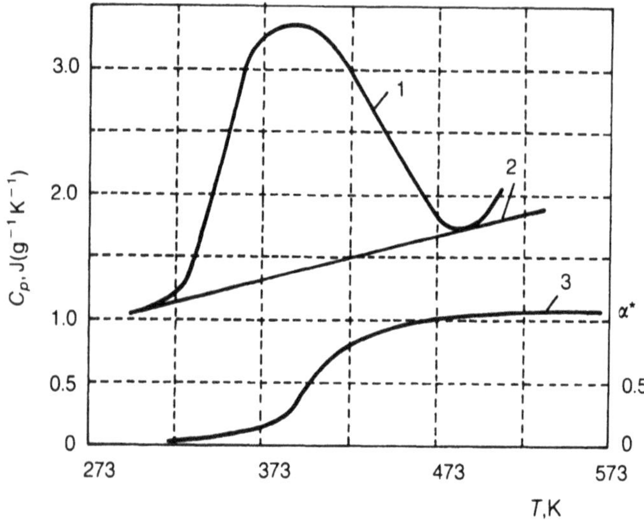

Fig. 7.62 Temperature dependence of specific heat capacity C_p (1, 2) and degree of conversion α^* (3) of aramid fibre: (1) first heating; (2) second heating. Degree of conversion α^* indirectly characterizes current weight of specimen, determined through degree of conversion, which is calculated with respect to peak area relative to whole peak area.

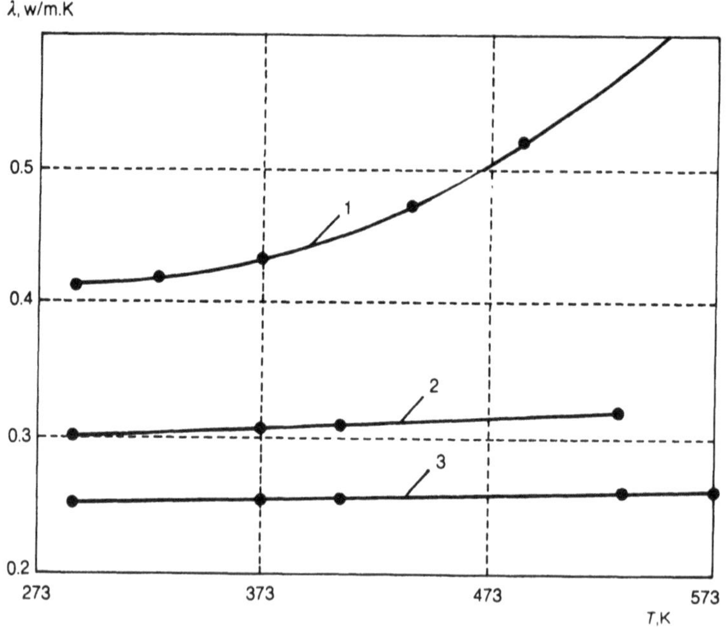

Fig. 7.63 Temperature dependence of coefficient of heat conductivity λ of composite materials based on epoxy matrix and multilayer structure cloth: (1) glass cloth made of silica fibres; (2) combined cloth of aramid and glass fibres (40:60); (3) cloth of aramid fibre SVM.

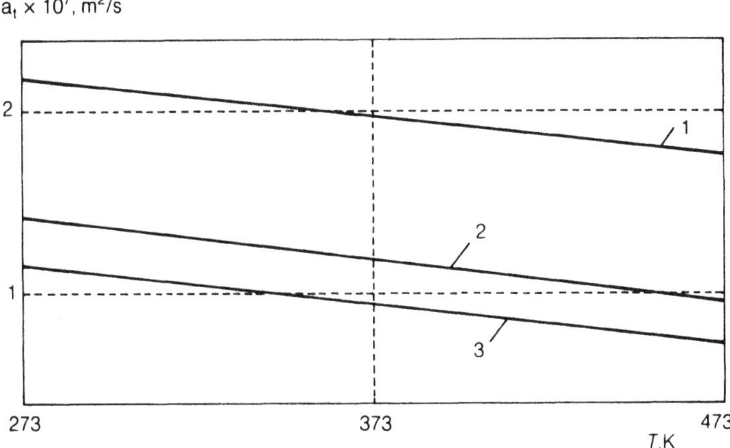

Fig. 7.64 Temperature dependence of coefficient of temperature conductivity a_t of composite materials based on epoxy matrix and multilayer structure cloth: (1) cloth of aramid fibre SVM; (2) combined cloth of aramid and glass; (3) glass cloth made of silica fibres.

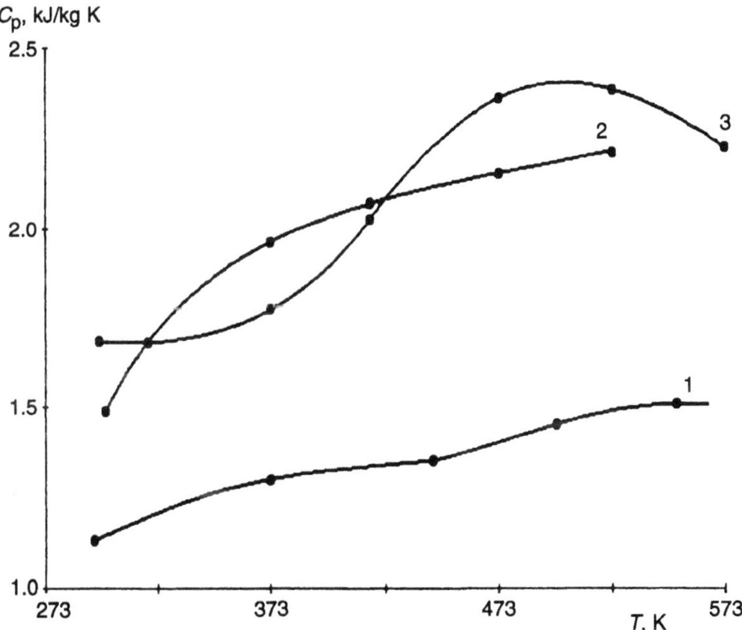

Fig. 7.65 Temperature dependence of coefficient of specific heat caracity C_p of composite materials based on epoxy matrix and multilayer structure cloth: (1) glass cloth of silica fibres; (2) combined cloth of aramid and glass fibres (40:60); (3) cloth of aramid fibre SVM.

Table 7.30 Comparative properties of epoxy composite materials based on multilayer structure cloth

	Reinforced fibres		
Properties	Aramid	Silica	Based on combination of aramid and silica fibres
Density (kg m^{-3})	1300	1650	1400
σ_b (MPa)	300	130	235
σ_{-b} (MPa)	170	150	200
E (MPa)	13500	17000	22000
Thermal conductivity, λ (W m^{-1}K^{-1})	0.25	0.41	0.29
Thermal diffusivity, $a_t \times 10^7$ (m^{2}s^{-1})	1.1	2.13	1.4
Specific heat capacity, C_p (kJ kg^{-1}K^{-1})	1.7	1.17	1.45
Impact elasticity (kg J m^{-2})	155	67	190

7.12 APPLICATIONS OF ORGANOPLASTICS IN AIRCRAFT ENGINEERING

A number of factors contribute to the high effectiveness of organoplastic application in various fields of engineering: the high level of physicomechanical properties and some unique characteristics of these materials (for example, tensile strength, resistance to impact loads); great possibilities for varying properties by changing the chemical composition and the structure of components; optimization of schemes for reinforcing fibrous filler and technological conditions for obtaining and processing semifinished products; and wide possibilities for hybridization – both monophase, due to changing only the structure of reinforcing fillers and the compositon of composite material, and heterophase, including modification of organoplastic properties due to introducing another composition in the composition of reinforcing fibres (glass, carbon, metal, boron, etc.).

Among modern structural materials, organoplastics are characterized by the lowest density, and the specific tensile strength (reaching 220–240 km) of unidirectional organoplastics exceeds that of all known structural materials. Specific stiffness of organoplastics is 2 times more than that of fibreglass, but 1.5–2.0 times less than the stiffness of carbon plastics. The fact that fibreglass specific indices of strength and stiffness are behind those of organo- and carbon plastics predetermined their limited use in aviation industry and replacement with more effective materials. In our opinion, composite materials based on aramid fibres do not compete with carbon plastics. At present, optimal areas of application are quite clearly determined for each of these materials, and the common area is overlapped with hybrid composites based on combination of carbon and aramid fibres.

Besides high specific characteristics of strength and stiffness, organoplastics, as structural materials, are able to provide durability, longevity and safety fracture of parts and units in case of partial (local) damage of constructions.

Analysis of properties and peculiarities of organoplastics' fracture showed that these materials are characterized by high work and toughness of fracture: those indices of organic composites exceed those for epoxies by 2–4 times. Low fatigue crack growth rate (several orders lower than that for metals), ability for long-term accumulation of fatigue damage without critical crack formation and low sensitivity to stress concentrators predetermine high service life characteristics of organoplastics, especially under the action of tensile stresses. The combination of high indices of vibration damping and resistance to fatigue loads ensures high swinging strength of organoplastics: the swinging strength of organoplastics is 5 times higher than that of aluminium alloys, and their vibration stiffness is at the level of that for carbon plastics.

High durability and longevity determine the interest in organoplastics as both a single material and as a material for creation of combined metal-containing polymer composites, providing a decrease of fatigue crack growth rate, by tens compared with ordinary materials and, in particular, with aluminium alloys.

The physical properties of organoplastics (thermophysical, dielectric), high chemical resistance in contact with aggressive media and retention of properties under the action of both high (up to 473 K) and cryogenic temperatures, etc., predetermine the wide possibilities for application of organoplastics for solving a wide range of various problems.

Providing high bearing ability, organoplastics as multipurpose structural materials are at the same time able to provide radio-transparency, armouring protection, erosion resistance, resistance to acoustic, mechanical and hydraulic impact and intensive vibration, thermal protection, heat and sound insulation, etc. Modification of organoplastics by adding metals and other fibres expands the range of effective use of these materials, ensuring non-electrifiability, resistance to lightning strikes, resistance to radio waves and other electromagnetic radiation, and stability in case of high-temperature contact with melted metal (for example, under conditions of 'titanium' fire) or hard bodies (friction pair), etc.

An essential disadvantage of organoplastics, limiting their application in a number of critical constructions, is comparatively low strength and stiffness in the case of compression. When choosing areas for application of organoplastics, it is necessary to take into account this disadvantage or to use certain methods for its compensation: in particular, combination of aramid fibres with glass or carbon fibres, which are more stable to compressive loads. It should be noted that combination of aramid fibres with rigid carbon fibres is considered the most effective one.

The aramid component in the composition of such hybrid composites performs various functions: it contributes to 1.5–2 times increase of material longevity due to increased fracture toughness and to functioning of the organoplastic component as a propagating crack stopper; it protects the carbon plastic from erosion attack; and it increases the resistance of constructions to the action of impact mechanical loads by 30–40% due to an increase of impact toughness and destruction of carbon plastic, which is modified by means of adding organoplastic. An increase of fracture toughness of hybrid carbon organoplastics considerably improves technological properties of these materials, in particular resistance of construction elements to impact riveting and other types of mechanical action, unavoidable when large-scale epoxy parts are transferred along the technological chain.

7.12.1 Organoplastics in construction of airframes

Being characterized by the valuable complex of physicomechanical properties, organoplastics are very useful materials for many branches of industry. However, it is preferable to use them in cases when the basic reason for choosing the material is the decrease of developed construction weight, i.e. first of all in the aviation and space industry. At present there is no single trend for creating various flying vehicles in which organoplastics could not be used effectively: for transport and passenger aircraft, such as the powerful 'Ruslan' or modern liners TU-204 and IL-96-300, helicopters, sporting aircrafts of type SU-26, 'small' aircraft, dirigibles, etc.

Organoplastics are used in airframe construction elements of modern aircraft and helicopters for manufacture of monolithic parts and cellular panels. Application of organoplastics as cellular panels is especially profitable for these materials, as this technical solution gives the possibility to compensate, to a certain extent, their main disadvantage – comparatively low resistance and inclination to losing stability under the action of compressive loads.

The construction of three-layer panels with thin shells and application of sufficiently stiff, strain-resistant cellular filler gives a possibility to realize high strength and stiffness of organoplastics under the action of tensile and shear stresses, arising in shells in case of complex loading of cellular panels. Besides the above-mentioned advantages three-layer constructions of such types are characterized by high corrosion resistance, high heat and sound insulation properties and radiotransparency.

Cellular panels with shells made of organocloth laminate are manufactured is accordance with adhesive, adhesive-free or combined technology. In the adhesive-free technology the preference is for use of prepregs with high (up to 60%) coating of binder, ensuring formation of a shell and, at the same time, its joining with cell faces. Combined formation is usually used for obtaining articles of intricate configuration, where it is very difficult to ensure good mating of elements to be joined with adhesive: a preliminarily formed shell and a cellular filler. In this case adhesive is applied on cell faces (or layed on cells as a film), followed by laying prepreg and simultaneous joining and forming a three-layer construction shell during one technological operation. Forming cellular panels and units is usually performed by means of an autoclave with specific formation pressure of 0.2–0.5 MPa.

Glass cellular plastic of solid or hollow glass fibres or polymeric cellular plastic made of paper based on Fenilon fibre (analogue of Nomex fibre, USA) (Table 7.31) are usually used as a cellular filler.

Combining cellular plastic as a light filler with shells made of organocloth laminate has become a standard solution, inevitably resulting in high technical and economical effect.

For the first time in our practice organoplastics as structural materials were used

Table 7.31 Strength in the case of uniform separation of cellular panels with shells made of organoplastics (binder – epoxy with modified urea)

Cellular filler	Material of shell	Method for combining shell with cells	Strength in case of uniform separation
Polymeric cellular plastic PSP-1, cell 2.5 mm, density 60 kg m^{-3}, height 9 mm	Organocloth laminate (cloth-satin, SVM fibre, 14.3 tex)	Combined[a]	3.3
		Adhesive[b]	2.6
	Organocloth laminate (combined fabric, containing 60% SVM fibre)	Combined	3.3
Glass-cloth-based laminate SSP-1, cell 2.5 mm, density 100 kg m^{-3}, height 9 mm	Organocloth laminate (cloth-satin, SVM fibre, 14.3 tex)	Adhesive-free	1.6
		Combined	5.1
		Adhesive	3.2
	Organo-glass-cloth laminate (combined cloth, containing 60% SVM fibre)	Adhesive-free	2.5
		Combined	7.0

[a] Prepreg of shell material is laid on cell faces through adhesive film VK-51.
[b] Film adhesive VK-41; roughing of organocloth laminate surface before joining is effected by removing a sacrificial layer.

for manufacturing Antonov OK transport aircraft. In 1975 organoplastics were used for manufacture of reinforced and stringer constructions of flight aircraft AN-28. This material was used for production of shell and fastening elements and the wing section behind the last longeron.

Constructions with polymeric cellular filler in which organoplastics are used as shells and bearing elements of outline profiles are widely used owing to their lightness (mass of 1 m^2 of panel varies within the limits of 0.9–2.9 kg, depending upon the shell thickness and load-bearing ability). These panels are used for manufacture of various partitions, doors, hatch covers, shields, electrical equipment tunnels, casings, protecting various units, heat insulating elements of battery bay constructions, etc.

High resistance to mechanical damage gave a possibility to use organoplastics as protective layers for epoxies, which are more sensitive to erosion and mechanical impact. External layers, reinforced with SVM fabric with thickness of about 0.12 mm, joined with the basic material in process of forming parts, ensured their high longevity and service reliability. The number of such construction elements in heavy transport aircraft exceeds 350. In particular, carbon plastic shells with a protective layer of aramid fabric are used for manufacture of gear doors, fillet panels, swing panels of fin and stabilizer, flag mechanism shields, etc.

Accumulation of experience in manufacture and operation has made it possible to use organoplastics as basic materials for manufacturing large-scale medium-load constructions of airframes for wide-body transport aircraft. The manufacture of gear shields with area of 155 m² is an example in this sphere. When manufacturing cellular panels of gear shields, use has been made of the method of combined formation of shells from woven prepregs directly on the cell surface, joining of cells with cellular plastic being effected by means of film adhesive laid preliminarily on cell faces (under prepregs). Application of organoplastic in combination with polymeric cellular filler resulted in a decrease of 1 m² construction mass by 2 times – to 5.25 kg.

Organoplastics were used even more widely in construction elements of passenger airplane TU-204. For standard low-load constructions, such as mounting hatch covers, covers of maintenance hatches, etc., organocloth laminate shell is used in construction with polymeric cells. Replacement of traditional metal covers with composite articles gives the possibility to decrease their mass by 26%. Cellular panels with shells made of organocloth laminate reinforced with stiffer cells of glass fabrics are used for manufacture of engine nacelle doors, pilot shield, forefin, wing fillet with gear shield, tail shield of fuselage and front wing edges. Compared with the metal version, the mass decrease of these parts varies from 12 to 22%.

More heavily loaded large-scale constructions such as doors of front and main gears, ailerons, flaps and wing panels behind the longeron, as well as panels of interceptor, rudder and elevator are manufactured of hybrid shells based on alternating layers of prepregs of SVM cloth and carbon-fibre unidirectional tape. The mass ratio of layers of composites, reinforced with aramid and carbon fibres, varies for hybrid shells within the limits from 1:3 to 1:1. The typical scheme of reinforcing such shells usually includes: one layer of full-strength satin cloth of SVM fibre, 14.3 tex, two layers of unidirectional carbon tape, a layer of SVM fabric, two layers of carbon tape and one layer of SVM cloth.

The amount of organoplastics in 4.4 tonnes of polymeric composites used in passenger airplane TU-204 construction elements is equal to about 3 tonnes, which ensured a decrease of mass of parts and units of modern airplane TU-204 by 1200 kg.

7.12.2 Organoplastics in helicopter construction

Organoplastics and hybrid composites based on them were successfully introduced and operated in several types of helicopters. The widespread application of polymeric composites and organoplastics in helicopter-building was caused by high rates of modernization of this class of machines.

Previous wide usage of fibreglass contributed to the comparatively quick introduction of organoplastics in helicopter construction elements. Experience in design, technology, materials technology and methods accumulated during manufacture of units and assembly units made of fibreglass was of principal value when transferring from all-metal constructions to constructions with increasing use of organoplastics in modern helicopters.

Owing to wide use of organoplastics (compared with fibreglass) it has become possible to transfer from creation of single parts made of composites, preferably included in a metal construction, to manufacture of units consisting of polymeric panels, built in a metal frame, and further on to develop all-plastic modular constructions. Cloth laminates based on long-service-life epoxy binders combined with aramid fibres of satin or serge structure are used for manufacture of helicopter parts, as well as with combined unidirectional cloth of aramid and glass fibres. Preliminarily impregnated cloth prepregs are cut by templates into blanks and laid out by volumetric production fixtures in accordance with the specified stack structure.

The purpose of the part and its loading factor determine the choice of its manufacture technology. The method of straight pressing under specific pressure of 50–800 MPa is used for obtaining flat sheets of ordinary thickness of 0.15–0.40 mm, which are then used for manufacture of part shells of three-layer design with cellular filler (tail sections of bearing screw, panels of rudder, fin, covers, etc.). Assembly of such panels is performed in an autoclave with use of epoxy adhesives in the form of films or compositions that are applied on cellular plastic blank faces before joining.

Low-load parts of single or double curvature are manufactured by the vacuum method.

Introduction of organoplastics in helicopter construction coincided in time with transfer from traditional metal stringer-beam (frame-type) construction of fuselage to the laminated cellular one.

Replacement of frame-type construction of fuselage units for panel type with organoplastic panels gave a possibility to increase specific strength, stiffness and stability of constructions in case of work under conditions of complex loading, to increase stability to impact action, and the heat and sound insulating characteristics of construction, to simplify its design and to decrease labour consumption for its manufacture. For example, the labour consumption required for assembly of tail fins made of cellular panels with organocloth laminate shells is decreased by 43–50% compared with that required for manufacturing a frame-type unit of similar purpose.

In the above-mentioned case replacement of frame construction with panel construction made possible the essential decrease of mass of units: on changing aluminium alloys for organoplastic, the mass of a cellular panel shell will be decreased by 23%. Application of organocloth laminates instead of fibreglass permits the mass of units to be decreased by 15–20%, providing and even increasing their specified service life.

Especially effective is use of light organoplastics instead of fibreglass for bearing screw blades, which require transverse balancing of sections in order to exclude vibrations of flutter type. So application of organocloth laminate shells gave a possibility to improve blade centring, decrease mass of section and at the same time to increase by more than 10 times the service life, compared with those of sections whose shells are made of fibreglass.

At present more than 400 units and parts made of organoplastics are used in helicopter constructions, and more than 1600 can be expected instead of metal ones. Organoplastics combined with stiffer materials (for example epoxies) may be most effectively used for highly loaded units, such as the bearing screw longeron. In this case it turns out to be possible to obtain the required reserve of strength and even provide control of the frequency characteristics of the longeron in order to detune it from resonance frequencies, exclude the danger of flutter and decrease the loads acting on the screw control pulling rods.

Organoplastics shells occupy the dominant place in construction of helicopter fuselage. The mass of $1\,m^2$ of non-load-bearing panels is 1.45–2.0 kg, that of load-bearing ones is 2.2–2.5 kg. Load-bearing panels with shells made of organoplastics are used in load-bearing units, such as cabin, fin, flaps and fuel tanks. Polymeric cellular plastic with cell of 2.5 and 4.2 mm is usually used as a filler for three-layer panels. Development of helicopter constructions with wide use of organoplastics has permitted the following:

1. Decreased mass of construction elements by 20–30% compared with metal version and ensured decrease of airframe unit mass by 14%.
2. Increased safety in respect of failure and survivability of the article.
3. Prolonged service life of some airframe units by 2–3 times and bearing screen end sections by 10–12 times.
4. Decreased the labour input and energy consumption for manufacture of intricate elements of construction due to decrease of number of parts, increase of module use factor for constructions and reduction of assembly cycle by 1–3 times.
5. Reduced duration of unit manufacture cycle by factor of 3.
6. Increased the material utilization coefficient to 0.8–0.9.
7. Decreased labour and metal consumption for manufacture of jigs and fixtures.
8. Reduced expenses for technological servicing by 3–5 times and provided exchangeability and repairability of parts and units.

The performed design and technological studies and the existing experience of operation testify to the usefulness and effectiveness of wide application of organoplastics in dynamically loaded constructions of bearing screws of helicopters, as well as for medium-loaded units and aggregates of airframe – tail fins, wing, fuselage, etc. Combination of organic SVM fibres with glass and carbon in laminated materials gives a possibility to apply these materials for high-load units of the frame and load-bearing system of helicopters.

Owing to wide use of constructions made of organoplastics and hybrid composites based on them, there has been a transfer from partial use of these materials in metal constructions to wide use of them (up to 30–35%), as well as to creation of constructions with preferential use of polymeric composites for new helicopter units (more than 50–60%).

7.12.3 Organoplastics in construction of airplane and helicopter interiors

The low density of organoplastics combined with sufficient stiffness gives a possibility to obtain the lightest constructions using them, which are of especial interest when designing interior details. Besides, organoplastics based on aramid fibres are characterized by high impact resistance, stability to action of aggressive liquids, detergents and mycological factors, low combustion ability and self-extinguishment, and limited release of smoke and toxic products during burning. This complex of properties ensures high operational reliability of interior constructions made of organoplastics.

Load-bearing cellular constructions for interiors are manufactured basically by means of two technological methods: with use of adhesive or without. In the second case one layer of prepreg with high technological stiffness is applied on both faces of a cellular plastic with subsequent vacuum forming. Prepregs of such type are characterized by high shelf-life (up to 4–7 months) and may be obtained on the basis of modified epoxy or phenol–formaldehyde binders. These semifinished products are produced from aramid cloths of satin or serge braiding and are well laid out on surfaces of intricate curvature. They are of technological grade and in the event of vacuum forming ensure the strength for uniform tear-out from polymeric cellular plastic with a 2.5 mm cell within the limits of 1.7–2.5 MPa.

Adhesive-free cellular panels with shells of one or two organocloth laminate layers are widely used for manufacture of elements of interior constructions of airplanes and helicopters. Replacement of fibreglass shells made of capillary fibres by organoplastic shells in the interior of passenger airplane IL-96-300 allowed a decrease of mass of 500 kg. Light cellular panels with organocloth laminate shells are used for parts of salon, cabin, auxiliary section and other interior rooms of passenger and transport airplanes, and for the spacecraft 'Buran'.

Light intricate curvature constructions, for example, elements of ceiling panels, may be manufactured by combining woven prepregs with light filler of knitted structure of aramid fibres. Blanks for such panels are obtained by means of two methods. According to the first one, knitted fabric is impregnated on a special impregnating machine with a solution of binder, and after removal of binder the obtained knitted prepreg is used for assembly of a stack in combination with woven prepregs, decorating the article surface. In accordance with the second method, the source knitted fabric is laid on the rewinding machine between two layers of aramid fabric and such stack is rolled into a roll. Impregnation of the prepared stack is effected on usual impregnating machines. The stack laid on the technological fixture is pressed by means of the vacuum or autoclave methods.

The properties of three-layer organoplastic with knitted filler depend upon the forming pressure (Table 7.32). Such materials including one or several layers of prepreg made of knitted linen are used for manufacturing low-loaded parts, such as ceiling panels or intricate configuration parts, for example the illuminator panel of TU-204 airplane. Application of knitted filler in construction of illuminator and ceiling panels ensured a decrease of mass by 50% and labour input for their manufacture by a factor of 3–4 (compared with cellular panels).

Sheet organocloth laminates with surfaces roughened for joining with adhesive formed during removal of a sacrificial layer are used for load-bearing cellular constructions – flat and single-curvature – as shells. Sheet semifinished products are manufactured on a multiplaten press, permitting pressing of 10–30 sheets on each press plate. Dimensions of sheets are usually equal to $1.0 \times 2.5 \, m^2$ and the maximum sheet

Table 7.32 Dependence of properties of three-layer organo-plastic[a] with knitted filler upon forming pressure

	Specific forming pressure (MPa)	
Properties of organoplastic	*0.09*	*0.02*
Thickness (mm)	1.4–1.8	0.9–1.5
Mass of 1 m^2 (kg)	0.8–1.7	0.8–1.7
Density (kg m^{-3})	550–1000	860–1200
σ_{bb} (MPa)	110–260	185–320
$\sigma_\perp$ (MPa)[b]	3–8	9–10

[a] Composition of three-layer organoplastic: aramid cloth of satin braiding of 14.3 tex fibres, knitted linen of SVM fibre, 58.4 tex.
[b] $\sigma_\perp$ = strength in the case of uniform separation.

thickness is determined by the thickness of monolayer of reinforcing cloth (0.13 or 0.2 mm).

Sheet (0.3–0.8 mm thick) organocloth laminates are most widely used. In the event of increase of prepreg layer number in the stack, the strength of organocloth laminate is increased and stabilized when 0.6–0.8 mm thickness is reached (σ_b = 800–900 MPa, E = 35–40 GPa).

Sheet organocloth laminate shells are used for manufacturing standard panels of floors and various partitions for internal constructions of airplanes and helicopters. Both polymeric and fibreglass cells may be used as fillers. Comparative tests of cellular floor panels with shells made of various polymeric composites – epoxy glass cloth laminates and epoxy organocloth laminates – showed the advantage of the latter when testing for bending under the action of concentrated load, i.e. under cyclic action of 100 kgf load applied through 20 mm diameter rod. Application of sheet organocloth laminate shells considerably improved the service life of floor panels and gave a possibility to decrease their mass compared with panels whose upper shell was manufactured of fibreglass cloth laminate and the lower one from epoxy.

Cellular panels with sheet organocloth laminate shells are widely used in airplanes IL-86, IL-96–300, TU-204, YaK-42 and others. So 360 m^2 of such panels are used for IL-96–300, and 205 m^2 for TU-204. The mass of 1 m^2 of such panels with polymeric cells PSP-1-2.5–60 is equal to 2.5 kg, that for fibreglass cells SSP-1-2.5 is equal to 3.1 kg when shell thickness is 0.7 mm (top) and 0.5 mm (bottom).

The low mass and sufficiently high strength and stiffness characteristics are a required but not a sufficient condition for successful use of structural organoplastics in interior parts. When heating polymeric material to a temperature that is equal to or higher than the decomposition temperature, it may release combustible or incombustible gases, liquids and solid particles, being carbon residues, coke or polymeric fragments. The gaseous products of decomposition may be flammable in the presence of an oxidizer (atmospheric oxygen). Combustion of material is a dangerous source of flames or flame-free burning, as well as of smoke and toxic products.

The analysis of emergency situations that occur during operation of passenger airplanes shows that 95% of lethal cases are bound up with fires, 40% being determined by the action of toxic gases and smoke. In this connection non-metallic materials used for airplane interiors should comply with the standards of flight fitness for civil airplanes in Russia (Unified Airworthiness Regulations for Civil Aircraft, UARCA).

The testing of materials for combustibility is carried out in accordance with methods corresponding to UARCA and the International Civil Aviation Organization (ICAO).

Tests of reinforcing fillers of various structures of aramid fibres showed that SVM and Terlon aramid fibre are in the group of materials highly resistant to combustion. Burning of these materials is characterized by the presence of an alight area, the rate of propagation of which is increased with increase of oxygen content. When the flame is spread from above downwards, samples of aramid fibres burn without a visible flame (burn incompletely), and in the event of increase of oxygen concentration in the mixture the burning rate is increased insufficiently. When burning from the bottom upwards, SVM fabric burns with a visible flame; the flame propagation rate is three times higher than in the opposite case and is considerably increased with increase of oxygen concentration. The oxygen index of aramid fibres is 24–39% and may be increased to 66% subject to respective processing.

Fibres of aromatic polyamide Fenilon (poly(m-phenylene isophthalamide)) – the base for paper used in manufacture of cells – are somewhat inferior to aramid fibres in respect of combustibility and are self-extinguishing materials.

Table 7.33 Combustibility of organocloth laminates based on SVM cloth

Matrix	Sample thickness (mm)	Time (s) Flame action	Time (s) Residual burning	Length of carbonized part of sample (mm)	Loss of mass (%)	Maximum height of flame (cm)	Group of combustibility[a]
Epoxyiso-	0.6	60	0	150	–	16	HB
cyanate	2.5	60	0	100	–	10	HB
Phenol–	0.17	60	0	60	3.8	8	HB
formal-	0.44	60	0	45	2.2	8	HB
dehyde							
Epoxyani-	0.5	60	0	35	3.0	20	HB
line	2.2	60	11	60	–	8	SE
phenol–							
formal-							
dehyde							
Epoxy-	2.0	60	9	7	3.1	13	SE
novolac							

[a]HB, hardly burning; SE, self-extinguishing.

Combustability of organoplastics, in the case of the same filler, depends upon the composition of binder and the sample thickness (Table 7.33). For interior constructions organocloth laminates based on epoxyisocyanate and phenol–formaldehyde resins, referred to as hardly burning materials, are employed.

On burning, the majority of polymeric materials release a considerable amount of smoke, which causes a harmful situation and hinders life-saving work. The maximum

Table 7.34 Indices of smoke formation for SVM cloth and organoplastics based on it

Material	Test mode	D_2	D_4	D_{max}	Smoke formation category[b]
			Indices of smoke formation[a]		
Cloth SVM, one layer, mass 90 g m^{-2}	Pyrolysis	1	1	~ 1	I
		1	1	~ 3	–
Organocloth laminate, thickness 1.9 mm	Pyrolysis	1	5	28	–
	Burning	3	14	100	III
Requirements of flight fitness standards		$\leqslant 100$	$\leqslant 200$	$\leqslant 200$	$< IV$

[a] D_2 and D_4, specific optical density of smoke during 2 and 4 min of tests; D_{max}, maximum specific optical density.
[b] Categories of smoke formation: I, practically non-releasing smoke; II, weakly smoking; III, medium smoking; IV, considerably smoking; V, strongly smoking materials.

Table 7.35 Fire hazard properties of organocloth laminates

Index	Requirements of flight fitness	Organocloth laminates	
		Epoxy	Phenol–for-maldehyde
Combustibility	Hardly combustible or self-extinguishing	Hardly combustible	Hardly combustible
Flame-forming ability	$\leqslant$ Group IV	III–IV	II–III
Toxicity of combustion products	$\geqslant$ Group II	II–III	II

Table 7.36 Air pollution/health characteristics of organocloth laminates used for constructions of passenger airplane salons[a]

Airplane	Area of surfaces of parts made of organo-cloth laminates, used in salon (m^2)	Saturation of salon volume with organoplastics $(m^2 m^{-3})$	
		Actual	Permissible
TU-204	224	1.2	> 2.5
IL-96-300	363	0.67	10
IL-114	112	1.5	10

[a] Under conditions of 1, 10 and 20-fold air exchange per hour, at a temperature of 293 K.

value of mass optical density of smoke for burning of organoplastics depends upon the sample thickness.

SVM aramid fibres are characterized by high coke residue (53%). Combination of these fibres with phenol–formaldehyde resins, which are also characterized by high residue of coke during burning and pyrolysis, gives a possibility to obtain materials with low flame-forming ability (Table 7.34).

The most important characteristic determining the possibility of applying polymeric material in an airplane interior is the toxicity of its oxidation products. Organocloth laminates based on epoxyisocyanate and phenol–formaldehyde are in the II–III category in respect of toxicity, which corresponds to UARCA requirements (Table 7.35). The analysis of volumes of use of these materials for salons shows a considerable reserve by permissible saturation and excuses their wide application (Table 7.36).

7.12.4 Organoplastics in construction of propulsion units

The rates of introducing polymeric composites in construction of aircraft units and engines are far behind those of introducing these materials in airframe construction. The maximum volume of use of polymeric composites in engines does not exceed 2–4%, but in the past few years one could note the beginning of technological and design work aimed at expanding the volume of both polymeric and metallic components into aircraft constructions.

It would be well to separate out the following main trends of creating and introducing polymeric composites, including organoplastics and hybrid polymeric and metal–polymeric composites, in aviation engine construction elements:

1. Shells of air intakes with high resistance to dust and rain corrosion.
2. Vibration-loaded thin shells of cold passage of engine.
3. Sound-absorbing constructions.
4. Impact-resistant protective fan shields.
5. Blade of rotor-fan engines.
6. Fire-fighting shields.

Structural aramid organoplastics are able to withstand long-term action of temperatures below 453–473 K; in this connection it is possible to use these materials for manufacture of large-scale construction elements of cold passage of modern gas-turbine engines (for example, shells of air intake, cowling, reverse flaps, nose cone, pylon, etc.).

Constructions of such details are usually three-layer, cellular, bonded ones; the materials and technology of their manufacture do not differ principally from those considered above in respect of airframe construction elements.

Application of cellular panels with organoplastic shells for construction of the gas–air channel of propulsion units is very effective owing to the highly crash proof nature and erosion resistance of these materials. The resilience of organoplastics is 3–5 times higher than that of carbon plastics, and the wear resistance of high-modulus aramid fibres is 20 times more than that of glass fibres. These advantages increase the reliability and operation safety of organoplastic parts in the case of the action of solid- and liquid-phase eroding flows.

The rain erosion intensity for organoplastics is almost an order lower than that for carbon plastics with a similar matrix (Table 7.37). As for resistance to dust erosion, when the speed of collision with solid particles of ~ 0.04 mm is equal to $200\,\mathrm{m\,s^{-1}}$ organoplastics exceed not only carbon plastics but fibre plastics as well. At the same speed of collision ($100\,\mathrm{m\,s^{-1}}$), the fibre plastic erosion depth is three times more than that for organoplastics; when the thickness of ablation layer is equal, organoplastic withstands the action of an eroding flow whose speed exceeds by two times the speed of the flow acting on fibre plastic with the same result.

The erosion resistance of aramid organoplastics may be increased by 2–4 times depending upon the chemical composition of the polymeric matrix. So, for example,

Table 7.37 Erosion resistance of orthotropic polymeric composites based on epoxyaniline phenol–formaldehyde matrix (test regimes shown in footnotes)

	Rain erosion[a]		Dust erosion	
Material	*Erosion intensity* $(cm^3\,kg^{-1})$	*Period till the beginning of destruction (s)*	*Erosion depth for two cycles[b] (mm)*	*Erosion intensity[c]* $(cm^3\,kg^{-1})$
Organoplastic	1.7	22	0 (loss of gloss)	0.51/1.18
Fibreglass plastic	4.3	35	–	0.48/4.58
Carbon plastic	11–19	9	600–800	0.7 to 1.2
				4.8 to 6.0

[a] Drop diameter, 1.5 mm; flow velocity, $170\,\mathrm{m\,s^{-1}}$; collision angle, $90°$; flow, 1.15 drop/cm^2.
[b] Diameter of particles of Al_2O_3, 0.5–0.8 mm; flow velocity, $50–60\,\mathrm{m\,s^{-1}}$; collision angle, $40–45°$.
[c] Diameter of particles of Al_2O_3, 0.04 mm; flow velocity, $200\,\mathrm{m\,s^{-1}}$; (in numerator, collision angle data; in denominator, $70°$).

the relative erosion resistance of an organoplastic based on epoxyaniline phenol–formaldehyde matrix is four times lower than that of an organoplastic in which an epoxyurethane polymer is used as the matrix.

Low sensitivity of organoplastics to stress concentrators provides considerably higher stability of these materials under conditions of simultaneous action of fatigue loads in two-phase aerodynamic flows. When the cycle stress $\sigma^{max} = 0.5\sigma_b$ the durability of fibre plastic specimens after erosion is an order lower, and under these conditions organoplastics exceed fibre-filled composites by more than 40 times.

Combination of high erosion resistance and longevity in damaged state with high resistance to mechanical impact with hard particles (stones, pebble, gravel) predetermined the effective use of organoplastics for manufacture of construction shells that are subjected to intensive erosion and impact action during operation (shells of intake channels of engines, as well as airframe construction elements, radio-transparent nose cones, front edges of wing and tail, undercarriage fairings, lower panels of wing and airframe).

Thin shells of air passage construction elements of engines operate under high vibration and acoustic loads. The high damping properties of organoplastics under the action of bending stresses combined with sufficiently high characteristics of durability and stiffness provide high effectiveness of their use for construction of bands in the form of outer reinforcing layer for thin mechanical shells. Application of organoplastics in thin metal–plastic shells gives the possibility to decrease their mass by 20% and at the same time to increase their vibration resistance by 3–4 times and the vibration rigidity by a factor of 1.5.

The resistance of constructions to vibrations may be increased due to the increase of swinging strength and vibration ridigity, as well as by making it possible to provide detuning from dangerous resonance frequencies and forms of vibration.

The advantages of aramid-based organoplastic compared with those of polymeric composites based on more brittle fibres are especially clearly seen when estimating the erosion resistance of these materials by variation of resistance of specimens to pulsed tension after the action of eroding flow.

Organoplastics are characterized by high durability in the acoustic range of frequencies, which makes it possible in a number of cases to use these materials effectively not only in metal shells but in cellular panels with shells made of other polymeric composites. For example, destruction of the tail part of a helicopter airframe made of organoplastics takes place in a zone subjected to acoustic action of engine exhaust gas stream. Replacement of fibre plastic with aramid organoplastic gave the possibility to increase the tail section wall service life by 1.5 times and to decrease its mass by 30%.

An important task is decreasing engine noise on the airport runway, as well as acoustic protection of the cabin crew and passengers. This problem becomes an important one among the traditional problems of increasing weight perfection, service life and reliability due to prescribing new stricter international requirements to noise.

Under the general concepts, the determining factors for designing noise-absorbing shields are high density, low elasticity and high damping of vibrations. Increasing the masses of aviation and particularly auxiliary constructions is extremely undesirable, especially when it does not give considerable effect for improvement of soundproofing characteristics. The sound insulating ability of low-surface-density partitions may be increased by increasing their elasticity, but for the cases when only low frequencies (30 Hz and lower) should be isolated. The decrease of partition elasticity not only decreases natural frequencies, but shortens the length of bending wave, decreases the insulating properties of the panel and thereby increases the critical frequency at which 'coincidence effect' takes place, equalization of longitudinal and bending wave lengths. In the case of low elasticity of the wall, resonances of vibrations and coincidences are spread to the ends of the frequency band, where these penomena are usually less essential ones.

For example, application of organoplastics (the modulus of elasticity of which is 2–3 times lower than that of aluminium alloys owing to optimizing the scheme of reinforcement or choice of special quasi-isotropic filler) gives the possibility, subject to equal thickness of partition, not only to increase its mass by a factor of 2, but to increase essentially the intensity of passed acoustic wave. The increase of the transmitted sound signal intensity at the expense of increase of resonance and 'coincidence effect' favours the use of organoplastics characterized by high damping properties.

The logarithmic decrement of damping oscillations of thin sheet organoplastics is an order higher than that for aluminium alloys and depends upon the structure, composition and method of composite manufacture (13–34%). The logarithmic decrement of damping oscillations of cellular constructions with organoplastic shells amounts to 11–18%.

The best soundproofness for a specified total mass is achieved by applying two or more separate partitions, interconnected by as many as possible pliable ties. The modulus of elasticity for three-layer cellular constructions is on average 20–30% lower than that for thin-sheet shells, which contributes to improvement of their sound-absorbing characteristics, but is not able to decrease the construction stiffness.

But for sound-absorbing cellular organoplastic panels there is a possibility for production of a volume sound suppressor, which is a light material with regulated porosity. In contrast to cellular ones, volume sound suppressors are less labour-intensive, can be manufactured by means of one technological operation and are characterized by at least double the range of effective absorption of acoustic waves. Utilization of such suppressors decreases the noise level by 10–13 dB.

The problem of sound decreasing may be solved not only by means of setting special sound-absorbing shields on engines. For the past years in connection with the sound-decreasing trend, as well as with the necessity to decrease fuel consumption, designers have met the problem of designing non-reactive engines with relatively low rotation speed of propellers with comparatively big diameters.

Proceeding from general ideas it is known that an increase of diameter subject to simultaneous decreasing of propeller speed, when the blade and M do not exceed 1.0, causes a decrease of noise level by up to 30% and an increase of propeller efficiency. However, as we know, the centrifugal force is quickly increasing with increase of blade length, which shifts the problem to one of selecting a light material, characterized by high relative tension strength and stiffness.

From the standpoint of this basic requirement, the high elastic strength properties of organoplastics in case of tension under conditions of durable static and fatigue loading combined with high resistance to mechanical impact and abrasive action, and ability to work reliably in case of local damage and to damp vibrations transferred from the propeller to the aircraft frame give the possibility to consider these materials as very useful ones for manufacture of propeller and propeller-fan engine blades.

Especially effective is the use of organoplastics for construction of a thin sabre-like blade with sufficiently flexible elastic longeron.

The chosen reinforcement schemes of a packet made of unidirectional prepreg taking into account the conditions of blade feather loading gave the possibility to obtain material with stiff high-strength unoriented central layers, receiving effect of centrifugal force and misoriented at angles of $\pm 45°$ by external layers, providing sufficiently high torsion stiffness and high level of aerodamping in case of flexural vibrations of blade.

Such a blade is subjected during operation to the action of unambiguous extension loads of variable value due to action of centrifugal forces, i.e. it works under the most favourable conditions for organoplastics, under conditions of extension.

The thin blade made of organoplastic is able to unload and damp vibrations in case of action of bending loads from aerodynamic flow.

Considerable bending moments and alternating fatigue loads arise and act in 'extension–compression' mode in traditional blades made of organoplastics with relatively thick root part, which forces the material to work under unfavourable conditions (Table 7.38).

Manufacture of thin and strong sabre-like organoplastic blades ensure obtaining high aerodynamic characteristics of propeller due to decrease of head resistance. Decreasing the blade thickness, especially in the root part, decreases also their resistance to air flow at the air intake inlet.

Organoplastic blades may be effectively used at airplane speeds of up to $800 \, \mathrm{km \, h^{-1}}$, an increase of efficiency being ensured, as well as decrease of noise in surrounding

Table 7.38 Comparative properties of unidirectional epoxy organo- and fibreglass plastics for blades of propeller and fan engines

Property	Material	
	Organoplastic	*Fibreglass plastic*
Tensile strength (MPa)	1800–2000	1000–1200
Tensile modulus of elasticity (GPa)	96	52
Fatigue strength after 10^7 cycles (MPa)	1400[a]	300[a]

[a] Properties are given by results of fatigue tests in the mode of this material operation in the blade set: for organoplastics, 'extension–extension'; for fibreglass plastics, 'extension–compression'.

places and in the airplane salon on take-off and in cruising flight. The reliability of such a blade during operation is predetermined by the high resistance of organoplastics to abrasive action, mechanical and ballistic impact. The interest in application of organoplastics has grown considerably in recent times, which is bound up with extension of work for creation of small airplanes with propeller and propeller-fan engines, characterized by economical fuel consumption.

High strength and impact resistance of organoplastics is effectively used for manufacture of shields of these materials, protecting essentially important units of airplane from destruction and the passenger salon and pilot's cabin from depressurization. One of the most widely met causes of such flight events is collisions of passenger and transport airplanes with birds.

Some 1500 such collisions are registered in Russia annually. Such collision results in destruction of ventilator blades, with subsequent destruction of other elements of constructions of engine, airframe and units of airplane. Velocity of formed fractures is usually equal to $300-400\,\mathrm{m\,s}^{-1}$ and mass may reach several kilograms, owing to which the impact energy of such fractures is serious for destruction of aviation constructions and causing emergency situations.

In order to localize zones of probable destruction within the limits of the destroyed engine overall dimensions, they use protective armoured shells made of organoplastics.

Circular protective shields with diameter of 1.5–2.5 m and width of 200–400 mm, depending upon the type of engine, are manufactured by the method of winding woven unidirectional prepregs of aramid fibres. Such shields are used in ventilator body constructions of airplanes YaK-42, IL-86, TU-204, IL-96-300, Ruslan and others. For example, the ventilator body protective ring with thickness of 19 mm and the protective band on sound-absorbing construction body ensured reliable protection of IL-86 airplane from non-localized fractures (mass up to 370 g) of working blades of fan and first-stage compressor, when the kinetic energy of fracture reached 29 kJ.

Protective organoplastic rings of fans are successfully used instead of fibreglass plastic used for these purposes.

Application of organoplastics for this purpose in Ruslan airplane gave the possibility to decrease its mass by 200 kg (compared with fibreglass plastics).

High resistance of organoplastics to high-energy mechanical impact is combined with high protective properties of these materials in case of damage with high-speed fractures. On being combined with metals, ceramics and other materials, aramid fibres and organoplastics based on them are, at present, standard materials, widely used for armouring combat machines, collector's cars and other special automobiles, as well as for manufacturing individual means of protection: helmets, jackets, shields, etc.

It should be noted that effective performance of protective functions in various constructions is typical for organoplastics. The above-mentioned functions of abrasive, anti-impact, armoured and vibration protection of constructions, as well as functions of sound absorbing partitions, should be supplemented with high effectiveness of organoplastics use for manufacturing shields of gas-turbine engines resistant to factors of 'titanium fire'.

Increased use of titanium alloys, providing an increase of specific characteristics of modern gas-turbine engines, compared with heavier steels and less heat-resistant aluminium alloys, is the cause of this dangerous phenomenon. In the case of operation of gas-turbine engines, parts of which are made of titanium alloys, sometimes (for example, in case of destruction of compressor blades) self-ignition of titanium parts can take place, converting into titanium fire. In this case a compressor body made of titanium alloy burns through for less than 1 s, a steel body for several seconds, owing to the combined action of high temperature (~ 3173 K), arising under effect of burning melted titanium alloy, and air pressure in the compressor, amounting to 5–12 atm in fire-dangerous zone, depending upon the stage. Titanium fire is accompanied with spraying of melted metal, as a result of which the fire is quickly propagated to other elements of the engine, which may cause a crash.

To localize 'titanium fire' in emergency situations, it is necessary to mount protective fire-fighting shields in the way of probable fire propagation to vital units of the airplane. These shields should be able to resist the action of melted burning metal for 40 s (until switching on the fire extinguishing system).

The results of studying climatic peculiarities of thermal destruction of aramid organoplastics permit use of these materials for constructions of shields that are resistant to titanium fire factors. The tests showed that protective organoplastic and hybrid organoplastic shields are two times lighter compared with traditional shields made of aluminium alloy sheet, protected with a layer of heat-resistant sealant. The fire hazard factor for shields made of the above-mentioned polymeric composite materials is 3.5–6 s kg^{-1} m^{-2}, which is 1.5–2 times higher compared with heat-resistant polyamide fibreglass plastic. The effectiveness of alternating layers of aramid and carbon fibres in the shield structure is predetermined by the combination of absorbing a large quantity of heat during thermal destruction and the high heat-insulating properties by thickness of wall made of organoplastic layers with high heat removal in the wall plane by carbon layers.

The most effective application of organoplastic in various units is provided in those cases when the unit construction permits realization of high characteristics of relative tensile strength and stiffness of unidirectional organoplastics.

Such units include articles subjected during operating to critical tensile loads, the origin of which may be bound up with the action of centrifugal forces (flywheels of accumulators of kinetic energy, shells of electric machine rotors and other high-speed systems) and forces of internal pressure in vessels for compressed liquids and gases (in bodies of engines and boosters, gas cylinders, hydraulic accumulators, booster cylinders and other systems).

A classical example of such an application is the use of organoplastics for manufacture of solid-propellant rocket engine bodies for space rocket complexes. When manufacturing these units by winding bundles or unidirectional tapes preliminarily impregnated with binder, it is possible to realize high relative strength and stiffness of aramid fibres. In this case the realized tensile strength of armouring fibre may reach 3200 MPa and parameter of mass effectiveness of the article is 40–44 km.

Under comparable conditions the strength of glass fibres realized in the article does not exceed 2200 MPa and parameter of mass effectiveness amounts to 18–23 km.

High-pressure vessels with working load up to 600 atm may be manufactured in both full-plastic and as combined metal–plastic structures with external reinforcement with unidirectional organoplastic with tensile strength of 2200–2500 MPa. The full-plastic cylinders are characterized by minimum mass, but during long operation such cylinders do not ensure the required reliability due to destruction of sealing shell under action of temperature difference and cyclic variation of pressure.

Metal alloys characterized by high crack resistance with the average level of tensile strength (steel 1100–1400 MPa, titanium alloy 850–1050 MPa) are used for manufacture of more reliable metal–plastic cylinders. Combination of thin metal shell with higher-tensile-strength external reinforcing layer of organoplastic ensures the creation of a metal–plastic vessel, the bearing ability of which is equally provided with mutually working metal and polymeric shells. External reinforcement of this cylinder permits increase of small cycle fatigue effectiveness factor and decrease of mass, as well as to exclude formation of dangerous fragments in case of destruction.

The optimal combination of metals and external reinforcing organoplastic layer under the specified technological conditions provides high reliability of cylinders under a working pressure of up to 600 atm, their mass being decreased by a factor of 4 compared with an all-metal analogue. Owing to high strength of unidirectional organoplastic external layer, the equivalent structural strength of a cylindrical balloon made of titanium alloy reaches 2270 MPa, that of a steel spherical balloon up to 2000 MPa and that of a cylindrical balloon made of aluminium alloy 700 MPa. The service life of such balloons in the case of cyclic loading amounts to no less than 500 cycles, and the strength reserve reaches 2.6.

When combining titanium with unidirectional organoplastic in high-pressure balloons, the specific strength may exceed 45 km. Thermocycling of titanium and organoplastic shells within the range of temperatures from 143 to 423 K does not practically affect their bearing ability: after 75 cycles of temperature difference the decrease of bearing ability did not exceed 8%.

The weight effectiveness of cylindrical metal and organoplastic balloons with metal shell made of steel and titanium alloy (Table 7.39) was 25% and 75% respectively higher than the weight effectiveness of a spherical balloon, designed for the same parameters (volume 28 litres, working pressure 350 atm). The titanium and organoplastic balloon was successfully used in the seat of a flying armchair of a cosmonaut.

Cylindrical single union balloons with volume of 3 litres and internal shell made of aluminium alloys and external one made of organoplastic in the case of working pressure of 280 atm and strength reserve of 2.6 have mass of 1.05 kg; in the case of increase of working pressure to 350 atm, the balloon mass is increased to 1.25 kg. Such balloons are of interest for use in medical and sporting equipment (Table 7.40).

Besides high-pressure vessels, the idea of external reinforcement of units with high-strength unidirectional organoplastic was realized for manufacture of armature banding and rotor casing of high-speed electrical machines. Application of organoplastic gave the possibility to increase unit rotation speed by 20% ensuring reliability and safety, to decrease labour consumption for the unit manufacture by 20–30%, to increase the coefficient of metal utilization up to 90% and to exclude use of deficient silver solder.

Table 7.39 Weight effectiveness of metal–organoplastic cylinder balloons for working pressure of 350 atm[a]

Material of metal shell	Mass of balloon (kg)	Share of organoplastic layer, by mass[b]	Coefficient of weight effectiveness (10^{-3} m)	
			$K_1{}^c$	$K_2{}^c$
Titanium alloy ($\sigma_b = 1050\,\mathrm{MPa}$)	15.0	0.107	6.5	13.0
Steel ($\sigma_b = 1200\,\mathrm{MPa}$)	22.0	0.070	4.5	10.5

[a] Volume of cylinders is 28 litres; strength reserve 2.25. Strength of external reinforcing organoplastic layer is 2000–2200 MPa.
[b] Ratio of organoplastic reinforcing layer mass to total mass of balloon.
[c] $K_1 = P_1 V/G$ and $K_2 = P_2 V/G$, where P_1 and P_2 are working pressure and destructive pressure respectively, V is balloon volume, and G is balloon mass.

Table 7.40 Comparative characteristics of all-metal balloons made of aluminium alloy with external reinforcing layer of unidirectional organo- and fibreglass plastics (volume of balloon 3 litres, mass of metal shell 0.85 kg, working pressure 280 atm[a])

Material	External reinforcing layer		Mass (kg)		Destructive pressure (atm)	Coefficient of effectiveness, $K_2{}^b$ (km)
	Thickness (mm)	σ_b (MPa)	Reinforcing layer	Balloon		
Organoplastic reinforced with:						
bundle	6.0	2000	0.2	1.05	730	20.0
fibre	3.0	4000	0.1	0.95	730	23.0
Fibreglass plastic, reinforced with bundle	9.0	1000	0.6	1.45	730	15.0

[a] Metal shell, fully formed (without welding), thickness 2 mm.
[b] $K_2 = PV/G$, where P is destructive pressure.

Experience of operation of parts and units made of organoplastics based on high-strength aramid fibres showed that application of these lightest fibrous polymeric composites was characterized by high specific strength and tensile stiffness, high fatigue and long-time strength, crack resistance and resistance to mechanical impacts and abrasive effects. These materials are characterized by high physical properties, and are sufficiently stable under conditions of durable action of various climatic factors and aggressive media, as well as high temperatures of up to 473 K.

Organoplastics are multipurpose materials. Depending upon the composition and the structure of the initial components, organoplastics may be used for manufacturing parts and units of various purposes: structural, electrical and radiotechnical, thermal and sound insulation, protection from mechanical and ballistic damage, action of aggressive media, etc.

It is useful to apply organoplastics as structural materials for manufacture of articles subjected to high tensile loads (static and fatigue ones); in elements of constructions subjected to the action of considerable vibrational, impact, acoustic and abrasive actions; in low-loaded light articles of one layer or laminated structure with cellular or other fillers. For parts working under medium loads in complicated stress states, organoplastics are usually used in combination with other materials, for example with thin sheets of light metal alloys, or are modified with additions of other fibres, characterized with higher fatigue strength during compression (carbon, fibreglass).

An increase of compression strength and stiffness of organoplastics is a major and a complicated problem. In years to come, aramid organoplastics of the first generation will be replaced by new high-strength organoplastics with tensile strength of up to 3500 MPa, as well as by high-modulus organoplastics, the modulus of elasticity of which is not lower than that of carbon plastics.

Considerable progress will also be reached in the field of increase of heat resistance of organoplastics; these materials will be able to operate for a long time at temperatures of 573–623 K. But tensile elastic and strength properties of organoplastics will be increased considerably more slowly. Solution of this problem will probably require a radical change of approach to choice of the structure of polymeric fibres and the technology for obtaining them.

REFERENCES

1. Perov V.V., Mashinskaya G.P., Some aspects of developing polymeric fibre-based composites, *Composite Materials: Reports of the First Soviet–Japanese Symposium, Moscow, May 1977*, Moscow, 1979, pp. 169–82.
2. Baranovsky V.V., Dulitskaya G.M., *Laminated Plastics for Electrotechnical Applications*, Energia, Moscow, 1976.
3. Mashinskaya G.P., *Organofibres, Plastiki Konstruktsionnogo Naznacheniya*, Energia, Moscow, 1976.
4. Pavlov V.V., Mashinskaya G.P., Teterev L.A., Organofibres, *Encyclopaedia of Polymers*, Sovetskaya Entsiklopedia, Moscow, 1974, vol. 2, pp. 510–13.
5. Lelinkov O.S., Perepelkin K.E., Utevski L.E., Mashinskaya G.P., Pavlov V.V., Structural resin-dipped fabric laminates based on PVC-fibre material, *New Chemical Fibres for Industrial Applications*, Khimiya, Leningrad, 1973, pp. 136–40.
6. Volokhina A.V., Kalmykova V.D., Manufacture of high-strength and heat-resistant synthetic fibres, *Itogi Nauki i Tekhniki. Khimiya i Tekhnologia Vysoko molekulyarnykh Soedineniy*, VINITI, Moscow, 1981, pp. 3–71.
7. Avrorova L.V., Volokhina A.V., Glazunov A.B., Kudryavtsev G.I., Oprits L.G., Tokarev A.V., Semenova A.S., Third-generation chemical fibres produced in the USSR. *Khimicheskie Volokna*, 1989, **4**, 21–6.

8. Material of the Jubilee Conference celebrating the 60th anniversary of the Khimvolokno Production Facility, *Khimicheskie volokna*, 1991, **2**, 3–63, **5**, 4–18.

9. Kuperman A.M., Snigireva N.A., Rodozinsky A.K., Gorbatkina Yu.A., Zelenski E.S., Vladimirov L.V., Berlin A.A., Highly oriented polyethylene fibres for reinforcing plastics, *Conference of the Department of Polymers and Composite Materials at the Institute of Physics, USSR Academy of Sciences, Chernogolovka*, Moscow, 1991, vol. 2, pp. 69–73.

10. Kudryavtsev G.I., Tokarev A.V., Avrorova L.V., Konstantinov V.A., High-strength high-modulus synthetic fibre SVM, *Khimicheskie volokna*, 1974, **6**, 70–1.

11. Deev I.S., Zherdev Yu.V., Mashinskaya G.P., Korolev A.Ya., Influence of various factors on the surface structure of organic fibres, *Aviatsionnye Materialy (Organoplastiki)*, ONTI VIAM, Moscow, 1984, pp. 8–18.

12. Deev I.S., Zherdev Yu.V., Mashinskaya G.P., Korolev A.Ya., Structure of the fibre–matrix boundary in organoplastics, *Aviatsionnye Materialy. Kompozitsionnye Materialy (Organoplastiki)*, ONTI VIAM, Moscow, 1984, pp. 19–24.

13. Kurzemnieks A.Kh., Deformational properties of the structure of organic fibres based on parapolyamides, *Mechanika Kompozitnykh Materialov*, 1979, **1**, 10–14.

14. Zarin A.E., Andreev A.S., *High-Strength Reinforced Chemical Fibres*, NIITEKHIM, Moscow, 1983.

15. Gal' A.E., Zakrevsky V.A., Perepelkin K.E., Study of the initial stage of the mechanism of mechanical breakdown of aromatic polyamides, *Vysokomolekulyarnye Soedineniya*, 1987, **26A**(11), 2326–31.

16. Badaev A.S., Perepechko I.I., Sorokin V.E., Viscoelastic behaviour of high-modulus polymer fibres in the temperature interval 20–900 K, *Doklady Akademii Nauk SSSR*, 1984, **278**, 387–9.

17. Tsobkalo E.S., Technical Sciences Dissertation, LITLP, Leningrad, 1982.

18. Badaev A.S., Perepechko I.I., Sorokin V.E., Comparative analysis of the dynamic properties of reinforcing polymer fibres *Mekhanika Kompozitnykh Materialov*, 1986, **4**, 579–84.

19. Lebedev L.B., Aleksashin V.M., Mashinskaya G.P., Dilatometric properties of high-strength fibre SVM, *Summaries of Papers from Seminar on Polymer-Material Fillers, Moscow, 1983*, MDNP, Moscow, 1983, pp. 9–13.

20. Lebedev L.B., Zhelezina G.F., Mashinskaya G.P., Kirilov V.N., Basov A.A., Dilatometric properties of SVM fibres and of organoplastics based on then, *Aviatsionnye Materialy Kompozitsionnye Materialy (Organoplastiki)*, ONTI VIAM, Moscow, 1984, pp. 31–8.

21. Gal' A.E., Leksovsky N.P., Vogman S.D., Influence of intermolecular interactions on the strength characteristics of polyheteroarylenes, *Vysokomolekulyarnya Soedineniya*, 1979, **21A**(10), 2241–7.

22. Kurzemieks A.Kh., Influence of moisture on the structure and properties of organofibres, *Mekhanika Kompozitnykh Materialov*, 1980, **5**, 919–22.

23. Pichugin V.S., Protasov V.D., Stepanychev E.I., Deformability and bearing strength of sheaths made on an expanding mandrel, *Abstracts of Papers of All-Union Conference on the Mechanics of Polymer Composite Materials, Riga, October 1983*, Riga, pp.148–9.

24. Stalevich A.M., Romanov V.A., Mesheryakov G.P., Mashinskaya G.P., Makarov A.V., Influence of vibration on the process of relaxation of tensile forces in aromatic polyamide fibres, *Khimicheskie Volokna*, 1982, **3**, 35–7.

25. Lebedev L.B., Aleksashin V.M., Mashinskaya G.P., Study of the formation of thermosetting polymer composites by thermal analysis, *Summaries of Papers from 8th All-Union Conference on Thermal Analysis, Kuibyshev, 1982*, Kuibyshev, p. 165.

26. Zhukova Z.N., Telegin F.Yu., Lebedev G.A., Mashinskaya G.P., Aleksandrova L.B., Voloshinova R.Z., Study of the interactions in an oligomer–polymer system, *Summaries of Papers from 1st All-Union Conference on Polymer Blends, Ivanovo, October 1986*, Ivanovo, p. 41.

27. Zhukova Z.N., Lebedev G.A., Aleksandrova L.B., Voloshinova R.Z., Mashinskaya G.P., Optimizing the composition of organoplastics on the basis of the thermodynamic and

kinetic characteristics of the interaction of its components, *Summaries of Papers from Moscow International Conference on Composites, November 1990*, Moscow, pp. 103–4.

28. Shul' G.S., Bonding Zhukova Z.N., Mashinskaya G.P., Influence of consistency parameters on the bonding strength of components in organoplastics, *Summaries of Papers from Moscow International Conference on Composities, November 1990*, Moscow, pp. 104–5.

29. Lebedev L.B., Aleksashin V.M., Fiedorova V.N., Mashinskaya G.P., Study of the formation of composites based on SVM aramid fibre and epoxy multicomponent binders, *Summaries of Papers from Moscow International Conference on Composites, November 1990*, Moscow p. 104.

30. Aleksashin V.M., Lebedev L.B., Ul'yanenko S.N., Mashinskaya G.P., Voloshinova R.Z., Study of curing and relaxation transitions in organofibres, *Summaries of Papers from the All-Union Conference on Composite Materials, Moscow, October 1981*, MGU, Moscow, pp. 101–2.

31. Voyutsky S.S. *Autoadhesion and Adhesion of High Polymers*, Rostekhizdat, Moscow, 1960.

32. Deryagin B.V., Krotova N.A., Smilga V.P., *Adhesion of Solid Bodies*, Nauka, Moscow, 1973.

33. Lipatov Yu.S., *Interphase Phenomena in Polymers*, Naukova Dumka, Kiev, 1980.

34. Kulezniev V.N., Voyutsky S.S., On local diffusion and segmental solubility of polymers, *Kolloidnyi Zhurnal*, 1973, **35**(1), 40–3.

35. Kulezniev V.N., Berlin A.A., Basin V.E., *Principles of Polymer Adhesion*, Khimiya, Moscow, 1969.

36. Basin V.E., *Adhesive Strength*, Khimiya, Moscow, 1981.

37. Gorbatkina Yu.A., *Adhesive Strength in Polymer–Fibre Systems*, Khimiya, Moscow, 1987.

38. Berlin A.A., Basin V.E., *Principles of Polymer Adhesion*, Khimiya, Moscow, 1969.

39. Plueddemann, E.P. (ed.), *Composite Materials*, vol. 6, *Interfaces in Polymer Composites*, Academic Press, New York, 1974; Mir, Moscow, 1978.

40. Paul, D.R., Newman, S. (eds), *Polymer Blends*, Academic Press, New York, 1978; Mir, Moscow, 1981.

41. Shul' G.S., Technical Sciences Dissertation, Obninsk, 1986.

42. Shul' G.S., Shchukina L.A., Shkirkova L.M., Gorbatkina Yu.A., Lebedev L.B., Voloshinova R.Z., Mashinskava G.P., Bonding strength of epoxy matrices with organic SVM fibre, *Aviatsionnye Materialy. Kompozitsionnye Materialy (Organoplastiki)* ONTI VIAM, Moscow, 1984, pp. 25–31.

43. Garanina S.D., Shul' G.S., Mashinskaya G.P., Lebedev L.B., Shirkova L.M., Shchukina L.A., Ermolaeva M.A., Influence of water on the properties of organoplastics, *Mekhanika Kompozitnykh Materialov*, 1984, **4**, 652–6.

44. Ul'yanenko S.N., Magomedov G.M., Mashinskaya G.P., Zelenev Yu.V., Relaxation processes and structural organisation in thermosetting organoplastics, *Summaries of Papers from Moscow International Conference on Composites, November 1990*, Moscow, p. 202.

45. Ul'yanenko S.N., Magomedov G.M., Lebedev L.B., Mashinskaya G.P., Zelenev Yu.K., The role of the interphase layer in the formation of the ductile elastic properties of high-strength organoplastics, *Mekhanika Kompozit Nykh Materialov*, 1987, **3**, 414–19.

46. Ul'yanenko S.N., Magomedov G.M., Mashinskaya G.P., Zelenev Yu.V., Evaluation of the interphase interaction in organoplastics using a dynamic mechanical method, *Plasticheskie massy*, 1987, **1**, 39–40.

47. Startsev O.V., Perepechko I.I., Voloshinova R.Z., Mashinskaya G.P., Active influence of fillers on the process of structural formation in the epoxy binder of a polymer composite material, *Doklady AN SR*, 1982, **267**(6), 1412–15.

48. Perov B.V., Skudra A.M., Mashinskaya G.P., Bulavs F.Ya., Peculiarities of organic fibre reinforced plastic failure and their influence on strength, *Fracture of Composite Materials*, Alphen aan den Rign, The Netherlands, 1979, pp. 279–89; *Mekhanika Kompozitnykh Materialov*, 1979, **2**, 317–21.

49. Skudra A.M., Mashinskaya G.P., Elastic properties of plastics reinforced with fabric, *Mekhanika Kompozitnykh Materialov*, RPI, Riga, 1979, pp. 9–16.

50. Slutsker G.Ya., Technical Sciences Dissertation, LITLP, Leningrad, 1982.

51. Zaitsev G.P., Arkhipov G.V., Mashinskaya G.P., Durability of an orthotropic polymer composite containing a crack under constant tensile load, *Mekhanika Kompozitnykh Materialov*, 1985, **5**, 834–42.

52. Suvorova Yu.V., Vasil'ev A.E., Mashinskaya G.P., Long-term rupture of non-elastic composites, *Mekhanika Kompozitnykh Materialov*, 1979, **5**, 794–8.

53. Suvorova Yu.V., Viktorova I.V., Mashinskaya G.P., Long-term strength and rupture of organoplastics, *Mekhanika Kompozitnykh Materialov*, 1980, **6**, 1010–13.

54. Suvorova Yu.V., Viktorova I.V., Vasil'ev A.E., Mashinskaya G.P., Study of the behaviour of organoplastics under different loading and temperature conditions, *Mashinovedenie*, 1980, **2**, 67–71.

55. Suvorova Yu.V., Viktorova I.V., Mashinskaya G.P., Lebedev L.B., Accumulation of faults in organoplastic under quasi-static and cyclic loading, *Mekhanika Kompozitnykh Materialov*, 1983, **4**, 614–18.

56. Garanina S.D., Basov A.A., Korolev A.Ya., Mashinskaya G.P., Water absorption by organoplastics, *Aviatsionnye Materialy. Kompozitsionnye Materialy (Organoplastiki)*, ONTI VIAM, Moscow, 1984, pp. 119–31.

57. Startseva L.T., Perepechko I.I., Mashinskaya G.P., Averkina N.K., Multiplet peaks of mechanical losses in the main relaxation region of organoplastics plasticized with moisture, *Mekhanika Kompozitnykh Materialov*, 1981, **6**, 1117–20.

58. Startsev O.V., Jartsev V.A., Mashinskaya G.P., Molecular mobility and relaxation processes in the epoxy matrix of a composite. 2. The effects of aging in a damp subtropical climate, *Mekhanika Kompozitnykh Materialov*, 1984, **4**, 593–7.

59. Bulmanis V.N., Krivonos V.V., Mashinskaya L.P., Chersky I.N., The manufacturing climatology of polymer fibre composites, *Summaries of Papers from Moscow International Conference on Composites, November 1990*, Moscow pp. 174–5.

60. Kobets L.P., Mashinskaya L.P., Deev I.S., Pyrolization processes and the structure of high-strength aramid fibres, *Summaries of Papers from Moscow International Conference on Composites, November 1990*, Moscow pp. 174–5.

61. Polyakova E.N., Zherdev Yu.V., Koroliev A.Yu., Thermal destruction of polymers with additional pyrolysis by volatiles, *Plasticheskie Massy*, 1973, **2**, 71–2.

62. Venger A.E., Fraiman Yu.E., Balashov A.Ya., Mashinskaya L.P., Study of the thermal breakdown and thermal capacity of organic fibres, *Teplofizika Vysokikh Temperatur*, 1981, **XIX**(1), 98–101.

Index

GPSR Compliance
The European Union's (EU) General Product Safety Regulation (GPSR) is a set
of rules that requires consumer products to be safe and our obligations to
ensure this.

If you have any concerns about our products, you can contact us on

ProductSafety@springernature.com

In case Publisher is established outside the EU, the EU authorized
representative is:

Springer Nature Customer Service Center GmbH
Europaplatz 3
69115 Heidelberg, Germany